GRAPE AND WINE BIOTECHNOLOGY

Edited by **Antonio Morata** and **Iris Loira**

Grape and Wine Biotechnology
http://dx.doi.org/10.5772/61694
Edited by Antonio Morata and Iris Loira

Contributors
Santiago Benito, Fengmei Zhu, Bin Du, Jun Li, Miguel Ángel Olego, Juan José Rubio Coque, José Enrique Garzón Jimeno, José Manuel Álvarez Pérez, Miguel Javier Quiroga Martínez, Rebeca Cobos, Mario Sánchez García, Jesús Esteban Medina, Sandra González García, Dimitris Tsaltas, Milena Lambri, Donato Colangelo, Fabrizio Torchio, Roberta Dordoni, Dante Marco De Faveri, Jorge Domínguez, Hugo Martínez-Cordeiro, Marta Lores, Antonio Morata, Iris Loira, Jose Antonio Suárez Lepe, Guillermo Cristian Martinez-Avila, Romeo Rojas, Cecilia Castro-López, Ernesto Sánchez-Alejo, Alice Vilela, Fernanda Cosme, Berta Gonçalves, António Inês, António M. Jordão, Jun Wang, Hong-Wei Xiao, Arun S Mujumdar, Xiaoshuan Zhang, Weisong Mu, Qian Zhang, Xiao-Ming Fang, Zhenjiang Gao, Mizuho Nita, Patricio Arce-Johnson, Grace Armijo, Carmen Espinoza, Rodrigo Loyola, Franko Restovic, Claudia Santibáñez, Rudolf Schlechter, Mario Agurto, Shunji Suzuki, Shinichi Enoki, Alina Kunicka-Styczyńska, Agata Czyżowska, Katarzyna Rajkowska, Agnieszka Wilkowska, Piotr Dziugan, Diego Lijavetzky, Claudio Muñoz, Constanza Chialva, Estefania Eichler, Silvana María Del Mónaco, Yolanda Curilen, Ramona Del Carmen Maturano, Adriana Beatriz Simes, Adriana Catalina Caballero, Sebastián Mario Ezequiel Bravo, Rosane Gomez, Denise Lacerda, Pedro C. Costa, Claudia Funchal, Caroline Dani, Arroyo Teresa, García Margarita, Esteve-Zarzoso Braulio, Giovanna Fia, Niina J Ronkainen, Elissavet Dotsika

CBS, Edition **2017**

Published by InTech
Janeza Trdine 9, 51000 Rijeka, Croatia

Publishing Process Manager Andrea Koric

Technical Editor SPi Global

Cover Designer InTech Design Team

Additional hard copies can be obtained from orders@intechopen.com

Grape and Wine Biotechnology, Edited by Antonio Morata and Iris Loira
p. cm.
Print ISBN 978-953-51-2692-8
Online ISBN 978-953-51-2693-5

INTECH

open science | open minds

Contents

Preface

This book shows a multidisciplinary approach on many current and essential topics of vine and wine biotechnology in its 21 chapters. Authors are experts in their respective research fields at international level and work in research centers or universities from 13 wine-producing countries. The topics in the chapters describe the innovative technologies or processes in wine production, but these are presented at a technical/reader-understandable level. Grape production and wine industry are evolving to improve sensory quality, in a safe, sustainable, and eco-friendly way. Innovations in cultivar selection or vine plant biotechnology are directed to improve grape sensory quality and at the same time to obtain cultivars with improved resistance to diseases and pests and to adapt to environmental conditions as a way to reduce the use of chemical pesticides. Wine biotechnology is focused on the improvement of both alcoholic and malolactic fermentations, as well as the microorganisms respectively involved in them—yeasts and malolactic bacteria. The goal is on the improvement of sensory quality trying to adapt fermentative quality to the specific aromatic and flavor profile of each variety and region. Also it is essential to improve wine safety in order to provide consumers with healthy wines. Reduction of alcohol content in beverages is another major concern in the framework of making wines. Moreover, this can be essential in a scenario of increasing global climatic changes, which caused higher contents of sugar in grapes, resulting in the increase of alcohol content. The use of non-*Saccharomyces* yeasts is also a current trend to improve wine quality at sensory and technological level. Moreover aging of wine is the traditional way of polishing its structure to increase its complexity and roundness. Wine aging in barrel is an emerging technique to modulate wine sensory profile.

The book has been divided into three sections according to specific topics: grape and vine biotechnology, fermentation and wine biotechnology, and analysis and origin authentication.

Chapters 1–4 are devoted to the management of fungal diseases at molecular level and pathogenic processes. Dr. Arce-Johnson et al. focus on molecular biotechnology in vine and how to manage biotic and abiotic stress in vine plants to reduce economic losses and in turn to get a good quality. Particular attention is given to fungal diseases and UV radiation and the effect of these stresses in plant development. Dr. Suzuki et al. describe the pathogenic alteration related to proteins in plants explaining the main plant defense mechanisms, proteins involved, and their role in pathogen control. Dr. Nita explains the use of cultivar selection to control fungal diseases in grape and vine from an integrated management with the aim to minimize the use of chemical pesticides. In Chapter 4 by Dr. Lijavetzky et al., the main molecular tools to obtain transgenic vines with improved properties are studied. New technologies of gene edition are explained highlighting their application in functional study of grapevine.

In Chapter 5, Dr. Martínez-Ávila et al. describe the main available technologies to extract and recover phenolic compounds from grapes and by-products to obtain bioactive compounds to be used as pigments, organic acids, or antioxidants. In a similar way, Dr. Gómez et al. (Chapter 6) describe the extraction and potential therapeutic applications of vine leaves' polyphenols in human diseases. Grape drying is used to produce raisins, another expansive grapevine production at world level. In Chapter 7, Dr. Xiao et al. explain the current status and future trends of grape drying technologies. In Chapter 8, Dr. Domínguez et al. explain the use of earthworms as a biological and eco-friendly process to produce a high-quality biofertilizer to be used in soil management.

Chapters 9–15 are focused on wine-grape microorganisms and the management of fermentations. Dr. Vilela et al. describe many metabolites that can be found in grapes but also those produced in alcoholic and malolactic fermentations. Their main compounds and average concentrations in grapes or wines and analytical techniques to elucidate them are exhaustively described. Dr. Tsaltas et al. pay attention to the natural grape microbiome and the distribution and ecology of microorganism species according to geographical origin, cultural practices, varieties, and climatic conditions. They also describe the potential applications of wine strains as starter cultures in wine fermentations. Dr. Arroyo et al. focus the reader's attention on the role of nonconventional yeasts or non-*Saccharomyces* yeasts in wine production, describing new applications to improve wine sensory quality with yeast species that were traditionally eliminated in wine fermentations. Dr. Zhu et al. describe the main volatile compounds in wines explaining the main factors affecting wine aromatic properties. Dr. Morata et al. explain the main application of the use of selected *Saccharomyces* and non-*Saccharomyces* yeast species to improve the formation of stable pigments and color stability in wines. Dr. Benito et al. describe the main applications of *S. pombe*, a non-*Saccharomyces* yeast that recently is gaining importance in wine fermentation for its specific properties and metabolism. Dr. del Monaco et al. explain the current application of the use of selected wild yeasts in Argentinian wines in Patagonian region.

Chapters 16–19 are focused on wine technology, aging, and stabilization. Dr. Fia explains the use of wine lees as a tool for biological aging of wines and describes a new method to apply it and to facilitate the process. Reduction of alcohol content in wines is a hot topic, and Dr. Olego et al. describe several viticultural and biotechnological techniques to achieve it. Bentonite is an additive used in enology for settling and clarification; Dr. Lambri et al. explain the main properties of this silicate and the interaction of wines with protein colloids, polyphenols, and aromatic compounds to understand better the applications in wine stabilization. Dr. Kunicka-Styczyńska explains the main technologies and processes used in wine industry in Poland.

Finally, Chapters 20–21 are dedicated to the use of instrumental analytical techniques in wine analysis and origin authentication. Dr. Ronkainen et al. focus the attention on trace elements analyzed by atomic spectroscopy and electroanalytical techniques, and Dr. Chantzi et al. highlight the application of NMR, stable isotopes, 14C radiocarbon, and isotopic techniques in wine authentication and the determination of geographical origin.

Antonio Morata, Associate Professor
Department of Chemistry and Food Technology
Technical University of Madrid, Spain

Iris Loira, Assistant Professor
Department of Chemistry and Food Technology
School of Agricultural Engineering
Technical University of Madrid, Spain

Section 1

Grape and Vine Biotechnology

Chapter 1

Grapevine Biotechnology: Molecular Approaches Underlying Abiotic and Biotic Stress Responses

Grace Armijo, Carmen Espinoza, Rodrigo Loyola,
Franko Restovic, Claudia Santibáñez,
Rudolf Schlechter, Mario Agurto and
Patricio Arce-Johnson

Additional information is available at the end of the chapter

http://dx.doi.org/10.5772/64872

Abstract

Grapevine is one of the most abundant crops worldwide, with varieties destined for fresh and dry consumption, as well as wine production. Unfortunately, grapevine plants are affected by both biotic and abiotic stresses, generating significant economic losses. These conditions can negatively impact grape cultivation at different stages: plant and berry development during pre- and post-harvest, production, fresh fruit processing and export, along with wine quality. Most of the grapevine varieties are susceptible to several pathogens and within this chapter, particular attention is given to fungi (*Botrytis cinerea* and *Erysiphe necator)* and viruses, since they are a worldwide concern. Within the latter, special focus is given to the grapevine leafroll disease, a complex and destructive infection. On the other hand, abiotic stress is also relevant in grapevine, and in this chapter it will be exemplified by UV-B radiation and its impact on growth and fruit development, plant adaptive responses and its relationship with the quality of grape berries for winemaking. The main biotic and abiotic grapevine stress factors are reviewed in this chapter, considering a special focus on biotechnological approaches carried out in order to address them and minimize their detrimental consequences.

Keywords: grapevine fungal diseases, *Erysiphe necator*, *Botrytis cinerea*, grapevine viruses, UV-B radiation, grapevine biotechnology

1. Introduction

Grapevine (*Vitis vinifera L.)* is one of the most important crops worldwide. Within the Vitaceae family, the *Vitis* genus has a major agronomic importance. Among them, *V. vinifera* is the only species extensively used in the global industry, dominating the market with only a few cultivars, generally classified according to their final production: wine grapes, table grapes and raisins [1]. This low variability is directly related to grapevine's high susceptibility to biotic and abiotic stresses, which is associated with significant economic losses.

Most of the grapevine varieties are susceptible to several biotic agents, such as phytoplasma, bacteria, fungi, oomycetes, viruses and nematodes, which dramatically reduce plant yield and fruit quality, and negatively impact plant development. In vineyards, the most important diseases are caused by microorganisms such as fungi, oomycetes and viruses. Within pathogenic fungi, *Botrytis cinerea* and *Erysiphe necator* are the most important ones, producing the grey mould and powdery mildew diseases, respectively [2]. *V. vinifera* is classified as a susceptible species, where low or no resistant phenotypes have been described in economically significant cultivars until now.

On the other hand, nearly 60 viruses can infect grapevine plants, a much higher number than the ones affecting other perennial crops. Under natural conditions, grapevine viruses are transmitted by insect or nematode vectors. However, since grapevine is usually propagated by grafting, viruses can also disseminate within plants through these cuttings [3]. It is noteworthy to mention that unlike to other pathogens, grapevine plants present no virus resistance, meaning that they can establish compatible interactions where viral pathogens can spread throughout all tissues, generating a global cellular stress and developmental defects [4–6]. Regarding these infections, the leafroll disease is one of the most complex viral diseases known and is considered one of the most destructive in grapevines. In addition to their economic detriment to grapevine cultures, all viruses are relevant when the sanitary status of the vineyard is considered.

Abiotic stress factors, particularly water availability, temperature and light are also relevant in grapevine. Among them, ultraviolet (UV)-B radiation impacts on grapevine plants growth and normal fruit development. *V. vinifera* is often cultivated in Mediterranean climates with varied UV-B radiation dosages [7]. Grapevines are considered as well adapted to solar radiation due to a variety of physiological adaptive responses mainly based on antioxidant enzyme activities and secondary metabolites [8], which besides their role in defence against abiotic stress are relevant for colour, taste and aroma of grapes. These responses are triggered by UV-B perception and signalling pathway, which was recently identified and characterized in grapes [9, 10]. The increase of flavonols in response to UV-B has been reported in grapevine berry skins [9]. As a consequence, the quality of grape berries for winemaking is correlated with the accumulation of UV-B-induced phenolic compounds. Hence, wines with the highest concentrations of phenolic compounds are generally considered of excellence [11]. Therefore, understanding the mechanisms of perception, signalling and response of the grapevine to UV-B and using this knowledge to improve both productivity and fruit quality by genetic modification are attractive targets for the wine industry.

http://dx.doi.org/10.5772/64872

Biotechnological approaches aimed to solve grapevine stress affections, in areas regarding grapevine physiology and genetics, are a main requirement for optimizing and improving quality of this species through biotechnological tools.

2. Grapevine biotic stress

As mentioned above, biotic stress is related to infection caused by phytopathogenic organisms such as bacteria, nematodes, fungi, oomycetes and viruses, among others. These pathogens get the necessary elements for growth and reproduction from its hosts. According to their infection strategies, plant pathogens can be classified as necrotrophics, biotrophics and hemibiotrophics. Necrotrophic pathogens on one hand, extract nutrients from dead cells during colonization, secreting lytic enzymes and phytotoxins in order to promote necrosis in the host plant. Biotrophic pathogens, on the other hand, feed on living tissue maintaining the viability of the host in order to obtain metabolism products. Finally, hemibiotrophic pathogens start with a biotrophic infection phase, followed by a late necrotrophic one [12].

2.1. Fungal diseases in grapevine: a biotrophic and necrotrophic model

Nowadays, most of the wine, table grape and dried fruit cultivars have the Eurasian grape species *V. vinifera* as a common ancestor, mostly due to its distinctive flavour and aroma. However, another similar trait is their limited genetic resistance against fungal pathogens, making these cultivars highly dependent on the use of fungicides [13].

The most common fungal grape diseases are the powdery mildew and grey mould caused by the biotrophic pathogen *E. necator* and the necrotrophic *B. cinerea*, respectively [2].

2.1.1. Powdery mildew: E. necator

E. necator Schwein (synonyms: *Uncinula necator* Burr., *E. tuckeri* Berk., *U. americana* Howe and *U. spiralis* Berk. and Curt.; anamorph: *Oidium tuckeri* Berk.) is a biotrophic and filamentous fungus that belongs to the Erisiphaceae family. *E. necator* is the etiologic agent of the powdery mildew disease in species of the *Vitis, Cissus, Parthenocissus* and *Ampelopsis* genera, being *V. vinifera* one of its most economically important hosts [14].

The powdery mildew disease is associated with large production losses as it reduces yield and fruit quality, mainly affecting the sugar content and acidity of the berries, although it can also infect other green tissues. This pathogen can be found in all grape-growing regions, especially in dry and warm weathers [15].

Being an obligated biotrophic pathogen, *E. necator* depends on its host for growth and development, as photosynthesis-active tissues are necessary to complete its life cycle. The infective process begins with the attachment of the asexual spore (conidia) on plant tissues, followed by the formation of a primary germ tube which differentiates in a specialized infective structure called appressorium. The latter then generates a mechanical pressure in order to penetrate and invade the host cell (**Figure 1**). Germination involves the secretion of fungal lytic enzymes which leads to the release of compounds that enhance fungal germination and development [13]. The successful invasion results in cell membrane invagination and the haustorium formation, which is a specialized hypha that facilitates the dynamic exchange of molecules derived from both fungal and host cells. The fungus retrieves nutrients from the host cells, while at the same time it secretes proteins to suppress host defences. After this establishment, secondary hyphae proliferate in order to colonize a larger area of tissue across the surface, producing more appressoria and haustoria at regular intervals. The overall process culminates with conidiation, which involves the formation and release of asexual spores to distal tissues [16]. The main symptom of *E. necator* colonization is the appearance of a white powder in the infected host tissue, corresponding to mycelial proliferation and conidiophores development [14].

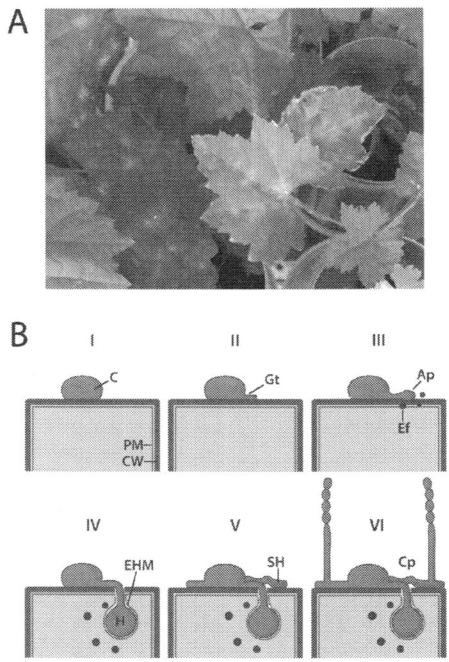

Figure 1. Grapevine powdery mildew and *E. necator* asexual life cycle. (A) Grape leaves infected with *E. necator* exhibit a white powder on the infected tissue surface. (B) Asexual life cycle stages. I: Conidium (C) attachment; II: Conidium germination and germ tube (Gt) formation; III: Appresorium differentiation (Ap); IV: Development of haustorium (H), extra-haustorial membrane (EHM) and secretion of virulence factors or effectors (Ef); V: Colonization and secondary hyphae (SH) formation; VI: Production of asexual reproductive organs or conidiophores (Cp).

When environmental or nutritional conditions become unfavourable, *E. necator* develops a structure of sexual reproduction that contains ascospores, called cleistothecia. Within this structure ascospores can remain dormant for months until favourable conditions allow

http://dx.doi.org/10.5772/64872

germination and, like asexual spores, appressorium formation in order to begin a new infective process [14].

2.1.2. Grey mould: B. cinerea

The necrotrophic fungus *B. cinerea* (Persoon: Fries; teleomorph *Botryotinia fuckeliana*) is widely distributed in nature and it lacks a specific host. This fungus is capable to infect vegetables, fruits and ornamental plants, among others, making it a great problem for many plant species [17]. However, its importance relies on its ability to infect crops of commercial interest, such as grapevine. It can cause soft rotting of all aerial plant tissues, and rotting of post-harvest fruits; production of grey conidiophores and (macro)conidia are typical signs of the disease [18].

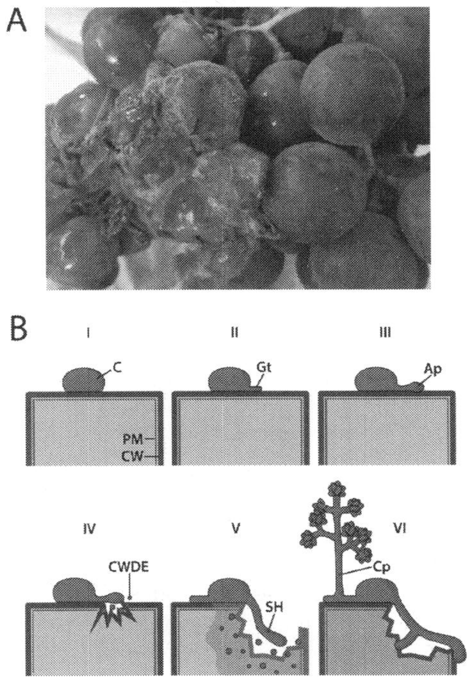

Figure 2. Grey mould and *Botrytis cinerea* asexual life cycle. (A) Grape berry cluster severely infected with *B. cinerea*. (B) Asexual life cycle stages. I: Conidium (C) attachment; II: Conidium germination and germ tube (Gt) formation; III: Appresorium (Ap) differentiation; IV: Secretion of cell wall degrading enzymes (CWDE); V: Colonization and secondary hyphae (SH) formation; VI: Development of asexual reproductive organs or conidiophores (Cp).

Grey mould disease causes heavy yield losses in table and wine grapes all around the world. As a consequence of the increase of the international trade of cold-stored products, this fungus has gained great importance because it can grow effectively over long periods of time at just above freezing temperatures [18]. In the field, it can spread to other grapes by insects which can carry viable conidia and generate mechanical damage [19]. Although, *B. cinerea* shows a remarkable flexibility to germinate in different host environments, several factors influence the germination of a conidia, such as temperature, surface water, relative humidity, among others [20].

Once the conidium attaches to its host, it can germinate and develop to an infective structure called appressorium, which is able to breach the cuticle by means of a penetration peg (**Figure 2**). The underlying cells are killed by the fungus, and the primary lesion is established. After the skin barrier is damaged, *B. cinerea* causes decomposition of plant tissues in order to consume the plant biomass. At this point it secretes cell wall degrading enzymes (CWDE), toxins and oxalic acid. Subsequently, the hyphal growth is induced in order to begin the sporulation cycle and infection of adjacent cells [17].

In some tissues, *B. cinerea* causes long-lasting quiescent infections, in which no symptoms are discernible at first. It can also penetrate floral tissue of grapes (petals, stigmas, styles or stamens) and remain dormant, often for several weeks, until it resumes activity and invades the fruit later in the season or during ripening. It has been postulated that high levels of phytoalexins in immature fruits contribute to quiescence, acting as fungitoxic or fungistatic compounds; and that post-harvest host physiological and biochemical responses might activate the pathogen [21].

2.2. Grapevine responses to fungal diseases

Plants are considered to have two types of immunity: a general one against a broad spectrum of microorganisms, and other specific one against a particular pathogen. Both responses are characterized by their ability to recognize pathogen components, transduce the stress signal and induce a defence response. However, the main difference between the both is considered to be the robustness and duration of the response [22].

The first type of immunity is known as PTI (pathogen-associated molecular patterns (PAMP) triggered immunity) and is activated by PAMP recognition receptors (PRR) that detect structural pathogen components and transduce the signal for the induction of a basal response. This type of immunity is mainly related to the prevention of pathogen entry into plant cells [23]; however, it is not completely effective against biotrophic and necrotrophic fungi. On the latter case, the response is activated by damage-associated molecular patterns (DAMP) recognition mainly derived from the host cell wall fragments generated by CWDE [24].

The second line of defence is known as ETI or effector triggered immunity, capable of directly or indirectly recognizing specific pathogen effectors through the expression of resistance proteins (R proteins). This recognition induces a more robust and efficient response, mainly against biotrophic pathogens, by preventing them to complete their life cycle in the host, interrupting nutrient uptake and eliminating the infected cells along with the pathogen [23]. Since this response against biotrophic pathogens (and hemibiotrophic too) generally ends with programmed cell death (PCD) of infected tissue, some necrotrophic pathogens induce this mechanism during infection in order to bypass plant defences and rapidly kill tissue for nutritional benefits [24].

Plant defence mechanisms are finely regulated by plant hormones, mainly jasmonic acid (JA), ethylene (Et) and salicylic acid (SA), which communicate synergistically or antagonistically depending on the type of pathogen. Generally speaking, the defence against necrotrophic pathogens are considered to be mediated by JA and Et, while the defence against biotrophic pathogens by SA [12]. However, *V. vinifera* cultivars are very susceptible to fungal pathogens, likely due to insufficient defence responses to contain these pathogens.

2.2.1. Grapevine defences against E. necator

E. necator corresponds to the only powdery mildew species adapted to *V. vinifera*. Nevertheless, several species from the Vitaceae family have been identified as resistant. In the latter, plants are able to restrict *E. necator* invasion and growth by means of two strategies: penetration resistance and programmed cell death (PCD)-mediated resistance (observed as a hypersensitive response). The first blocks the breach of the cell wall and membrane and thus prevents the formation of the haustorium. On the other hand, the PCD-mediated resistance is exerted once the epidermal cell is penetrated and induces the death of it, thereby interrupting the supply of nutrients required by the biotrophic fungus for further growth and development [13]. This type of resistance is related to the detection of pathogen effectors by the plant due to specific resistance genes (R genes) [25, 26]. Different loci have been found in several species of the Vitaceae family which confer resistance to *E. necator*; carrying resistance gene analogues (RGA) and in some cases associated to complete resistance to powdery mildew mostly related with PCD induction [27]. However, very few candidate R-genes have been identified to date and molecular defence mechanisms triggered by these resistance loci are being studied. A number of genes have been implicated in resistance in certain wild *Vitis* species showing increased transcription during infection or differential expression levels between resistant and susceptible plants [13], but the identification of key components in PTI and ETI responses against *E. necator* in grapevines is still pending.

Resistance to powdery mildew in the Vitaceae family is closely related to its evolutionary history. *V. vinifera* is native to Eurasia and developed evolutionarily isolated from *E. necator*, a pathogenic fungus from North America, until the 1840s. This is the reason why nearly all *V. vinifera* cultivars lack the genetic protection mechanisms against the fungus and are highly susceptible to infection [14, 26, 28]. Even though *V. vinifera* susceptible plants are able to initiate a basal defence response, they are unable to restrict fungal growth and arrest the disease [29].

Ontogenic, or age-related resistance, also has a role in the defence against *E. necator*. It may operate at a whole plant level or at specific organs or tissues. Grape berries develop ontogenic resistance to *powdery mildew* within 4 weeks after fruit set. However, adhesion of conidia, germination and appressorium formation were not impeded on older berries [30]. Ontogenically resistant berries respond rapidly to infection by synthesis of a germin-like protein that has been previously shown to play a role in the host defence against barley powdery mildew. This type of defence, which conditions ontogenic resistance, operates in the earliest stages of the infection process prior to the formation of a functional haustorium [30].

2.2.2. *Grapevine defences against B. cinerea*

Low or no resistant phenotypes to grey mould have been described in most common table grape *V. vinifera* cultivars, whereas high level of resistance has only been found in the *Muscadinia rotundifolia* (*Vitis rotundifolia*), *Vitis labrusca* and other grape hybrids species. This resistance appears to be related to mechanical barriers, such as cuticle and wax contents [31]. Pre-existing or basal defences seem to be an important part of the machinery against *B. cinerea*, along with the activation of inducible defence mechanisms mediated by SA or JA/Et pathways, which in turn depend on the developmental stage, and an appropriate kinetics between ROS production and the generation of antioxidant compounds [32, 33].

Structural barriers are related to the fungal primary infection process (i.e. appressoria formation and plant tissue penetration), while inducible responses are associated with subsequent infection ones [34]. In this case, PTI is mainly activated by DAMPs, host cell wall fragments generated by fungal CWDE and PAMPs such as chitin fragments of fungal cell walls, among others. These are identified by specific PRR receptors, such as cell-wall-associated kinases, which in turn activate the defence signalling cascade, culminating in hormones and transcription factors biosynthesis [12, 24]. This response induces protease inhibitors generation and secondary metabolite biosynthesis (i.e. anthocyanins and phytoalexins). The flavonoid phytoalexin plays an important role in the defence response of grapes. The rapid production of resveratrol, major compound of the stilbene family, and its transformation into Viniferins enhance resistance to fungal pathogens in grapevine cultivars [35]. Resveratrol and pterostilbene (two grapevine phytoalexins) produce malformation or growth inhibition of germ tubes, cytoplasmic granulation of the cellular content and the disruption of the plasma membrane in *B. cinerea* conidia [36].

2.3. Biotechnological strategies for fungal control in grapevine

Regarding control of fungal pathogens, major improvement efforts have been directed towards enhancing fungal-disease resistance in table and wine grape cultivars. Development and optimization of alternative strategies to reduce the use of classic chemical inputs for protection against diseases in vineyard is becoming a necessity. Nowadays, fungal-related diseases are controlled through fungicide applications of organic and inorganic composition. The most used compounds are sulphurs, petroleum-based oils, inorganic salts, benzimidazoles and ergosterol biosynthesis inhibitors, among others [14]. However, these management practices usually generate negative impacts on the environment and have elevated health and safety hazards. Various sources have speculated that sulphur, the most heavily used agricultural chemical, can cause respiratory illnesses and other adverse health effects [37]. In soil, sulphur is slowly converted by bacteria to sulphate, which generally does not cause harm. Other synthetic compounds used for treatment and prevention, such as sterol inhibitors have not been reported as having negative environmental or human health effects.

2.3.1. Genetic improvement

Genetic improvement is an agronomic practice widely used to confer interest features to a crop through hybridization between different cultivars or even species, in order to obtain new varieties. In *V. vinifera*, most interesting features vary depending on the use the fruit will be given. Nevertheless, in all cases, the importance of introducing fungal disease resistance is a priority, in order to reduce pathogen management and to minimize environmental impact [38].

Many North American *Vitis* species show various levels of resistance to *E. necator* but lack productive and commercial qualities; however, they represent a valuable germplasm to be used as natural sources of resistance in grapevine breeding programs. Among the resistant North American species identified to date we can find *V. aestivalis, V. cinerea, V. riparia, V. berlandieri, V. labrusca* and also *Muscadinia rotundifolia* [14, 26, 39]. However, the powdery mildew resistance character is not restricted to North America. The Central Asian *V. vinifera* cvs. 'Kishmish vatkana' and 'Dzhandzhal kara' have also been identified as resistant genotypes [26, 40–42].

Genetic knowledge of the resistance trait is crucial to achieve a significant improvement of grapevine through breeding. Several powdery mildew resistance loci have been identified and mapped to date. The Run1 locus was described in *M. rotundifolia* and has been successfully introgressed into *V. vinifera*. According to the closest SSR markers VMC4f3.1 and VMC8g9, this locus was mapped to a region in chromosome 12 and co-segregates with the *Plasmopara viticola* resistance locus Rpv1 [14, 26, 43, 44]. The *MrRUN1* and *MrRPV1* genes, which code for TIR-NBS-LRR proteins (a class of R proteins), are the first cloned and functionally characterized resistance genes from grapevine [27]. The Ren1 locus, on the other hand, belongs to the *V. vinifera* cvs. 'Kishmish vatkana' and 'Dzhandzhal kara' from Central Asia. It has been mapped in linkage group 13 with the closest linked SSR markers VMC9H4-2, VMCNG4E10-1 and UDV-020. To date, this gene has not been fully identified, although near the SRR markers, an NBS-LRR and a CAD gene have been recognized, being both probably part of the hypersensitive response [40, 42]. Other identified powdery mildew resistance loci are Run2.1, Run2.2, Ren5 and Ren.4 from *M. rotundifolia* cvs. 'Magnolia', 'Trayshed', 'Regale' and *V. romanetii*, respectively. All of these loci have been mapped in chromosome 18, which have a significant higher density of NBS-LRR genes compared to the other linkage groups, except for Ren5 mapped in linkage group 14. The resistance mechanism mediated by Ren4 may differ from the other loci since extremely low penetration and secondary hyphae development rates with no cell death have been observed at the infection site [45–47].

One of the main concerns about using pathogen resistance genes in plant breeding is the potential appearance of new pathogen strains that could breakdown the resistance. To overcome this latent problem, actual breeding efforts are focusing on stacking or pyramiding two or more resistance genes within a single cultivar to increase the durability of the resistance in the field. In this scenario, pathogen reproduction will be restricted even if infection by a new pathogen strain with a modified or lost effector molecule occurs. Thus, biotechnological tools have become essential for the development of new resistant cultivars. Marker-assisted gene pyramiding is one of the main applications of DNA markers in plant breeding. The use of molecular marker-assisted selection allows the identification of segregants that may exhibit the same phenotype but carry multiple resistant genes [26]. A grapevine progeny with individuals carrying both Run1 and Ren1 loci was developed in 2010, where Run1 was introgressed from a *M. rotundifolia* × *V. vinifera* hybrid plant derived from a pseudo-backcrossing breeding scheme, while Ren1 was introgressed from the resistant *V. vinifera* cv 'Kishmish vatkana' [48].

Unlike to what happens with *E. necator*, no genetic resistance components against *B. cinerea* have been identified until now, being this the main reason why no breeding programs against this fungus have been reported. All efforts in this area have been developed within the transgenic field, which will be described below.

2.3.2. Genetic manipulation

The development of highly reproducible genetic engineering protocols for grapevine cultivars and rootstocks now allows the identification, screening and/or introduction of grapevine-derived genes related to desirable traits, such as disease resistance.

Pathogenesis-related (PR) proteins were screened for their response to fungal pathogen infection. Genetically modified (GM) grapevines constitutively expressing rice chitinase genes exhibited enhanced resistance to powdery mildew [49, 50]; however, no resistance was observed when plants expressed barley chitinase genes [51]. Other non-grapevine-derived genes, such as the polygalacturonase inhibiting protein (PGIP) and other lytic peptides, were demonstrated to improve fungal disease resistance [49].

Two endochitinase (*ECH42* and *ECH33*) genes and a *N*-acetyl-β-D-hexosaminidase (*NAG70*) gene related to *Trichoderma* spp. were used to develop a set of genetically modified 'Thompson Seedless' lines in order to evaluate fungal tolerance against *B. cinerea*. The highest resistant plants were the ones expressing the *ECH42–NAG70* double gene construct and the *ECH33* gene [52].

Genetic manipulation of phytoalexins has been done in order to increase disease resistance of plants. Use of modern molecular biology tools for elucidating the control mechanisms of phytoalexin synthesis and for engineering disease-resistant plants is based on the expression of stress- or disease-related genes. Few reports attempting the manipulation of phytoalexins biosynthesis by genetic engineering have been published, with most of them related to resveratrol, the major phytoalexin from Vitaceae. STS, the key enzyme in resveratrol synthesis, uses as substrates precursor molecules that are present throughout the plant kingdom. Therefore, the introduction of a single gene is sufficient to synthesize resveratrol in heterologous plant species [53].

The grapevine rootstock 41-B, overexpressing the grapevine *VST1* stilbene synthase gene under the control of the fungus inducible promoter PR 10.1, produced high stilbene levels and exhibited *in vitro* resistance to *B. cinerea* [54]. Stilbene phytoalexin resveratrol levels in grapes have been directly correlated with grey mould resistance [55].

All the aforementioned results demonstrate that improved fungal tolerance can be accomplished through transgene expression. In addition, they support the use of iterative molecular and physiological phenotyping in order to select tolerant individuals from GM grapevine populations.

2.3.3. *Biological control*

Biological control of fungal pathogens is based on the use of microorganisms to prevent or reduce the damage produced during infection. Among the best studied biocontrol agents we can find are the filamentous fungi of the *Trichoderma* genus [56], bacteria of the *Bacillus, Pseudomonas* and *Serratia* genera [57] and yeasts of the *Pichia* and *Candida* genera [58]. Within the proposed mechanisms of how biological control of plant pathogenic fungi work, we can describe the competition for ecological niches, especially for nutrient utilization and elements obtaining such as nitrogen and carbon and/or the secretion of toxic molecules for the fungus [59, 60].

Another biocontrol mechanism is the activation of the induced systemic resistance (ISR) in plants; this mechanism can be induced by elicitors released by the biocontrol agent (ranging a wide variety of molecules), and it has been attributed to non-pathogenic microorganisms associated to plants, such as saprophytes [61]. Generally speaking, microorganisms exhibit a combination of the mentioned mechanisms, thus reducing the risk of pathogen resistance [62].

Among the bacteria able to synthesize and secrete anti-fungal molecules, those belonging to the genus *Bacillus* are the most important. These bacteria are characterized by their ability to secrete a wide range of bioactive molecules, including anti-fungal, anti-microbial, insecticides, plant growth promoters and ISR-inducing ones [63]. In addition, these molecules have a low toxicity to animals and humans, and are highly biodegradable, so they do not represent a hazard to the environment unlike chemical fungicides [64]. The best characterized molecules secreted by bacteria of the genus *Bacillus* with anti-fungal activity, are the cyclic lipopeptides. These molecules are mainly classified into three families: iturins, fengicines and surfactins. They are formed by a non-ribosomal peptide synthesis ring, which is attached to a fatty acid chain [65]. An important feature of these molecules is that within each family there are different counterparts, differing in the amino acid composition of the polypeptide in the ring and the chain length of the fatty acid [65].

2.3.4. *Elicitors*

Another control strategy consists in the stimulation and/or potentiation of the grapevine defence responses by the means of elicitors [66]. Elicitors are defined as a more specific class of purified molecules originated from microorganisms or plants which are able to stimulate an innate immune response in plants [67].

Elicitor perception also increases the level of plant resistance against future pathogen attack [12]. Induced resistance is often related to the 'priming' or potentiation phenomenon, and some molecules perceived by plants have also been shown to induce these effects [66, 68]. The definition of priming is related to the physiological state of the plant after an initial biotic or abiotic stimulus. This priming allows the plant to respond in a faster and/or stronger way to following biotic and/or abiotic challenges, often resulting in an improved tolerance in comparison to non-primed plants [68]. The mechanism of this phenomenon remains relatively unknown to date, but recent hypotheses suggest that accumulation of dormant MAPKs, chromatin modifications and alterations of primary metabolism could be involved in the process [66, 68].

Bacterial elicitors were recently shown to stimulate innate immunity in grapevine cultivars through cytoskeleton re-organization, early signalling event activation and defence gene induction [69]. Fungal elicitors have also been proved to be very efficient in stimulating innate immunity in grapevines. The deacetylated derivative of chitin (chitosan) elicitor triggered defence responses and protection against *B. cinerea* [70]. Also, ergosterol has been found to trigger defence gene expression in grapevine plants [71]. However, there are few references that show positive and effective results against pathogens under vineyard conditions [66].

Few of these products have shown acceptable effectiveness against biotrophic pathogens. Therefore, until now, there is not an elicitor-based product that can be used instead of conventional agrochemicals in order to successfully fight *B. cinerea*. Additional research needs to be pursued in order to fully understand the defence mechanisms under vineyard conditions.

2.4. Viral infections in grapevine: an example of compatible host-pathogen interaction

Viral diseases in grapevine are highly complex. This complexity is due to the large amount of different viruses that can infect grapevine plants, occurring most of the time as multiple infections, and because of the nature of the compatible pathogen-host interactions that is

Figure 3. Characteristic symptomatology of the main viruses affecting grapevine. Leaves showing different virus-triggered symptomatology, including reddish areas (A, B, E, H and I), leaf thickness and downward rolling (B), ringspots (D and G), chlorosis (C and F) and yellow veins (H).

established. Viral infections in grapevine plants affect vegetative organs inducing foliar deformations, alterations in leaf colour and, in some cases, graft rejection (**Figure 3**) [72].

Severe infections also reduce berry setting and cause irregular and delayed ripening [72, 73]. Currently, more than 60 viruses have been described in grapevine [73], which together with viroids, phytoplasmas and insect-transmitted xylematic bacteria, correspond to the highest number of intracellular pathogen described for a single crop. Grapevine infecting viruses are classified according to several parameters, including size particle, genome structure, replication strategies, transmission vector and serological information [73]. In general, grapevine infecting viruses exhibit single-stranded RNA (ssRNA) genomes, and the most relevant belong to the Nepovirus, Ampelovirus, Closterovirus and Vitivirus genera.

Viruses belonging to the Nepovirus genus are widely disseminated and are responsible for the degeneration disease. The most representative are the GFLV (Grapevine Fanleaf Virus), ArMV (Arabis Mosaic Virus), SLRSV (Strawberry Latent Ringspot Virus), ToRSV (Tomato Ringspot Virus) and TRSV (Tobacco Ringspot Virus) [72–75]. Most viruses of these groups are not serologically related but share physical and biological attributes [75]. Regarding their infection vectors, Nepoviruses can be transmitted by one or more nematodes species [76, 77]. Moreover, it has been established that GFLV is transmitted by *Xiphinema index*, ArMV by *Xiphinema diversicaudatum* and SLRSV by *Paralongidorus maximus*. Nepoviruses induce grapevine degeneration and leaf decline and produce serious yield losses [74]. However, leaf, steam and berries symptoms vary according to the graft and scion combination, virus strains and environmental conditions. These symptoms include delay in bud break, irregular bud growth, leaf deformity and reduced berry size. Together with the virus-induced decay, a reduction in vegetative growth takes place and even plant death can occur [78, 79]. Decay disease is a major threat for the grape industry since vigour reduction triggered by GFLV infection can reduce yields in about 80% or more [73, 80].

Grapevine leafroll disease (GLD) is one of the most important viral diseases affecting grapevines worldwide [3, 81–83]. It is generally accepted that this disease is caused by 11 viral agents, named GLRaV-1 to GLRaV-11 [3], and according to specific genome sequences, their taxonomic classification includes members of the Ampelovirus genus (GLRaV-1, -3, -4, -5, -6 and -9), the Closterovirus genus (GLRaV-2) and the Velarivirus genus (GLRaV-7). Besides the diversity of viral agents associated with GLD, it is widely assumed that GLRaV-3 is the main etiological factor contributing to the disease. Viral agents responsible for GLD are flexuous filaments, 1800 × 12 nm in size with the unique Closterovirus architecture [3]. These particles are responsible for the characteristic GLD symptom, expressed as red colour leaves with green vein pattern, often curled downwards and brittle [83]. In red cultivars, GLD symptomatology is much more evident in comparison to white cultivars, where the disease can be asymptomatic [84]; nevertheless, white cultivars can show inter-veinal yellowing of leaves and leaf rolling [83].

Grapevine virus A and B (GVA and GVB), which belong to the Vitivirus genus, are also relevant [85]. GVA is related with the Kober stem grooving symptom, where severe grooving on the grafted stems occurs [86, 87], while GVB is associated with the corky bark syndrome consisting of soft, rubbery and abnormal swelling of the basal internodes of the canes, longitudinal cracks and cork forming, typical of the rugose wood complex [88]. Vitivirus genus

such as GLRaV-3. Therefore, it is reasonable to think that the infection may alter the molecules flow towards the berries, and that this effect may vary according to the asymptomatic or symptomatic phases of the infection and grapevine phenological stages [3, 83].

The effects of a chronical infection with GLRaV-3 during berry ripening in grapevine have been studied in the red cultivar Cabernet Sauvignon [97]. Interestingly, this virus affects the normal fruit ripening process, resulting in incomplete berry ripening in terms of gene expression patterns. Genes associated with anthocyanin biosynthesis and sugar metabolism are down-regulated in berries from infected plants, consistent with a decrease in up to 40% in total anthocyanin content. These changes are observed specifically at ripening, where the infection has a greater impact in comparison with other stages of berries development. These authors also suggest the presence of viral particles in berries, probably colonizing the organ through the vasculature during fruit development.

Lately, the effect of GLRaV-3 on the chemical properties of fruit, juice and wine from *V. vinifera* L. cv. Sauvignon blanc was assessed, allowing comparisons between recent and established infections [84]. Authors propose that the duration of the infection is significant to this comparison, and that established infections modify berry development at later stages. The pathogen causes a delay in grape ripening, with a concomitant delay in harvest date. However, when berries from uninfected and infected plants reached similar ripeness, minimal effects on juice and wine chemistry were observed.

2.6. Diagnostic and control methods for grapevine viruses

2.6.1. Diagnostic methods

Since grapevine viruses can show detrimental effects on plant physiology, it is necessary to have appropriate and reliable diagnosis methods to achieve an efficient control of pathogens propagation. So far, several techniques have been applied to identify infected plant material, including biological indexing, serology and molecular assays [3, 83, 99].

Biological indexing, mostly performed as part of certification programs, refers to grafting of candidate vine on woody indicators of the Vitis genus. Later on, the indicator plant is observed for the development of virus disease symptoms. However, this approach is time-consuming, labour intensive and dependent on virus titer, the success of the viral inoculation, strain variations and skilled personnel [83]. Serological methods are based on the recognition of viral proteins by specific antibodies. Of these, the enzyme-linked immunosorbent assay (ELISA) is the most widely applied [83, 99], and commercial kits are available. Serological approaches are robust and scalable, although less sensitive than nucleic acid-based techniques. Special attention must be given to sampling, considering differences in virus accumulation through plant tissues and seasonal variations in virus titer, and it has been described that genetic variants can affect the robustness of these methods [3, 83]. Nucleic acid-based methods, on the other hand, detect the genomic components of the viruses. These methods are commonly used due to their high sensitivity, in comparison with other diagnostic approaches. They can detect the presence of viral genomes even at low viral titer, are rapid, allow the scaling and the simultaneous analysis of a high number of samples or several viruses at once [100]. Since most of grapevine viruses have RNA genomes, reverse transcription-polymerase chain reaction (RT-PCR) is the selected molecular assay for the detection of these pathogens [83, 101]. Several techniques have been developed based on PCR variants [83, 102], but the use of real-time PCR allows quantification of virus titer [103]. Recently, new generation sequencing (NGS) has been used for rapid identification and sequencing of all putative viruses present in a candidate sample, allowing the identification of new viral agents as well [3, 81, 99, 104]. The use of NGS technologies as diagnostic tool requires no prior knowledge of the pathogens present in the sample, but is still expensive in order to be used as a routine procedure.

2.6.2. Multiplex PCR to detect complex viral infections

As it was mentioned before, viral diseases in grapevine often occurs as multiple infections, where several viral agents are present simultaneously and can contribute to the overall symptoms development. Several papers describe viral detection by molecular approaches, which are reviewed in [83, 99]. However, a simple and efficient commercial method for the

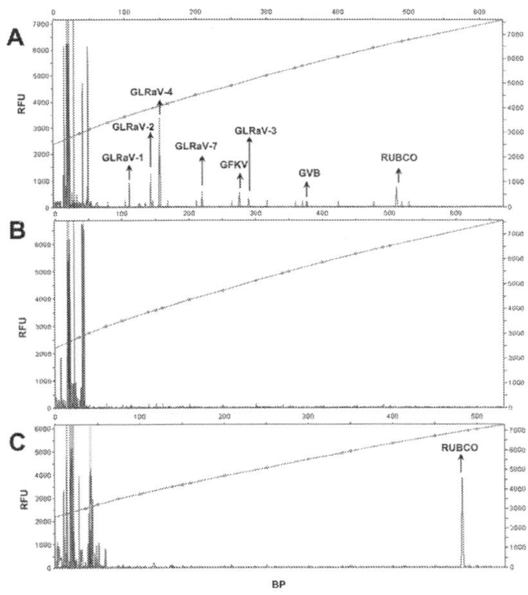

Figure 5. Multiplex PCR detection of grapevine viruses in complex samples. PCR fragments were analysed by capillary electrophoresis in order to detect the different amplicons. (A) Detection of the specific fragments corresponding to GFKV, GVB, GLRaV-1, GLRaV-2, GLRaV-3, GLRaV-4 and GLRaV-7, in a sample containing a mix of cDNA from plants infected with these viruses. (B) Negative control of the PCR. (C) PCR using a virus-free plant. The ROX500 standard was used to estimate the size of each fragment. Blue peaks correspond to the amplified fragments obtained in each reaction. Red line represents the linear regression obtained with the Peak Scanner Software V.1.0 to estimate fragment size. Base pairs of each fragment (*X*-axis), fluorescence intensity (RFU) of amplified fragments (*Y*-axis, left) or RFU of the standard (*Y*-axis, right) are shown.

detection of several grapevine viruses at once is currently not available. A method for virus detection must fulfil several criteria, such as sensitivity, specificity, accuracy, number of samples that can be tested simultaneously and cost, among others. Therefore, to have reliable diagnosis methods is a permanent challenge for grapevine growers.

In our laboratory (unpublished work), we have designed a system for simultaneous virus detection in vines, consisting in a multiplex PCR that can detect up to seven RNA viral genomes, in addition to the detection of the gene coding for the small sub-unit of grapevine Rubisco enzyme as a plant positive control. Using bioinformatics tools, specific primers against different viruses were designed to generate products of different sizes. Then, primers were labelled at the 5′ end with 6-FAM fluorophore, in order to be detected by capillary electrophoresis. This method allows the specific and simultaneous detection of GFKV, GVB, GLRaV-1, GLRaV-2, GLRaV-3, GLRaV-4 and GLRaV-7, in a quick, efficient and single PCR (**Figure 5**).

This type of multiplex PCR can be used to generate commercial kits that can serve to detect viral agents present in a vineyard or to test the plant material that will be later used in clonal propagation. With the proper bioinformatics analysis, more viruses can be added to the system, allowing a much more versatile detection kit.

2.6.3. *Control methods*

There are several control methods that are routinely applied in order to prevent virus dissemination. For instance, sanitary selection and certification of propagation material helps to reduce potential virus dispersion [99]. Since viruses are transmitted by vectors, control of viral diseases can be achieved by the restriction of such vectors with the use of agrochemicals [105]. However, agrochemicals utilization increases production costs, and additionally are associated with detrimental effects to the environment and human health, while most of modern agronomical practices tend to reduce its use. Sanitation techniques, on the other hand, are aimed to treat infected material and eliminate viral titer. Among these techniques, thermotherapy is the most frequently applied although it is not effective for all grapevine viruses [106]. The *in vitro* culture of meristems, somatic embryos and shoot tips allows the regeneration of virus-free plantlets [99], an approach that is probably based on the unequal distribution of viruses along plant tissues. Other sanitation techniques include chemotherapy, very often applied as an alternative to eliminate more recalcitrant viruses and cryotherapy, a highly efficient method which is effective when the treatment of high number of samples is needed, but its implementation is difficult as some genotypes are refractory to it [99].

2.6.4. *Inducing virus resistance in grapevine by transgenesis*

Biotechnology arises as an alternative to allow the generation of virus-resistant grapevine plants by transgenesis, mainly involving the expression of viral components and exploiting the naturally occurring gene silencing [107–117]. This strategy requires plant transformation with a short sequence of the pathogen genome in a way that a double-strand RNA structure is formed during transcription, initiating gene silencing in the host. In our lab, induction of virus silencing was accomplished in grapevine rootstocks in order to be used for grafting [118].

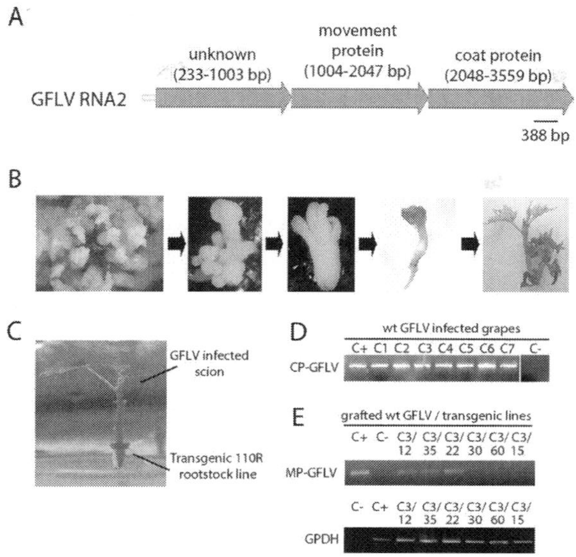

Figure 6. GFLV silencing induced by grafting strategy. (A) RNA 2 of GFLV genome showing the unknown, movement and coat protein corresponding ORFs; the 388 bp region of coat protein used to produce the inverted duplicate that triggers PTGS is denoted. (B) To obtain transgenic rootstock, globular embryos were used for transformation. (C) *In vitro* grafting of GFLV infected plants on transgenic rootstock expressing the *cp-gflv* transgene. (D) Presence of GFLV in infected wild-type (wt) plants (C1–C7) showed by amplification using CP-GFLV primers. (E) Evaluation of GFLV presence in upper leaves of the grafts with MP-GFLV primers, 30 days after grafting. MP-GFLV was undetected in C3 plants grafted on transgenic lines 30, 60 and 15. Expression of the GPDH gene was used as a housekeeping control.

It is expected that the mobile signal-inducing virus silencing in the rootstock will also be able to reach the scion, and as a consequence, trigger virus silencing in the non-transgenic scion. This approach is very versatile, since the resistance against a specific virus can be obtained in all the varieties used as scion with a particular virus-resistant transgenic rootstock. We have transformed rootstocks plants (110 Richter and Harmony) by co-culture of embryogenic and organogenic tissues with *Rhizobium radiobacter* carrying a vector containing a silencing sequence of the coat protein of the grapevine fanleaf virus (GFLV) cloned as an inverted duplicate in a way that triggers post-transcriptional gene silencing (PGTS) at transcription. Twenty-six transgenic plants of the 110 Richter rootstock have been recovered, analysed by RT-PCR against the GFLV sequence, and lines properly expressing the construction were propagated to obtain several plants of each line. The transgenic rootstocks have been grafted with GFLV-infected plants that were positive for virus presence by RT-PCR analysis. Once the grafts were set, the GFLV detection was made in the scion using primers for the viral movement protein. After 1 month of grafting, the detection of the virus has been abolished in the scion, in three of the six analysed rootstocks lines (**Figure 6**).

Therefore, a viral infection of a non-transgenic scion could be silenced if it is grafted on a transgenic rootstock carrying sequences that triggers PTGS. This strategy is an interesting alternative to considerate in virus-free breeding programs because the infection in non-transgenic grapevines from any cultivar could be abolished using a transgenic rootstock, keeping cultivars and, more important, the fruit produced non-transgenic.

3. Grapevine abiotic stress

Grapevine crops are often exposed to sub-optimal growing conditions which cause several abiotic stresses, as they are constantly exposed to different water regimes, nutrient deficiency or excess, extreme heat or low temperatures and deficit or excess of light [8]. All plants, including grapevine, need Sun energy in order to produce organic compounds through photosynthesis, but sunlight is a sum of different wavelengths. Among them, the ultraviolet radiation (UVR) plays an important role, however, the main problem with UV light is that as the wavelength declines, its energy content increases, mainly as UV-B radiation, and therefore its potential to cause photo-biological damage increases. UV-B is not only potentially harmful, but it also serves as an environmental information source, though information about it is still scarce. As general abiotic stresses have been extensively reviewed [119–125], we will have special focus on UV-B-mediated perception and signalling responses of grapevine and photo-biotechnological approaches to improve fruit quality for winemaking.

3.1. Solar ultraviolet B levels, ozone layer depletion and increase of UV-B radiation

Solar energy is the primary source of energy for all surface phenomena, especially autotrophic organisms. Among them, plants use solar radiation not only as an energy source, but also as a key signal containing vital information about the environment in which they live [126, 127]. Solar radiation not only includes the visible spectrum (400–700 nm) necessary for photosynthesis, but also other types of radiation. Near 7% of the electromagnetic radiation emitted by the Sun is within the ultraviolet radiation (UVR) spectrum (200–400 nm) [128–130]. UVR has been divided into three different bands: UV-A (315–400 nm), UV-B (280–315 nm) and UV-C (200–280 nm) [130, 131]. As it passes through the atmosphere, the total transmitted radiation flux is considerably reduced, and the composition of UVR is modified. Shortwave UV-C is completely absorbed by atmospheric gases, while UV-B is partially absorbed by the stratospheric ozone (O_3), leaving only a small fraction (<0.5% of total sunlight energy) transmitted to the Earth surface. UV-A, on the other hand, is not absorbed by ozone [130, 132]. Over the last 50 years, the ozone concentration has diminished by 5%, mainly due to the release of anthropogenic pollutants, such as chlorofluorocarbons (CFCs) and other halogenated ozone-depleting substances [126, 133]. As a direct consequence of the ozone reduction, an increase in the flux of UV-B radiation has been registered during the last years [126, 128, 132]. Although, UV-B radiation is only a minor component of solar radiation, due to its high energy, its potential for causing biological damage is exceptionally high [133]. Besides the regulation of solar UV-B by the ozone layer, there are several factors influencing UV-B radiation levels, such as latitude, altitude, season, time of the day, weather conditions, surface reflection, atmospheric pollution and shading by plant canopies [126, 133]. From a wine producer's point of view, the establishment, planning and vineyard management are additional factors to take into account that can influence UV-B levels on plants. These factors including climate, presence and slope aspect, site elevation, trellis and training system and vine vigour, among others, could be directly influencing both intercepted light in canopies and fruit zone [8].

3.2. Effects of UV-B in plants

Due to the sessile lifestyle, plants are forced to adapt to changes in environmental conditions while achieving an equilibrium between optimal photosynthetically active radiation (PAR) capture and UV-B protection [131, 132]. The UV-B radiation has several detrimental effects but it also serves as a key regulator of plant morphology and physiological, biochemical and genetic mechanisms [127, 129, 133, 134]. Plants actively respond to irradiation with high or low UV-B doses, either by the activation of repair mechanisms or by stimulation of photomorphogenic processes [128, 129]. In general, low UV-B doses reduces growth and expansion of leaves, produces leaf thickness, increases epicuticular waxes, trichomes number and axillary branching, reduces stem elongation, suppresses both hypocotyl extension and root growth and enhances flavonoid biosynthesis, mostly flavonol [127–129, 132, 134]. Plants under UV-B radiation present a compact architecture, although different phenotypes have been reported. This may relate to UV-induced morphological changes being underpinned by different mechanisms at high and low UV-B doses [127].

In grapevine, the high UV-B doses reduce shoot length and leaf area, increase both leaf thickness [135] and accumulation of terpenes with antioxidant properties [136]. On the other hand, flavonol biosynthesis is dramatically activated under both high and low UV-B exposures in the berry skin [9, 10]. Also, membrane-related terpenes are increased in low fluence of UV-B in grapevine leaves [136].

3.3. UVR8-mediated photomorphogenic mechanisms in response to UV-B in plants

In order to maximize its growth and survival, plants detect, respond and adapt to UV-B rays. This type of radiation is a key environmental cue, which initiates diverse pathways affecting metabolism, development and viability. Many of the UV-B radiation effects involve differential regulation of gene expression. This response depends on the exposition nature (high or low UV-B doses), the degree of adaptation and acclimation to the radiation, and the interaction with other environmental factors. UV-B radiation responses are mediated by two signalling pathways in *Arabidopsis thaliana*, (1) the non-specific signalling pathway, which involves DNA damage, accumulation of reactive oxygen species (ROS) and synthesis of defence-related molecules in response to high levels of UV-B radiation; and (2) the specific signalling pathway on the other hand, which is mediated by photomorphogenic responses to low levels of UV-B radiation [128]. It is important to note that photomorphogenic signalling promotes the expression of genes involved in the protection and acclimation against UV-B radiation and, hence, promotes the survival of exposed plants. Photomorphogenic signalling implies the participation of a specific component in *Arabidopsis*, the UV-B photoreceptor UVR8 (UV Resistance Locus 8) with specific tryptophan residues which act as intrinsic chromophores [137]. UVR8 perceives radiation, triggering the dissociation from its non-active homodimer configuration [137, 138]. Following monomerization, UVR8 accumulates in the nucleus and interacts with the positive regulator Constitutively Photomorphogenic 1 (COP1) [139–142], a WD40/RING-E3 ubiquitin ligase that in non-inductive conditions targets HY5 (Elongated Hypocotyl 5) for proteosome-dependent degradation [143]. HY5 is a key effector of UV-B protection and light photomorphogenic responses [144, 145], and it is transcriptionally activated by UV-B in a UVR8- and COP1-dependent manner [146–148]. Other components of the UVR8 signalling pathway are repressor of UV-B photomorphogenesis 1 (RUP1) and RUP2. Both RUP1 and RUP2 act as feedback inhibitors of UVR8 signalling by facilitating UVR8 redimerization after exposure to UV-B and thus preserve responsiveness to changing levels of the input signal [149, 150].

3.4. Elucidating the grapevine UV-B signalling pathway

Grapevine (*Vitis vinifera* L.) is a woody species often cultivated in Mediterranean climates with varied UV-B radiation dosages, generally ranging between moderate (5 kJ m^{-2} d^{-1}) to high (12 kJ m^{-2} d^{-1}) levels [7]. Grapevines are considered well-adapted plants to solar radiation due to a variety of physiological adaptive responses, mainly based in antioxidant enzyme activities and secondary metabolites production [7]. The most common protective mechanisms against potentially harmful radiation are the synthesis of phenolic compounds that absorb UV-B radiation [128, 132]. Among the versatile range of functions flavonoids possesses, the most important related to UV-B include the ability to attenuate radiation by filtering, an antioxidant activity capable of scavenging free radicals, and the modulation of reactive oxygen signalling cascades involved in growth and development [151]. These secondary metabolites, phenolic compounds, flavonoids and cinnamate esters, among others, accumulate in vacuoles of epidermal cells in response to UV-B radiation and attenuate the further diffusion of solar UV-B in deeper cell layers [128, 132]. In grapevine, it has been reported that in leaf epidermis and fruit berry skins, anthocyanins and flavonols increase in response to UV-B [11, 152–154]. Considered as a relevant model for studying adaptive responses, several approaches have been conducted for understanding the effects of UV-B in grapevine. Analysis of the transcriptomic variations caused by a particular UV-B radiation dose (4.75 kJ m^{-2} d^{-1}) given at high and low

Gene	ID	Expression[a]	Reference
VvUVR1	VIT_07s0031g02560	No change	Carbonell-Bejerano et al. (2014)
VvHY5	VIT_04s0008g05210	Up-regulated	Loyola et al. (2016)
VvHYH	VIT_05s0020g01090	Up-regulated	
VvCOP1-1	VIT_12s0059g01420	No change	
VvCOP1-2	VIT_10s0523g00030	No change	
VvRUP	VIT_16s0050g00020	Up-regulated	

[a]Differential expression in UV-B radiation treatments compared to the control.

Table 1. UV-B perception and signalling grape homologues.

fluence rates demonstrated that DNA repair, synthesis of UV-B sunscreens and general multiple-stress pathways were the main activated processes [155]. Additionally, has been reported the identification and characterization of the grapevine homologues of the Arabidopsis UV-B signalling components (**Table 1**) and that this UV-B radiation-specific signalling cascade is activated in berry skin along with the accumulation of secondary metabolites, mainly flavonols [9, 10].

3.5. Manipulation of UV-B perception and signalling components to improve plant shape and fruit quality in grapevine

It is known that grapevine is a vigorous growing plant; hence, one of the main objectives for viticulture practices is to reduce the size of the canopies and alter the shape of the vine, in order to increase field plant density and improve fruit organoleptic qualities, among others [8, 156]. Moreover, a higher plant density means greater productivity per area unit. To meet these objectives, conventional genetic improvement of most fruit crops, including grapevine, has been extensively done, with several obstacles in the way. Among the latter we can find long juvenility periods, seedlessness, self-incompatibility, high heterozygosity and sterility. Therefore, conventional breeding techniques are difficult, expensive and time consuming [156]. Because of this, genetic improvement through genetic engineering techniques offers an attractive alternative in order to overcome these problems.

Gene name	Role in UV-B signalling[a]	Grape homologue	Experimental approach	Phenotype or trait of interest[a]	Reference
AtUVR8	Photoreceptor	VvUVR1	Over-expression	UV-B tolerance, dwarfing, increased flavonoids levels, enhanced *B. cinerea* resistance.	Rizzini et al. [137]; Demkura and Ballaré [157]
AtCOP1	Positive regutator	VvCOP1-1 VvCOP1-2	Over-expression	UV-B tolerance, dwarfing, increased flavonoids levels.	Oravecz et al. [142]
AtHY5	Positive regulator	VvHY5	Over-expression	UV-B tolerance, dwarfing, increased flavonoids levels.	Ulm et al. [148]
AtHYH	Positive regulator	VvHYH	Over-expression	Moderated UV-B tolerance, increased flavonoids levels.	Brown and Jenkins [158]
AtRUP2	Negative regulator	VvRUP	Silencing or down-regulation	UV-B tolerance, extreme dwarfing, increased flavonoids levels.	Heijde and Ulm [150]

[a]Derived from studies in *Arabidopsis*.

Table 2. UV-B signalling target genes that could be genetically engineered in grapevine.

In vine growing, the production of dwarf and semi-dwarf canopies with short and numerous shoots, in order to increase field vine density, are normally used for both dwarfing rootstocks and spur varieties [8, 156]. However, rootstocks and spur varieties are available for only a few species and graft compatibility is often a problem. Therefore, an alternative to this is the use of photo-biotechnology techniques which may contribute to the creation of dwarf varieties by genetic engineering, modifying, for example, UV-B perception and/or signalling components (see **Table 2**).

Photo-biotechnology refers to over- or down-expressing genes with photo-biological relevance [159]. Since photoreceptors and/or light signalling cascade components regulate the expression of critical development and plant growth genes, genetic manipulation of these is viewed as a promising strategy to develop fruit crops with improved agronomic traits [159]. Therefore, photo-biotechnology offers a promising approach for studying the influence of UV-B signal transduction components on plant development and may be used to improve crop yield, shade tolerance, growth and fruit ripening, canopies shape, hormone synthesis and biosynthesis of metabolites and pigments. For example, a promising study in tomato showed that down-regulation of *LeHY5* by RNAi-mediated gene repression exhibited defects in light photomorphogenesis response, loss of thylakoid organization and reduced carotenoid accumulation. In contrast, repression of *LeCOP1LIKE* expression resulted in plants with an exaggerated photomorphogenesis response, high levels of chlorophyll and elevated fruit carotenoid levels [160]. These results demonstrate that genes encoding components of UV-B signalling cascade represent a promising genetic tool for manipulation of fruit quality. Additionally, several studies summarized by [159] in various plant species show that modulation of the expression of phytochromes (mainly PhyA and PhyB) can be used to produce high-yielding crops.

The quality of grape berries for winemaking integrates various aspects, but as for red wines, the accumulation of phenolic compounds by UV-B is highly necessary [161]. Wines with the highest phenolic concentrations are generally considered of excellence, therefore, these molecules are said to play a significant role in winemaking since they are key determinants of wine quality [161]. All of the aforementioned evidence suggests that UV-B protective mechanisms may potentially lead to important industrial applications, relevant to the wine industry. UVR8 may prove to be an attractive and suitable target to manipulate plant growth and/or plant tolerance to abiotic stress, generating UV-B-resistant grapevines with enhanced secondary metabolites levels (i.e. phenolic compounds).

In summary, the elucidation of the UV-B signalling pathway and the role of photomorphogenesis, in addition to advances in genetic manipulation of grapes, are unique biotechnological tools that could be used to improve grapevines in order to meet and surpass market expectations.

Acknowledgements

This work was supported by the National Commission for Science and Technology CONICYT (FONDECYT grant number 1150220) and the Millennium Nucleus for Plant Synthetic Biology and Systems Biology NC130030. G. Armijo, R. Loyola and F. Restovic were supported by FONDECYT postdoctoral grant number 3140324, 3150578 and 3150259, respectively.

Author details

Grace Armijo[1], Carmen Espinoza[1], Rodrigo Loyola[1], Franko Restovic[1], Claudia Santibáñez[1,2], Rudolf Schlechter[1,3], Mario Agurto[1] and Patricio Arce-Johnson[1*]

*Address all correspondence to: parce@bio.puc.cl

1 Department of Molecular Genetics and Microbiology, Faculty of Biological Sciences, Pontifical Catholic University of Chile, Santiago, Chile

2 Université de Bordeaux, ISVV, INRA, Ecophysiology and Functional Genomic of Grapevine, Villenave d'Ornon, France

3 Department of Biology, Institute of Agricultural Sciences, ETH Zurich, Zurich, Switzerland

References

[1] This, P., T. Lacombe, and M.R. Thomas: Historical origins and genetic diversity of wine grapes. Trends Genet. 2006; 22(9): p. 511–519. DOI: 10.1016/j.tig.2006.07.008.

[2] Ferreira, R.B., S.S. Monteiro, M.A. Picarra-Pereira, and A.R. Teixeira: Engineering grapevine for increased resistance to fungal pathogens without compromising wine stability. Trends Biotechnol. 2004; 22(4): p. 168–173. DOI: 10.1016/j.tibtech.2004.02.001.

[3] Naidu, R.A., H.J. Maree, and J.T. Burger: Grapevine leafroll disease and associated viruses: A unique pathosystem. Ann Rev Phytopathol. 2015; 53: p. 613–634. DOI: 10.1146/annurev-phyto-102313-045946.

[4] Ehrenfeld, N., P. Canon, C. Stange, C. Medina, and P. Arce-Johnson: Tobamovirus coat protein CPCg induces an HR-like response in sensitive tobacco plants. Mol Cells. 2005; 19(3): p. 418–427.

[5] Espinoza, C., A. Vega, C. Medina, K. Schlauch, G. Cramer, and P. Arce-Johnson: Gene expression associated with compatible viral diseases in grapevine cultivars. Funct Integr Genom. 2007; 7(2): p. 95–110. DOI: 10.1007/s10142-006-0031-6.

[6] O'Donnell, P.J., E.A. Schmelz, P. Moussatche, S.T. Lund, J.B. Jones, and H.J. Klee: Susceptible to intolerance – a range of hormonal actions in a susceptible Arabidopsis pathogen response. Plant J. 2003; 33(2): p. 245–257. DOI: 10.1046/j.1365-313X.2003.01619.x.

[7] Martinez-Luscher, J., F. Morales, S. Delrot, M. Sanchez-Diaz, E. Gomes, J. Aguirreolea, and I. Pascual: Short- and long-term physiological responses of grapevine leaves to UV-B radiation. Plant Sci. 2013; 213: p. 114–122. DOI: 10.1016/j.plantsci.2013.08.010.

[8] Keller, M. Environmental constraints and stress physiology. In M. Keller, Editor. The Science of Grapevines, Second ed. Academic Press: San Diego; 2015. Pp. 267–341. DOI: 10.1016/B978-0-12-419987-3.00007-8

[9] Carbonell-Bejerano, P., M.P. Diago, J. Martinez-Abaigar, J.M. Martinez-Zapater, J. Tardaguila, and E. Nunez-Olivera: Solar ultraviolet radiation is necessary to enhance grapevine fruit ripening transcriptional and phenolic responses. BMC Plant Biol. 2014; 14: p. 183. DOI: 10.1186/1471-2229-14-183.

[10] Liu, L., S. Gregan, C. Winefield, and B. Jordan: From UVR8 to flavonol synthase: UV-B-induced gene expression in Sauvignon blanc grape berry. Plant Cell Environ. 2015; 38(5): p. 905–919. DOI: 10.1111/pce.12349.

[11] Berli, F.J., M. Fanzone, P. Piccoli, and R. Bottini: Solar UV-B and ABA are involved in phenol metabolism of *Vitis vinifera L.* increasing biosynthesis of berry skin polyphenols. J Agric Food Chem. 2011; 59(9): p. 4874–4884. DOI: 10.1021/jf200040z.

[12] Pieterse, C.M., A. Leon-Reyes, S. Van der Ent, and S.C. Van Wees: Networking by small-molecule hormones in plant immunity. Nat Chem Biol. 2009; 5(5): p. 308–316. DOI: 10.1038/nchembio.164.

[13] Qiu, W., A. Feechan, and I. Dry: Current understanding of grapevine defense mechanisms against the biotrophic fungus (*Erysiphe necator*), the causal agent of powdery mildew disease. Hort Res. 2015; 2: p. 15020. DOI: 10.1038/hortres.2015.20.

[14] Gadoury, D.M., L. Cadle-Davidson, W.F. Wilcox, I.B. Dry, R.C. Seem, and M.G. Milgroom: Grapevine powdery mildew (*Erysiphe necator*): A fascinating system for the study of the biology, ecology and epidemiology of an obligate biotroph. Mol Plant Pathol. 2012; 13(1): p. 1–16. DOI: 10.1111/j.1364-3703.2011.00728.x.

[15] Gadoury, D.M., R.C. Seem, A. Ficke, and W.F. Wilcox: Epidemiology of powdery mildew on concord grapes. Phytopathology. 2001; 91(10): p. 948–955.

[16] Micali, C., K. Gollner, M. Humphry, C. Consonni, and R. Panstruga: The powdery mildew disease of arabidopsis: A paradigm for the interaction between plants and biotrophic fungi. Arabidopsis Book. 2008; 6: p. e0115. DOI: 10.1199/tab.0115.

[17] van Kan, J.A.: Licensed to kill: The lifestyle of a necrotrophic plant pathogen. Trends Plant Sci. 2006; 11(5): p. 247–253. DOI: 10.1016/j.tplants.2006.03.005.

[18] Williamson, B., B. Tudzynski, P. Tudzynski, and J.A. van Kan: *Botrytis cinerea*: The cause of grey mould disease. Mol Plant Pathol. 2007; 8(5): p. 561–580. DOI: 10.1111/j.1364-3703.2007.00417.x.

[19] Machota, R., Jr., L.C. Bortoli, F.R. Cavalcanti, M. Botton, and A.D. Grutzmacher: Assessment of injuries caused by *Anastrepha fraterculus* (Wied.) (Diptera: Tephritidae) on the incidence of bunch rot diseases in table grape. Neotrop Entomol. 2016. DOI: 10.1007/s13744-016-0377-y.

[20] Williamson, B., G. Duncan, J. Harrison, and G. Zimand: Effect of humidity on infection of rose petals by dry-inoculated conidia of *Botrytis cinerea*. Mycol Res. 1995; 99(11): p. 1303–1310. DOI: 10.1016/S0953-7562(09)81212-4.

[21] Prusky, D.: Pathogen quiescence in postharvest diseases. Annu Rev Phytopathol. 1996; 34: p. 413–434. DOI: 10.1146/annurev.phyto.34.1.413.

[22] Tsuda, K. and F. Katagiri: Comparing signaling mechanisms engaged in pattern-triggered and effector-triggered immunity. Curr Opin Plant Biol. 2010; 13(4): p. 459–465. DOI: 10.1016/j.pbi.2010.04.006.

[23] Jones, J.D. and J.L. Dangl: The plant immune system. Nature. 2006; 444(7117): p. 323–329. DOI: 10.1038/nature05286.

[24] Mengiste, T.: Plant immunity to necrotrophs. Annu Rev Phytopathol. 2012; 50: p. 267–294. DOI: 10.1146/annurev-phyto-081211-172955.

[25] Bent, A.F. and D. Mackey: Elicitors, effectors, and R genes: The new paradigm and a lifetime supply of questions. Annu Rev Phytopathol. 2007; 45: p. 399–436. DOI: 10.1146/annurev.phyto.45.062806.094427.

[26] Dry, I., Feechan, A., Anderson, C., Jermakow, A., Bouquet, A., Adam-Blondon, A., and Thomas, M.: Molecular strategies to enhance the genetic resistance of grapevines to powdery mildew. Aust J Grape Wine R. 2010; 16: p. 94–105. DOI: 10.1111/j.1755-0238.2009.00076.x.

[27] Feechan, A., C. Anderson, L. Torregrosa, A. Jermakow, P. Mestre, S. Wiedemann-Merdinoglu, D. Merdinoglu, A.R. Walker, L. Cadle-Davidson, B. Reisch, S. Aubourg, N. Bentahar, B. Shrestha, A. Bouquet, A.F. Adam-Blondon, M.R. Thomas, and I.B. Dry: Genetic dissection of a TIR-NB-LRR locus from the wild North American grapevine species Muscadinia rotundifolia identifies paralogous genes conferring resistance to major fungal and oomycete pathogens in cultivated grapevine. Plant J. 2013; 76(4): p. 661–674. DOI: 10.1111/tpj.12327.

[28] Glawe, D.A.: The powdery mildews: a review of the world's most familiar (yet poorly known) plant pathogens. Annu Rev Phytopathol. 2008; 46: p. 27–51. DOI: 10.1146/annurev.phyto.46.081407.104740.

[29] Marsh, E., S. Alvarez, L.M. Hicks, W.B. Barbazuk, W. Qiu, L. Kovacs, and D. Schachtman: Changes in protein abundance during powdery mildew infection of leaf tissues of Cabernet Sauvignon grapevine (Vitis vinifera L.). Proteomics. 2010; 10(10): p. 2057–2064. DOI: 10.1002/pmic.200900712.

[30] Ficke, A., D.M. Gadoury, and R.C. Seem: Ontogenic resistance and plant disease management: A case study of grape powdery mildew. Phytopathology. 2002; 92(6): p. 671–675. DOI: 10.1094/PHYTO.2002.92.6.671.

[31] Gabler, F.M., J.L. Smilanick, M. Mansour, D.W. Ramming, and B.E. Mackey: Correlations of morphological, anatomical, and chemical features of grape berries with

resistance to Botrytis cinerea. Phytopathology. 2003; 93(10): p. 1263–1273. DOI: 10.1094/PHYTO.2003.93.10.1263.

[32] Aziz, A., B. Poinssot, X. Daire, M. Adrian, A. Bezier, B. Lambert, J.M. Joubert, and A. Pugin: Laminarin elicits defense responses in grapevine and induces protection against Botrytis cinerea and Plasmopara viticola. Mol Plant Microbe Interact. 2003; 16(12): p. 1118–1128. DOI: 10.1094/MPMI.2003.16.12.1118.

[33] Verhagen, B., P. Trotel-Aziz, P. Jeandet, F. Baillieul, and A. Aziz: Improved resistance against Botrytis cinerea by grapevine-associated bacteria that induce a prime oxidative burst and phytoalexin production. Phytopathology. 2011; 101(7): p. 768–777. DOI: 10.1094/PHYTO-09-10-0242.

[34] Armijo, G., R. Schlechter, M. Agurto, D. Muñoz, C. Núñez, and P. Arce: Grapevine pathogenic microorganisms: understanding infection strategies and host response scenarios. Front Plant Sci. 2016. DOI: 10.3389/fpls.2016.00382.

[35] Dai, R., H. Ge, S. Howard, and W. Qiu: Transcriptional expression of Stilbene synthase genes are regulated developmentally and differentially in response to powdery mildew in Norton and Cabernet Sauvignon grapevine. Plant Sci. 2012; 197: p. 70–76. DOI: 10.1016/j.plantsci.2012.09.004.

[36] Delaunois, B., S. Cordelier, A. Conreux, C. Clement, and P. Jeandet: Molecular engineering of resveratrol in plants. Plant Biotechnol J. 2009; 7(1): p. 2–12. DOI: 10.1111/j.1467-7652.2008.00377.x.

[37] Lee, K., J.L. Smith, and J.A. Last: Absence of respiratory inflammatory reaction of elemental sulfur using the California Pesticide Illness Database and a mouse model. J Agromed. 2005; 10(3): p. 41–47.

[38] Burger, P., A. Bouquet, and M.J. Striem. Grape breeding. In S.M. Jain and P.M. Priyadarshan, Editors. Breeding Plantation Tree Crops: Tropical Species. Springer: New York, EE.UU.; 2009. Pp. 161–189.

[39] Pauquet, J., A. Bouquet, P. This, and A.-F. Adam-Blondon: Establishment of a local map of AFLP markers around the powdery mildew resistance gene Run1 in grapevine and assessment of their usefulness for marker assisted selection. Theor Appl Genet. 2001; 103(8): p. 1201–1210. DOI: 10.1007/s001220100664.

[40] Coleman, C., D. Copetti, G. Cipriani, S. Hoffmann, P. Kozma, L. Kovacs, M. Morgante, R. Testolin, and G. Di Gaspero: The powdery mildew resistance gene REN1 co-segregates with an NBS-LRR gene cluster in two Central Asian grapevines. BMC Genet. 2009; 10: p. 89. DOI: 10.1186/1471-2156-10-89.

[41] Feechan, A., S. Kabbara, and I.B. Dry: Mechanisms of powdery mildew resistance in the Vitaceae family. Mol Plant Pathol. 2011; 12(3): p. 263–274. DOI: 10.1111/j.1364-3703.2010.00668.x.

[42] Hoffmann, S., G. Di Gaspero, L. Kovacs, S. Howard, E. Kiss, Z. Galbacs, R. Testolin, and P. Kozma: Resistance to Erysiphe necator in the grapevine 'Kishmish vatkana' is

controlled by a single locus through restriction of hyphal growth. Theor Appl Genet. 2008; 116(3): p. 427–438. DOI: 10.1007/s00122-007-0680-4.

[43] Barker, C.L., T. Donald, J. Pauquet, M.B. Ratnaparkhe, A. Bouquet, A.F. Adam-Blondon, M.R. Thomas, and I. Dry: Genetic and physical mapping of the grapevine powdery mildew resistance gene, Run1, using a bacterial artificial chromosome library. Theor Appl Genet. 2005; 111(2): p. 370–377. DOI: 10.1007/s00122-005-2030-8.

[44] Molnár, S., Galbács, Z., Halász, G., Hoffmann, S., Kiss, E., Kozma, P., Veres, A., Galli, Z., Szőke, A., and Heszky, L.: Marker assisted selection (MAS) for powdery mildew resistance in a grapevine hybrid family. Vitis. 2007; 46(4): p. 212–213.

[45] Blanc, S., S. Wiedemann-Merdinoglu, V. Dumas, P. Mestre, and D. Merdinoglu: A reference genetic map of Muscadinia rotundifolia and identification of Ren5, a new major locus for resistance to grapevine powdery mildew. Theor Appl Genet. 2012; 125(8): p. 1663–1675. DOI: 10.1007/s00122-012-1942-3.

[46] Ramming, D.W., F. Gabler, J. Smilanick, M. Cadle-Davidson, P. Barba, S. Mahanil, and L. Cadle-Davidson: A single dominant locus, ren4, confers rapid non-race-specific resistance to grapevine powdery mildew. Phytopathology. 2011; 101(4): p. 502–508. DOI: 10.1094/PHYTO-09-10-0237.

[47] Riaz, S., A.C. Tenscher, D.W. Ramming, and M.A. Walker: Using a limited mapping strategy to identify major QTLs for resistance to grapevine powdery mildew (*Erysiphe necator*) and their use in marker-assisted breeding. Theor Appl Genet. 2011; 122(6): p. 1059–1073. DOI: 10.1007/s00122-010-1511-6.

[48] Katula-Debreceni, D., A.K. Lencsés, A. Szőke, A. Veres, S. Hoffmann, P.K. Kozma, L.G. , L. Heszky, and E. Kiss: Marker-assisted selection for two dominant powdery mildew resistance genes introgressed into a hybrid grape population. Sci Hortic. 2010; 126(4): p. 448–453. DOI: 10.1016/j.scienta.2010.08.012.

[49] Gray, D.J., Z.T. Li, and S.A. Dhekney: Precision breeding of grapevine (*Vitis vinifera L.*) for improved traits. Plant Sci. 2014; 228: p. 3–10. DOI: 10.1016/j.plantsci.2014.03.023.

[50] Nirala, N.K., D.K. Das, P.S. Srivastava, S.K. Sopory, and K.C. Upadhyaya: Expression of a rice chitinase gene enhances antifungal potential in transgenic grapevine (*Vitis vinifera* L.). 2015; 49(4).

[51] Bornhoff, B.A., M. Harst, E. Zyprian, and R. Topfer: Transgenic plants of Vitis vinifera cv. Seyval blanc. Plant Cell Rep. 2005; 24(7): p. 433–438. DOI: 10.1007/s00299-005-0959-3.

[52] Rubio, J., C. Montes, A. Castro, C. Alvarez, B. Olmedo, M. Munoz, E. Tapia, F. Reyes, M. Ortega, E. Sanchez, M. Miccono, L. Dalla Costa, L. Martinelli, M. Malnoy, and H. Prieto: Genetically engineered Thompson seedless grapevine plants designed for fungal tolerance: selection and characterization of the best performing individuals in a field trial. Transgenic Res. 2015; 24(1): p. 43–60. DOI: 10.1007/s11248-014-9811-2.

[53] Jeandet, P., C. Clement, E. Courot, and S. Cordelier: Modulation of phytoalexin biosynthesis in engineered plants for disease resistance. Int J Mol Sci. 2013; 14(7): p. 14136–14170. DOI: 10.3390/ijms140714136.

[54] Coutos-Thevenot, P., B. Poinssot, A. Bonomelli, H. Yean, C. Breda, D. Buffard, R. Esnault, R. Hain, and M. Boulay: In vitro tolerance to Botrytis cinerea of grapevine 41B rootstock in transgenic plants expressing the stilbene synthase Vst1 gene under the control of a pathogen-inducible PR 10 promoter. J Exp Bot. 2001; 52(358): p. 901–910.

[55] Bavaresco, L., C. Fregoni, E. Cantu, and M. Trevisan: Stilbene compounds: from the grapevine to wine. Drugs Exp Clin Res. 1999; 25(2-3): p. 57–63.

[56] Nawrocka, J. and U. Małolepsza: Diversity in plant systemic resistance induced by Trichoderma. Biol Control. 2013; 67(2): p. 149–156. DOI: 10.1016/j.biocontrol.2013.07.005.

[57] Elmer, P. and T. Reglinski: Biosuppression of Botrytis cinerea in grapes. Plant Pathol. 2006; 55(2): p. 155–177. DOI: 10.1111/j.1365-3059.2006.01348.x.

[58] Buck, J.W.: Combinations of fungicides with phylloplane yeasts for improved control of Botrytis cinerea on geranium seedlings. Phytopathology. 2004; 94(2): p. 196–202. DOI: 10.1094/PHYTO.2004.94.2.196.

[59] Lindow, S.E. and M.T. Brandl: Microbiology of the phyllosphere. Appl Environ Microbiol. 2003; 69(4): p. 1875–1883.

[60] Sharma, R.R., D. Singh, and R. Singh: Biological control of postharvest diseases of fruits and vegetables by microbial antagonists: A review. Biol Control. 2009; 50(3): p. 205–221. DOI: http://dx.doi.org/10.1016/j.biocontrol.2009.05.001.

[61] El Oirdi, M., T.A. El Rahman, L. Rigano, A. El Hadrami, M.C. Rodriguez, F. Daayf, A. Vojnov, and K. Bouarab: *Botrytis cinerea* manipulates the antagonistic effects between immune pathways to promote disease development in tomato. Plant Cell. 2011; 23(6): p. 2405–2421. DOI: 10.1105/tpc.111.083394.

[62] Alamri, S., M. Hashem, and Y.S. Mostafa: In vitro and in vivo biocontrol of soil-borne phytopathogenic fungi by certain bioagents and their possible mode of action. Biocontr Sci. 2012; 17(4): p. 155–167.

[63] Bacon, C.W., E.R. Palencia, and D.M. Hinton. Abiotic and biotic plant stress-tolerant and beneficial secondary metabolites produced by endophytic *Bacillus* species. In K.N. Arora, Editor. Plant Microbes Symbiosis: Applied Facets. Springer India: New Delhi; 2015. Pp. 163–177. DOI: 10.1007/978-81-322-2068-8_8

[64] Stein, T.: Bacillus subtilis antibiotics: structures, syntheses and specific functions. Mol Microbiol. 2005; 56(4): p. 845–857. DOI: 10.1111/j.1365-2958.2005.04587.x.

[65] Arguelles-Arias, A., M. Ongena, B. Halimi, Y. Lara, A. Brans, B. Joris, and P. Fickers: Bacillus amyloliquefaciens GA1 as a source of potent antibiotics and other secondary

metabolites for biocontrol of plant pathogens. Microb Cell Fact. 2009; 8: p. 63. DOI: 10.1186/1475-2859-8-63.

[66] Delaunois, B., G. Farace, P. Jeandet, C. Clement, F. Baillieul, S. Dorey, and S. Cordelier: Elicitors as alternative strategy to pesticides in grapevine? Current knowledge on their mode of action from controlled conditions to vineyard. Environ Sci Pollut Res Int. 2014; 21(7): p. 4837–4846. DOI: 10.1007/s11356-013-1841-4.

[67] Boller, T. and G. Felix: A renaissance of elicitors: perception of microbe-associated molecular patterns and danger signals by pattern-recognition receptors. Annu Rev Plant Biol. 2009; 60: p. 379–406. DOI: 10.1146/annurev.arplant.57.032905.105346.

[68] Conrath, U.: Molecular aspects of defence priming. Trends Plant Sci. 2011; 16(10): p. 524–531. DOI: 10.1016/j.tplants.2011.06.004.

[69] Chang, X. and P. Nick: Defence signalling triggered by Flg22 and Harpin is integrated into a different stilbene output in Vitis cells. PLoS One. 2012; 7(7): p. e40446. DOI: 10.1371/journal.pone.0040446.

[70] Aziz, A., P. Trotel-Aziz, L. Dhuicq, P. Jeandet, M. Couderchet, and G. Vernet: Chitosan oligomers and copper sulfate induce grapevine defense reactions and resistance to gray mold and downy mildew. Phytopathology. 2006; 96(11): p. 1188–1194. DOI: 10.1094/PHYTO-96-1188.

[71] Laquitaine, L., E. Gomes, J. Francois, C. Marchive, S. Pascal, S. Hamdi, R. Atanassova, S. Delrot, and P. Coutos-Thevenot: Molecular basis of ergosterol-induced protection of grape against botrytis cinerea: induction of type I LTP promoter activity, WRKY, and stilbene synthase gene expression. Mol Plant Microbe Interact. 2006; 19(10): p. 1103–1112. DOI: 10.1094/MPMI-19-1103.

[72] Martelli, G.P.: Graft-transmissible diseases of grapevines. Handbook for Detection and Diagnosis. Rome: FAO; 1993.

[73] Martelli, G.P.: Directory of virus and virus-like diseases of the grapevine and their agents. J Plant Pathol. 2014; 96(1): p. 1–136.

[74] Goheen, A.C.: Virus and viruslike diseases of grapes. Hortscience. 1977; 12(5): p. 465–469.

[75] Oliver, J.E., E. Vigne, and M. Fuchs: Genetic structure and molecular variability of Grapevine fanleaf virus populations. Virus Res. 2010; 152(1–2): p. 30–40. DOI: 10.1016/j.virusres.2010.05.017.

[76] Forer, L.B., C.A. Powell, and R.F. Stouffer: Transmission of tomato ringspot virus to apple rootstock cuttings and to cherry and peach seedlings by Xiphinema-Rivesi. Plant Dis. 1984; 68(12): p. 1052–1054. DOI: 10.1094/Pd-69-1052.

[77] Sturhan, D., W.M. Wouts, G.S. Grandison, and C.J. Barber: Nematode vectors of plant viruses in New Zealand. NZ J Zool. 1997; 24(4): p. 309–322.

[78] Gilmer, R.M. and J.K. Uyemoto: Tomato ringspot virus in Baco Noir grapevines in New York. Plant Dis Rep. 1972; 56: p. 133–135.

[79] Gilmer, R.M., J.K. Uyemoto, and L.J. Kelts: A new grapevine disease induced by tobacco ringspot virus. Phytopathology. 1970; 60: p. 619–627.

[80] Andret-Link, P., C. Laporte, L. Valat, C. Ritzenthaler, G. Demangeat, E. Vigne, V. Laval, P. Pfeiffer, C. Stussi-Garaud, and M. Fuchs: Grapevine fanleaf virus: Still a major threat to the grapevine industry. J Plant Pathol. 2004; 86(3): p. 183–195.

[81] Coetzee, B., M.J. Freeborough, H.J. Maree, J.M. Celton, D.J. Rees, and J.T. Burger: Deep sequencing analysis of viruses infecting grapevines: Virome of a vineyard. Virology. 2010; 400(2): p. 157–163. DOI: 10.1016/j.virol.2010.01.023.

[82] Fiore, N., S. Prodan, J. Montealegre, E. Aballay, A.M. Pino, and A. Zarnorano: Survey of grapevine viruses in Chile. J Plant Pathol. 2008; 90(1): p. 125–130.

[83] Maree, H.J., R.P. Almeida, R. Bester, K.M. Chooi, D. Cohen, V.V. Dolja, M.F. Fuchs, D.A. Golino, A.E. Jooste, G.P. Martelli, R.A. Naidu, A. Rowhani, P. Saldarelli, and J.T. Burger: Grapevine leafroll-associated virus 3. Front Microbiol. 2013; 4: p. 82. DOI: 10.3389/fmicb.2013.00082.

[84] Montero, R., D. Mundy, A. Albright, C. Grose, M.C.T. Trought, D. Cohen, K.M. Chooi, R. MacDiarmid, J. Flexas, and J. Bota: Effects of Grapevine Leafroll associated Virus 3 (GLRaV-3) and duration of infection on fruit composition and wine chemical profile of *Vitis vinifera* L. cv. Sauvignon blanc. Food Chem. 2016; 197: p. 1177–1183. DOI: 10.1016/j.foodchem.2015.11.086.

[85] Adams, M.J., J.F. Antoniw, M. Bar-Joseph, A.A. Brunt, T. Candresse, G.D. Foster, G.P. Martelli, R.G. Milne, S.K. Zavriev, and C.M. Fauquet: The new plant virus family Flexiviridae and assessment of molecular criteria for species demarcation. Arch Virol. 2004; 149(5): p. 1045–1060. DOI: 10.1007/s00705-004-0304-0.

[86] Alabi, O.J., M. Al Rwahnih, T.A. Mekuria, and R.A. Naidu: Genetic diversity of Grapevine virus A in Washington and California vineyards. Phytopathology. 2014; 104(5): p. 548–560. DOI: 10.1094/PHYTO-06-13-0179-R.

[87] Goszczynski, D.E., J. du Preez, and J.T. Burger: Molecular divergence of Grapevine virus A (GVA) variants associated with Shiraz disease in South Africa. Virus Res. 2008; 138(1–2): p. 105–110. DOI: 10.1016/j.virusres.2008.08.014.

[88] Bonavia, M., M. Digiaro, D. Boscia, A. Boari, G. Bottalico, V. Savino, and G.P. Martelli: Studies on "corky rugose wood" of grapevine and on the diagnosis of grapevine virus B. Vitis. 1996; 35(1): p. 53–58.

[89] Al Rwahnih, M., M.R. Sudarshana, J.K. Uyemoto, and A. Rowhani: Complete genome sequence of a novel vitivirus isolated from grapevine. J Virol. 2012; 86(17): p. 9545. DOI: 10.1128/JVI.01444-12.

[90] Rosa, C., M. Polek, B.W. Falk, and A. Rowhani: Improved efficiency for quantitative and qualitative indexing for Citrus tristeza virus and Citrus psorosis virus. Plant Dis. 2007; 91(9): p. 1089–1095. DOI: 10.1094/Pdis-91-9-1089.

[91] Jooste, A.E.C., N. Molenaar, H.J. Maree, R. Bester, L. Morey, W.C. de Koker, and J.T. Burger: Identification and distribution of multiple virus infections in Grapevine leafroll diseased vineyards. Eur J Plant Pathol. 2015; 142(2): p. 363–375. DOI: 10.1007/s10658-015-0620-0.

[92] Prosser, S.W., D.E. Goszczynski, and B. Meng: Molecular analysis of double-stranded RNAs reveals complex infection of grapevines with multiple viruses. Virus Res. 2007; 124(1–2): p. 151–159. DOI: 10.1016/j.virusres.2006.10.014.

[93] Maule, A., V. Leh, and C. Lederer: The dialogue between viruses and hosts in compatible interactions. Curr Opin Plant Biol. 2002; 5(4): p. 279–284. DOI: 10.1016/S1369-5266(02)00272-8.

[94] Mittler, R.: Abiotic stress, the field environment and stress combination. Trends Plant Sci. 2006; 11(1): p. 15–19. DOI: 10.1016/j.tplants.2005.11.002.

[95] Espinoza, C., C. Medina, S. Somerville, and P. Arce-Johnson: Senescence-associated genes induced during compatible viral interactions with grapevine and Arabidopsis. J Exp Bot. 2007; 58(12): p. 3197–3212. DOI: 10.1093/jxb/erm165.

[96] Gutha, L.R., L.F. Casassa, J.F. Harbertson, and R.A. Naidu: Modulation of flavonoid biosynthetic pathway genes and anthocyanins due to virus infection in grapevine (Vitis vinifera L.) leaves. BMC Plant Biol. 2010; 10: p. 187. DOI: 10.1186/1471-2229-10-187.

[97] Vega, A., R.A. Gutierrez, A. Pena-Neira, G.R. Cramer, and P. Arce-Johnson: Compatible GLRaV-3 viral infections affect berry ripening decreasing sugar accumulation and anthocyanin biosynthesis in Vitis vinifera. Plant Mol Biol. 2011; 77(3): p. 261–274. DOI: 10.1007/s11103-011-9807-8.

[98] Hung, K.T. and C.H. Kao: Hydrogen peroxide is necessary for abscisic acid-induced senescence of rice leaves. J Plant Physiol. 2004; 161(12): p. 1347–1357. DOI: 10.1016/j.jplph.2004.05.011.

[99] Maliogka, V.I., G.P. Martelli, M. Fuchs, and N.I. Katis: Control of viruses infecting grapevine. Adv Virus Res. 2015; 91: p. 175–227. DOI: 10.1016/bs.aivir.2014.11.002.

[100] Lopez-Fabuel, I., T. Wetzel, E. Bertolini, A. Bassler, E. Vidal, L.B. Torres, A. Yuste, and A. Olmos: Real-time multiplex RT-PCR for the simultaneous detection of the five main grapevine viruses. J Virol Methods. 2013; 188(1–2): p. 21–24. DOI: 10.1016/j.jviromet.2012.11.034.

[101] Ward, E., S.J. Foster, B.A. Fraaije, and H.A. McCartney: Plant pathogen diagnostics: immunological and nucleic acid-based approaches. Ann Appl Biol. 2004; 145(1): p. 1–16. DOI: 10.1111/j.1744-7348.2004.tb00354.x.

[102] Masuta, C. and I. Uyeda: Plant Virology Protocols New Approaches to Detect Viruses and Host Responses Third Edition Preface. Plant Virology Protocols: New Approaches to Detect Viruses and Host Responses, 3rd edition. 2015; 1236: p. V–VI. DOI: 10.1007/978-1-4939-1743-3.

[103] Bester, R., P.T. Pepler, J.T. Burger, and H.J. Maree: Relative quantitation goes viral: An RT-qPCR assay for a grapevine virus. J Virol Methods. 2014; 210: p. 67–75. DOI: 10.1016/j.jviromet.2014.09.022.

[104] Motooka, D., S. Nakamura, K. Hagiwara, and T. Nakaya: Viral Detection by High-Throughput Sequencing. Plant Virology Protocols: New Approaches to Detect Viruses and Host Responses, 3rd edition. 2015; 1236: pp. 125–134. DOI: 10.1007/978-1-4939-1743-3_11.

[105] Almeida, R.P., K.M. Daane, V.A. Bell, G.K. Blaisdell, M.L. Cooper, E. Herrbach, and G. Pietersen: Ecology and management of grapevine leafroll disease. Front Microbiol. 2013; 4: p. 94. DOI: 10.3389/fmicb.2013.00094.

[106] Panattoni, A., A. Luvisi, and E. Triolo: Review. Elimination of viruses in plants: Twenty years of progress. Spanish J Agric Res. 2013; 11(1): p. 173–188. DOI: 10.5424/sjar/2013111-3201.

[107] Gambino, G., I. Gribaudo, S. Leopold, A. Schartl, and M. Laimer: Molecular characterization of grapevine plants transformed with GFLV resistance genes: I. Plant Cell Rep. 2005; 24(11): p. 655–662. DOI: 10.1007/s00299-005-0006-4.

[108] Gambino, G., I. Perrone, A. Carra, W. Chitarra, P. Boccacci, D. Torello Marinoni, M. Barberis, F. Maghuly, M. Laimer, and I. Gribaudo: Transgene silencing in grapevines transformed with GFLV resistance genes: analysis of variable expression of transgene, siRNAs production and cytosine methylation. Transgenic Res. 2010; 19(1): p. 17–27. DOI: 10.1007/s11248-009-9289-5.

[109] Jardak-Jamoussi, R., P. Winterhagen, B. Bouamama, C. Dubois, A. Mliki, T. Wetzel, A. Ghorbel, and G.M. Reustle: Development and evaluation of a GFLV inverted repeat construct for genetic transformation of grapevine. Plant Cell Tissue Org Cult. 2009; 97(2): p. 187–196. DOI: 10.1007/s11240-009-9514-1.

[110] Krastanova, S., M. Perrin, P. Barbier, G. Demangeat, P. Cornuet, N. Bardonnet, L. Otten, L. Pinck, and B. Walter: Transformation of grapevine rootstocks with the coat protein gene of grapevine fanleaf nepovirus. Plant Cell Rep. 1995; 14(9): p. 550–554.

[111] Maghuly, F., S. Leopold, A.D. Machado, E.B. Fernandez, M.A. Khan, G. Gambino, I. Gribaudo, A. Schartl, and M. Laimer: Molecular characterization of grapevine plants transformed with GFLV resistance genes: II. Plant Cell Rep. 2006; 25(6): p. 546–553. DOI: 10.1007/s00299-005-0087-0.

[112] Martinelli, L., E. Candioli, D. Costa, and A. Minafra: Stable insertion and expression of the movement protein gene of Grapevine Virus A (GVA) in grape (Vitis rupestris S.). Vitis. 2002; 41(4): p. 189–193.

[113] Mauro, M.C., S. Toutain, B. Walter, L. Pinck, L. Otten, P. CoutosThevenot, A. Deloire, and P. Barbier: High efficiency regeneration of grapevine plants transformed with the GFLV coat protein gene. Plant Sci. 1995; 112(1): p. 97–106. DOI: 10.1016/0168-9452(95)04246-Q.

[114] Radian-Sade, S., A. Perl, O. Edelbaum, L. Kuznetsova, R. Gafny, I. Sela, and E. Tanne: Transgenic Nicotiana benthamiana and grapevine plants transformed with grapevine virus A (GVA) sequences. Phytoparasitica. 2000; 28(1): p. 79–86. DOI: 10.1007/Bf02994025.

[115] Spielmann, A., S. Krastanova, V. Douet-Orhant, and P. Gugerli: Analysis of transgenic grapevine (Vitis rupestris) and Nicotiana benthamiana plants expressing an Arabis mosaic virus coat protein gene. Plant Sci. 2000; 156(2): p. 235–244. DOI: 10.1016/S0168-9452(00)00259-4.

[116] Winterhagen, P., C. Dubois, M. Sinn, T. Wetzel, and G.M. Reustle: Gene silencing and virus resistance based on defective interfering constructs in transgenic Nicotiana benthamiana is not linked to accumulation of siRNA. Plant Physiol Biochem. 2009; 47(8): p. 739–742. DOI: 10.1016/j.plaphy.2009.03.012.

[117] Xue, B., K.S. Ling, C.L. Reid, S. Krastanova, M. Sekiya, E.A. Momol, S. Sule, J. Mozsar, D. Gonsalves, and T.J. Burr: Transformation of five grape rootstocks with plant virus genes and a virE2 gene from Agrobacterium tumefaciens. In Vitro Cell Dev Biol Plant. 1999; 35(3): p. 226–231.

[118] Torres, E., C. Santibanez, F. Rubio, F. Godoy, A. Cadavid-Labrada, C. Bruno, C. Medina, and P. Arce-Johnson: Gene Silencing as a Strategy to Induce Grapevine FanLeaf Virus (GFLV) Resistance in Grapevine Rootstocks. X International Conference on Grapevine Breeding and Genetics. 2014; 1046: pp. 187–193.

[119] Chaves, M.M., O. Zarrouk, R. Francisco, J.M. Costa, T. Santos, A.P. Regalado, M.L. Rodrigues, and C.M. Lopes: Grapevine under deficit irrigation: hints from physiological and molecular data. Ann Bot. 2010; 105(5): p. 661–676. DOI: 10.1093/aob/mcq030.

[120] Gupta, B. and B. Huang: Mechanism of salinity tolerance in plants: Physiological, biochemical, and molecular characterization. Int J Genom. 2014; 2014: p. 18. DOI: 10.1155/2014/701596.

[121] Golldack, D., C. Li, H. Mohan, and N. Probst: Tolerance to drought and salt stress in plants: Unraveling the signaling networks. Front Plant Sci. 2014; 5. DOI: 10.3389/fpls.2014.00151.

[122] Rocheta, M., J.D. Becker, J.L. Coito, L. Carvalho, and S. Amâncio: Heat and water stress induce unique transcriptional signatures of heat-shock proteins and transcription

factors in grapevine. Funct Integr Genom. 2013; 14(1): p. 135–148. DOI: 10.1007/s10142-013-0338-z.

[123] Carvalho, L.C., J.L. Coito, S. Colaço, M. Sangiogo, and S. Amâncio: Heat stress in grapevine: The pros and cons of acclimation. Plant Cell Environ. 2015; 38(4): p. 777–789. DOI: 10.1111/pce.12445.

[124] Janmohammadi, M., L. Zolla, and S. Rinalducci: Low temperature tolerance in plants: Changes at the protein level. Phytochemistry. 2015; 117: p. 76–89. DOI: http://dx.doi.org/10.1016/j.phytochem.2015.06.003.

[125] Guo, M., J.-H. Liu, X. Ma, D.-X. Luo, Z.-H. Gong, and M.-H. Lu: The plant heat stress transcription factors (HSFs): Structure, regulation and function in response to abiotic stresses. Front Plant Sci. 2016; 7. DOI: 10.3389/fpls.2016.00114.

[126] Wargent, J.J. and B.R. Jordan: From ozone depletion to agriculture: Understanding the role of UV radiation in sustainable crop production. New Phytol. 2013; 197(4): p. 1058–1076. DOI: 10.1111/nph.12132.

[127] Robson, T.M., K. Klem, O. Urban, and M.A.K. Jansen: Re-interpreting plant morphological responses to UV-B radiation. Plant Cell Environ. 2015; 38(5): p. 856–866. DOI: 10.1111/pce.12374.

[128] Jenkins, G.I.: Signal transduction in responses to UV-B radiation. Ann Rev Plant Biol. 2009; 60(1): p. 407–431. DOI: 10.1146/annurev.arplant.59.032607.092953.

[129] Jenkins, G.I.: The UV-B photoreceptor UVR8: From structure to physiology. Plant Cell. 2014; 26(1): p. 21–37. DOI: 10.1105/tpc.113.119446.

[130] Parihar, P., S. Singh, R. Singh, V.P. Singh, and S.M. Prasad: Changing scenario in plant UV-B research: UV-B from a generic stressor to a specific regulator. J Photochem Photobiol B: Biol. 2015; 153: p. 334–343. DOI: http://dx.doi.org/10.1016/j.jphotobiol.2015.10.004.

[131] Ulm, R. and G.I. Jenkins: Q&A: How do plants sense and respond to UV-B radiation? BMC Biol. 2015; 13(1): p. 1–6. DOI: 10.1186/s12915-015-0156-y.

[132] Frohnmeyer, H. and D. Staiger: Ultraviolet-B radiation-mediated responses in plants. Balancing damage and protection. Plant Physiol. 2003; 133(4): p. 1420–1428. DOI: 10.1104/pp.103.030049.

[133] Bornman, J.F., P.W. Barnes, S.A. Robinson, C.L. Ballare, S.D. Flint, and M.M. Caldwell: Solar ultraviolet radiation and ozone depletion-driven climate change: effects on terrestrial ecosystems. Photochem Photobiol Sci. 2015; 14(1): p. 88–107. DOI: 10.1039/C4PP90034K.

[134] Tilbrook, K., A.B. Arongaus, M. Binkert, M. Heijde, R. Yin, and R. Ulm: The UVR8 UV-B Photoreceptor: Perception, Signaling and Response. The Arabidopsis Book. 2013: p. e0164. DOI: 10.1199/tab.0164.

[135] Berli, F.J., R. Alonso, R. Bressan-Smith, and R. Bottini: UV-B impairs growth and gas exchange in grapevines grown in high altitude. Physiol Plant. 2013; 149(1): p. 127–140. DOI: 10.1111/ppl.12012.

[136] Gil, M., M. Pontin, F. Berli, R. Bottini, and P. Piccoli: Metabolism of terpenes in the response of grape (Vitis vinifera L.) leaf tissues to UV-B radiation. Phytochemistry. 2012; 77: p. 89–98. DOI: http://dx.doi.org/10.1016/j.phytochem.2011.12.011.

[137] Rizzini, L., J.J. Favory, C. Cloix, D. Faggionato, A. O'Hara, and E. Kaiserli: Perception of UV-B by the Arabidopsis UVR8 protein. Science. 2011; 332. DOI: 10.1126/science.1200660.

[138] Wu, D., Q. Hu, Z. Yan, W. Chen, C. Yan, and X. Huang: Structural basis of ultraviolet-B perception by UVR8. Nature. 2012; 484. DOI: 10.1038/nature10931.

[139] Cloix, C., E. Kaiserli, M. Heilmann, K.J. Baxter, B.A. Brown, and A. O'Hara: C-terminal region of the UV-B photoreceptor UVR8 initiates signaling through interaction with the COP1 protein. Proc Natl Acad Sci U S A. 2012; 109. DOI: 10.1073/pnas.1210898109.

[140] Favory, J.J., A. Stec, H. Gruber, L. Rizzini, A. Oravecz, M. Funk, A. Albert, C. Cloix, G.I. Jenkins, E.J. Oakeley, H.K. Seidlitz, F. Nagy, and R. Ulm: Interaction of COP1 and UVR8 regulates UV-B-induced photomorphogenesis and stress acclimation in Arabidopsis. EMBO J. 2009; 28(5): p. 591–601. DOI: 10.1038/emboj.2009.4.

[141] Lau, O.S. and X.W. Deng: The photomorphogenic repressors COP1 and DET1: 20 years later. Trends Plant Sci. 2012; 17. DOI: 10.1016/j.tplants.2012.05.004.

[142] Oravecz, A., A. Baumann, Z. Máté, A. Brzezinska, J. Molinier, E.J. Oakeley, É. Ádám, E. Schäfer, F. Nagy, and R. Ulm: Constitutively Photomorphogenic 1 is required for the UV-B response in Arabidopsis. Plant Cell. 2006; 18(8): p. 1975–1990. DOI: 10.1105/tpc.105.040097.

[143] Saijo, Y., J.A. Sullivan, H. Wang, J. Yang, Y. Shen, V. Rubio, L. Ma, U. Hoecker, and X.W. Deng: The COP1–SPA1 interaction defines a critical step in phytochrome A-mediated regulation of HY5 activity. Genes Dev. 2003; 17(21): p. 2642–2647. DOI: 10.1101/gad.1122903.

[144] Brown, B.A., C. Cloix, G.H. Jiang, E. Kaiserli, P. Herzyk, D.J. Kliebenstein, and G.I. Jenkins: A UV-B-specific signaling component orchestrates plant UV protection. Proc Natl Acad Sci U S A. 2005; 102. DOI: 10.1073/pnas.0507187102.

[145] Lee, J., K. He, V. Stolc, H. Lee, P. Figueroa, Y. Gao, W. Tongprasit, H. Zhao, I. Lee, and X.W. Deng: Analysis of transcription factor HY5 genomic binding sites revealed its hierarchical role in light regulation of development. Plant Cell. 2007; 19(3): p. 731–749. DOI: 10.1105/tpc.106.047688.

[146] Shin, D.H., M. Choi, K. Kim, G. Bang, M. Cho, S.-B. Choi, G. Choi, and Y.-I. Park: HY5 regulates anthocyanin biosynthesis by inducing the transcriptional activation of the

MYB75/PAP1 transcription factor in Arabidopsis. FEBS Lett. 2013; 587(10): p. 1543–1547. DOI: 10.1016/j.febslet.2013.03.037.

[147] Stracke, R., J.-J. Favory, H. Gruber, L. Bartelniewoehner, S. Bartels, M. Binkert, M. Funk, B. Weisshaar, and R. Ulm: The Arabidopsis bZIP transcription factor HY5 regulates expression of the PFG1/MYB12 gene in response to light and ultraviolet-B radiation. Plant Cell Environ. 2010; 33(1): p. 88–103. DOI: 10.1111/j.1365-3040.2009.02061.x.

[148] Ulm, R., A. Baumann, A. Oravecz, Z. Mate, E. Adam, and E.J. Oakeley: Genome-wide analysis of gene expression reveals function of the bZIP transcription factor HY5 in the UV-B response of Arabidopsis. Proc Natl Acad Sci U S A. 2004; 101. DOI: 10.1073/pnas.0308044100.

[149] Gruber, H., M. Heijde, W. Heller, A. Albert, H.K. Seidlitz, and R. Ulm: Negative feedback regulation of UV-B-induced photomorphogenesis and stress acclimation in Arabidopsis. Proc Natl Acad Sci U S A. 2010; 107. DOI: 10.1073/pnas.0914532107.

[150] Heijde, M. and R. Ulm: Reversion of the Arabidopsis UV-B photoreceptor UVR8 to the homodimeric ground state. Proc Natl Acad Sci U S A. 2013; 110. DOI: 10.1073/pnas.1214237110.

[151] Hatier, J. and K. Gould. Anthocyanin function in vegetative organs. In K. Gould and K.W. Davies, Editors. Anthocyanins: Biosynthesis, Functions, and Applications. Springer New York: New York; 2008. Pp. 1–19. DOI: 10.1007/978-0-387-77335-3

[152] Berli, F., J. D'Angelo, B. Cavagnaro, R. Bottini, R. Wuilloud, and M.F. Silva: Phenolic composition in grape (Vitis vinifera L. cv. Malbec) ripened with different solar UV-B radiation levels by capillary zone electrophoresis. J Agric Food Chem. 2008; 56(9): p. 2892–2898. DOI: 10.1021/jf073421+.

[153] Kolb, C.A., M.A. Käser, J. Kopecký, G. Zotz, M. Riederer, and E.E. Pfündel: Effects of natural intensities of visible and ultraviolet radiation on epidermal ultraviolet screening and photosynthesis in grape leaves. Plant Physiol. 2001; 127(3): p. 863–875. DOI: 10.1104/pp.010373.

[154] Pollastrini, M., V. Di Stefano, M. Ferretti, G. Agati, D. Grifoni, G. Zipoli, S. Orlandini, and F. Bussotti: Influence of different light intensity regimes on leaf features of Vitis vinifera L. in ultraviolet radiation filtered condition. Environ Exp Bot. 2011; 73: p. 108–115. DOI: http://dx.doi.org/10.1016/j.envexpbot.2010.10.027.

[155] Pontin, M.A., P.N. Piccoli, R. Francisco, R. Bottini, J.M. Martinez-Zapater, and D. Lijavetzky: Transcriptome changes in grapevine (Vitis vinifera L.) cv. Malbec leaves induced by ultraviolet-B radiation. BMC Plant Biol. 2010; 10(1): p. 1–13. DOI: 10.1186/1471-2229-10-224.

[156] Baldoni, L. and E. Rugini. 3-Genetic modification of agronomic traits in fruit crops A2 — Valpuesta, Victoriano. Fruit and Vegetable Biotechnology. Woodhead Publishing; 2002. Pp. 25–113. DOI: http://dx.doi.org/10.1533/9781855736412.1.25

[157] Demkura, P. and C. Ballaré: UVR8 Mediates UV-B-Induced Arabidopsis Defense Responses against Botrytis cinerea by Controlling Sinapate Accumulation. Molecular Plant. 2012. 5(3): p. 642–652.

[158] Brown, B. A. and G. I. Jenkins: UV-B Signaling Pathways with Different Fluence-Rate Response Profiles Are Distinguished in Mature Arabidopsis Leaf Tissue by Requirement for UVR8, HY5, and HYH. Plant Physiol. 2008. 146(2): p. 576–588.

[159] Gururani, M.A., M. Ganesan, and P.-S. Song: Photo-biotechnology as a tool to improve agronomic traits in crops. Biotechnol Adv. 2015; 33(1): p. 53–63. DOI: http://dx.doi.org/10.1016/j.biotechadv.2014.12.005.

[160] Liu, Y., S. Roof, Z. Ye, C. Barry, A. van Tuinen, J. Vrebalov, C. Bowler, and J. Giovannoni: Manipulation of light signal transduction as a means of modifying fruit nutritional quality in tomato. Proc Natl Acad Sci U S A. 2004; 101(26): p. 9897–9902. DOI: 10.1073/pnas.0400935101.

[161] Berli, F. and R. Bottini: UV-B and abscisic acid effects on grape berry maturation and quality. J Berry Res. 2013; 3(1): p. 1–14. DOI: 10.3233/JBR-130047.

Chapter 2

Pathogenesis-Related Proteins in Grape

Shinichi Enoki and Shunji Suzuki

Additional information is available at the end of the chapter

http://dx.doi.org/10.5772/64873

Abstract

An overview of major pathogens and their control, plant defense mechanisms, and pathogenesis-related (PR) proteins and their roles in pathogen control is presented herein. *Vitis vinifera*, including wine grape and table grape, is one of the most valuable horticultural crops in the world because of its commercial use. However, *V. vinifera* cultivars are extremely susceptible to pathogens, particularly fungi and oomycetes, such as *Botrytis cinerea* and *Plasmopara viticola*, respectively. Plants have various defense mechanisms to counter these pathogens. One example is induced resistance, which involves the induction of the immune system in the event of a pathogen attack, including the generation of PR proteins. Some PR proteins possess antimicrobial activity. PR proteins are classified into 17 families, some of which are found in grape. Thus, their roles in grape have been actively studied. A new strategy to increase plant resistance to pathogens has been developed. A good understanding of grape defense mechanism through PR proteins is expected to open new doors to improve grape quality and yield by efficiently controlling pathogens in the future.

Keywords: pathogen, plant defense mechanism, induced resistance, pathogenesis-related (PR) protein, pathogen control

1. Introduction

Vitis vinifera is one of the most important grape species for wine making. However, it is extremely susceptible to such pathogens as fungi and oomycetes. An understanding of grape defense mechanism is important because infection by pathogens leads to marked loss of fruit quality and yield.

Higher plants possess a variety of defense mechanisms against pathogens. Pathogenesis-related (PR) proteins are induced in response to infection by pathogens. PR proteins are classified into 17 families according to molecular structure and enzyme activity. The functions of all the families have not been reported in grape.

In this chapter, we review PR proteins in *V. vinifera* by presenting an overview of the latest knowledge of PR proteins in grape.

2. Major pathogens and their control in grapevine

Grapevine pathogens are roughly divided into fungi, bacteria, virus, and others. *V. vinifera* cultivars in particular are extremely susceptible to diseases caused by fungi and oomycetes, which result in huge economic losses worldwide. Of those diseases, the most prevalent are powdery mildew, gray mould, and downy mildew. Their characteristics, the resistance of grapevine to these diseases, and pathogen control in grapevine are described below.

2.1. Major pathogens

2.1.1. Powdery mildew

Powdery mildew caused by ascomycete *Uncinula necator* (syn. *Erysiphe necator*) is one of the most well-known fungal diseases in viticulture. In recent years, the decline of the effects of chemical fungicides and the emergence of climate suitable for fungal growth have led to the development of powdery mildew epidemic in European vineyards [1]. The symptoms are white powdery spots on the entire grapevine. Infected berries become brown and crack, resulting in loss of yield and seriously altering wine quality. Most *V. vinifera* cultivars are susceptible to *U. necator*, although some *Vitis* species show various levels of resistance [1–3].

2.1.2. Gray mould

Gray mould caused by ascomycete *Botrytis cinerea* is the major fungal disease in humid and temperate regions of the world [4]. *B. cinerea* invades a variety of agricultural crops and has a broad host range. In grape, the symptoms appear as 'gray mould' on the lesioned part of infected leaves, flower clusters, and ripening grape berries. *B. cinerea* infection likely occurs at all stages of grape growth. The development of gray mould in ripening berries results in loss of yield and berry quality.

2.1.3. Downy mildew

Downy mildew is caused by oomycete *Plasmopara viticola*. It is one of the most harmful diseases in grape grown in Europe and in the US [5]. Downy mildew appears as oily yellow spots on the surface of infected leaves. Loss of yield and lowered berry quality occur due to weakening of young shoot and death of leaf tissue. Boso et al. [6] investigated the susceptibility of *V. vinifera* cultivars and other *Vitis* species to downy mildew. *V. vinifera* 'Cabernet Sauvignon' was the least susceptible, whereas non-*vinifera* cultivars did not show symptoms of downy mildew in the field.

2.2. Pathogen control by chemical fungicide

The above diseases caused by ascomycete and oomycete are generally controlled by spraying chemical fungicides, such as quinone outside inhibiting (QoI) fungicides, in the vineyard. QoI fungicides, which act by inhibiting fungal mitochondrial respiration, is one of the most widely used agents against pathogens in viticulture. However, in addition to the adverse effects of these fungicides, the emergence of fungi resistant to these fungicides has been reported [7, 8]. Therefore, restrictions and laws for use have been set by individual countries. The resistance of *P. viticola* to QoI fungicide is acquired by G143A mutation in cytochrome b that constitutes mitochondrial electron transport chain complex III [8, 9].

3. Plant defense mechanism

Plant defense mechanism is roughly classified into two categories: constitutive (static) resistance and induced (active) resistance (**Table 1**).

Category	Feature	Reference
Constitutive (static) resistance [resistance inherent in plants]		
Physical resistance	Thickness and hardness of cell wall	[10]
	Hydrophobic environment created by cuticle layer	
Chemical resistance	Antimicrobial substances, such as phenol and saponin	[11]
Induced (active) resistance [resistance newly induced by pathogen attack]		
1 Formation of papilla	Physical and chemical barrier against penetration	[12]
2 Hardening of cell wall	Lignification	[13]
	Crosslinked polymers with glycoprotein	
3 Hypersensitive reaction	Containment of pathogen by autocide activity of cells	[14]
	Generation of ROS	
4 Production of phytoalexins	Low molecular weight antimicrobial substance	[15]
5 Production of PR proteins	With antimicrobial activity	[16]

Table 1. Categories of plant defense mechanisms against pathogens.

Constitutive resistance is a prophylactic resistance mechanism inherent in plants and is divided into physical resistance and chemical resistance. The former is the first barrier against pathogens created by the cell wall. The latter is realized by antimicrobial substances present in plants, such as polyphenols.

On the other hand, induced resistance involves the induction of the immune system by pathogen attack and is roughly divided into five types: (1) formation of papilla, (2) hardening of cell wall, (3) hypersensitive response (HR), (4) production of phytoalexins, and (5) production of PR proteins. (1) and (2) are resistance acquisition through the formation of a physical barrier. The generation of reactive oxygen species (ROS) by HR induces another type of defense mechanisms. Phytoalexins and some PR proteins, on the other hand, have antimicrobial activity. All the above-mentioned types of induced resistance are accompanied by changes in the metabolic system and take place in not only infected cells localized acquired resistance (LAR), but also the whole plant systemic acquired resistance (SAR).

These types of defense mechanisms in induced resistance operate by sensing a substance called elicitor on the receptor. The elicitors include abiotic substances, such as heavy metals and synthetic compounds in the form of fungicides, and biotic substances, such as proteins, lipids, oligosaccharides, and antibiotics of biological origin. In fact, elicitors are found in the cell wall of pathogens, and PR proteins, such as chitinase and β-1,3-glucanase, function indirectly by releasing oligosaccharide elicitors from the cell wall of pathogens [17, 18].

4. Definition and classification of PR proteins

In this section, we define and classify in detail the PR proteins described above. PR proteins were discovered for the first time in tobacco leaves, indicating the plant's hypersensitive reaction to tobacco mosaic virus (TMV) [14, 19]. These proteins are found in many plant species [20], including grape.

4.1. Definition

PR proteins are proteins encoded but not expressed in host plant in the absence of interaction with a pathogen. They are also defined as proteins generally induced in an infection [21, 22]. PR proteins are also induced under conditions of nonpathogenic origin, such as stress. Examples include cytoplasm separation [23] and high concentrations of plant hormones [24].

Family	Property	Function/target site	Reference
PR-1	Antifungal	Unknown	[31]
PR-2	β-1,3-Glucanase	Cell wall (β-1,3-glucan)	[35–39, 56]
PR-3	Chitinase (types I, II, IV, V, VI, and VII)	Cell wall (chitin)	
PR-4	Chitinase (types I and II)	Cell wall (chitin)	
PR-5	Thaumatin-like	Plasma membrane	[37, 40–42]
PR-10	Ribonuclease (like)	RNA	[43–46, 57]
PR-14	Lipid-transfer protein	Involvement in defense signaling pathway	[47–54, 58–61]
PR-15	Oxalate oxidase	Production of H_2O_2 with	[55]
PR-16	Oxalate oxidase-like protein	Antimicrobial activity	

Table 2. Classification of PR proteins in *Vitis*.

4.2. Classification

PR proteins share many biochemical properties that render them easily distinguishable. They have relatively low molecular weights, are stably extractable at low pH [25, 26], are highly resistant to proteases [27], and have extreme isoelectric points. Most of them are located in the apoplast [28, 29]. In general, acidic PR proteins are located in the apoplast and basic ones, in the vacuole. PR proteins are classified on the basis of amino acid sequences, serological reaction, enzymatic activity, and others. Five groups of PR proteins (PR-1 to PR-5) were initially characterized in tobacco. Currently, PR proteins are categorized into 17 families [30], but not all are found in grape (**Table 2**). In the next section, the roles and functions of PR proteins in grape (**Table 2**) are described in detail.

5. PR protein gene in *V. vinifera* grape

5.1. Pathological function

5.1.1. PR-1 (unknown)

Although PR-1 proteins exhibit antifungal activity, their functions remain unclear. *V. vinifera* PR-1 was identified and cloned [31]. However, in grapes, signaling pathway related to PR-1 is remains to be determined. The signaling network of LAR and SAR has been well studied in recent years in *Arabidopsis thaliana* (**Figure 1**).

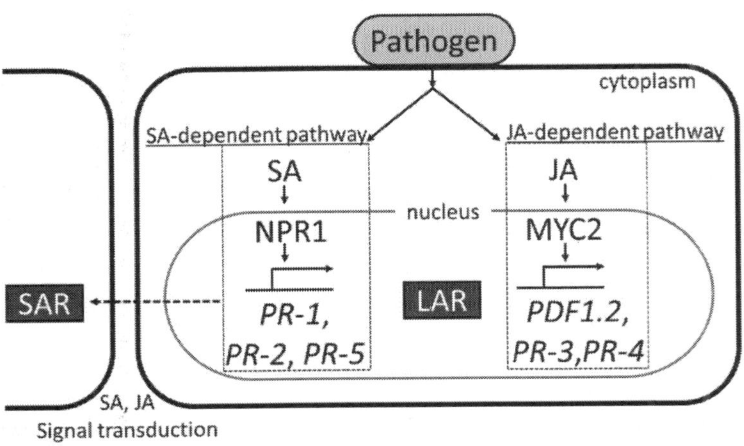

Figure 1. Predicted pathogen-induced signaling pathways leading to LAR and SAR in *Arabidopsis thaliana*. HR, hypersensitive reaction; SA, salicylic acid; NPR1, NON-EXPRESSER OF PATHOGENESIS-RELATED GENES 1; JA, jasmonic acid; MYC2, transcription factor MYC2; PDF1.2, plant defensin 1.2; LAR, localized acquired resistance; SAR, systemic acquired resistance.

LAR is induced by pathogen attack. SA and JA act as a second messenger [32]. In response to SA, positive regulator protein non-expresser of pathogenesis-related genes 1 (NPR1) is transported to the nucleus and activate the expression of PR protein genes, including the *PR-1* [33]. On the other hand, JA dependent pathway is upregulated by MYC2, transcription factor, and induces plant defense-related genes, such as the plant defensin 1.2 (PDF1.2) [34]. As a result, LAR is induced in pathogen-infected cells. Further, when SA and JA signaling are transported to other cells, SA and JA similarly induce the expression of plant defense-related genes through the NPR1 and MYC2. Finally, SAR is expressed in other noninfected cells. Some classes of PR proteins, including PR-1, are known to be expressed in association with SAR.

5.1.2. PR-2 (β-1,3-glucanases) and PR-3 and -4 (chitinases)

PR-2 proteins are β-1,3-glucanases and PR-3 and -4 proteins are chitinases. Because β-1,3-glucanases and chitinases were discovered as PR proteins early on, they had been widely studied for their roles in plant defense against pathogens in many species, including grape. They exert antimicrobial activity as a result of their ability to hydrolyze fungal cell wall components. The former hydrolyze β-1,3-glucan and the latter hydrolyze chitin. The synergistic effects of β-1,3-glucanases and chitinases inhibit the growth of fungal pathogens [35, 36]. These proteins function indirectly by releasing the elicitor of oligosaccharides from the cell wall of pathogens, thereby inducing various plant defense mechanisms. As shown in **Figure 1**, the expression of PR-2 gene along with the PR-1, PR-5 gene is dependent on SA [33], expression of PR-3, PR-4 gene depends on the JA in *Arabidopsis* [32].

Jacobs et al. [37] showed that the hydrolytic activity in grape directly affects the extent of infection by powdery mildew at the pathogen infection site. β-1,3-Glucanase and chitinase activities were strongly induced in leaves and pre-véraison berries by ethephon treatment. Moreover, PR protein expression was decreased during grape maturation, which explains the increased susceptibility of grape to pathogen attack at the final stage of maturation [38]. Apoplasmic β-1,3-glucanase gene (*VvGHF17*) is expressed constitutively in grape leaves, berry pulp, and skin, and VvGHF17-overexpressing Arabidopsis plants exhibited disease resistance to *B. cinerea* and *Colletotrichum higginsianum* [39].

5.1.3. PR-5 (thaumatin-like proteins)

PR-5 proteins include thaumatin-like proteins and osmotin. The amino acid compositions and the NH_2 terminal sequences of thaumatin-like proteins showed that thaumatin-like proteins are actually osmotins, which are known to accumulate in tobacco cells in response to osmotic stress [40]. PR-5 proteins are believed to be involved in enhancing fungal membrane permeability and causing osmotic rupture of fungal plasma membrane [41]. Jayasankar et al. [42] demonstrated *in vitro* and *in vivo* that the constitutive expression of *V. vinifera* thaumatin-like protein 1 (VVTL-1) in *V. vinifera* protected grape from anthrax. Treatment with ethephon, which effects ethylene generation, induced grape PR-5 gene families in leaves and berries [37].

5.1.4. PR-10 (ribonuclease (like))

PR-10 proteins exhibit ribonuclease (RNase) activity. RNase contributes to plant defense in programmed cell death during HR or acts directly against pathogens [43]. PR-10 gene (VpPR10.2) isolated from *Vitis pseudoreticulata* exhibited resistance to fungi and had high homology with VvPR10.2 from susceptible *V. vinifera* cultivar. In contrast to VvPR10.2, VpPR10.2 was induced at high levels in response to *P. viticola* infection [44]. On the other hand, PR-10 proteins with cytokinin-binding activity were identified in mung bean [45] and moss [46]. Questions persist regarding the roles of PR-10.

5.1.5. PR-14 (lipid transfer proteins)

PR-14 proteins are lipid transfer proteins (LTPs). Some LTP-like polypeptides show antifungal or antibacterial activity [47, 48]. Several isoforms are involved in the plant defense signaling pathway [49–51]. Type I LTP of tobacco binds jasmonic acid (JA), a signaling molecule, and the complex interacts with receptors on the cell membrane [52]. Some grape LTPs bind JA. The external application of the VvLTP4-JA complex to grape plantlets enhanced resistance to *B. cinerea* infection compared to the application of either VvLTP4 or JA alone [53]. Such elicitors as ergosterol triggered VvLTP1 upregulation by WRKY transcription factor and stilbene synthase gene expression in grape plantlets, and enhanced the production of resveratrol (grape phytoalexin) and the resistance to *B. cinerea* [54].

5.1.6. PR-15 (oxalate oxidases) and PR-16 (oxalate-oxidase-like proteins)

PR-15 and PR-16 include germins (oxalate oxidases) and germin-like proteins (oxalate-oxidase-like proteins), respectively. Many germin-like proteins exhibit oxalic acid ester oxidase (OXO) or superoxide dismutase (SOD) activity. They are involved in the production of H_2O_2 a ROS, which has antimicrobial activity. Seven germin-like protein (GLP) cDNA clones were isolated from *V. vinifera* 'Chardonnay' [55]. *V. vinifera* germin-like 3 (VvGLP3) is strongly induced by *U. necator* infection. Neither VvGLP1 nor VvGLP7 was induced by *U. necator*, but these were induced by *B. cinerea* or *P. viticola* infection.

5.1.7. Others

PR-6, PR-7, PR-8, PR-9, PR-11, PR-12, and PR-13 are proteinase inhibitors, endoproteases, chitinases (type III), peroxidases, chitinases (type I), defensins, and thionins, respectively. To the best of our knowledge, these proteins have not yet been detected in grape.

5.2. Physiological function

Some PR protein families have physiological functions. For example, PR-2 proteins (β-1,3-glucanases) hydrolyze β-1,3-glucan in fungal cell wall, but because β-1,3-glucan (called 'callose' in plants), the substrate of β-1,3-glucanase, is widespread in plants, PR-2 proteins must perform various physiological functions, such as flower formation [56]. Many examples of PR-10 and PR-14 proteins have been reported in grape. The overexpression of grapevine PR-10 gene (*Vvpr10*) in *Saccharomyces cerevisiae* conferred salt tolerance in the yeast [57]. Some LTP isoforms are involved in somatic embryo development and epidermal layer formation. LTP gene is expressed during zygotic and somatic embryogenesis [58–60]. The overexpression of VvLTP1 gene interferes with somatic embryo development and invalidates the bilateral symmetry of the embryo in grape [61].

6. Application of PR proteins to pathogen control in *V. vinifera* grapevine

Fungicides are used to control fungal diseases. However, their adverse effects on the environment and the appearance of fungi resistant to the fungicides have been reported. Therefore, the development of new plant disease control methods is desired. In recent years, new strategies to increase plant resistance have been examined. Among them, PR-protein-related pathogen control methods are described here.

6.1. Molecular breeding

For a long time, researchers and breeders have used conventional breeding methods for the development of disease-resistant cultivars from available resources in genus *Vitis*. However, conventional breeding methods are hampered by a major problem, namely, hybrid produced by crossing often exhibits undesirable traits from hybrid parent resistant to fungi. In order to remove these traits, backcrossing is performed many times. On the other hand, molecular breeding by a transformation method can use breeding resources other than those from genus *Vitis* and offers the possibility of improving only the objective trait, such as disease resistance, without modifying other viticulturally desirable traits in the target cultivars.

Target cultivar	Introduced gene	Gene source	Acquired resistance to	Reference
'Merlot'	Chitinase	*Trichoderma*	*B. cinerea* and *U. necator*	[63]
'Chardonnay'				
'Chardonnay'	Chitinase	*Trichoderma*	*B. cinerea* and *U. necator*	[65]
'Thompson Seedless'	Chitinase	*Trichoderma*	*B. cinerea*	[64]
'Neo Muscat'	Chitinase	Rice	*U. necator* and *Elsinoe ampelina*	[66]
'Pusa Seedless'	Chitinase	Rice	*U. necator*	[67]
'Crimson Seedless'	Chitinase and β-1,3-glucanase	Wheat	*P. viticola*	[68]
'Thompson Seedless'	TLP (cisgenic)	Chardonnay	*U. necator*	[70]
'Seyval blanc'	Chitinase and RIP	Barley	No effect	[69]

β-1,3-Glucanase (PR-2); chitinase (PR-3 and -4); TLP, thaumatin-like protein (PR-5); RIP, ribosome-inactivating protein.

Table 3. Disease resistance traits introduced into *V. vinifera* cultivars.

Most commercially valuable *V. vinifera* cultivars are susceptible to pathogens, particularly fungi, as described above. Many studies that have attempted to confer fungal resistance to *V. vinifera* cultivars, including wine grape and table grape, via the transformation technique have been reported. The major findings appear in **Table 3**. The first report of a transformation technique for *V. vinifera* was by Kikkert et al. [62, 63]. The introduction of chitinase from *Trichoderma* conferred disease resistance to *V. vinifera* cultivars and established a transformation system with embryogenic cultures of *V. vinifera* 'Merlot' and 'Chardonnay' by a biolistic method. Thereafter, examples of the *Agrobacterium*-mediated transformation method using transgenic chitinase were reported [64–68]. Nookaraju and Agrawal [68] also reported that the introduction of both β-1,3-glucanase (PR-2) and chitinase in transgenic plants increased resistance to downy mildew. Transgenic plants that displayed no resistance in field tests have also been reported [69]. The increase of disease resistance by chitinase introduction has been reported in other plants as well. Chitin is the main component of the cell wall in major pathogens, *U. necator, B. cinerea,* and *P. viticola*. As higher plants do not contain chitin, chitinase does not affect growth and development in plants. Therefore, this control strategy using chitinase is considered to be very effective.

Recently, PR proteins aside from the above have been used and strategies other than transgenic approaches have been attempted. *V. vinifera* 'Crimson Seedless' expressing cisgenic thaumatin-like proteins (TLP, PR-5) from *V. vinifera* 'Chardonnay' displayed resistance to powdery mildew [70]. Although the transformation method is adopted, cisgenic plants are produced by using a gene derived from a closely related species originally present in nature as genetic resources. Although cisgenic plants fall under regulations for transgenic plants, discussions are underway as to whether cisgenic plants should be excluded from the regulations in the EU and the US [71].

6.2. New chemical control method using elicitors

Transgenic grapevines are forbidden in French vineyards. Aziz et al. [72] proposed an alternative strategy for controlling pathogens, which is the activation of plant defense mechanism by elicitors. Defense reactions elicited by laminarin increase PR protein (chitinase and β-1,3-glucanase) activities and confer resistance to *B. cinerea* and *P. viticola* growth [72]. It is considered that the induction of natural plant defense mechanisms by elicitors is an attractive and powerful method compared with chemical fungicides [73]. Indeed, a number of positive results against downy and powdery mildew and gray mold have been reported in laboratories and greenhouses [74]. However, little research has been carried out in vineyards. In addition, results comparable to those obtained in laboratories have not been generated and its effect seems to be weaker than that of the conventional fungicide method. This variability is likely due to environmental, plant, pathogen, and external conditions. To quote Delaunois et al. [74], 'Additional research needs to be pursued to fully understand defense stimulation under vineyard conditions.'

7. Conclusions

Knowledge of the roles and functions of genes encoding PR proteins in grape has been accumulated from previous studies of other plants, such as tobacco, and used in the development of new methods for controlling pathogens in grape. However, questions remain, such as the presence of PR proteins of other classes in grape and the regulation PR protein genes involved in the plant defense mechanism. Elucidating the answers to these questions at the molecular level will help further our understanding of PR proteins in *V. vinifera* and create possibilities to improve grape quality and yield by efficiently preventing pathogen growth in the future.

Acknowledgements

Special thanks go to Ms. Kayo Arita, Mr. Masachika Mikami, and Mr. Yoshinao Aoki of the University of Yamanashi for their assistance.

Author details

Shinichi Enoki and Shunji Suzuki*

*Address all correspondence to: suzukis@yamanashi.ac.jp

Laboratory of Fruit Genetic Engineering, The Institute of Enology and Viticulture, University of Yamanashi, Yamanashi, Japan

References

[1] Staudt, G. (1997). Evaluation of resistance to grapevine powdery mildew (*Uncinula necator* [Schw.] Burr., anamorph *Oidium tuckeri* Berk.) in accessions of *Vitis* species. VITIS, 36(3), 151.

[2] Doster, M. A., & Schnathorst, W. C. (1985). Comparative susceptibility of various grapevine cultivars to the powdery mildew fungus *Uncinula necator*. American Journal of Enology and Viticulture, 36(2), 101–104.

[3] Mullins, M. G., Bouquet, A., & Williams, L. E. (1992). Biology of the grapevine. Cambridge University Press, Cambridge.

[4] Latorre, B. A., Elfar, K., & Ferrada, E. E. (2015). Gray mold caused by *Botrytis cinerea* limits grape production in Chile. Ciencia e investigación agraria: revista latinoamericana de ciencias de la agricultura, 42(3), 316–330.

[5] Clark, J. S. C., & Spencer-Phillips, P. T. N. 2000. Downy mildew. In: Lederberg, J., Alexander, M., Bloom, B. R., Hopwood, D., Hull, R., Iglewski, B. H., Laskin, A. I., Oliver,

S. G., Schaechter, M., & Summers, W. C. (Eds.), The Encyclopedia of Micro-biology, vol. 2. Academic Press, San Diego, USA, pp. 117e129.

[6] Boso, S., Alonso-Villaverde, V., Gago, P., Santiago, J. L., & Martínez, M. C. (2014). Susceptibility to downy mildew (*Plasmopara viticola*) of different *Vitis* varieties. Crop Protection, 63, 26–35.

[7] Delye, C., Laigret, F., & Corio-Costet, M. F. (1997). A mutation in the 14 alpha-demethylase gene of *Uncinula necator* that correlates with resistance to a sterol biosynthesis inhibitor. Applied and Environmental Microbiology, 63(8), 2966–2970.

[8] Chen, W. J., Delmotte, F., Cervera, S. R., Douence, L., Greif, C., & Corio-Costet, M. F. (2007). At least two origins of fungicide resistance in grapevine downy mildew populations. Applied and Environmental Microbiology, 73(16), 5162–5172.

[9] Ishii, H., Fraaije, B. A., Sugiyama, T., Noguchi, K., Nishimura, K., Takeda, T., Amano, T., & Hollomon, D. W. (2001). Occurrence and molecular characterization of strobilurin resistance in cucumber powdery mildew and downy mildew. Phytopathology, 91(12), 1166–1171.

[10] Mizuta, H., & Yasui, H. (2012). Protective function of silicon deposition in Saccharina japonica sporophytes (Phaeophyceae). Journal of applied phycology, 24(5), 1177–1182.

[11] Mabekoje, O. O., Bello, O. O., & Egberongbe, H. O. (2013). Antimicrobial efficacy of ocimum gratissimum andvernonia amygdalina on gastrointestinal bacteria. Canadian Journal of pure and applied sciences, 7(2), 2341.

[12] Aist, J. R., Structural responses as resistance mechanisms, in The Dynamics of Host Defense, Bailey, J. A., Ed., Academic Press, New York, 1983, 33.

[13] Brisson, L. F., Tenhaken, R., & Lamb, C. (1994). Function of oxidative cross-linking of cell wall structural proteins in plant disease resistance. The Plant Cell, 6(12), 1703–1712.

[14] Van Loon, L. C., & Van Kammen, A. (1970). Polyacrylamide disc electrophoresis of the soluble leaf proteins from *Nicotiana tabacum* var.'Samsun'and 'Samsun NN': II. Changes in protein constitution after infection with tobacco mosaic virus. Virology, 40(2), 199–211.

[15] Dixon, R. A. (1986). The phytoalexin response: elicitation, signalling and control of host gene expression. Biological Reviews, 61(3), 239–291.

[16] Linthorst, H. J., & Van Loon, L. C. (1991). Pathogenesis-related proteins of plants. Critical reviews in plant sciences, 10(2), 123–150.

[17] Keen, N. T., & Yoshikawa, M. (1983). β-1, 3-Endoglucanase from soybean releases elicitor-active carbohydrates from fungus cell walls. Plant Physiology, 71(3), 460–465.

[18] Yamaguchi, T., Yamada, A., Hong, N., Ogawa, T., Ishii, T., & Shibuya, N. (2000). Differences in the recognition of glucan elicitor signals between rice and soybean:

β-glucan fragments from the rice blast disease fungus *Pyricularia oryzae* that elicit phytoalexin biosynthesis in suspension-cultured rice cells. Plant Cell, 12(5), 817–826.

[19] Gianinazzi, S., Martin, C., & Vallee, J. (1970). Hypersensibilite aux virus, temperature et proteines solubles chez le *Nicotiana xanthi* n. c: apparition de nouvelles macromolecules lors de la repression de la synthese virale. Academic Press Paris CR Ser D, 270, 2382–2386.

[20] Van Loon, L. C. (1985). Pathogenesis-related proteins. Plant Molecular Biology, 4(2), 111–116.

[21] Antoniw, J. F., & White, R. F. (1980). The effects of aspirin and polyacrylic acid on soluble leaf proteins and resistance to virus infection in five cultivars of tobacco. Journal of Phytopathology, 98(4), 331–341.

[22] Van Loon, L. C., Pierpoint, W. S., Boller, T. H., & Conejero, V. (1994). Recommendations for naming plant pathogenesis-related proteins. Plant Molecular Biology Reporter, 12(3), 245–264.

[23] Wagih, E. E., & Coutts, R. H. (1981). Similarities in the soluble protein profiles of leaf tissue following either a hypersensitive reaction to virus infection or plasmolysis. Plant Science Letters, 21(1), 61–69.

[24] Antoniw, J. F., Kueh, J. S. H., Walkey, D. G. A., & White, R. F. (1981). The presence of pathogenesis-related proteins in callus of Xanthi-nc Tobacco. Journal of Phytopathology, 101(2), 179–184.

[25] Van Loon, L. C. (1976). Specific soluble leaf proteins in virus-infected tobacco plants are not normal constituents. Journal of General Virology, 30(3), 375–379.

[26] Gianinazzi, S., Pratt, H. M., Shewry, P. R., & Miflin, B. J. (1977). Partial purification and preliminary characterization of soluble leaf proteins specific to virus infected tobacco plants. Journal of General Virology, 34(2), 345–351.

[27] Pierpoint, W. S., Robinson, N. P., & Leason, M. B. (1981). The pathogenesis-related proteins of tobacco: their induction by viruses in intact plants and their induction by chemicals in detached leaves. Physiological Plant Pathology, 19(1), 85-IN29.

[28] Parent, J. G., & Asselin, A. (1984). Detection of pathogenesis-related proteins (PR or b) and of other proteins in the intercellular fluid of hypersensitive plants infected with tobacco mosaic virus. Canadian Journal of Botany, 62(3), 564–569.

[29] Carr, J. P., Dixon, D. C., Nikolau, B. J., Voelkerding, K. V., & Klessig, D. F. (1987). Synthesis and localization of pathogenesis-related proteins in tobacco. Molecular and Cellular Biology, 7(4), 1580–1583.

[30] Van Loon, L. C., Rep, M., & Pieterse, C. M. J. (2006). Significance of inducible defense-related proteins in infected plants. Annual Review of Phytopathology, 44, 135–162.

[31] Bertsch, C., Joly, D., & Walter, B. (2003). Sequence of a putative *Vitis vinifera* PR-1. Vitis, 42(2), 103–104.

[32] Thomma, B. P., Eggermont, K., Penninckx, I. A., Mauch-Mani, B., Vogelsang, R., Cammue, B. P., & Broekaert, W. F. (1998). Separate jasmonate-dependent and salicylate-dependent defense-response pathways in Arabidopsis are essential for resistance to distinct microbial pathogens. Proceedings of the National Academy of Sciences, 95(25), 15107–15111.

[33] Uknes, S., Mauch-Mani, B., Moyer, M., Potter, S., Williams, S., Dincher, S., Chandler, D., Slusarenko, A., & Ryals, J. (1992). Acquired resistance in Arabidopsis. The Plant Cell, 4(6), 645–656.

[34] Dombrecht, B., Xue, G. P., Sprague, S. J., Kirkegaard, J. A., Ross, J. J., Reid, J. B., Fit, G. P., Sewelam, N., Schenk, P. M., Manners, J. M., & Kazan, K. (2007). MYC2 differentially modulates diverse jasmonate-dependent functions in Arabidopsis. The Plant Cell, 19(7), 2225–2245.

[35] Mauch, F., Mauch-Mani, B., & Boller, T. (1988). Antifungal hydrolases in pea tissue II. Inhibition of fungal growth by combinations of chitinase and β-1,3-glucanase. Plant Physiology, 88(3), 936–942.

[36] Jongedijk, E., Tigelaar, H., Van Roekel, J. S., Bres-Vloemans, S. A., Dekker, I., van den Elzen, P. J., Cornelissen, B. J., & Melchers, L. S. (1995). Synergistic activity of chitinases and β-1,3-glucanases enhances fungal resistance in transgenic tomato plants. Euphytica, 85(1–3), 173–180.

[37] Jacobs, A. K., Dry, I. B., & Robinson, S. P. (1999). Induction of different pathogenesis-related cDNAs in grapevine infected with powdery mildew and treated with ethephon. Plant Pathology, 48(3), 325–336.

[38] Monteiro, S., Piçarra-Pereira, M. A., Loureiro, V. B., Teixeira, A. R., & Ferreira, R. B. (2007). The diversity of pathogenesis-related proteins decreases during grape maturation. Phytochemistry, 68(4), 416–425.

[39] Fujimori, N., Enoki, S., Suzuki, A., Naznin, H. A., Shimizu, M., & Suzuki, S. (2016). Grape apoplasmic β-1,3-glucanase confers fungal disease resistance in *Arabidopsis*. Scientia Horticulturae, 200, 105–110.

[40] Stintzi, A., Heitz, T., Kauffmann, S., Legrand, M., & Fritig, B. (1991). Identification of a basic pathogenesis-related, thaumatin-like protein of virus-infected tobacco as osmotin. Physiological and Molecular Plant Pathology, 38(2), 137–146.

[41] Roberts, W. J., & Selitrennikoff, C. P. (1990). Zeamatin, an antifungal protein from maize with membrane-permeabilizing activity. Journal of General Microbiology, 136(9), 1771–1778.

[42] Jayasankar, S., Li, Z., & Gray, D. J. (2003). Constitutive expression of *Vitis vinifera* thaumatin-like protein after in vitro selection and its role in anthracnose resistance. Functional Plant Biology, 30(11), 1105–1115.

[43] Liu, J. J., & Ekramoddoullah, A. K. (2006). The family 10 of plant pathogenesis-related proteins: their structure, regulation, and function in response to biotic and abiotic stresses. Physiological and Molecular Plant Pathology, 68(1), 3–13.

[44] He, M., Xu, Y., Cao, J., Zhu, Z., Jiao, Y., Wang, Y., Xin Guan, X., Yang, Y., Xu, W., & Fu, Z. (2013). Subcellular localization and functional analyses of a PR10 protein gene from *Vitis pseudoreticulata* in response to *Plasmopara viticola* infection. Protoplasma, 250(1), 129–140.

[45] Fujimoto, Y., Nagata, R., Fukasawa, H., Yano, K., & Azuma, M. others. (1998). Purification and cDNA cloning of cytokinin-specific binding protein from mung bean (*Vigna radiata*). European Journal of Biochemistry, 258, 794–802.

[46] Gonneau, M., Pagant, S., Brun, F., & Laloue, M. (2001). Photoaffinity labelling with the cytokinin agonist azido-CPPU of a 34 kDa peptide of the intracellular pathogenesis-related protein family in the moss *Physcomitrella patens*. Plant Molecular Biology, 46(5), 539–548.

[47] Cammue, B. P. A., Thevissen, K., Hendriks, M., Eggermont, K., Goderis, I. J., Proost, P., Van Damme, J., Osborn, R. W., Guerbette, F., Kader, J. C., & Broekaert, W. F. (1995). A potent antimicrobial protein from onion seeds showing sequence homology to plant lipid transfer proteins. Plant Physiology, 109(2), 445–455.

[48] Terras, F. R., Schoofs, H. M., De Bolle, M. F., Van Leuven, F., Rees, S. B., Vanderleyden, J., Cammue, B. P., & Broekaert, W. F. (1992). Analysis of two novel classes of plant antifungal proteins from radish (*Raphanus sativus* L.) seeds. Journal of Biological Chemistry, 267(22), 15301–15309.

[49] Maldonado, A. M., Doerner, P., Dixon, R. A., Lamb, C. J., & Cameron, R. K. (2002). A putative lipid transfer protein involved in systemic resistance signalling in Arabidopsis. Nature, 419(6905), 399–403.

[50] Blein, J. P., Coutos-Thévenot, P., Marion, D., & Ponchet, M. (2002). From elicitins to lipid-transfer proteins: a new insight in cell signalling involved in plant defense mechanisms. Trends in Plant Science, 7(7), 293–296.

[51] Grant, M., & Lamb, C. (2006). Systemic immunity. Current Opinion in Plant Biology, 9(4), 414–420.

[52] Buhot, N., Gomès, E., Milat, M. L., Ponchet, M., Marion, D., Lequeu, J., Delrot, S., Thévenot, P. C., & Blein, J. P. (2004). Modulation of the biological activity of a tobacco LTP1 by lipid complexation. Molecular Biology of the Cell, 15(11), 5047–5052.

[53] Girault, T., François, J., Rogniaux, H., Pascal, S., Delrot, S., Coutos-Thévenot, P., & Gomès, E. (2008). Exogenous application of a lipid transfer protein-jasmonic acid

complex induces protection of grapevine towards infection by *Botrytis cinerea*. Plant Physiology and Biochemistry, 46(2), 140–149.

[54] Laquitaine, L., Gomès, E., François, J., Marchive, C., Pascal, S., Hamdi, S., Atanassova, R., Delrot, S., & Coutos-Thévenot, P. (2006). Molecular basis of ergosterol-induced protection of grape against *Botrytis cinerea*: induction of type I LTP promoter activity, WRKY, and stilbene synthase gene expression. Molecular Plant-Microbe Interactions, 19(10), 1103–1112.

[55] Godfrey, D., Able, A. J., & Dry, I. B. (2007). Induction of a grapevine germin-like protein (VvGLP3) gene is closely linked to the site of *Erysiphe necator* infection: a possible role in defense? Molecular Plant-Microbe Interactions, 20(9), 1112–1125.

[56] Akiyama, T., Pillai, M. A., & Sentoku, N. (2004). Cloning, characterization and expression of OsGLN2, a rice endo-1,3-β-glucanase gene regulated developmentally in flowers and hormonally in germinating seeds. Planta, 220(1), 129–139.

[57] Jellouli, N., Jouira, H. B., Daldoul, S., Chenennaoui, S., Ghorbel, A., Salem, A. B., & Gargouri, A. (2010). Proteomic and transcriptomic analysis of grapevine PR10 expression during salt stress and functional characterization in yeast. Plant Molecular Biology Reporter, 28(1), 1–8.

[58] Vroemen, C. W., Langeveld, S., Mayer, U., Ripper, G., Jurgens, G., Van Kammen, A., & De Vries, S. C. (1996). Pattern formation in the Arabidopsis embryo revealed by position-specific lipid transfer protein gene expression. Plant Cell, 8(5), 783–791.

[59] Coutos-Thevenot, P., Jouenne, T., Maes, O., Guerbette, F., Grosbois, M., Caer, J. P., Boulay, M., Deloire, A., Kader, J. C., & Guern, J. (1993). Four 9-kDa proteins excreted by somatic embryos of grapevine are isoforms of lipid-transfer proteins. European Journal of Biochemistry, 217(3), 885–889.

[60] Sterk, P., Booij, H., Schellekens, G. A., Van Kammen, A., & De Vries, S. C. (1991). Cell-specific expression of the carrot EP2 lipid transfer protein gene. Plant Cell, 3(9), 907–921.

[61] François, J., Lallemand, M., Fleurat-Lessard, P., Laquitaine, L., Delrot, S., Coutos-Thévenot, P., & Gomès, E. (2008). Overexpression of the *VvLTP1* gene interferes with somatic embryo development in grapevine. Functional Plant Biology, 35(5), 394–402.

[62] Kikkert, J. R., Ali, G. S., Striem, M. J., Martens, M., Wallace, P. G., Molino, L., & Reisch, B. I. (1996). Genetic engineering of grapevine (*Vitis* sp.) for enhancement of disease resistance. In III International Symposium on In Vitro Culture and Horticultural Breeding, 447, 273–280.

[63] Kikkert, J. R., Ali, G. S., Wallace, P. G., Reisch, B., & Reustle, G. M. (1998). Expression of a fungal chitinase in *Vitis vinifera* L. 'Merlot' and 'Chardonnay' plants produced by biolistic transformation. In VII International Symposium on Grapevine Genetics and Breeding, 528, 299–306.

[64] Hinrichsen, P., Reyes, M. A., Castro, A., Araya, S., Garnier, M., Prieto, H., Reyes, F., Muñoz, C., Dell'Orto, P., & Moynihan, M. R. (2005). Genetic transformation of grapevines with *Trichoderma harzianum* and antimicrobial peptide genes for improvement of fungal tolerance. In VII International Symposium on Grapevine Physiology and Biotechnology, 689, 469–474.

[65] Reisch, B., Kikkert, J., Vidal, J., Ali, G., & Gadoury, D. (2002). Genetic transformation of *Vitis vinifera* to improve disease resistance. In VIII International Conference on Grape Genetics and Breeding, 603, 303–308.

[66] Yamamoto, T., Iketani, H., Ieki, H., Nishizawa, Y., Notsuka, K., Hibi, T., Hayashi, T., & Matsuta, N. (2000). Transgenic grapevine plants expressing a rice chitinase with enhanced resistance to fungal pathogens. Plant Cell Reports, 19(7), 639–646.

[67] Nirala, N. K., Das, D. K., Srivastava, P. S., Sopory, S. K., & Upadhyaya, K. C. (2010). Expression of a rice chitinase gene enhances antifungal potential in transgenic grapevine (*Vitis vinifera* L.). VITIS, 49(4), 181.

[68] Nookaraju, A., & Agrawal, D. C. (2012). Enhanced tolerance of transgenic grapevines expressing chitinase and β-1,3-glucanase genes to downy mildew. Plant Cell, Tissue and Organ Culture (PCTOC), 111(1), 15–28.

[69] Bornhoff, B. A., Harst, M., Zyprian, E., & Töpfer, R. (2005). Transgenic plants of *Vitis vinifera* cv. Seyval blanc. Plant Cell Reports, 24(7), 433–438.

[70] Dhekney, S. A., Li, Z. T., & Gray, D. J. (2011). Grapevines engineered to express cisgenic *Vitis vinifera* thaumatin-like protein exhibit fungal disease resistance. In Vitro Cellular & Developmental Biology-Plant, 47(4), 458–466.

[71] Holme, I. B., Wendt, T., & Holm, P. B. (2013). Intragenesis and cisgenesis as alternatives to transgenic crop development. Plant Biotechnology Journal, 11(4), 395–407.

[72] Aziz, A., Poinssot, B., Daire, X., Adrian, M., Bézier, A., Lambert, B., Joubert, J. M., & Pugin, A. (2003). Laminarin elicits defense responses in grapevine and induces protection against *Botrytis cinerea* and *Plasmopara viticola*. Molecular Plant-Microbe Interactions, 16(12), 1118–1128.

[73] Klarzynski, O., & Fritig, B. (2001). [Stimulation of plant natural defenses]. Comptes rendus de l'Academie des sciences. Series III, Sciences de la vie, 324(10), 953–963.

[74] Delaunois, B., Farace, G., Jeandet, P., Clément, C., Baillieul, F., Dorey, S., & Cordelier, S. (2014). Elicitors as alternative strategy to pesticides in grapevine? Current knowledge on their mode of action from controlled conditions to vineyard. Environmental Science and Pollution Research, 21(7), 4837–4846.

Chapter 3

Evaluation of the Cultivar Effect on Wine Grape Fungal Diseases with a Use of a Low-Input Fungicide Regimen in Southeastern Virginia, USA

Mizuho Nita

Additional information is available at the end of the chapter

http://dx.doi.org/10.5772/64845

Abstract

Wine grape cultivar selection was examined as a use of biotechnology and a part of integrated plant disease management in grape and wine production. The efficacy of 18 wine grape cultivars against various wine grape diseases was examined using a relatively low-input fungicide regimen to determine whether these diseases can be managed under hot and humid southeastern Virginia conditions over 5 years. Disease developments of black rot, Botrytis bunch rot, downy mildew, Phomopsis cane and leaf spot, and powdery mildew was evaluated. Although overall level of disease was low in each year, we observed a significant ($P \leq 0.05$) effect of cultivar in many cases, indicating the importance of cultivar selection. Cultivars such as 'Norton', 'Noire', 'Traminette', 'Vidal blanc', and 'Viognier' were found to be less susceptible to the major diseases under Virginia environmental conditions.

Keywords: grape cultivar selection, low input disease management, Southeastern US, Norton, Noiret

1. Introduction

The state of Virginia (VA) is in the mid-Atlantic region of the USA. Currently, VA ranks fifth in wine grape production in the US with 1100 hectares of wine grape acreage (*Vitis vinifera*, inter- or intraspecific hybrids, and *V. labrusca*) and over 230 wineries [1, 2]. An annual economic impact of VA wine industry was estimated to be $740 million in 2010 [2], which represents a 106% increase from 2005.

VA is known for its diverse climate with five different climate regions (**Figure 1**): Tidewater, Piedmont, Northern Virginia, Western Mountain, and Southwestern Mountain regions [3]. Some localities, such as Charlottesville in the Central VA, are preferred for commercial wine grape production due to long growing seasons and relatively lower risk of winter temperature extremes. On the other hand, winters on the north of the Blue Ridge mountains (which is a part of the Appalachian Mountains, the border between Northern and Western Mountain in **Figure 1**) in Northern VA can be very severe.

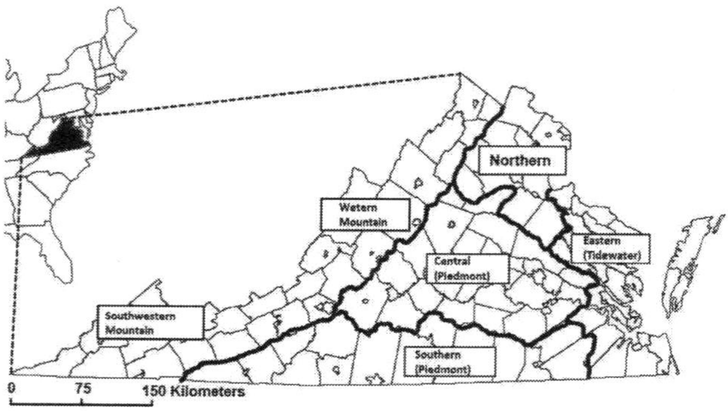

Figure 1. Map of the state of Virginia (adapted with permission from [4]).

Annual rainfall totals can vary from 838 mm in the Shenandoah Valley, which is located the west of the Blue Ridge mountains, to more than 1500 mm in the mountains of southwestern Virginia [3]. The Gulf Stream in the Atlantic Ocean and moist air coming from the Pacific Ocean are the primary drivers for precipitations in VA. Storms resulted from warm and humid air generally moves from the west to the east in the continental US. At the East coast, the storms are pushed toward the northeast by moist air that comes from the south, following the Gulf Stream. The meeting of two air masses often results in frontal storms coming from the west to produce rainfalls, especially along the western side of the Appalachian Mountains. In addition to the frontal storms, tropical storms, which typically come to VA in late August to early October, provide 10–40% of the annual rainfall [3]. When these storms become very intense and/or take more westward pass, the strong winds that hit the mountain range result in heavy rainfalls on the Blue Ridge. However, often time, these tropical storms come from the southern US, cross the East coast to the south of Virginia, and then exit out to the Atlantic. In such cases, the heaviest rain usually falls in southeastern VA.

Southeastern VA is traditionally known for tobacco production since the first tobacco planting in the US in 1612 near the James River, which runs through VA [5]. However, due to a recent increase in cheaper tobacco import from foreign countries, the tobacco industry is struggling [6]. Thus, many growers are seeking for alternative crops. Since labor and land cost is relatively low, cash crops such as wine grapes are often considered as a crop of choice [7]. Traditionally,

Central VA and Northern VA have been the major wine grape growing regions in VA; however, based on 2013 information, 64 farm wineries are located in the lower half of VA where relatively lower price of land allows growers to plant more vines (www.virginiawine.org).

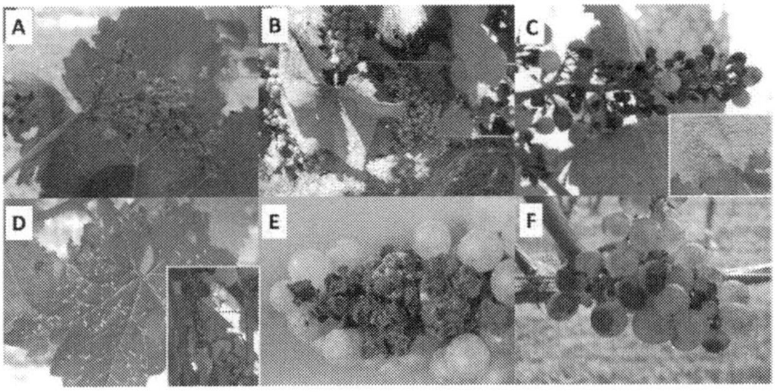

Figure 2. Symptoms commonly found grapevine fungal diseases in Virginia. Panel (A) Downy mildew on a cluster and leaf, (B) powdery mildew on clusters and leaves, (C) black rot on clusters (and on a leaf in the lower right insert), (D) phomopsis cane and leaf spot on a leaf, (and on a shoot and rachis in the lower right insert), (E) Botrytis bunch rot, and (F) ripe rot, pictures courtesy of the author.

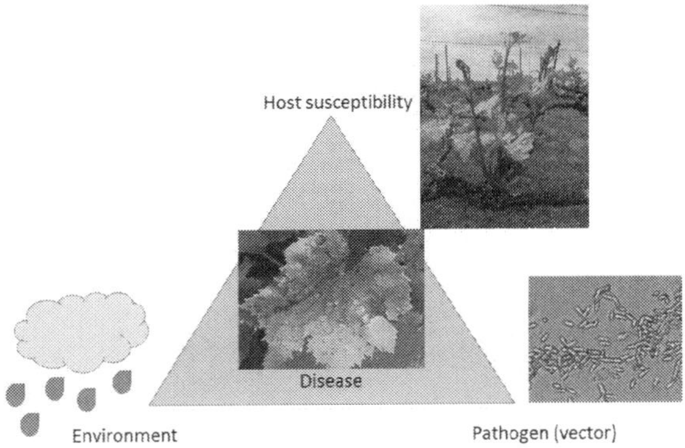

Figure 3. Disease triangle adapted with permission from Agrios [9], pictures courtesy of the author.

Frequent rainfalls in the spring and summer create environmental conditions that favor development of various diseases [8]. Common summer time diseases in Virginia are downy mildew (caused by *Plasmopara viticola*) (**Figure 2A**), powdery mildew (caused by *Erysiphe necator*) (**Figure 2B**), black rot (caused by *Guignardia bidwellii*) (**Figure 2C**), and Phomopsis cane and leaf spot (caused by *Phomopsis viticola*) (**Figure 2D**). As the berries mature, late season rots such as Botrytis bunch rot (also known as Botrytis gray mold, caused by *Botrytis cinerea*) (**Figure 2E**), ripe rot (caused by *Colletotrichum acutatum* and *C. gloeosporioides* species complexes) (**Figure 2F**), and sour rot (caused by several bacteria and yeast species) become more common.

Development of plant disease is often described with a concept called the disease triangle [9]. Three components (environment, susceptible host, and presence of pathogen) have to be met in order for any plant disease to develop (**Figure 3**). The frequent rain events in VA produce the environment for many summer diseases, as well as late season rots. In addition, growers often select cultivars based on their marketing needs, rather than viticultural decisions, such as disease resistance [10]. In VA, cultivars such as 'Chardonnay', 'Merlot', and 'Cabernet Sauvignon' are commonly grown commercially [8]. These cultivars are relatively susceptible to the diseases mentioned above, when you compared with some of intraspecific hybrid cultivars, such as 'Vidal blanc' or 'Traminette' [8].

The high risk of disease development often results in more frequent applications of fungicides [9, 10]. This decision is often made not based on the facts (i.e., components of the disease triangle), but based on the fear of losing the crop. The unnecessary fungicide applications not only cost growers as out of pocket expenses, but also cause significant environmental impacts, such as increased risk of fungicide drift, fungicide resistance, soil compaction due to more traffic, more CO_2 emission, etc. [9]. The over use of the materials often resulted in the misuse of them, which can increase the risk of fungicide resistance [10, 11]. The frequent application of fungicides also increases the concern on the over use of "hard" materials such as mancozeb and captan, which are used frequently because of their broad target pathogen ranges and low fungicide resistance risks [8, 9]. Heavy reliance on mancozeb can results in increased European red mite population [9], and captan is known as an eye irritant [12].

This chapter will focus on cultivar selection as a use of biotechnology and a part of integrated plant disease management in grape and wine production. The efficacy of 18 wine grape cultivars against various wine grape diseases was examined using a relatively low-input fungicide regimen to determine whether these diseases can be managed under hot and humid southeastern VA conditions over 5 years.

2. Materials and methods

A research vineyard at Virginia Polytechnic Institute and State University's Southern Piedmont Agricultural Research and Extension Center (SPAREC) (latitude = 37.0754205, longitude = −77.8670789) was utilized for the experiment. Research plots were completely randomized design with 6 replications of 18 cultivars (**Table 1**). Each replication consisted of three vines of the same cultivar. Disease incidence and severity were visually assessed twice in the season (approximately at bloom and veraison) on leaves and clusters for five major diseases for the area (black rot, Botrytis bunch rot, downy mildew, Phomopsis cane and leaf spot, and Powdery mildew). The vineyard was evaluated for 5 years (2009–2013).

Cultivar	Short names used in this chapter
Aleatico	Aleatico
Cabernet Franc clone #1	CabF1
Cabernet Franc clone #313	CabF313
Cabernet Sauvignon clone #337	CabS337
Chardonnay clone #96	Chard96
Mourvedre	Mourvedre
Muscat blanc	MuscatB
Norton	Norton
NY73.0136.17 (Noiret)	NY73 (Noiret)
Petit Manseng	PetitM
Petit Verdot	PetitV
Roussanne	Roussanne
Tannat	Tanna
Tinta Cao	TintaCao
Touriga nacional	TourigaN
Traminette	Traminet
Vidal blanc	VidalB
Viognier	Viognier

Table 1. Cultivars tested at SPAREC, Blackstone, VA, USA, 2009–2013.

Fungicides were applied 10–11 times per year (**Table 2**) with a low-cost and low-chemical input in mind. Applications were made based on a 14-day interval schedule with an exception of "at bloom" application that could be applied less than 14 days of the previous application. During 2009 and 2010, the first application was not applied, thus, there were a total of 10 applications. In general, mancozeb (Dithane Rainshield 75DF, Dow Agro Science, Indianapolis, IN, USA) and sulfur (Microthiol Disperss, United Phosphorous Inc., King of Prussia, PA, USA) were applied until prebloom. Boscalid + pyraclostrobin (Pristine, BASF Corporation, Research Triangle Park, NC, USA) were applied at bloom and bunch closure. After the second cover, mancozeb was replaced with a potassium phosphite (Prophyt, Helena Chemical, Collierville, TN, USA). All were applied with a 100 gallon (379 L) air blast sprayer (John Bean, Durand-Wayland Inc., GA, USA). The one exception was the cultivar 'Norton' that was not treated with any fungicides during the 5 years.

The effect of cultivar on the mean leaf and cluster disease incidence and severity was analyzed using a generalized linear mixed model (PROC GLIMMIX, SAS, ver. 9.4, SAS Institute, Cary, NC, USA). The GLIMMIX model utilized the log it link function for the mean leaf and cluster disease incidences and identity for the mean leaf and cluster disease severities. When the effect of cultivar was found to be significant, differences among cultivars were compared using Fisher's least square difference (LSD) method.

Application number	Growth stage	Material and rate per hectare[a]	
1	3″	Mancozeb (3.4 kg)	+ sulfur (3.4 kg)
2	10″	Mancozeb (3.4 kg)	+ sulfur (3.4 kg)
3	Prebloom	Mancozeb (3.4 kg)	+ sulfur (3.4 kg)
4	50% bloom	Pyraclostrobin + boscalid (0.88 kg)	
5	1st cover	Mancozeb (3.4 kg)	+ sulfur (3.4 kg)
6	Pea (2nd)	Mancozeb (3.4 kg)	+ sulfur (3.4 kg)
7	Pea (3rd)	Potassium phosphite (3.5 L)	+ sulfur (3.4 kg)
8	Berry touch	Pyraclostrobin + boscalid (0.88 kg)	
9	Veraison	Potassium phosphite (3.5 L)	+ sulfur (3.4 kg)
10	Preharvest	Potassium phosphite (3.5 L)	+ sulfur (3.4 kg)
11	Preharvest	Potassium phosphite (3.5 L)	+ sulfur (3.4 kg)

Note. [a]Materials used were Mancozeb (Dithane Rainshield 75DF, Dow Agro Science, Indianapolis, IN, USA), sulfur (Microthiol Disperss, United Phosphorous Inc., King of Prussia, PA, USA), pyraclostrobin + boscalid (Pristine, BASF Corporation, Research Triangle Park, NC, USA), and a potassium phosphite (Prophyt, Helena Chemical, Collierville, TN).

Table 2. Fungicide application program used at SPAREC, Blackstone, VA, USA 2009–2013.

3. Results

Weather data were obtained from a weather station (ET106, Campbell Scientific, Logan, UT, USA), located approximately 50 m east of the experimental vineyard. The average precipitation per year between 2009 and 2013 was 1250 mm. The average temperature from April to August varied from 13 to 23°C, and total precipitation varied from 378 to 646 mm (**Table 3**).

Year	Average temperature (C)	Total precipitation (mm)
2009	21.6	515.1
2010	13.0	438.4
2011	22.0	582.2
2012	21.1	378.7
2013	23.1	645.9

Table 3. Average temperature in Celsius and total precipitation in mm from April to August, SPAREC, Blackstone, VA, USA, 2009–2013.

Although a consistent level of precipitation was recorded each year, overall development of diseases during 2009–2013 was lower than expected (**Table 4**). Consistent development of both black rot and downy mildew was observed each year; however, other than 2011 when the mean downy mildew disease incidence and severity was 61% and 5%, respectively, we did not observe any major development of the other four diseases on leaves. In many cases, disease incidence was less than 10% and disease severity was less than 0.2%. There was a higher level of Phomopsis cane and leaf spot observed in 2013. These higher levels of downy mildew and Phomopsis cane and leaf spot probably were probably due to more precipitations observed in 2011 and 2013 (**Table 3**). Botrytis was not recorded because leaf symptom development was rarely observed.

Disease incidence	**2009**		**2010**		**2011**		**2012**		**2013**	
Black rot	1.72	(0.21)	8.41	(0.45)	4.81	(0.36)	1.72	(0.21)	0.47	(0.11)
Downy mildew	5.19	(0.36)	0.48	(0.11)	61.00	(0.81)	5.19	(0.36)	1.03	(0.17)
Phomopsis	0.11	(0.05)	0.24	(0.08)	0.19	(0.07)	0.11	(0.05)	2.75	(0.27)
Powdery mildew	0.03	(0.03)	7.23	(0.42)	2.58	(0.26)	0.03	(0.03)	0.19	(0.07)
Disease severity	**2009**		**2010**		**2011**		**2012**		**2013**	
Black rot	0.019	(0.003)	0.141	(0.011)	0.093	(0.009)	0.019	(0.003)	0.005	(0.001)
Downy mildew	0.084	(0.009)	0.008	(0.003)	5.367	(0.14)	0.084	(0.009)	0.01	(0.002)
Phomopsis	0.001	(0.001)	0.003	(0.002)	0.002	(0.001)	0.001	(0.001)	0.115	(0.017)
Powdery mildew	<0.001	(0)	0.162	(0.017)	0.081	(0.015)	<0.001	(0)	0.002	(0.001)

Table 4. The mean and standard error (in parentheses) of leaf disease incidence and severity at veraison, SPAREC, Blackstone, VA, USA, 2009–2013.

Disease incidence	**2009**		**2012**		**2013**	
Black rot	1.75	(0.38)	1.75	(0.38)	0.08	(0.08)
Botrytis	0.08	(0.08)	0.08	(0.08)	0.00	(0)
Downy mildew	0.17	(0.12)	0.17	(0.12)	0.08	(0.08)
Phomopsis	0.00	(0)	0.00	(0)	0.00	(0)
Powdery mildew	0.00	(0)	0.00	(0)	0.25	(0.15)
Disease severity	**2009**		**2012**		**2013**	
Black rot	0.027	(0.008)	0.027	(0.008)	0.004	(0.004)
Botrytis	0.001	(0.001)	0.001	(0.001)	0	(0)
Downy mildew	0.005	(0.004)	0.005	(0.004)	0.004	(0.004)
Phomopsis	0	(0)	0	(0)	0	(0)
Powdery mildew	0	(0)	0	(0)	0.003	(0.001)

Table 5. The mean and standard error (in parentheses) of cluster disease incidence and severity at veraison, SPAREC, Blackstone, VA, USA, 2009, 2012, and 2013.

		Disease incidence[a]			Disease severity[a]		
Year	Disease	*F*	*P* > *F*		*F*	*P* > *F*	
2009	Black rot	1.0	0.50		1.6	0.07	
	Downy mildew	10.6	<0.001	**	8.9	<0.001	**
	Phomopsis	0.7	0.40		2.0	0.01	**
	Powdery mildew	0.0	0.94		0.9	0.56	
2010	Black rot	15.0	<0.001	**	22.2	<0.001	**
	Downy mildew	1.7	0.11		4.1	<0.001	**
	Phomopsis	2.0	0.11		3.4	<0.001	**
	Powdery mildew	6.3	<0.001	**	4.6	<0.001	**
2011	Black rot	10.4	<0.001	**	18.7	<0.001	**
	Downy mildew	32.0	<0.001	**	61.5	<0.001	**
	Phomopsis	0.4	0.76		2.1	0.01	**
	Powdery mildew	3.2	<0.001	**	4.9	<0.001	**
2012	Black rot	1.0	0.50		1.6	0.07	
	Downy mildew	10.6	<0.001	**	8.9	<0.001	**
	Phomopsis	0.7	0.40		2.0	0.01	**
	Powdery mildew	0.0	0.94		0.9	0.56	
2013	Black rot	0.8	0.60		1.7	0.04	*
	Downy mildew	3.3	<0.001	**	5.8	<0.001	**
	Phomopsis	7.7	<0.001	**	14.7	<0.001	**
	Powdery mildew	0.3	0.92		1.7	0.04	*

Note. [a] *F* statistics and *P*-values from ANOVA (PROC GLIMMIX in SAS 9.4) results are shown. One and two asterisk(s) following the number represent 95 and 99% confidence level, respectively.

Table 6. Effect of cultivar on leaf disease incidence and severity at veraison, SPAREC, Blackstone, VA, USA, 2009–2013.

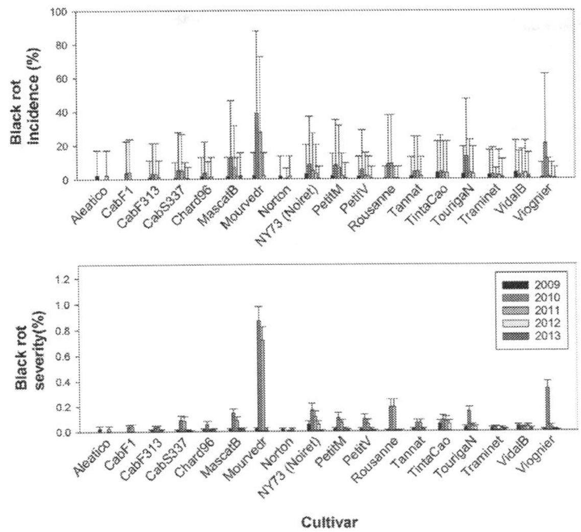

Figure 4. Leaf disease incidence and severity of black rot among 18 cultivars examined at SPAREC, Blackstone, VA, USA, 2009–2013 (note: The *Y*-axis scale of each panel is different).

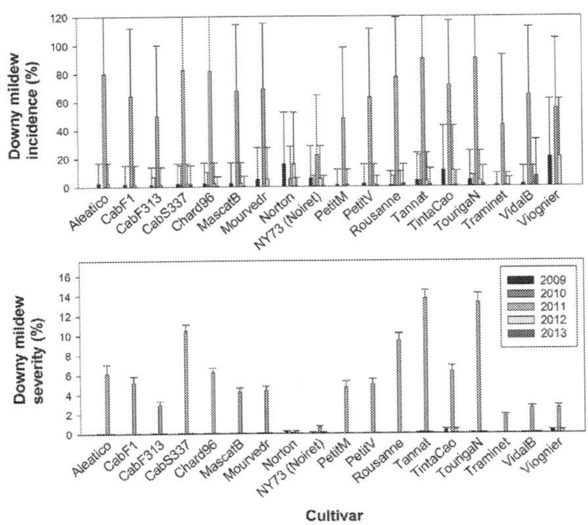

Figure 5. Leaf disease incidence and severity of downy mildew among 18 cultivars examined at SPAREC, Blackstone, VA, USA, 2009–2013 (note: The *Y*-axis scale of each panel is different).

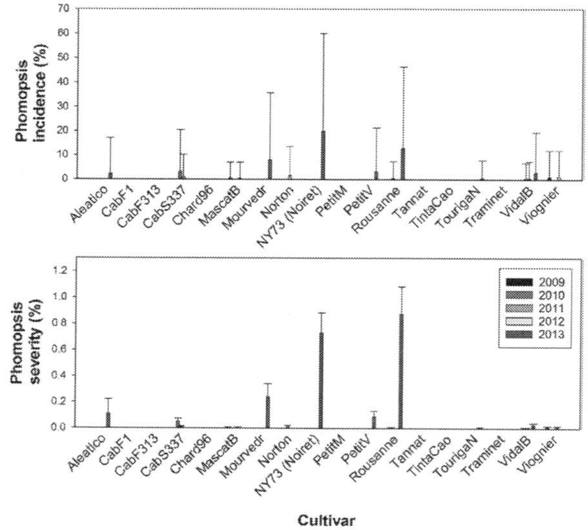

Figure 6. Leaf disease incidence and severity of Phomopsis cane and leaf spot among 18 cultivars examined at SPAR-EC, Blackstone, VA, USA, 2009–2013 (note: The *Y*-axis scale of each panel is different).

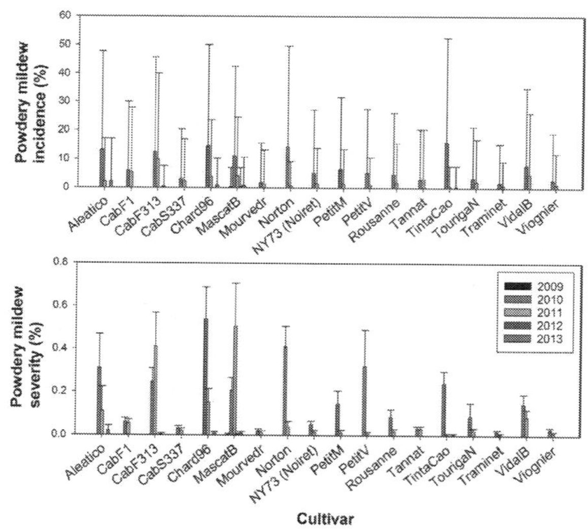

Figure 7. Leaf disease incidence and severity of powdery mildew among 18 cultivars examined at SPAREC, Blackstone, VA, USA, 2009–2013 (note: The *Y*-axis scale of each panel is different).

Cluster disease incidence and severity were measured in 2009, 2012, and 2013. The majority of clusters were lost due to bird damages in 2010 and 2011. As in leaf disease incidence and severity, the overall disease levels in these 3 years were very low (**Table 5**). The highest disease development was black rot on 2009 and 2012, but disease incidence was less than 2% and disease severity was less than 0.03%. To our surprise, there were very limited developments of Botrytis. Most likely due to very low development of cluster diseases, the effect of cultivar on cluster disease incidence and severity on all five diseases was not significant ($P > 0.05$) for all 3 years. Thus, data shown hereafter are on leaf disease incidence and severity.

Cultivara[a]	Black rot 2010[b]				Downy mildew 2011[b]			
	Incidence (%)		Severity (%)		Incidence (%)		Severity (%)	
Aleatico	0	G	0	CDEF	80.00	ABCD	6.13	CD
CabF1	3.64	EFG	0.04	EF	64.00	EF	5.21	CD
CabF313	3.33	G	0.03	F	50.00	GH	2.94	EF
CabS337	5.33	DEFG	0.09	CDEF	82.67	B	10.42	B
Chard96	3.56	FG	0.05	DEF	81.78	B	6.25	C
MascatB	12.89	C	0.15	CDE	67.62	DE	4.23	DE
Mourvedre	39.05	A	0.87	A	69.05	CDE	4.39	DE
Norton	0	G	0	F	5.56	J	0.07	H
NY73 (Noiret)	8.72	CDE	0.17	CD	22.56	I	0.65	GH
PetitM	8.00	CDEF	0.11	CDEF	48.15	GH	4.67	D
PetitV	5.71	DEFG	0.10	CDEF	62.86	EF	5.00	CD
Roussanne	9.05	CD	0.19	C	77.14	BC	9.41	B
Tannat	4.44	DEFG	0.06	DEF	90.22	A	13.7	A
Tinta Cao	4.62	DEFG	0.09	CDEF	71.79	CDE	6.33	C
TourigaN	13.33	C	0.16	CDE	90.56	A	13.26	A
Traminet	2.59	G	0.03	F	43.33	H	1.75	FG
VidalB	2.08	G	0.02	F	64.76	DE	2.54	F
Viognier	21.18	B	0.34	B	54.90	FG	2.59	F

Note. [a]See **Table 1** for actual names of cultivars.
Note. [b]Least square means from GLIMMIX results are shown. Different letters following the number indicate significant ($P \leq 0.05$) difference based on LSD.

Table 7. Effect of cultivars on black rot in 2010 and downy mildew in 2011, SPAREC, Blackstone, VA, USA.

In spite of low development of diseases, there are many cases where significant effect of cultivar ($P \leq 0.05$) on disease incidence and severity were observed (**Table 6**). In many cases, it was one or two cultivars that resulted in relatively higher or lower level of disease incidence or severity (**Figures 4–7**, **Tables 7** and **8**). For example, "Mourvedre" and "Viognier" resulted in higher disease incidences and severities of black rot than other cultivars examined (**Figure 4**, **Table 7**). Leaf downy mildew disease incidences and severities of 'Tannat', 'Touriga nacional', 'Cabernet Sauvignon clone 337', 'Chardonnay clone 96', and 'Roussanne' were higher than others in 2011 (**Figure 5**, **Table 7**). Mourvedre, NY73.0136.17 ('Noiret'), and Roussanne were more susceptible to Phomopsis cane and leaf spot than others in 2013 (**Figure 6**, **Table 8**). Interestingly, these three cultivars resulted in relatively lower powdery mildew incidence than other cultivars (**Figure 7**, **Table 8**). 'Aleatico', 'Cabernet Franc clone 313', Chardonnay clone 96, 'Muscat blanc', Norton, and 'Tinta Cao' resulted in relatively higher powdery mildew incidence in 2010 (**Figure 7**, **Table 8**).

Cultivara[a]	**Phomopsis 2013**[b]				**Powdery mildew 2010b**[b]			
	Incidence (%)		**Severity (%)**		**Incidence (%)**		**Severity (%)**	
Aleatico	2.22	AB	0.11	BC	13.33	ABCD	0.31	ABCDEF
CabF1	0	B	0	C	6.06	CDEF	0.06	DEF
CabF313	0	B	0	C	12.50	AB	0.25	BCD
CabS337	0	B	0	C	3.11	FGH	0.03	EF
Chard96	0	B	0	C	14.67	A	0.54	A
MascatB	0	B	0	C	11.11	ABC	0.20	CDE
Mourvedre	8.21	AB	0.24	B	1.90	GH	0.02	EF
Norton	0	B	0	C	14.44	A	0.41	AB
NY73 (Noiret)	20.00	A	0.73	A	5.13	DEFGH	0.05	DEF
PetitM	0	AB	0	C	6.67	BCDEF	0.15	CDEF
PetitV	3.33	AB	0.09	BC	5.24	DEFG	0.32	BC
Roussanne	12.86	AB	0.88	A	4.76	EFGH	0.09	DEF
Tannat	0	B	0	C	3.11	FGH	0.03	EF
Tinta Cao	0	B	0	C	15.90	A	0.24	BCD
TourigaN	0.56	B	0.01	C	3.33	EFGH	0.09	DEF
Traminet	0	B	0	C	1.85	H	0.02	F
VidalB	2.86	AB	0.03	C	7.92	BCDE	0.15	CDEF
Viognier	0	B	0	C	2.75	FGH	0.03	EF

Note. [a]See **Table 1** for actual names of cultivars.
Note. [b]Least square means from GLIMMIX results are shown. Different letters following the number indicate significant ($P \leq 0.05$) difference based on LSD.

Table 8. Effect of cultivars on Phomopsis cane and leaf spot in 2013 and powdery mildew in 2010, SPAREC, Blackstone, VA, USA.

4. Summary and concluding remarks

As noted earlier, plant disease requires three conditions to develop: host susceptibility, pathogen (or vector with pathogen) availability, and conducive environmental condition (**Figure 2**). This study examined the choice of cultivars as a use of biotechnology in the vineyard, and challenge it with a low-input fungicide regimen under severe weather conditions in the southeastern VA. The results showed that a proper selection of cultivar coupled with a relatively simple disease management plan can prevent the majority of foliar and cluster diseases of wine grapes.

There was significant cultivar effect on the development of each disease, and each cultivar resulted in different level of susceptibility to five of diseases we measured, with an exception of Botrytis that only showed limited development during the course of 5 years. Some cultivars, such as Norton and Noiret (NY73), resulted in very low level of diseases regardless of the very limited use of fungicide. Noiret is one of the newer red-fruited interspecific hybrid cultivar that was introduced in 1994 from the New York State Agricultural Experiment Station, Cornell University [13]. It is a cross of 'Chancellor' and 'Steuben', and it is rated moderately resistant to powdery mildew and Botrytis bunch rot. Our results also support that Noiret is more resistant to powdery mildew than other cultivar examined. Although a report from Cornell University [13] suggested that downy mildew can be an issue with Noiret, our results demonstrated that it is less susceptible than other tested cultivars. However, it seemed that Noiret is susceptible to Phomopsis cane and leaf spot.

Norton (syn. 'Cynthiana') is an intraspecific red-fruited hybrid cultivar, which was developed by Dr. Daniel Norborne Norton of Richmond, Virginia, around 1820 [14, 15]. It is considered a cross between *Vitis aestivalis* and *V. vinifera,* but with some traits of other *Vitis* species (*V. labrusca, V. cinerea,* and *V. cordifolia*) from previous crossing [15] were also speculated. It is also known for strong resistance level against many pathogens [15, 16]. In our study, Norton vines did not receive any fungicides over 5 years, yet, resulted in very low level of disease development. There was a relatively high level of powdery mildew development in 2010, and a low level of downy mildew development in 2011, but we did not observe any major outbreak of disease. Growers in VA recognize the advantage of the cultivar and a total of 51 hectares of Norton is grown in VA in 2015 [17]. One of growers produce more than 16 hectares of Norton, which is a very large hectarage for VA [15]. It is often described that Norton only requires 20–25% of pesticides when you compared with *V. vinifera* cultivars [15].

Five years of observation revealed characteristics of each cultivar. For instance, Traminette, Vidal blanc, and Viognier were another set of cultivars with low level of overall disease development. Viognier is designated as the state's signature grape in 2011 by the Virginia Wine Board, a state-sponsored trade association that promotes Virginia wine. Probably because of the promotion, more than 138 hectares of Viognier are grown in VA in 2015 [17]. Viognier is known to be susceptible to Phomopsis cane and leaf spot; however, results from this study did not show the same trend. Disease incidence of Phomopsis cane and leaf spot of grapes tend to aggregated among previously infected vines [18], thus, our results may be due to the lack of infected vines nearby Viognier subplots. Traminette and Vidal blanc are two intraspecific white-fruited hybrid cultivars that are commonly grown in VA. Traminette is a cross between Joannes Seyve 23-416 and Gewurztraminer and was introduced from New York State Agricultural Experimental Station in Geneva in 1996 [8]. Traminette is known for excellent wine quality as well as good disease resistance [8], as shown in our study. Vidal blanc is a cross between Ugni blanc and Seibel 4986, and known for good resistance against downy mildew, Botrytis and black rot [8, 16]. As of 2015, 46 and 65 hectares of Traminette and Vidal blanc, respectively, are grown in VA [17].

Other cultivars to be noted are 'Cabernet sauvignon clone #337', 'Tinta Cao', and 'Touriga nacional', which resulted in relatively low powdery mildew development. Powdery mildew is considered as one of the most important diseases among VA and mid-Atlantic regions [8], thus, it is important to know that these cultivars are less susceptible to it. Mourvedre was another cultivar with low powdery mildew; however, it was shown to be more susceptible to black rot and Phomopsis than other cultivars, therefore, Mourvedre may not be the best cultivar for the southeastern VA. Also, it should be noted that although the overall disease level was low with Noiret, it was very susceptible to Phomopsis cane and leaf spot, thus a specific protective application for this cultivar may be required.

Weather conditions varied among 5 years of experiment, and the development of diseases showed the influence of these weather events. For example, the lack of Botrytis development during the 5 years of experiment can be explained by the lack of conducive environmental conditions because *Botrytis* cinerea has a very wide host range, and some of cultivar, such as Chardonnay, is known to be susceptible to Botrytis. In 2011 and 2013 when the total amount of precipitation was higher than other years, there were more prominent developments of both downy mildew and Phomopsis cane and leaf spot. In this study, the fungicide application schedule was predetermined, and it was not altered based on weather conditions. In reality, any grower would change/modify their fungicide application schedule or chemicals based on weather conditions. For example, under rainy weather conditions we observed in 2011 or 2013, growers would have used downy mildew-specific material with curative activity (e.g., metalaxyl, potassium phosphite, etc.), in addition to the planned fungicides used in this study, or mix one of them with one of the applications. Thus, very low downy mildew incidences and severities shown with Noiret, Norton, Viognier, Traminette, and Vidal blanc with our fungicide program shows that these cultivars can be managed with lower fungicide inputs against downy mildew.

The advent of fungicide resistance has been documented in many modes of action groups in VA and surrounding states [19–21]. Moreover, risks of fungicide resistance development from the reliance on curative fungicide application have been discussed [11]. Therefore, the demonstrated efficacy of a relatively low-input protective fungicide program helps us to reduce unnecessary fungicide applications.

The information obtained from this study will be used to select proper cultivars for hot and humid growing conditions in the southeastern Virginia, and to develop a backbone fungicide application program to manage major fungal diseases with a relatively simple, and low-input program.

Acknowledgements

The author thanks Ed Jones (SPAREC) for applications of fungicides and general vineyard maintenance, Amanda Bly and Sabrina Hartley from AHS Jr. Agricultural Research and Extension Center of Virginia Polytechnic Institute and State University for disease assessment, data entry, and reviewing of the manuscript, and Akiko Nita for disease assessment and data entry.

Author details

Mizuho Nita

Address all correspondence to: nita24@vt.edu

Department of Plant Pathology, Physiology, and Weed Science, Alson H. Smith Jr. Agricultural Research and Extension Center, Virginia Polytechnic Institute and State University, Winchester, VA, USA

References

[1] J. Caldwell, *Virginia wine industry jobs grow by 50%; Economic impact doubles, new study finds*. 2012; Available from: http://www.governor.virginia.gov/News/viewRelease.cfm?id=1114. 24 June 2016.

[2] VWB, *The Economic Impact of Wine and Wine Grapes on the State of Virginia-2010*. 2012, A Frank, Rimerman + Co. LLP: St. Helena, CA.

[3] B.P. Hayden and P.J. Michaels, *Virginia's Climate*. University of Virginia Climatology Office, 2011.

[4] T. Jones, N. Rayapati, and M. Nita, *Occurrence of Grapevine leafroll associated virus-2, -3 and Grapevine fleck virus in Virginia, U.S.A., and factors affecting virus infected vines*. European Journal of Plant Pathology, 2015. 142(2): 209–222.

[5] B.W. Arnold, *History of the Tobacco Industry in Virginia from 1860 to 1894*. Johns Hopkins Press, 1897.

[6] H.F. Gale, L. Foreman, and T. Capehart, *Tobacco and the Economy: Farms, Jobs, and Communities*, in *Agricultural Economic Report*. 2000, Economic Research Service, U.S. Department of Agriculture.

[7] Anonymous, *From Leaf to Vine Tobacco Farmers Turn to Grapes*, in *Kentucky New Era*. 2000: Hopkinsville, KY.

[8] T.K. Wolf, ed. *Wine Grape Production Guide for Eastern North America*. 2008, Natural Resource, Agriculture, and Engineering Service (NRAES) Cooperative Extension: Ithaca, NY. 336.

[9] G.N. Agrios, *Plant Pathology*. 4th ed. San Diego: Academic Press, 1997.

[10] H. Ishii, *Impact of fungicide resistance in plant pathogens on crop disease control and agricultural environment*. Japan Agricultural Research Quarterly: JARQ, 2006. 40(3): 205–211.

[11] Beckerman, J. L., et al. (2015). "Do some IPM concepts contribute to the development of fungicide resistance? Lessons learned from the apple scab pathosystem in the United States". Pest Manag Sci 71(3): 331–342.

[12] Anonymous, *Occupational Safety and Health Guideline for Captan*. 1995, U.S. Department of Health and Human Services and US Department of Labor.

[13] B.I. Reisch, et al., 'Noiret'™ *Grape*, in *New York's Food and Life Sciences Bulletin*. 2006, Cornell University: New York Agricultural Experimenal Station, Geneva.

[14] R.K.R. Ambers and C.P. Ambers, *Daniel Norborne Norton and the origin of the Norton grape*. American Wine Society Journal, 2004. 36(3): 77–87.

[15] Ambers, C. P. (2013). "A historical hypothesis on the origin of the norton grape". Journal of Wine Research 24(2): 85–95.

[16] W.F. Wilcox, W.D. Gubler, and J.K. Uyemoto, eds. *Compendium of Grape Diseases, Disorders, and Pests*. 2015, APS Press: St. Paul, MN.

[17] V.W.B.M. Office, *Virginia 2015 Commercial Grape Report*. Richmond, VA, 2015.

[18] M. Nita, M.A. Ellis, and L.V. Madden, *Spatial pattern of phomopsis cane and leaf spot symptoms in commercial vineyards in Ohio*. Journal of Phytopathology, 2011: no-no.

[19] A. Baudoin, et al., *QoI resistance of Plasmopara viticola and Erysiphe necator in the mid-Atlantic United States*. Plant Health Progress, 2008: 0211–02.

[20] M. Rouxel, et al., *Geographic distribution of cryptic species of Plasmopara viticola causing downy mildew on wild and cultivated grape in Eastern North America*. Phytopathology, 2014. 104(7): 692–701.

[21] H. Ishii, et al., *Efficacy of SDHI fungicides, including benzovindiflupyr, against Colletotrichum species*. Pest Management Science, 2016: n/a-n/a.

Chapter 4

Development and Use of Biotechnology Tools for Grape Functional Analysis

Constanza Chialva, Estefanía Eichler,
Claudio Muñoz and Diego Lijavetzky

Additional information is available at the end of the chapter

http://dx.doi.org/10.5772/64915

Abstract

The aim of this chapter is to provide a description of the latest scientific advances in the field of gene functional analysis in grapevine. It provides general information about the studies conducted during the past decade to understand the natural variation of this plant and how this information has been exploited for the understanding of traits of interest. Likewise, it is exposed how the use of biotechnology tools have helped to characterize the mechanisms of gene expression and its regulation, as well as the subcellular localization of proteins and their interactions with other molecules. Finally, an approximation to the new technologies of gene editing and their potential application in the functional study of grapevine has been carried out.

Keywords: *Vitis vinifera*, QTL, gene transfer, genome editing, CRISPR/Cas system

1. Introduction

In recent years, studies of plant functional analysis are becoming increasingly relevant. Genome sequencing of a number of organisms is providing the scientific community with a vast resource of DNA sequence information that is revolutionizing the way science is being done. Thereby, progress has been made in the functional genomics of grapevine following the whole genome sequencing and assembling of *Vitis vinifera* PN40024 reference genome [1]. This has led to new interesting perspectives in genomic research and in functional analysis, providing insights into genetic regulation of grapevine genes and novel ways for isolating and characterizing genes, transcription factors, and proteins. The functional genomics methodology is also changing the experimental strategy from a forward genetics, that is, mutant to gene approach, to a reverse genetics or sequenced-gene to mutant and function approach [2]. Although functional analyzes are very relevant to basic plant biology, an important approach for crop plant improvement and commercial applications is expected. For instance, the current development of "clean" transformation techniques intend to obtain plants without insertion of antibiotic resistance genes and non-plant sequences [3].

Nowadays, there are several tools used by reverse genetics to induce variation into a gene and then used to infer its function. Genome editing is one of them [4]. This novel technique employs engineered nucleases that cut the DNA specifically generating targeted double-strand breaks (DSBs) [5]. Starting from the already known nucleases, such as zinc finger proteins and transcription activator-like effectors, and the recent discovered technology, the clustered regularly interspaced short palindromic repeat (CRISPR)/Cas system, a precise manipulation of gene sequences as well as the addition or deletion of DNA fragments on specific locus has recently become possible for some areas. For functional analysis, this means the possibility of making specific mutations in order to discover new gene functions, different relations between them, or even a synthetic pathway.

Another interesting challenge in grapevine investigation is the identification of those genes that contribute to the natural genetic variation of specific traits as well as understanding their main functions. The frequent quantitative nature of genetic variation requires the use of quantitative trait loci (QTLs) mapping approaches to understand the genetic structure of traits [6]. Interval mapping based on DNA markers could be used to genetically localize QTLs in natural and experimental populations [7, 8]. The development of new QTL mapping strategies and DNA research tools [9, 10], as well as the successful identification of few genes responsible of simple Mendelian traits [11, 12], have reinvigorated interest in grapevine QTL analysis. The present chapter summarizes the latest scientific advances in the field of grapevine functional analysis through the development and use of different biotechnological tools. It also describes diverse methods for the discovery and modification of genes that contribute to the finding of new sources of variability.

2. From QTL mapping to gene function

In model organisms, induced mutagenesis provides a powerful alternative for gene function discovery strategies derived from the knowledge of the phenotypic variation in plant (i.e., forward genetics). Because the prospects of gene identification are high and every gene affecting a trait is potentially a target, mutagenesis may present advantages with respect to natural, polygenic variation (i.e., quantitative trait loci) for identifying functional pathways and complex traits [13]. However, mutagenesis has not been extensively used in grapevine, although physical and chemical mutagens have been investigated [14, 15]. So far, the only published report on chemical mutagenesis of *in vitro*-grown grapevine buds comes from experiments using ethyl methane sulphonate and ethyl bromide on cultivar Pusa Seedless [16]. Likewise, the use of mutagenesis-induced variation for functional analysis has not been fully implemented. The main limitations are related to the almost inexistence of homozygous genotypes (besides the PN40024 line [1]) and the poor germination rate of most grapevine cultivars [6].

On the other hand, the extensive development of genetic resources including Amplified fragment length polymorphism (AFLP) [17], Random amplification of polymorphic DNA (RAPD) [18], Simple Sequence Repeat (SSR) [19], and single nucleotide polymorphism (SNP) [20] markers for grapevine genetics in the era before the availability of the grapevine genome [1] allowed the characterization of the genetic determinants for several grapevine traits by means of QTL mapping. The study of the genetic control of major agronomic traits in grapevine using QTL analysis allowed the elucidation of traits such as seedlessness and berry weight [12, 18, 21–23]; berry phenolic composition [24–28]; aroma [29, 30]; berry firmness [31, 32]; fertility [9, 33]; flower hermaphroditism [34]; cluster architecture [35]; pathogen resistance [34, 36–38]; plant phenology [39, 40]; and adaptation to abiotic stresses [41] and to climate change [42]. Despite several of the reports allowed the identification of candidate sequences, just few of them have finally characterized the genes responsible for a particular function or trait. Fortunately, those few paradigmatic examples are relate to two major grapevine quality traits: seedlessness and aroma.

Seedlessness, the absence of seeds in the berry [43], has been largely studied since the first genetic reports on the trait [18, 21, 44]. Those studies were mainly performed by the analysis of seed and berry size inheritance. In these works, the phenotypes segregate in experimental populations with a continuous distribution, an indicative of the quantitate nature of the trait. To date, the most accepted model proposed that seedlessness is under the control of a dominant regulator gene named seed development inhibitor (*SDI*) [18, 45, 46]. The further reports of a major QTL for seedlessness colocalizing with *SDI* on linkage group (LG) 18 confirmed the prediction of the model. These coincidental studies described a large-effect QTL by explaining between 50% and 70% of the phenotypic variation for seedlessness [18, 21, 22, 39].

The functional characterization of *SDI* begun with the identification of *VviAGL11*, a MADS-box gene putatively involved in grapevine ovule, seed, and berry development [47, 48], along with its in silico colocalization to the same contig containing the *SDI* locus in LG 18 [39]. Further, Mejía et al. [12] integrated multiple genetic, molecular, and genomic resources [1, 22, 39, 49] to elucidate the molecular basis underlying the *SDI* locus. In order to test the hypothesis for a possible role of *VviAGL11* in seedlessness, Mejía et al. [12] performed a comprehensive set of experiments providing additional genetic and transcriptional support for this hypothesis. These experiments consisted on the fine positional identification of *VviAGL11* as a candidate gene in a reduced confidence interval of 92 kb (**Figure 1**), and its additional characterization at the molecular, genetic, and transcriptional level. The authors showed that (i) the proportion of phenotypic variation in seedlessness explained by *VviAGL11* was higher than 70%; (ii) the promoter sequence comparison of the *VviAGL11* alleles at seedless and seeded genotypes showed several polymorphisms with putative functional effects, particularly two short insertions and deletions (INDELs); and (iii) the level of *VviAGL11* expression was associated with the *VviAGL11* genotype, since the homozygous genotypes for the seeded allele showed transcription 25-times higher than the homozygous genotypes for the seedless allele. In that

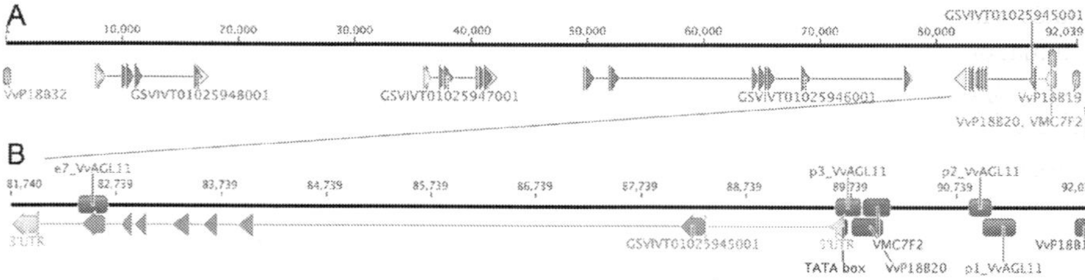

Figure 1. Annotation of the different genes found within the berry size and seedlessness QTL. (A) 92 kb window showing the structure of the four genes comprised between the VvP18B19 and VvP18B32 microsatellite markers. (B) Detailed view of the most probable candidate gene for seedlessness, VviAGL11 (VIT_218s0041g01880). Microsatellite and newly developed intragenic markers used for the fine mapping of the gene are indicated in red and blue, respectively. © Mejía et al.; licensee BioMed Central Ltd. 2011.

way, the genetic and transcriptional evidence suggested that seedlessness in table grapes might be due to missexpression of *VviAGL11* caused by short insertions and deletions (INDELs) in its regulatory elements. Together, all the presented information by Mejía et al. [12] pointed out to *VviAGL11* as being the major gene responsible for seedlessness.

On the other hand, it has been widely studied that aromas in grapevine arise from volatile compounds, such as terpenes, norisoprenoids, and thiols [50]. Particularly, the aroma of Muscat grapes is linked to the presence of the monoterpenes geraniol, linalool, nerol, and α-terpineol [51]. In plants, all isoprenoids are formed through two different and partially independent pathways, the mevalonic acid pathway (MEP, in the cytoplasm) and the methylerythritol phosphate pathway (in plastids). However, experimental reports have shown that most species mainly use one of the two pathways. Particularly in grapevine, the dominant pathway for monoterpene biosynthesis in leaves and berries is MEP route [52].

The route to the functional characterization of the genetic basis of aroma in grapevine starts with two simultaneous reports demonstrating the colocalization of a grapevine 1-deoxy-D-xylulose 5-phosphate synthase (*VviDXS*) gene with a major QTL for terpenol content in LG 5 [11, 53]. Moreover, by the analysis of the nucleotide diversity and linkage disequilibrium within the *VviDXS* gene, and testing for association between individual polymorphisms and Muscat flavor in different genetic backgrounds, Emanuelli et al. [54] identified significant single nucleotide polymorphisms. Further analysis corroborated that all those linked polymorphisms shared a particular SNP responsible for the substitution of a lysine with an asparagine at position 284 of the *VviDXS* protein. Finally, to test the functional relationship between *VviDXS* and Muscat flavor, Battilana et al. [29] compared the monoterpene profiles of cultivar Moscato Bianco with the expression of *VviDXS* alleles throughout berry development. It is worth to mention that the cultivar Moscato Bianco is heterozygous for the SNP mutation like most of the Muscat-flavored genotypes, thus containing both a "Muscat-type" allele (284N) and a "neutral" allele (284K) [54]. By comparing the transcription profile of *VviDXS* and free monoterpenol odorant variations during berry ripening, Battilana et al. [29] showed that monoterpene accumulation in Moscato Bianco berries correlates with *VviDXS* expression (at both the gene and the allele level). Despite all these accumulation evidences, the most dramatic demonstration of functional effect of the *VviDXS* gene polymorphisms were the biochemical experiments showing that the amino acid non-neutral substitution (K284N) influences the enzyme kinetics by increasing the catalytic efficiency. The kinetic analyses performed by Battilana et al. [29] clearly showed that the substrate affinities of the proteins encoded by the *VviDXS* alleles were similar but a major difference was found in the catalytic efficiencies of the enzymes, being *VviDXS* N284 twice as efficient as *VviDXS* K284.

Despite the robustness of the two presented examples for seedlessness and aroma, forthcoming gene functions determination in grapevine will be boosted by the use of new and more efficient tools. Recently, the availability of next-generation sequencing and whole-genome sequence information allowed the generation of a SNP-based genotyping array [55]. This array, developed for the grapevine international community, was built after the resequencing of 43 *V. vinifera ssp. vinifera,* four *V. vinifera ssp. sylvestris,* three *V. cinerea,* three *V. berlandieri,* three *V. aestivalis,* three *V. labrusca,* one *V. lincecumii,* and five *Muscadinia rotundifolia* genotypes using Illumina platforms. The use of this powerful tool in combination with the microvine model system [56] allowed an innovative study in grapevine genetics [9]. This work achieved the identification of 10 robust and stable QTLs for vegetative and reproductive traits as well as the first berry acidity QTLs reported so far in grapevine [9]. Additional initiatives like the recent development of large diversity panel including 279 cultivars from different uses (table and wine grapes) and geographical origin (eastern and western), and including most of the major founders of modern cultivars [10], will certainly increase the power and value of the 18K SNP chip for genome wide association studies and to gain more insight into the genetic control of many agronomic traits and their interaction with the environment.

3. Gene transfer technologies as a tool for functional analysis

In the plant biotechnology community, the term transformation is used to describe the insertion of engineered gene sequences into a plant cell, leading to a change in the genetic makeup of the target cell and its derivatives (i.e., reverse genetics). The foreign molecule can function for a short time in the nucleus as an extrachromosomal entity (transient transformation) [57], or the integration into the genetic material of the target cell can be necessary for long-term functionality and expression (stable transformation) [58, 59]. Therefore, transient expression has been used to evaluate factors that influence the stability or consistency of gene expression [60]. The final determination of factors that modulate transgene expression must ultimately be made only following introduction to plant cells for stable transformation. Methods for molecule transfer in grapevine involve biological methods (*Agrobacterium*-mediated transformation) for indirect gene transfer or chemical/physical methods (biolistics, electroporation, and protoplasts) for direct gene transfer to plant cells [61, 62] (**Figure 2**).

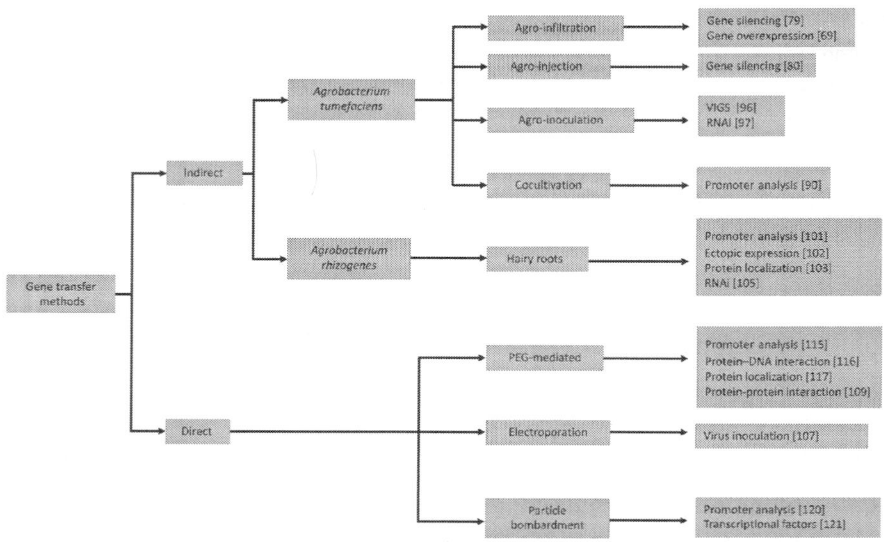

Figure 2. Schematic summary of direct and indirect transformation methods assayed in *Vitis vinifera* L. References regarding each functional analysis are presented in brackets.

3.1. Indirect gene transfer methods: *Agrobacterium*-mediated transformation of grapevine

Indirect transformation methods introduce plasmids, that is, independent circular molecules of DNA that are found in bacteria, into the target cell by means of bacteria capable of transferring genes to higher plant species [63]. The most popular used microorganisms are *Agrobacterium tumefaciens* and *Agrobacterium rhizogenes*. As wild-type bacterium, they can transfer a crown gall disease-inducing plasmid to its host, promoting tumor formation (Ti plasmid) [64], or a hairy root disease-inducing plasmid (Ri plasmid) promoting the formation of proliferative multibranched adventitious roots [65], respectively. However, the disease-inducing genes have been removed from current (disarmed) vectors and thus, they are not able to cause disease anymore. In their place, any genetic construction or gene of interest can be placed and those constructions introduced into plant cells.

Although other methods are suitable for plant transformation, such as protoplast or biolistic transformation, the *Agrobacterium*-mediated transformation is preferred since plants bearing single transgene copy can be more easily obtained (e.g., see [66]). The system is cheap and has been shown to work effectively in a variety of plant species [67]. However, it features some drawbacks. For instance, the host tissue must experiment some physical damage, the vectors are normally designed to infect the nucleus, the bacteria need to be eliminated using antibiotics, and the host range must be sensible to infection. Moreover, the presence of *Agrobacterium* may alter the activity of several plant proteins [68]. So, this aspect should be considered when using

Agrobacterium-mediated system to study stress signaling components [69]. In addition, this method is often associated with a high level of nonspecific autofluorescence and is difficult to be used for fluorescence-based analysis, such as subcellular localization [70]. In addition, the waxy cuticles of some plants organs can limit observations using a fluorescence microscope.

A. tumefaciens is a soilborne bacterium and a plant pathogen causing crown gall disease in angiosperms and gymnosperms [71]. *Agrobacterium*-mediated gene transfer methods were developed in grapevine in the early 1990s. Baribault et al. [72] first succeeded in the transformation of cell suspension cultures of Cabernet Sauvignon. Presently, transformation methods based on the Ti plasmid of *A. tumefaciens* are the predominantly employed protocols for grape transformation worldwide [73] and are compatible with the regeneration of transgenic plants from a variety of cultivars [74]. It has been widely used as a versatile tool for development of stably transformed model plants and crops. However, the development of *Agrobacterium*-based transient plant transformation methods attracted substantial attention in recent years.

The most significant applications of transformation with *A. tumefaciens* include functional genomics by reverse genetics. In grapevines, this technology has been recently applied to analyze the function of several genes such as *VviAdh* [75], *VviPIP2* [76], *VviCCD1* [77], and *VviWRKY33* [78]. Likewise, this method has been proved to be very useful for transgenic complementation [79–81], technical studies [82–87], promoter analysis [88–91], subcellular localization analysis [92], and protein production [69, 93] in grapevine. Nonetheless, the genetic background and plant growth conditions are key factors in performing successful *Agrobacterium*-mediated transformation. For instance, the compatibility of *A. tumefaciens* strains with the plant species represents an important variable to be considered in this kind of assay. Thus, the strains most efficiently used for gene transfer into grapevine are probably C58C1 (pCH32) [69, 78, 86] and EHA105 [75, 77, 82, 83, 85, 88, 90], which contain extra copies of *vir* genes that make them hypervirulent [61].

3.1.1. Agroinjection and agroinfiltration

Transformation and regeneration of grapevine plants have been achieved via organogenesis [84], embryogenesis [83, 93], or from nonembryogenic cell cultures [86, 94], demonstrating that grapevine is not recalcitrant to *Agrobacterium* infection but this approach is time consuming and takes several months to produce transgenic plants suitable for analyses. In some cases, the production of whole transgenic organisms may not be needed if a large number of cells within a plant can be uniformly and consistently transformed. Indeed, direct *Agrobacterium*-mediated transformation at the plant-organ level has proven very useful when the recovery of transgenic plants is not required [95]. This is particularly interesting for species like grapevine, where the regeneration of transgenic plants is difficult.

A. tumefaciens can be infiltrated into plant leaves using a syringe or vacuum, allowing different origins of the target material (greenhouse-grown plants, plantlets grown *in vitro*, green cuttings, or *in vitro* shoots). The bacteria enter the intercellular air spaces within the leaf and transform a very large percentage of the internal mesophyll cells. The agroinjection method involves a needleless syringe that can be filled with the bacterial suspension and then pressed against the underside of a leaf to infiltrate the suspension by active pushing through the stomata [92]. This method can be used to rapidly and simply generate a chimeric plant, where a large number of leaf cells contain the gene of interest. Using this technique, Urso et al. [80] developed an efficient agroinjection-based gene silencing assay of specific genes likely to be involved in resistance to powdery mildew in grapevine leaves of *in vitro* plantlets.

Similarly, the agroinfiltration method consists of plunging detached leaves [69, 79] or whole plants [81] into the bacterial suspension. As the flow of bacterial suspensions through stomata and across the epidermis is impeded by the high surface tension of aqueous solutions, transient transformation is obtained through the rapid release of a vacuum to introduce the bacterial suspension into the mesophyll cells. Nowadays, this method represents an easy and noninvasive technique that allows gene expression in the whole leaf [67]. Agroinfiltration is usually performed on tissues of young plantlets grown *in vitro*, as greenhouse grown plants have often been described as recalcitrant to this technique [67, 92]. Santos-Rosa et al. [69] transiently overexpressed stilbene synthase genes in detached grapevine leaves in order to study the influence of stilbenes on downy mildew infection. Bertazzon et al. [79] assayed the transient downregulation of a grapevine defense-related gene by the agroinfiltration of the constructs for the expression of dsRNA. Interestingly, Ben-Amar et al. [81] established a protocol to agroinfiltrate leaves of greenhouse-grown plants using a vacuum device. In their work, they delivered the first evidence of GFP gene silencing in grapevine achieved for the first time using *in planta* agroinfiltration method.

3.1.2. Cocultivation

This method is used to introduce bacterial suspension into intercellular spaces within plant tissue, simply by submerging above-ground parts of the plant into an *Agrobacterium* solution for a few seconds or inoculating the plant material (cell culture) with bacteria cocultivated on a solid medium. The technique involves the preparation of a diluted bacterial solution that incorporates the addition of a surfactant, the preparation of the explant, the dip of the material in the bacterial solution, and the cocultivation in the presence of the antibiotic. The method facilitates high-throughput transformation at ambient pressures and considers that vacuum infiltration or syringe pressure is unnecessary as long as a suitable surfactant is used. For instance, Lizamore and Winefield [87] used the organosilicone surfactant Silwet L-77 to increase transient transformation in grape without the need for vacuum- or syringe-based method for infiltration of leaves. The transformation efficiency was achieved by measuring red pigmentation of cells, transiently transformed with the transcriptional activator of anthocyanin biosynthesis, *Vvi*Myb*A1*.

Cocultivation with *A. tumefaciens* is normally used to perform transient expression assays [87]. Likewise, it is the most common method to obtain stably transformed grapevines [83, 86, 88, 90, 93]. The cocultivation of cell suspension cultures of Gamay Red with EHA105 strain of *Agrobacterium* was used for studying the expression of the grape dihydroflavonol 4-reductase gene (*VviDFR*) and the analysis of its promoter region [88]. In another work, Li et al. [90] cocultivated somatic embryos of Thompson Seedless with EHA105 strain of *Agrobacterium* harboring a construct of *Vvi*Myb*A1* as a reporter gene and a vast number of grapevine constitutive promoters from various genotypes. Cocultivation has been the transformation method chosen for stable transformation of embryogenic and nonembryogenic cell cultures as well. Cheng et al. [93] used proembryogenic masses of the grapevine cv. Thompson Seedless for genetic transformation via *Agrobacterium* with a gene involved in the defense system. Martínez-Márquez et al. [86] applied the same method to stably transform nonembryogenic cell cultures of two *V. vinifera* cell lines with the GFP gene under the control of pCaMV35S. This study was relevant for those interested in bioactive metabolite production.

3.1.3. Agroinoculation

Agroinfiltration was classically used for introducing gene constructs driven by a Ti plasmid. Additionally, this technique has enabled the introduction of virus-derived vectors into several *V. vinifera* cultivars [96, 97]. Agroinoculation, first developed as a simple tool to study plant-virus interactions, is a popular method of choice for functional gene analysis of viral genomes. It also serves as a mean for disease control via RNA interference (RNAi)-enabled vaccination against pathogens or invertebrate pests. It uses a live virus to attain desirable traits via either expressing a protein of interest or knocking down gene expression via RNAi. The latter is an attractive approach of virus-induced gene silencing (VIGS). Muruganantham et al. [96] developed a VIGS vector based on the *Grapevine virus A* (GVA) that is a member of the genus *Vitivirus*, family *Flexiviridae.* Their described an *Agrobacterium*-mediated method for inoculating *in vitro*-propagated *V. vinifera* plantlets via their roots with the GVA-derived vector for silencing the endogenous phytoene desaturase (PDS) gene. Similarly, Kurth et al. [97] generated a virus-derived gene expression and regulation vector based on *Grapevine leafroll-associated virus-2* (GLRaV-2). This relatively benign virus of the family *Closteroviridae* is spread throughout grape-growing areas worldwide. The GLRaV-2 vector expresses recombinant proteins in the phloem tissue that is involved in sugar transport throughout the plant, from leaves to roots to berries. This avenue provided a tool to track virus infection through the entire pathway of sugar transport. Furthermore, the vector provides a powerful RNAi capability of regulating the expression of endogenous genes via virus-induced gene silencing for disease protection.

3.1.4. Sonication-assisted Agrobacterium-mediated transformation (SAAT)

A number of grapevine cultivars have been stably transformed using *Agrobacterium*-mediated procedures and most progress has been achieved using embryogenic cell masses [82, 83, 85, 93]. Sonication-assisted *Agrobacterium*-mediated transformation (SAAT) was first reported by Trick and Finer [98], who used this technique for the production of transgenic soybean (*Glycine max*) and Ohio buckeye (*Aesculus glabra*) plants from SAAT-treated embryogenic cell suspensions. Chu et al. [94] recently reported the effect of different times in a sonication bath while infecting dedifferentiated nonembryogenic grapevine suspension cell cultures with *A. tumefaciens*. Plant tissue damaged by sonication allows the tissue to be much more easily transformed by *A. tumefaciens.* This pioneering protocol significantly increased the number of cells expressing the reporter gene that consistently produces transgenic microcalli that can be converted into transgenic cellular lines.

3.1.5. Hairy roots

A. rhizogenes is a soil pathogen that elicits adventitious and genetically transformed roots. This leads to the production of so-called "composite plants" comprising a transgenic hairy root system attached to nontransformed shoots and leaves. While grapevine roots have been successfully transformed with *A. rhizogenes* [99], the regeneration of transgenic plants via *A. rhizogenes*-mediated transformation was only obtained using embryogenic calluses [100].

Hairy root technology has already been used in several functional studies relative to grapevine. Secondary metabolism investigation associated to flavonoid analysis of hairy roots overexpressing *Vvi*MybA1-2 determined that this transcription factor is specifically involved in the last steps of anthocyanin biosynthesis and transport [101]. Moreover, ectopic expression of either *Vvi*MybPA1 or *Vvi*MybPA2 in grapevine hairy roots induced qualitative and quantitative changes of the proanthocyanidin profile [102]. Localization studies, also using this technology, showed that anthoMATE transporters play their role in the tonoplast [103]. Regulation studies of stilbene biosynthesis demonstrated that ectopic expression of MYB15 in grapevine hairy roots resulted in increased stilbene synthase gene expression and in the accumulation of glycosylated stilbenes in planta [104]. Other studies dealing with pathogen interaction used hairy root methodology to enhance nematode resistance in transgenic grape. For this purpose, a biotech-based solution was designed for controlling root-knot nematodes (RKNs) by introducing RNA interference to silence RKN effector gene [105].

3.2. Direct gene transfer methods in grapevine

Due to the difficulty of transforming plants through the use of *Agrobacterium*, direct gene transfer methods were developed based on chemical, physicochemical, and mechanical procedures. The principle of passing DNA molecules through large pores or holes in the cell wall or membrane are very efficient in the introduction of DNA but inefficient for the recovery of transgenic plants. Taking this into consideration, direct gene transfer methods in plants require one of the following techniques for transformation: permeation of protoplast membranes by a chemical (PEG) treatment or by electroporation to allow direct DNA uptake, or cell bombardment of plant tissues with microparticles coated with the DNA of interest.

3.2.1. Polyethylene glycol (PEG) treatment and electroporation of protoplasts

Due to the rigid cellulose wall, it has been relatively difficult to handle plant cells. Several methods based on mechanical removal of cell walls and on the use of solvents have been used to degrade cell wall for the obtention of protoplasts. However, the methods involving the use of hydrolytic enzymes have been the most popular ones [106]. The protoplasts used for transformation are usually isolated by enzymatic digestion of mesophyll cells from leaves [107–109], berry mesocarp [106, 110], roots [111], stems [112], embryogenic tissue [108, 113, 114], and from fast-grown suspension-cultured cells [115–117].

Although the generation of transgenic lines represents a powerful research tool for characterizing plant gene function, protoplast-based protocols for grapevine stable transformation have some drawbacks. The production of polyphenols and phytoalexins is induced at a high level during the digestion process and the corresponding genes remain activated during the culture of *Vitis* spp. protoplasts, diminishing its viability [118, 119]. Despite this, some progress has been reached in obtaining whole plants from protoplast regeneration [113, 114]. However, plant protoplasts constitute a versatile system for transient gene expression and have been widely used for the functional characterization of genes [109], virus inoculation [107, 108], protein subcellular localization [109, 117], promoter analysis [115], protein-DNA interaction [116], and protein-protein interaction [109].

PEG-mediated transient assay utilizing protoplasts has become a powerful tool for rapid gene functional analysis that can be readily carried out using ordinary lab supplies and usually has high transformation efficiency. Wang et al. [117] described a rapid and efficient transient expression system for PEG-mediated transformation of protoplasts derived from grape berry suspension-cultured cells. The system was applied for subcellular localization studies of flavonoid biosynthesis enzymes using GFP as a reporter gene. In the same way, Zhao et al. [109] reported a simplified and highly efficient method for the isolation of mesophyll protoplasts from grapevine leaves and a modified transfection protocol using PEG. This transient transformation of protoplasts was developed to characterize the function of a heterologous plant defense gene through its gene expression, and was regarded as suitable for the study of protein expression, protein subcellular localization, and protein-protein interaction.

In electroporation, cells are permeabilized by the application of very short, high-voltage electric pulses to introduce DNA into cells. As mentioned before, the use of electroporation is restricted to stable transformation in species whose protoplasts are regenerable. Thus, the primary application of electroporation to plants has been for DNA uptake for studies of transient gene expression. As an example, protoplast electroporation has been used for virus inoculation [107, 108]. Valat et al. [107] used the mesophyll protoplast electroporation as a rapid screening technique of transgenic grapevine clones expressing the viral capsid gene or the movement protein gene of grapevine fanleaf virus (GFLV) to identify material that reduces or inhibit the accumulation of viral proteins at the cell level.

3.2.2. *Particle bombardment (biolistics)*

The most commonly used method for direct DNA uptake (or naked DNA introduction) is particle bombardment. Also known as biolistics, the technique consists on the acceleration of high-density carrier particles covered with genes that pass through the cells, leaving the DNA inside. Although it is mainly reported as a stable transformation method, it is also convenient for transient expression assays. However, it requires expensive equipment and causes severe tissue damage, and usually yields low transformation efficiency. The advantages of this technique are that it has no limitation on species ranges, genotypes, or subcellular organelles. In grapevine, it has been assayed with the use of circular plasmids [89, 91, 104, 120] and minimal cassettes [121, 122]. Moreover, it has let cotransformation with multiple genes [61, 121]. Likewise, particle bombardment has allowed the manipulation of organ sections such as leaves [91], embryos [91], and suspension-cultured cells [89, 104, 120–122] as targets for grape transformation.

Due to its versatility, the technique has been applied for several functional studies such as promoter analysis [89, 91, 120] and the regulatory function of some transcription factors [104]. In their work, Höll et al. [104] demonstrated via transient gene reporter assays that the cotransformation of cell suspensions with transcription factors (MYB14 and MYB15) and promoter sequences specifically activate the promoters of *STS* genes. Without neglecting the technical aspects, the work of Vidal et al. [121] is worth mentioning. In their work, they compared the efficiency of the method when an embryogenic cell suspension culture was cotransformed via biolistics using a minimal gene cassette and a traditional circular plasmid. The stability of the plant phenotype compared to nontransgenic lines after its regeneration, confirmed the effectiveness of the minimal cassette technology for genetic transformation of grapevine cultivars. Later studies [122] demonstrated the importance of 3′-end cassette protection for successful protein expression using the minimal cassette technology. Protection of the minimal cassette upstream promoter and downstream terminator may be necessary due to the nuclease activity of target plant material.

4. New tools for genome editing

The dominant genome editing tools before 2013 were zinc finger nucleases (ZFNs) and transcription activator-like effector nucleases (TALENs) [4]. Both are artificial fusion proteins comprising an engineered DNA-binding domain fused to the nonspecific nuclease domain of the restriction enzyme *Fok*I. While these nucleases have generated efficient targeted mutagenesis and other genome editing applications [123, 124], the design and construction of large modular proteins are both laborious and expensive. For instance, the utility of ZFNs is limited by their long synthesis time and their high rate of failure [125]. These nucleases generate the introduction of targeted DNA double-strand breaks, stimulating cellular DNA repair mechanisms. Two different DSBs repair pathways have been defined: nonhomologous end joining (NHEJ) and homologous recombination (HR) [126]. Subsequent cellular DNA repair process generates desired insertions, deletions, or substitutions at the loci of interest. The newest technology for genome editing is based on RNA-guided engineered nucleases, which seems to have a great future due to their simplicity, efficiency, and versatility.

4.1. The CRISPR/Cas9 system

Clustered regularly interspaced short palindromic repeat-associated Cas is an adaptive bacterial and archaeal immune system that uses antisense RNAs to control invasions of phages and plasmids [127]. CRISPR loci are short variable spacers separated by short repeats that are transcribed into noncoding RNAs. The noncoding RNAs form a functional complex with CRISPR-associated Cas proteins and guide the complex to cleave complementary invading DNA [128] (**Figure 3**).

There are three CRISPR/Cas system types (I, II, and III) that uses distinct molecular mechanisms to achieve nucleic acid recognition and cleavage [129]. In genome editing, type II CRISPR/Cas system has been developed as a new gene-targeting tool. The Cas9 endonuclease, from *Streptococcus pyogenes*, forms a complex with two short RNA molecules called CRISPR RNA (crRNA) and transactivating crRNA (transcrRNA) that guide the nuclease to cleave DNA on both strands at a specific site. A prerequisite for cleavage is the presence of a conserved protospacer adjacent motif (PAM) downstream of the target DNA, which usually has the sequence 5′-NGG-3′ [4, 130, 131]. The dual tracrRNA:crRNA was then engineered as a single guide RNA (sgRNA). This structure maintains the 20-nucleotide sequence at the 5′ end of the sgRNA, which determines the DNA target site by Watson-Crick base pairing (crRNA), and the double-stranded structure at the 3′ side of the guide sequence, which binds to Cas9 (tracrRNA) [131]. Given the sequence specificity conferred by a 20-nt sequence in the sgRNA, CRISPR/Cas system can be retargeted to cleave virtually any DNA sequence by redesigning the sgRNA. These techniques could induce several DSBs and generates genomic modifications such as deletions, insertions, or gene replacement [132–135]. These genomic modifications will depend on the repair pathway. If NHEJ is taken, small deletion or insertion will occur; if homology-directed repair (HDR) is chosen, DNA seg-

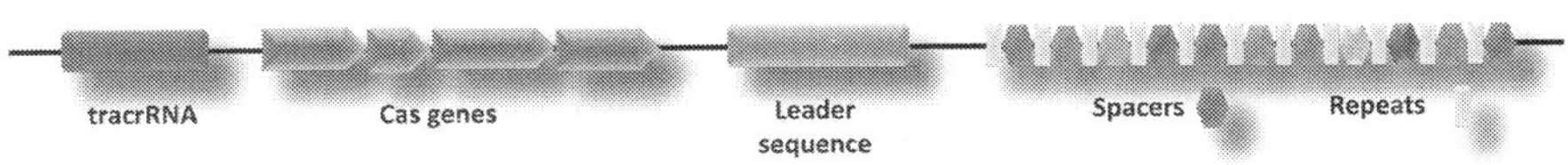

Figure 3. A typical structure of CRISPR/Cas9 type II locus. It includes a tracrRNA section (pink box), a family of Cas genes (light blue boxes), a CRISPR section that is an array of alternating nonrepetitive spacer (green hexagons), a short palindromic direct repeats (beige rectangle), and a leader sequence (orange box) that is an AT-rich not conserved sequence. The leader sequence always precedes the clustered regularly interspaced short palindromic repeats.

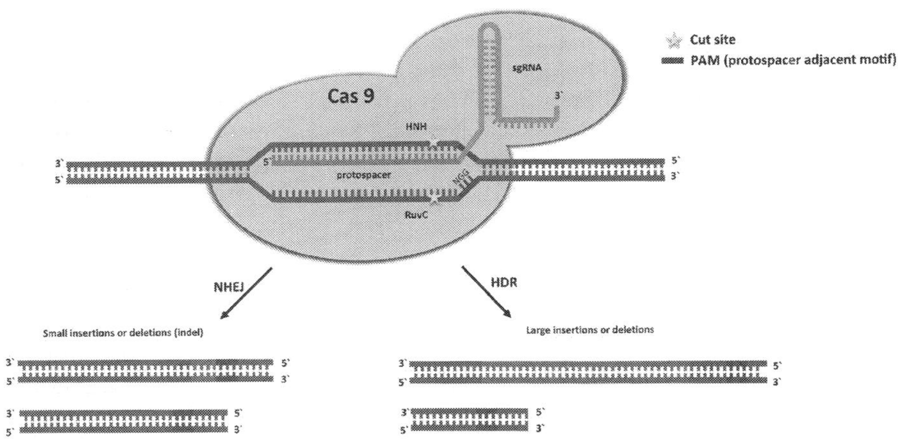

Figure 4. The mechanism of genome editing using CRISPR/Cas9. The genomic DNA target must lie adjacent to a protospacer adjacent motif. Cas9 protein associates with the sgRNA and binds to the target sequence, cleaving both strands of the DNA upstream of the PAM. Cleavage results in a DSB that becomes substrate for endogenous cellular DNA repair machinery that could catalyze nonhomologous end joining or homology-directed repair.

ments with sequences homologous to the break site will be inserted (large insertion) or gene replacement could happen [136] (**Figure 4**).

4.2. Applications in plants

Although CRISPR/Cas system has not been applied on *V. vinifera*, it could be harnessed to achieve a great progress in grapevine functional genomics. This technique is an interesting alternative tool to induce DSBs in plant genomes. In 2013, several reports were published discussing the first applications of CRISPR/Cas9-based genome editing in plants [137, 138]. Different species were used for these studies such as *Arabidopsis thaliana* [139], tobacco [140], rice [141], wheat [132], maize [142], sorghum [133], and tomato [143]. Jiang et al. [133] demonstrated that three slightly different versions of the Cas9/sgRNA system delivered by *A. tumefaciens* or polyethylene glycol-mediated transfection are functional in four plant types: *A. thaliana* and tobacco (dicots), and rice and sorghum (monocots). Jia and Wang [144] made the first report on targeted genome modification in citrus using the Cas9/sgRNA system. They used agroinfiltration to deliver Cas9 along with a synthetic sgRNA targeting the CsPDS gene into sweet orange. Another important finding was made by Woo et al. [145]. In their studies, they achieved the edition of plant genomes without the introduction of foreign DNA into cells. Instead, they transfected protoplasts of *A. thaliana*, tobacco, lettuce, and rice with preassembled complexes of purified Cas9 protein and guide RNA.

With CRISPR/Cas9 system, efficient NHEJ-mediated targeted mutagenesis was detected in *A. thaliana* and tobacco protoplasts. Positive HDR events were documented in tobacco protoplasts as well [146]. Rice and wheat protoplasts were also studied exhibiting efficient mutagenesis frequencies [137, 138]. Besides of being a very promising tool for generating modifications to the genome, the CRISPR/Cas9 system could generate genome modifications that could be present in the germ line and be segregated normally to the next generation of plants without new mutation or reversion [139, 143, 147]. This encourage the system to be a very promising tool for generating modifications in the genome that can be present in the germ line and be segregated normally to the next generation of plants without new mutation or reversion [139, 143, 147].

The fact that the CRISPR/Cas system allows multiplexing gene editing is particularly valuable [141, 148]. Xing et al. [149] developed a system where several sgRNA could be generated from one construct. Subsequently, simultaneous multiplex mutageneses were analyzed in maize protoplasts, transgenic maize lines, and transgenic *Arabidopsis* lines exhibiting targeted mutations. This kind of studies could be very helpful in the future to understand different gene functions or biosynthetic pathways. An additional potential use for CRISPR/Cas9 system is to confer molecular immunity against DNA plant viruses [150, 151].

The CRISPR/Cas9 system can be used for several purposes in addition to genome editing. Disabled nucleases (catalytically inactive version of Cas9, dCas9) can still bind to their target DNA sequence, so they can be expressed as a fusion protein with the transactivation or transrepression domain of a transcription factor and be used to regulate gene expression [152, 153].

Besides all the great applications of this system, a major concern when using an RNA-guided Cas9 is the off-site target activity. Off-site targeting is defined by the tolerance of Cas9 to mismatch in the RNA guide sequence and it is dependent on the number, position, and distribution of mismatches throughout the entire guide sequence [154–156]. The technical decision of using a 20-nucleotide motif complementary to the target DNA in the sgRNA may leads to the possibility that it cross-hybridizes to highly similar DNA sequences in other genomic regions. So, the use of genome-specific designing tools for guide RNAs [157] in CRISPR-Cas experiments is a strongly desirable prerequisite.

Although the CRISPR/Cas9 system has been thoroughly investigated in the past three years, there are no experiments made on grapevine yet. The first advances need to mention that Wang et al. [157] computationally identified and characterized five different types of CRISPR/Cas9 target sites and developed a user-friendly database for upcoming editing projects of grapevine genomes. These novelties provide an encouraging future perspective for genome editing by the use of the CRISPR/Cas9 system.

5. Conclusions

A major challenge for grapevine research is the identification of genes and gene variants responsible for important agronomic traits and to assign biological function to annotated sequences. Demonstration of biological function requires genetic approaches that deal with genetic variation. During the last two decades, different molecular techniques have allowed the fine characterization of the natural genetic variation underlying QTLs for traits of agronomic interest, and in few cases, identified the responsible genes. Along with QTLs, the development of new methodologies of gene editing such as CRISPR/Cas and gene transfer methods, the detailed genomic, transcriptomic, and proteomic studies have been performed. Nowadays, these techniques are constantly evolving and becoming more and more simple, efficient, and precise. These set of tools will soon help promoting the progress in knowledge, both in functional genomics and biotechnology, for its subsequent application.

Author details

Constanza Chialva, Estefanía Eichler, Claudio Muñoz and Diego Lijavetzky*

*Address all correspondence to: dlijavetzky@conicet.gov.ar

Instituto de Biología Agrícola de Mendoza (IBAM), Consejo Nacional de Investigaciones Científicas y Técnicas (CONICET), Universidad Nacional de Cuyo (FCA-UNCuyo), Chacras de Coria, Mendoza, Argentina

References

[1] Jaillon O, Aury JM, Noel B, Policriti A, Clepet C, Casagrande A, et al. The grapevine genome sequence suggests ancestral hexaploidization in major angiosperm phyla. Nature. 2007;449(7161):463–467.

[2] Bouchez D, Ho H. Update on genomics functional genomics in plants. Plant Physiol. 1998;(118):725–732.

[3] Gambino G, Gribaudo I. Genetic transformation of fruit trees: current status and remaining challenges. Transgenic Res. 2012;21(6):1163–1181.

[4] Bortesi L, Fischer R. The CRISPR/Cas9 system for plant genome editing and beyond. Biotechnol Adv. Elsevier B.V.; 2015;33(1):41–52.

[5] Gaj T, Gersbach CA, Barbas CF. ZFN, TALEN, and CRISPR/Cas-based methods for genome engineering. Trends Biotechnol. Elsevier; 2013;31(7):397–405.

[6] Martínez-Zapater JM, Carmona MJ, Díaz-Riquelme J, Fernández L, Lijavetzky D. Grapevine genetics after the genome sequence: challenges and limitations. Aust J Grape Wine Res. 2010;16(s1):33–46.

[7] Lander E, Kruglyak L. Genetic dissection of complex traits: guidelines for interpreting and reporting linkage results. Nat Genet. 1995;11(3):241–247.

[8] Lander ES, Botstein D. Mapping mendelian factors underlying quantitative traits using RFLP linkage maps. Genetics. 1989;121:185–189.

[9] Houel C, Chatbanyong R, Doligez A, Rienth M, Foria S, Luchaire N, et al. Identification of stable QTLs for vegetative and reproductive traits in the microvine (*Vitis vinifera* L.) using the 18 K Infinium chip. BMC Plant Biol. 2015;15(1):205.

[10] Nicolas SD, Péros J-P, Lacombe T, Launay A, Le Paslier M-C, Bérard A, et al. Genetic diversity, linkage disequilibrium and power of a large grapevine (*Vitis vinifera* L.) diversity panel newly designed for association studies. BMC Plant Biol. BioMed Central; 2016;16(1):74.

[11] Battilana J, Costantini L, Emanuelli F, Sevini F, Segala C, Moser S, et al. The 1-deoxy-d-xylulose 5-phosphate synthase gene co-localizes with a major QTL affecting monoterpene content in grapevine. Theor Appl Genet. 2009;118(4):653–669.

[12] Mejía N, Soto B, Guerrero M, Casanueva X, Houel C, Miccono M de LÁ, et al. Molecular, genetic and transcriptional evidence for a role of VvAGL11 in stenospermocarpic seedlessness in grapevine. BMC Plant Biol. 2011;11(1):57.

[13] Nadeau JH, Frankel WN. The roads from phenotypic variation to gene discovery: mutagenesis versus QTLs. Nat Genet. 2000;25(4):381–384.

[14] Alleweldt G, Possingham JV. Progress in grapevine breeding. TAG Theor Appl Genet. 1988;75(5):669–673.

[15] Kuksova VB, Piven NM, Gleba YY. Somaclonal variation and in vitro induced mutagenesis in grapevine. Plant Cell Tissue Organ Cult. Springer; 1997;49(1):17–27.

[16] Khawale RN, Singh SK, Yerramilli V. Molecular marker-assisted selection of in vitro chemical mutagen-induced grapevine mutants. Curr Sci. 2007;92:1056–1060.

[17] Cabezas JA, Cervera MT, Arroyo-Garcia R, Ibanez J, Rodriguez-Torres I, Borrego J, et al. Garnacha and Garnacha Tintorera: genetic relationships and the origin of teinturier varieties cultivated in Spain. Am J Enol Vitic. 2003;54(4):237–245.

[18] Doligez A, Bouquet A, Danglot Y, Lahogue F, Riaz S, Meredith CP, et al. Genetic mapping of grapevine (*Vitis vinifera* L.) applied to the detection of QTLs for seedlessness and berry weight. Theor Appl Genet. 2002;105(5):780–795.

[19] Adam-Blondon AF, Roux C, Claux D, Butterlin G, Merdinoglu D, This P. Mapping 245 SSR markers on the *Vitis vinifera* genome: a tool for grape genetics. Theor Appl Genet. 2004;109(5):1017–1027.

[20] Lijavetzky D, Cabezas JA, Ibáñez A, Rodríguez V, Martínez-Zapater JM. High throughput SNP discovery and genotyping in grapevine (Vitis vinifera L.) by combining a resequencing approach and SNPlex technology. BMC Genomics. BioMed Central; 2007;8(1):1.

[21] Cabezas JA, Cervera MT, Ruiz-Garcia L, Carreno J, Martinez-Zapater JM. A genetic analysis of seed and berry weight in grapevine. Genome. 2006;49(12):1572–1585.

[22] Mejia N, Gebauer M, Munoz L, Hewstone N, Munoz C, Hinrichsen P. Identification of QTLs for seedlessness, bBerry size, and ripening date in a seedless x seedless table grape progeny. Am J Enol Vitic. 2007;58(4):499–507.

[23] Doligez A, Bertrand Y, Farnos M, Grolier M, Romieu C, Esnault F, et al. New stable QTLs for berry weight do not colocalize with QTLs for seed traits in cultivated grapevine (*Vitis vinifera* L.). BMC Plant Biol. 2013;13(1):217.

[24] Lijavetzky D, Ruiz-Garcia L, Cabezas JA, De Andres MT, Bravo G, Ibanez A, et al. Molecular genetics of berry colour variation in table grape. Mol Genet Genomics. 2006;276(5):427–435.

[25] Walker AR, Lee E, Bogs J, McDavid DAJ, Thomas MR, Robinson SP. White grapes arose through the mutation of two similar and adjacent regulatory genes. Plant J. 2007;49(5):772–785.

[26] Fournier-Level A, Le Cunff L, Gomez C, Doligez A, Ageorges A, Roux C, et al. Quantitative genetic bases of anthocyanin variation in grape (*Vitis vinifera* L. ssp. sativa) berry: a qtl to QTN integrated study. Genetics. 2009;183(3):1127–1139.

[27] Carrier G, Huang YF, Le Cunff L, Fournier-Level A, Vialet S, Souquet JM, et al. Selection of candidate genes for grape proanthocyanidin pathway by an integrative approach. Plant Physiol Biochem PPB/ Socie´te´ Fr Physiol ve´ge´tale. 2013;72:87–95.

[28] Huang YF, Doligez A, Fournier-Level A, Le Cunff L, Bertrand Y, Canaguier A, et al. Dissecting genetic architecture of grape proanthocyanidin composition through quantitative trait locus mapping. BMC Plant Biol. BioMed Central; 2012;12(1):30.

[29] Battilana J, Emanuelli F, Gambino G, Gribaudo I, Gasperi F, Boss PK, et al. Functional effect of grapevine 1-deoxy-D-xylulose 5-phosphate synthase substitution K284N on Muscat flavour formation. J Exp Bot. 2011;62(15):5497–5508.

[30] Guillaumie S, Ilg A, Réty S, Brette M, Trossat-Magnin C, Decroocq S, et al. Genetic analysis of the biosynthesis of 2-methoxy-3-isobutylpyrazine, a major grape-derived aroma compound impacting wine quality. Plant Physiol. 2013;162(2):604–615.

[31] Carreño I, Cabezas JA, Martínez-Mora C, Arroyo-García R, Cenis JL, Martínez-Zapater JM, et al. Quantitative genetic analysis of berry firmness in table grape (*Vitis vinifera* L.). Tree Genet Genomes. 2014;11(1):818.

[32] Correa J, Mamani M, Munoz-Espinoza C, Gonzalez-Aguero M, Defilippi BG, Campos-Vargas R, et al. New stable QTLs for berry firmness in table grapes. Am J Enol Vitic. 2016;67(2):212–217.

[33] Doligez A, Bertrand Y, Dias S, Grolier M, Ballester JF, Bouquet A, et al. QTLs for fertility in table grape (*Vitis vinifera* L.). Tree Genet Genomes. 2010;6(3):413–422.

[34] Riaz S, Krivanek AF, Xu K, Walker MA. Refined mapping of the Pierce's disease resistance locus, PdR1, and sex on an extended genetic map of Vitis rupestris x V. arizonica. Theor Appl Genet. 2006;113(7):1317–1329.

[35] Correa J, Mamani M, Muñoz-Espinoza C, Laborie D, Muñoz C, Pinto M, et al. Heritability and identification of QTLs and underlying candidate genes associated with the architecture of the grapevine cluster (*Vitis vinifera* L.). Theor Appl Genet. 2014;127(5): 1143–1162.

[36] Hwang CF, Xu K, Hu R, Zhou R, Riaz S, Walker MA. Cloning and characterization of XiR1, a locus responsible for dagger nematode resistance in grape. Theor Appl Genet. 2010;121(4):789–799.

[37] Barba P, Cadle-Davidson L, Harriman J, Glaubitz JC, Brooks S, Hyma K, et al. Grapevine powdery mildew resistance and susceptibility loci identified on a high-resolution SNP map. Theor Appl Genet. 2014 ;127(1):73–84.

[38] Zyprian E, Ochßer I, Schwander F, Šimon S, Hausmann L, Bonow-Rex M, et al. Quantitative trait loci affecting pathogen resistance and ripening of grapevines. Mol Genet Genomics. 2016; 291(4):1573–1594.

[39] Costantini L, Battilana J, Lamaj F, Fanizza G, Grando MS. Berry and phenology-related traits in grapevine (*Vitis vinifera* L.): from quantitative trait loci to underlying genes. BMC Plant Biol. 2008;8(1):38.

[40] Grzeskowiak L, Costantini L, Lorenzi S, Grando MS. Candidate loci for phenology and fruitfulness contributing to the phenotypic variability observed in grapevine. Theor Appl Genet. 2013;126(11):2763–2776.

[41] Coupel-Ledru A, Lebon É, Christophe A, Doligez A, Cabrera-Bosquet L, Péchier P, et al. Genetic variation in a grapevine progeny (*Vitis vinifera* L. cvs Grenache×Syrah) reveals inconsistencies between maintenance of daytime leaf water potential and response of transpiration rate under drought. J Exp Bot. 2014;65(21):6205–6218.

[42] Duchêne E, Butterlin G, Dumas V, Merdinoglu D. Towards the adaptation of grapevine varieties to climate change: QTLs and candidate genes for developmental stages. Theor Appl Genet. 2012;124(4):623–635.

[43] Ledbetter C, Ramming D. Seedlessness in grapes. Hort Rev. 1989;11:159–184.

[44] Ledbetter CA, Burgos L. Inheritance of stenospermocarpic seedlessness in *Vitis vinifera* L. J Hered. 1994;85(2):157–160.

[45] Bouquet A, Danglot Y. Inheritance of seedlessness in grapevine (*Vitis vinifera* L.). VITIS —Journal of Grapevine Research. 1996;35:35.

[46] Adam-Blondon AF, Lahogue-Esnault F, Bouquet A, Boursiquot JM, This P. Usefulness of two SCAR markers for marker-assisted selection of seedless grapevine cultivars. VITIS—Journal of Grapevine Research. 2001;40:147.

[47] Boss PK, Sensi E, Hua C, Davies C, Thomas MR. Cloning and characterisation of grapevine (*Vitis vinifera* L.) MADS-box genes expressed during inflorescence and berry development. Plant Sci. 2002;162(6):887–895.

[48] Diaz-Riquelme J, Lijavetzky D, Martinez-Zapater JM, Carmona MJ. Genome-wide analysis of MIKCC-type MADS box genes in grapevine. Plant Physiol. 2008/11/11 ed. 2009;149(1):354–369.

[49] Moroldo M, Paillard S, Marconi R, Fabrice L, Canaguier A, Cruaud C, et al. A physical map of the heterozygous grapevine "Cabernet Sauvignon" allows mapping candidate genes for disease resistance. BMC Plant Biol. 2008;8:66.

[50] Lund ST, Bohlmann J. The molecular basis for wine grape quality—a volatile subject. Am Assoc Adv Sci. 2006;311: 804–805.

[51] Mateo JJ, Jiménez M. Monoterpenes in grape juice and wines. J Chromatogr A. 2000;881(1–2):557–567.

[52] Luan F, Wust M. Differential incorporation of 1-deoxy-d-xylulose into (3S)-linalool and geraniol in grape berry exocarp and mesocarp. Phytochemistry. 2002;60(5):451–459.

[53] Duchêne E, Butterlin G, Claudel P, Dumas V, Jaegli N, Merdinoglu D. A grapevine (*Vitis vinifera* L.) deoxy-d-xylulose synthase gene colocates with a major quantitative trait loci for terpenol content. Theor Appl Genet. 2009;118(3):541–552.

[54] Emanuelli F, Battilana J, Costantini L, Le Cunff L, Boursiquot JM, This P, et al. A candidate gene association study on Muscat flavor in grapevine (*Vitis vinifera* L.). BMC Plant Biol. BioMed Central; 2010;10(1):241.

[55] Le Paslier MC, Choisne R, Bacilieri R, Boursiquot JM, Bras M. The GrapeReSeq 18 K Vitis genotyping chip. In: Ninth International Symposium on Grapevine Physiology and Biotechnology, La Serena, Chili, 2013.

[56] Chaïb J, Torregrosa L, Mackenzie D, Corena P, Bouquet A, Thomas MR. The grape microvine—a model system for rapid forward and reverse genetics of grapevines. Plant J. 2010;62(6):1083–1092.

[57] Lacroix B, Citovsky V. The roles of bacterial and host plant factors in Agrobacterium-mediated genetic transformation. Int J Dev Biol. 2013;57(6–8):467–481.

[58] Janssen BJ, Gardner RC. Localized transient expression of GUS in leaf discs following cocultivation with Agrobacterium. Plant Mol Biol. 1990;14(1):61–72.

[59] Finer JH. Plant nuclear transformation. pp.3–18. In: F. Kempken and C. Jung (eds.), Genetic Modification of Plants, Biotechnology in Agriculture and Forestry 64, Springer-Verlag Berlin, Heidelberg.

[60] Dhillon T, Chiera JM, Lindbo JA, Finer JJ. Quantitative evaluation of six different viral suppressors of silencing using image analysis of transient GFP expression. Plant Cell Rep. 2009;28(4):639–647.

[61] Vidal JR, Gomez C, Cutanda MC, Shrestha BR, Bouquet A, Thomas MR, et al. Use of gene transfer technology for functional studies in grapevine. Aust J Grape Wine Res. 2010;16(Suppl. 1):138–151.

[62] Jelly NS, Valat L, Walter B, Maillot P. Transient expression assays in grapevine: a step towards genetic improvement. Plant Biotechnol J. 2014;12(9):1231–1245.

[63] Broothaerts W, Mitchell HJ, Weir B, Kaines S, Smith LM, Yang W, et al. Gene transfer to plants by diverse species of bacteria. Nature. 2005;433:629–633.

[64] Meyers B, Zaltsman A, Lacroix B, Kozlovsky S V., Krichevsky A. Nuclear and plastid genetic engineering of plants: Comparison of opportunities and challenges. Biotechnol Adv. Elsevier Inc.; 2010;28(6):747–756.

[65] Chandra S. Natural plant genetic engineer Agrobacterium rhizogenes: role of T-DNA in plant secondary metabolism. Biotechnol Lett. 2012;34(3):407–415.

[66] Bartlett JG, Alves SC, Smedley M, Snape JW, Harwood WA. Agrobacterium-mediated barley transformation. Acta Biol Szeged. 2008;52(1):49–51.

[67] Wroblewski T, Tomczak A, Michelmore R. Optimization of Agrobacterium-mediated transient assays of gene expression in lettuce, tomato and Arabidopsis. Plant Biotechnol J. Wiley Online Library; 2005;3(2):259–273.

[68] Pitzschke A, Hirt H. New insights into an old story: agrobacterium-induced tumour formation in plants by plant transformation. EMBO J. Nature Publishing Group; 2010;29(6):1021–1032.

[69] Santos-Rosa M, Poutaraud A, Merdinoglu D, Mestre P. Development of a transient expression system in grapevine via agro-infiltration. Plant Cell Rep. 2008;27(6):1053–1063.

[70] Zhang Y, Su J, Duan S, Ao Y, Dai J, Liu J, et al. A highly efficient rice green tissue protoplast system for transient gene expression and studying light/chloroplast-related processes. Plant Methods. BioMed Central Ltd; 2011;7(1):30.

[71] Smith EF, Townsend CO. A plant-tumor of bacterial origin. Science. 1907;25(643):671–673.

[72] Baribault TJ, Skene KGM, Steele Scott N. Genetic transformation of grapevine cells. Plant Cell Rep. 1989;8:137–140.

[73] Perl A, Ofra L, Abu-Abied M, Holland D. Establishment of an Agrobacterium-mediated transformation system for grape (*Vitis vinifera* L.): the role of antioxidants during grape-Agrobacterium interactions. Nature. 1996;14:624–628.

[74] Iocco P, Franks T, Thomas MR. Genetic transformation of major wine grape cultivars of *Vitis vinifera* L. Transgenic Res. 2001;10(2):105–112.

[75] Torregrosa L, Pradal M, Souquet JM, Rambert M, Gunata Z, Tesniere C. Manipulation of VvAdh to investigate its function in grape berry development. Plant Sci. 2008;174(2): 149–155.

[76] Perrone I, Gambino G, Chitarra W, Vitali M, Pagliarani C, Riccomagno N, et al. The grapevine root-specific aquaporin VvPIP2;4N controls root hydraulic conductance and leaf gas exchange under well-watered conditions but not under water stress. Plant Physiol. 2012;160:965–977.

[77] Lashbrooke JG, Young PR, Dockrall SJ, Vasanth K, Vivier MA. Functional characterisation of three members of the Vitis vinifera L. carotenoid cleavage dioxygenase gene family. BMC Plant Biol. 2013;13:156.

[78] Merz PR, Moser T, Höll J, Kortekamp A, Buchholz G, Zyprian E, et al. The transcription factor VvWRKY33 is involved in the regulation of grapevine (*Vitis vinifera*) defense against the oomycete pathogen Plasmopara viticola. Physiol Plant. 2015;153(3):365–380.

[79] Bertazzon N, Raiola A, Castiglioni C, Gardiman M, Angelini E, Borgo M, et al. Transient silencing of the grapevine gene VvPGIP1 by agroinfiltration with a construct for RNA interference. Plant Cell Rep. 2012;31(1):133–143.

[80] Urso S, Zottini M, Ruberti C, Lo Schiavo F, Stanca AM, Cattivelli L, et al. An Agrobacterium tumefaciens-mediated gene silencing system for functional analysis in grapevine. Plant Cell Tissue Organ Cult. 2013;114(1):49–60.

[81] Ben-Amar A, Cobanov P, Buchholz G, Mliki A, Reustle G. In planta agro-infiltration system for transient gene expression in grapevine (*Vitis* spp.). Acta Physiol Plant. 2013;35(11):3147–3156.

[82] Franks T, Gang He D, Thomas M. Regeneration of transgenic shape *Vitis vinifera* L. Sultana plants: Genotypic and phenotypic analysis. Mol Breed. 1998;4:321–333.

[83] Torregrosa L, Iocco P, Thomas MR. Influence of Agrobacterium strain, culture medium, and cultivar on the transformation efficiency of *Vitis vinifera* L. Am J Enol Vitic. 2002;53(3):183–190.

[84] Mezzetti B, Pandolfini T, Navacchi O, Landi L. Genetic transformation of *Vitis vinifera* via organogenesis. BMC Biotechnol. 2002;2(1):18.

[85] Li ZT, Dhekney S, Dutt M, Aman M, Tattersall J, Kelley KT, et al. Optimizing Agrobacterium-mediated transformation of grapevine. Vitro Cell Dev Biol Plant. 2006;42(3): 220–227.

[86] Martínez-Márquez A, Morante-Carriel J, Ramírez-Estrada K, Cusido R, Sellés-Marchart S, Palazon J, et al. A reliable protocol for the stable transformation of non-embryogenic cells cultures of grapevine (*Vitis vinifera* L.) and Taxus x media. J Biol Methods. 2015;2(2):21.

[87] Lizamore D, Winefield C. The addition of an organosilicone surfactant to Agrobacterium suspensions enables efficient transient transformation of in vitro grapevine leaf tissue at ambient pressure. Plant Cell Tissue Organ Cult. 2014;120(2):607–615.

[88] Gollop R, Even S, Colova-Tsolova V, Perl A. Expression of the grape dihydroflavonol reductase gene and analysis of its promoter region. J Exp Bot. 2002;53(373):1397–1409.

[89] Torregrosa L, Verriès C, Tesnière C. Grapevine (*Vitis vinifera* L.) promoter analysis by biolistic-mediated transient transformation of cell suspensions. Vitis. 2002;41(1):27–32.

[90] Li ZT, Kim KH, Jasinski JR, Creech MR, Gray DJ. Large-scale characterization of promoters from grapevine (*Vitis* spp.) using quantitative anthocyanin and GUS assay systems. Plant Sci. Elsevier Ireland Ltd; 2012;196:132–142.

[91] Joubert DA, de Lorenzo G, Vivier MA. Regulation of the grapevine polygalacturonase-inhibiting protein encoding gene: expression pattern, induction profile and promoter analysis. J Plant Res. 2013;126(2):267–281.

[92] Zottini M, Barizza E, Costa A, Formentin E, Ruberti C, Carimi F, et al. Agroinfiltration of grapevine leaves for fast transient assays of gene expression and for long-term production of stable transformed cells. Plant Cell Rep. 2008;27(5):845–853.

[93] Cheng S, Xie X, Xu Y, Zhang C, Wang X, Zhang J, et al. Genetic transformation of a fruit-specific, highly expressed stilbene synthase gene from Chinese wild Vitis quinquangularis. Planta. Springer, Berlin Heidelberg; 2016;243(4):1041–1053.

[94] Chu M, Quiñonero C, Akdemir H, Alburquerque N, Pedreño M, Burgos L. Agrobacterium-mediated transformation of Vitis cv. Monastrell suspension-cultured cells: Determination of critical parameters. Biotechnol Prog. Wiley Online Library; 2016. DOI 10.1002/btpr.2246

[95] Orzaez D, Mirabel S, Wieland WH, Granell A. Agroinjection of tomato fruits. A tool for rapid functional analysis of transgenes directly in fruit. Plant Physiol. Am Soc Plant Biol; 2006;140(1):3–11.

[96] Muruganantham M, Moskovitz Y, Haviv S, Horesh T, Fenigstein A, Preez J du, et al. Grapevine virus A-mediated gene silencing in Nicotiana benthamiana and Vitis vinifera. J Virol Methods. 2009;155(2):167–174.

[97] Kurth EG, Peremyslov V V, Prokhnevsky AI, Kasschau KD, Miller M, Carrington JC, et al. Virus-derived gene expression and RNA interference vector for grapevine. J Virol. 2012;86(11):6002–6009.

[98] Trick HN, Finer JJ. SAAT: Sonication-assisted Agrobacterium-mediated transformation. Transgenic Research. 1997;6: 329–336.

[99] Guellec V, David C, Branchard M, Temp J. Agrobacterium rhizogenes-mediated transformation of grapevine (*Vitis vinifera* L.). Plant Cell Tissue Organ Cult. 1990;(20): 211–215.

[100] Nakano M, Hoshino Y, Mii M. Regeneration of transgenic plants of grapevine (*Vitis vinifera* L.) via Agrobacterium rhizogenes-mediated transformation of embyronic calli. J Exp Bot. 1994;45(274):649–656.

[101] Cutanda-Perez MC, Ageorges A, Gomez C, Vialet S, Terrier N, Romieu C, et al. Ectopic expression of VlmybA1 in grapevine activates a narrow set of genes involved in anthocyanin synthesis and transport. Plant Mol Biol. 2009;69(6):633–648.

[102] Terrier N, Torregrosa L, Ageorges A, Vialet S, Verries C, Cheynier V, et al. Ectopic expression of VvMybPA2 promotes proanthocyanidin biosynthesis in grapevine and suggests additional targets in the pathway. Plant Physiol. 2008;149(2):1028–1041.

[103] Gomez C, Terrier N, Torregrosa L, Vialet S, Fournier-Level A, Verries C, et al. Grapevine MATE-type proteins act as vacuolar H^+-dependent acylated anthocyanin transporters. Plant Physiol. 2009;150(1):402–415.

[104] Höll J, Vannozzi A, Czemmel S, D'Onofrio C, Walker AR, Rausch T, et al. The R2R3-MYB transcription factors MYB14 and MYB15 regulate stilbene biosynthesis in Vitis vinifera. Plant Cell. 2013;25(10):4135–4149.

[105] Yang Y, Jittayasothorn Y, Chronis D, Wang X, Cousins P, Zhong GY. Molecular characteristics and efficacy of 16D10 siRNAs in inhibiting root-knot nematode infection in transgenic grape hairy roots. PLoS One. 2013;8(7). e69463.

[106] Fontes N, Delrot S, Gerós H. A method for the isolation of protoplasts from grape berry mesocarp tissue. Recent Pat Biotechnol. 2010;4(2):125–129.

[107] Valat L, Fuchs M, Burrus M. Transgenic grapevine rootstock clones expressing the coat protein or movement protein genes of grapevine fanleaf virus: Characterization and reaction to virus infection upon protoplast electroporation. Plant Sci. 2006;170(4):739–747.

[108] Valat L, Toutain S, Courtois N, Gaire F, Decout E, Pinck L, et al. GFLV replication in electroporated grapevine protoplasts. Plant Sci. 2000;155(2):203–212.

[109] Zhao FL, Li Y-J, Hu Y, Gao Y-R, Zang X-W, Ding Q, et al. A highly efficient grapevine mesophyll protoplast system for transient gene expression and the study of disease resistance proteins. Plant Cell Tissue Organ Cult. Springer, Netherlands; 2016;125(1): 43–57.

[110] Fontes N, Silva R, Vignault C, Lecourieux F, Gerós H, Delrot S. Purification and functional characterization of protoplasts and intact vacuoles from grape cells. BMC Res Notes. 2010;3:19.

[111] Reustle G, Alleweldt G. Isolation and culture of grapevine protoplasts. Vitis. JKI-Institut Rebenzuchtung Geilweilerhof, D-76833 Siebeldingen, Germany; 1990; pp. 423–431.

[112] Reustle G, Alleweldt G. Callus formation from protoplasts of grapevine (*Vitis* spp). Phys Plant. 1991;82:97.

[113] Reustle G, Harst M, Alleweldt G. Plant regeneration of grapevine (*Vitis* sp.) protoplasts isolated from embryogenic tissue. Plant Cell Rep. 1995;10:238–241.

[114] Zhu YM, Hoshino Y, Nakano M, Takahashi E, Mii M. Highly efficient system of plant regeneration from protoplasts of grapevine (*Vitis vinifera* L.) through somatic embryogenesis by using embryogenic callus culture and activated charcoal. Plant Sci. 1997;123(1–2):151–157.

[115] Saumonneau A, Laloi M, Lallemand M, Rabot A, Atanassova R. Dissection of the transcriptional regulation of grape ASR and response to glucose and abscisic acid. J Exp Bot. 2012;63(3):1495–1510.

[116] Marchive C, Léon C, Kappel C, Coutos-Thévenot P, Corio-Costet MF, Delrot S, et al. Over-expression of VvWRKY1 in grapevines induces expression of jasmonic acid pathway-related genes and confers higher tolerance to the downy mildew. PLoS One. 2013;8(1):1–8.

[117] Wang H, Wang W, Zhan J, Huang W, Xu H. An efficient PEG-mediated transient gene expression system in grape protoplasts and its application in subcellular localization

studies of flavonoids biosynthesis enzymes. Sci Hortic. Elsevier B.V., Amsterdam; 2015;191:82–89.

[118] Commun K, Mauro MC, Chupeau Y, Boulay M, Burrus M, Jeandet P. Phytoalexin production in grapevine protoplasts during isolation and culture. Plant Physiol Biochem. 2003;41(4):317–323.

[119] Reustle G, Natter I. Effect of polyvinylpyrrolidone and activated charcoal on formation of microcallus from grapevine protoplasts (*Vitis* sp.). Vitis. 1994;33(3):117–121.

[120] Harris NN, Luczo JM, Robinson SP, Walker AR. Transcriptional regulation of the three grapevine chalcone synthase genes and their role in flavonoid synthesis in Shiraz. Aust J Grape Wine Res. 2013;19(2):221–229.

[121] Vidal JR, Kikkert JR, Donzelli BD, Wallace PG, Reisch BI. Biolistic transformation of grapevine using minimal gene cassette technology. Plant Cell Rep. 2006;25(8):807–814.

[122] Sanjurjo L, Vidal JR, Segura A, De la Torre F. Genetic transformation of grapevine cells using the minimal cassette technology: The need of 3′-end protection. J Biotechnol. 2013;163(4):386–390.

[123] Palpant NJ, Dudzinski D. Zinc finger nucleases: looking toward translation. Gene Ther. Nature Publishing Group; 2013;20(2):121–127.

[124] Joung JK, Sander JD. TALENs: A widely applicable technology for targeted genome editing. Nat Rev Mol Cell Biol. Nature Publishing Group; 2013;14(1): 49–55.

[125] Puchta H, Fauser F. Synthetic nucleases for genome engineering in plants: Prospects for a bright future. Plant J. 2014;78(5):727–741.

[126] Wyman C, Kanaar R. DNA double-strand break repair: all's well that ends well. Annu Rev Genet. 2006;40:363–383.

[127] Makarova KS, Grishin N V, Shabalina SA, Wolf YI, Koonin EV. A putative RNA-interference-based immune system in prokaryotes: computational analysis of the predicted enzymatic machinery, functional analogies with eukaryotic RNAi, and hypothetical mechanisms of action. Biol Direct. 2006;1(1):7.

[128] Zhang F, Wen Y, Guo X. CRISPR/Cas9 for genome editing: Progress, implications and challenges. Hum Mol Genet. 2014;23(R1):40–46.

[129] Makarova KS. Evolution and classification of the CRISPR–Cas systems. Nat Publ Gr. Nature Publishing Group; 2011;9(6):467–477.

[130] Gasiunas G, Barrangou R, Horvath P, Siksnys V. Cas9-crRNA ribonucleoprotein complex mediates specific DNA cleavage for adaptive immunity in bacteria. Proc Natl Acad Sci USA. 2012;109(39):E2579–2586.

[131] Jinek M, Chylinski K, Fonfara I, Hauer M, Doudna JA, Charpentier E. A programmable dual-RNA-guided DNA endonuclease in adaptive bacterial immunity. Science. 2012;337(6096):816–821.

[132] Upadhyay SK, Kumar J, Alok a., Tuli R. RNA-guided genome editing for target gene mutations in wheat. G3 Genes | Genomes | Genetics. 2013;3:2233–2238.

[133] Jiang W, Zhou H, Bi H, Fromm M, Yang B, Weeks DP. Demonstration of CRISPR/Cas9/sgRNA-mediated targeted gene modification in Arabidopsis, tobacco, sorghum and rice. Nucleic Acids Res. 2013;41(20):1–12.

[134] Zhou H, Liu B, Weeks DP, Spalding MH, Yang B. Large chromosomal deletions and heritable small genetic changes induced by CRISPR/Cas9 in rice. Nucleic Acids Res. 2014;42(17):10903–10914.

[135] Zhao Y, Zhang C, Liu W, Gao W, Liu C, Song G, et al. An alternative strategy for targeted gene replacement in plants using a dual-sgRNA/Cas9 design. Sci Rep. Nature Publishing Group; 2016;6:23890.

[136] Voytas DF. Plant genome engineering with sequence-specific nucleases. Annu Rev Plant Biol. 2013;64:327–350.

[137] Shan Q, Wang Y, Li J, Zhang Y, Chen K, Liang Z, et al. Targeted genome modification of crop plants using a CRISPR-Cas system. Nat Biotechnol. 2013;31(8):686–688.

[138] Xie K, Yang Y. RNA-guided genome editing in plants using a CRISPR-Cas system. Mol Plant. 2013;6(6):1975–1983.

[139] Feng Z, Mao Y, Xu N, Zhang B, Wei P, Yang DL, et al. Multigeneration analysis reveals the inheritance, specificity, and patterns of CRISPR/Cas-induced gene modifications in Arabidopsis. Proc Natl Acad Sci USA. 2014;111(12):4632–4637.

[140] Nekrasov V, Staskawicz B, Weigel D, Jones JDG, Kamoun S. Targeted mutagenesis in the model plant Nicotiana benthamiana. Nat Biotechnol. 2013;31(8):688–691.

[141] Miao J, Guo D, Zhang J, Huang Q, Qin G, Zhang X, et al. Targeted mutagenesis in rice using CRISPR-Cas system. Cell Res. 2013;23(10):1233–1236.

[142] Liang Z, Zhang K, Chen K, Gao C. Targeted mutagenesis in Zea mays using TALENs and the CRISPR/Cas system. J Genet Genomics. Elsevier Ltd; 2014;41(2):63–68.

[143] Brooks C, Nekrasov V, Lippman ZB, Van Eck J. Efficient gene editing in tomato in the first generation using the clustered regularly interspaced short palindromic repeats/CRISPR-associated9 system. Plant Physiol. 2014;166(3):1292–1297.

[144] Jia H, Wang N. Targeted genome editing of sweet orange using Cas9/sgRNA. PLoS One. 2014;9(4). e93806.

[145] Woo JW, Kim J, Kwon S Il, Corvalán C, Cho SW, Kim H, et al. DNA-free genome editing in plants with preassembled CRISPR-Cas9 ribonucleoproteins. Nat Biotechnol. 2015;33(11):1162–1164.

[146] Li JF, Norville JE, Aach J, McCormack M, Zhang D, Bush J, et al. Multiplex and homologous recombination-mediated genome editing in Arabidopsis and Nicotiana benthamiana using guide RNA and Cas9. Nat Biotechnol. 2013;31(8):688–691.

[147] Pan C, Ye L, Qin L, Liu X, He Y, Wang J, et al. CRISPR/Cas9-mediated efficient and heritable targeted mutagenesis in tomato plants in the first and later generations. Sci Rep. Nature Publishing Group; 2016;6:24765.

[148] Lowder LG, Zhang D, Baltes NJ, Paul JW, Tang X, Zheng X, et al. A CRISPR/Cas9 toolbox for multiplexed plant genome editing and transcriptional regulation. Plant Physiol. 2015;169(2):971–985.

[149] Xing HL, Dong L, Wang ZP, Zhang HY, Han CY, Liu B, et al. A CRISPR/Cas9 toolkit for multiplex genome editing in plants. BMC Plant Biol. 2014;14(1):327.

[150] Ali Z, Abulfaraj A, Idris A, Ali S, Tashkandi M, Mahfouz MM. CRISPR/Cas9-mediated viral interference in plants. Genome Biol. 2015;16(1):238.

[151] Chandrasekaran J, Brumin M, Wolf D, Leibman D, Klap C, Pearlsman M, et al. Development of broad virus resistance in non-transgenic cucumber using CRISPR/Cas9 technology. Mol Plant Pathol. Wiley Online Library; 2016.

[152] Gilbert LA, Larson MH, Morsut L, Liu Z, Brar GA, Torres SE, et al. XCRISPR-mediated modular RNA-guided regulation of transcription in eukaryotes. Cell. Elsevier Inc.; 2013;154(2):1–10.

[153] Piatek A, Mahfouz MM. Targeted genome regulation via synthetic programmable transcriptional regulators. Crit Rev Biotechnol. 2016;8551:1–12.

[154] Fu Y, Foden J a, Khayter C, Maeder ML, Reyon D, Joung JK, et al. High-frequency off-target mutagenesis induced by CRISPR-Cas nucleases in human cells. Nat Biotechnol. Nature Publishing Group; 2013;31(9):822–826.

[155] Hsu PD, Scott D a, Weinstein J a, Ran FA, Konermann S, Agarwala V, et al. DNA targeting specificity of RNA-guided Cas9 nucleases. Nat Biotechnol. Nature Publishing Group; 2013;31(9):827–832.

[156] Doench JG, Fusi N, Sullender M, Hegde M, Vaimberg EW, Donovan KF, et al. Optimized sgRNA design to maximize activity and minimize off-target effects of CRISPR-Cas9. Nat Biotechnol. Nature Publishing Group; 2016;1–12.

[157] Wang Y, Liu X, Ren C, Zhong GY, Yang L, Li S, et al. Identification of genomic sites for CRISPR/Cas9-based genome editing in the Vitis vinifera genome. BMC Plant Biol. 2016;16(1):96.

Chapter 5

Phenolic Compounds Recovery from Grape Fruit and By-Products: An Overview of Extraction Methods

Cecilia Castro-López, Romeo Rojas,
Ernesto J. Sánchez-Alejo,
Guillermo Niño-Medina and
Guillermo C.G. Martínez-Ávila

Additional information is available at the end of the chapter

http://dx.doi.org/10.5772/64821

Abstract

Phenolic compounds are considered as bioactive compounds having beneficial effects on human health. Because of their biological properties, they have wide applications on pharmaceutical and food industries, and for this reason, it is important to identify most appropriate procedures, which permits the standardization for recovery of these compounds from several plant materials including grapes. Grape fruit and by-products are excellent sources of bioactive compounds such as pigments, organic acids, and phenolic compounds. Several convectional and emerging technologies have been evaluated in order to recover phenolic compounds from grape fruits and wastes such as chemical, physical, and biotechnological techniques, which offer different advantages related to economic, environmental, time-saving, and yield aspects. Nowadays, there is no updated information, which provides an overview about the techniques applied of these bioactive compound recovery in order to obtain high-quality and high-activity extracts rich in phenolic compounds from grape fruit and by-products. This chapter offers relevant aspects related to the techniques employed during the last five years by researches for phenolic compound recovery from grapes.

Keywords: phenolic compounds, grape fruit and by-products, extraction methods, emerging technologies

1. Introduction

Bioactive compounds are extra-nutritional components that naturally occur in small quantities in plant and food products. Most common bioactive compounds include secondary metabolites such as antibiotics, mycotoxins, alkaloids, food grade pigments, plant growth factors, and phenolic compounds [1]. Phenolic compounds are considered as bioactive compounds having beneficial effects on human health by decreasing the incidence of some degenerative diseases, such as cancer, diabetes, and reducing the risk factors of cardiovascular diseases. In addition, phenolic compounds have other biological properties such as inhibitors of cellular proliferation [2]. Because of their biological properties, phenolic compounds have wide applications on pharmaceutical, chemical, and food industries, and for this reason, it is important to identify the most appropriate procedures, which permits the standardization and/or optimization for recovery from these compounds, which are considered as the most abundant antioxidants in berries including grapes. Grape is the most widely cultivated fruit crop in the world with a global production of around 69 million tons, being Europe the biggest producer [3]. Grape fruit and by-products are an excellent source of bioactive compounds [4] such as pigments, organic acids, and phenolic compounds. Several emerging or conventional technologies have been evaluated in order to recover phenolic compounds from grape fruits and wastes such as chemical, physical, and biotechnological techniques, which offer different advantages related to economic, environmental, time-saving, and yield aspects. These techniques including ultrasound, microwave, micro- and ultra filtration, supercritical fluids, and electric fields assisted extraction. In addition, Soxhlet method, pressurized hot water, and the use of different organic solvents had been reported for this proposal. Moreover, enzyme technology and solid-state fermentation have been successfully applied for phenolic extraction from grape samples with important environmental advantages. Nowadays, there is no updated information, which provides an overview about the techniques applied of this bioactive compound recovery in order to obtain high-quality and high-activity phenolic compounds from grape fruit and by-products. This chapter offers relevant aspects related to the techniques employed during the last five years by researches around the world for phenolic compound recovery.

2. Soxhlet method

Soxhlet is equipment for extracting bioactive compounds, generally from lipid nature. It was invented by Franz von Soxhlet in 1879. Nowadays, the Soxhlet extraction represents the classical methodology for lipophilic compounds extraction [5]. For more of one century, this methodology has been used for different purposes and is described as the universal chemical extraction process [6]. Nevertheless, by itself, it is an optimized extraction process, but the literature offers a high amount of practical examples from bioactive compound extraction using different Soxhlet extraction conditions. However, this methodology requires large extraction times and quantities of solvents. The solvents more used are methanol [7, 8], etano [9, 10], n-hexane [7, 11], petroleum ether [12], toluene, chloroform [13], benzene, diethyl ether, dichloromethane, acetone, isooctane, cyclohexane [14], isopropanol [15], and water (for comparison only).

Grape variety	Solvent	Conditions	Yield	Antioxidant activity	Reference
Ruby Cabernet	Methanol	1:6 w/v 16 h	8.2% g/100 g d.b.	11.62 μg/mL (IC_{50} for AAPH)	[8]
Agiorgitiko	Ethanol	3:100 w/v 2–3 h	15% d.b. (approximately)	1.35 ± 0.02 (IC_{50} mg/mL)	[9]
	Water	3:100 w/v 5–6 h	Up to 24.35 ± 0.34% d.b.	2.02 ± 0.02 (IC_{50} mg/mL)	
NR (from Serralunga d' Alba, Italy)	Ethanol	6:85 w/w 18 h	7.7 ± 0.2 mg_{GAE}/g from seed fraction 11.9 ± 0.3 mg_{GAE}/g from skin fraction	NR	[10]
NR (from Friuli Venezia-Giulia, Italy)	n-hexane Methanol	1:12 w/v 6 h at 70°C	15.6 ± 1.2% d.b.	678 ± 15.5 $mg_{\alpha\text{-tocopherol}}$ 100 g^{-1}	[7]
Raboso Piave	n-hexane	1:12 w/v 6 h at 70°C	14.64 ± 0.29% d.b.	97.24 ± 0.35 Eq a Toc/g flour	[11]
Tempranillo (GSEJ)	Water	5 h 80–90°C	6.04 ± 0.69 g_{GAE}/L^{-1}	57.48 ± 3.61% Inhibition (DPPH) 36.57 ± 2.26 mg TROLOX L^{-1} (FRAP)	[65]
Tempranillo (GSEW)			2.41 ± 0.34 g_{GAE}/L^{-1}	40.35 ± 4.64% Inhibition (DPPH) 23.89 ± 5.55 mg TROLOX L^{-1} (FRAP)	
Gamay			2.09%	151.8 μg/g for IC_{50}, DPPH	[66]
Kalecik Karasi	Ethanol: water (95:5)	1:1 w/v 60°C for 8 h	2.49%	189.6 μg/g for IC_{50}, DPPH	
Okuzgozu			2.63%	109.8 μg/g for IC_{50}, DPPH	

NR, non-reported; GAE, gallic acid equivalents; d.b., dry basis; TPC, total phenolic compounds; GSEJ, grape seed extracts from juice; GSEW, grape seeds extracts from wine.

Table 1. Comparison of different organic solvents for phenolic compounds extraction from different varieties of grape residues.

Soxhlet method is based on the separation of a specific fraction from several food or plant materials with the use of a polar solvent depending on the solubility characteristic of the target compounds and the physicochemical nature of source, which can determine the surface contact and diffusivity of the solvent into the samples. Grape pomace is a waste product of grape juice and wine industry. These by-products contain high phenolic compounds because of poor extraction during the winemaking processes; hence, it makes their utilization worthwhile. In the last years, a large number of investigations have been conducted in order to find the best conditions for extraction of bioactive compounds from agroindustrial waste including grape waste as shown in **Table 1**. However, only this extraction method is used as comparison to replace the use of organic solvents.

3. Pressurized hot water extraction (PHWE)

As an alternative to use solvents in the several extraction methods, PHWE promotes the reduction or elimination of organic solvents into extraction processes. It improves the extraction process due to the water is non-flammable, non-toxic, available, and eco-friendly solvent [16]. High-pressure processing is a technology that has shown good prospects to extract bioactive compounds from several agroindustrial wastes [17]. PHWE is a non-conventional extraction method based on the extraction of molecules using hot liquid water as solvent. This technique is based on the use of temperatures above 100°C and 0.1 MPa, but lower than its critical point (374°C and 22.1 MPa) [18]. In addition, it is a highly promising energy-efficient and eco-friendly technique for recovering phenolic compounds from several sources [19]. However, in the beginning, it was not well received as analytical extraction solvent because the water is too polar to efficiently dissolve most target compounds. But, with PHWE, the water properties (polarity, viscosity, and surface tension) can be manipulated to optimize the phenolic extraction [20]. This manipulation improves the mass transfer rate and disrupts the water surface equilibrium, thereby lowering the activation energy required for desorption process [16]. When water is used as solvent, PHWE technology could also be designated as subcritical water extraction (SWE), superheated liquid extraction (SHLE), and pressurized liquid extraction (PLE) or accelerated solvent extraction (ASE) [21]. In recent years, several methodologies have been applied to the extraction of bioactive compounds of grape waste. Among them, the PHWE is a viable alternative, economic, with low-energy consumption, and eco-friendly. Most recently, possible scale up has been proposed by [9] with the use of a 10 l reactor to extract bioactive compounds from *Withania somnifera*. More recently, a new technique called high hydrostatic pressure has been successfully applied for phenolics recovery and microbial control, which could considerate to improve wine quality [22]. **Table 2** summarizes the last 5 years of investigation in the recovery of bioactive compounds from grape pomace. These investigations show that it has been possible to replace organic solvents with high yields and high antioxidant capacity. Therefore, it is an alternative extraction technique for application in pharmaceutical, food, and biotechnological industries.

Grape variety	Solvent	Conditions	Yield	Antioxidant activity	Reference
Cabernet Sauvignon	Hot water	1:10 w/v 5 min at 100°C	65.58 mg/g d.e	10.2 mg AAE/g d.e.	[19]
		1:10 w/v 5 min at 200°C	20.38 mg/g d.e	15 mg AAE/g d.e	
NR	Ethanol/water (70:30 vol%)	100 bars 140°C 5 cycles × 0 min each	7.28 g GAE/100 g d.b.	Up to 49.12% (DPPH) compared to resveratrol 50 μg/mL	[67]
Cabernet Sauvignon	Hot water	1:10 w/v 50–200°C 5–30 min	Up to 4.1 mg GAE/g dp	Up to 4.4 mg AAE/g dp (FRAP) Up to 184 mg TE/g dp (DPPH)	[68]
White Zinfandel	Hot water	Lower than 60 psi 140°C 9 mL/min water flow rate	130 mg/100 g d.b. (anthocyanins) 2077 mg/100 g d.b. (procyanidins)	NR	[69]
Sunbelt	80% aqueous ethanol	10.3 MPa 124°C 1 min	9.65 mg/100 g d.b. (Flavonols)	NR	[70]

NR, non-reported; GAE, gallic acid equivalents; d.b., dry basis; TPC, total phenolic compounds; d.e., dry extract.

Table 2. Summary of conditions and solvents for phenolic compounds extraction from different varieties of grape residues.

4. Ultrasound-assisted extraction (UAE)

Ultrasounds are sound waves of very high frequency (20 kHz to 100 MHz), which are propagated via compression and rarefaction, and require a medium (i.e., tissue) in which to travel. Ultrasound-assisted extraction makes use of high-intensity ultrasonic energy created by the implosion of cavitation bubbles. The bubbles collapse can produce physical, chemical, and mechanical effects. When this energy reaches the surface of the raw material through the extraction solvent, it is transformed into mechanical energy that is equivalent to several thousand atmospheres of pressure [23]. The high-pressure breaks the material particles, destroys cell membranes, improves penetration of solvent, and increases the contact surface area between the solid and liquid faces resulting in the release of phenolics to the extraction solvent in a relative short time [23, 24]. UAE is a simple, environmentally friendly, and efficient alternative to conventional extraction techniques [25]. The method's main advantages are simplicity of use and low instrumental requirements [23, 24, 26]. Ultrasonic devices include an ultrasonic bath, mainly used for small-scale extractions, or an ultrasonic probe system for large-scale industrial extractions (**Figure 1**) [26–29]. There are some previous applications of

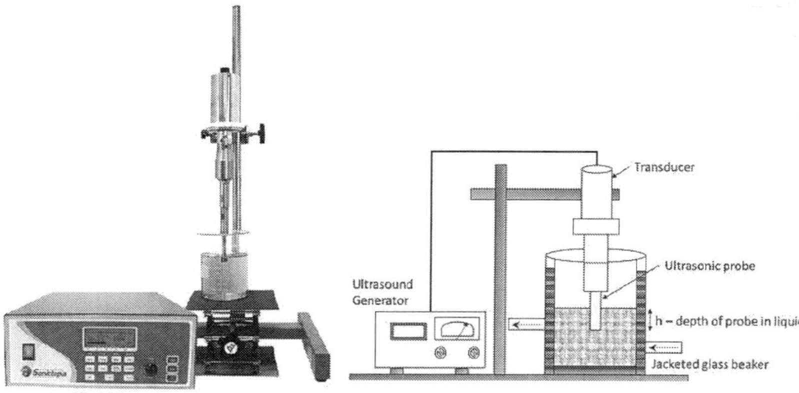

Figure 1. Schematic representation of an ultrasound-assisted extraction equipment.

UAE in the determination of phenolic compounds on specific parts of grapes. Ghassempour et al. [30] worked with red grape skin and the recuperation of anthocyanins. They compared UAE and microwave-assisted extraction (MAE), and the results showed that UAE has a slight lower recovery than MAE. Moreover, in their study, González-Centeno et al. [31] evaluated UAE as an extraction method for the quantification of total phenolic content and antioxidant capacity from grape pomace using an ultrasonic water bath. According to the results, the UAE resulted to aqueous extracts with phenolic and antioxidant characteristics similar to those obtained with conventional extraction, working under lower temperature conditions, and during less operating time (eight times less time than the conventional method).

Grape seeds have also been an interesting sample for UAE-based methods. Tao et al. [32] evaluated the effects of acoustic energy density (6.8–47.4 W/L) and temperature (20–50°C) on the extraction yields of total phenolics and tartaric esters during UAE from grape marc and demonstrated that ultrasound is an effective and promising technology to extract these kind of bioactive substances from this source. Ultrasound energy for extraction also facilitates more effective mixing, faster energy transfer, reduced thermal gradients and extraction temperature, selective extraction, reduced equipment size, faster response to process extraction control, quick start-up, increased production, and eliminates process steps among others [33]. Thus, the advantages of UAE include reduction in extraction time, energy, and use of solvent, which is reflected on economic and environmental aspects in the recovery of bioactive compounds.

5. Microwave-assisted extraction (MAE)

The fundamentals of the microwave extraction (MAE) process are different from those of conventional methods (solid–liquid or simply extraction) because the extraction occurs as the result of changes in the cell structure caused by electromagnetic waves. Microwave energy is a non-ionizing radiation that covers a third order of magnitude scale from 300 MHz to 300 GHz [34]. The principle of heating using microwaves is based on its direct effects on molecules of the material. Electromagnetic energy is converted to heat following ionic conduction and dipole rotation mechanisms (**Figure 2**) [35]. MAE process is assumed to involve three sequential steps [36]: (i) separation of the solutes from the active sites of the sample matrix under increased temperature and pressure, (ii) diffusion of solvent across the sample matrix, and (iii) release of the solutes from the sample matrix to the solvent. The operating conditions could be related to the success on the efficiency of this process; in this sense, parameters such as solvent extraction, temperature and time of the extraction, microwave power, and the physicochemical characteristics of the material should have special attention as they could potentially influence the recovery of target compounds [37, 38].

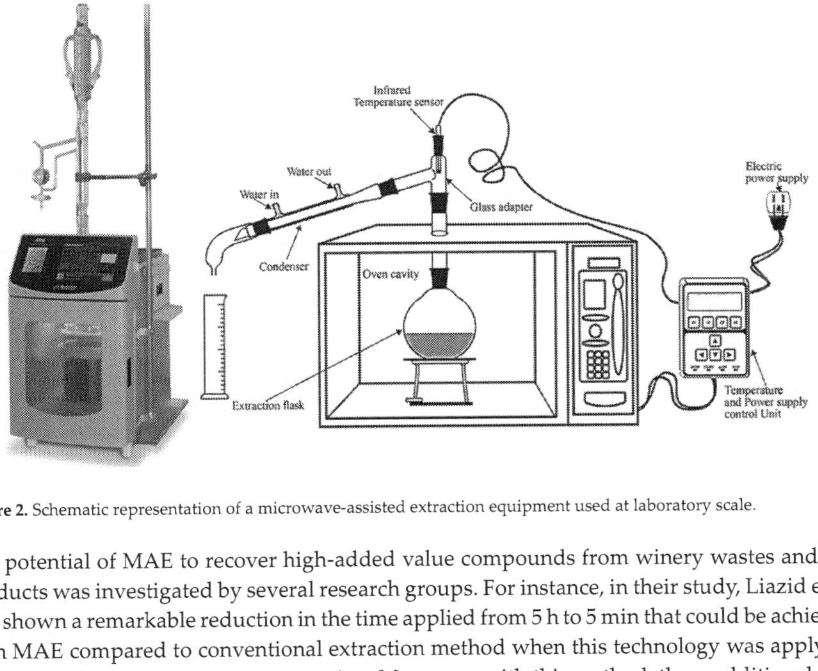

Figure 2. Schematic representation of a microwave-assisted extraction equipment used at laboratory scale.

The potential of MAE to recover high-added value compounds from winery wastes and by-products was investigated by several research groups. For instance, in their study, Liazid et al. [39] shown a remarkable reduction in the time applied from 5 h to 5 min that could be achieved with MAE compared to conventional extraction method when this technology was apply on grape skins for anthocyanins recuperation. Moreover, with this method, three additional acyl derivatives were extracted and quantified, while with the conventional method, it was impossible to measure. Bittar et al. [40] produced grape juice rich in polyphenols by MAE. Microwave-assisted process was evidenced to possess the highest values of TPC (21.41 ± 0.04 mg GAE/g DW) and TAC (4.49 ± 0.01 1 g MVGE/g DW). In addition, Li et al. [41] developed a microwave-assisted extraction (MAE) method for the extraction of phenolic compounds from grape seeds of *Vitis vinifera*. To optimize the extraction, it was considered the ethanol concentration in the extraction solvent, liquid: solid ratio, time, power, and temperature. The results obtained revealed that the optimal extraction conditions were ethanol concentration (47.2%), liquid: solid ratio (45.3:1), and time (4.6 min). The total phenolic content also was determined. Sequential application of the optimal conditions to one sample revealed that approximately 92% of the total phenolics were extracted in the first instance. Concluding that, MAE provides comparable or better extraction and it was very much quicker than other extraction methods.

6. Membrane separation

Membranes can be defined as semipermeable barriers that separate two phases and restrict the transport of defined components in a selective manner. The transport of components through the membrane is achieved by applying a driving force (concentration gradient, pressure, temperature, or electric potential). Thus, membrane separation processes use semipermeable membrane of definite nature to separate the components of a solution based on molecular size differences. In every membrane separation process, there is a membrane that is placed between two phases. One phase is called feed and the other is called permeate. When the feed consists of equal to or more than two components, and some of those components flow faster than others through the membrane, separation of the feed mixture takes place (**Figure 3**) [42, 43].

Microfiltration (MF) and ultrafiltration (UF) have already been widely used in the recovery, concentration, and fractionation of value-added products from agroindustrial wastes [44, 45]. Nevertheless, the use of membrane separation processes for the recovery of value-added products from wine lees is still a matter of research. In general, MF membranes rate according to nominal pore sizes, which are in the range of approximately 0.1–10 μm and operate at very low pressure, typically 10 psi or less, while UF membranes have molecular weight cut-off values between 1000 and 300,000 Da and pore diameters in the range of ≤10 nm–0.1 μm and typically operate at pressures ranging from 15 to 100 psi [28, 46].

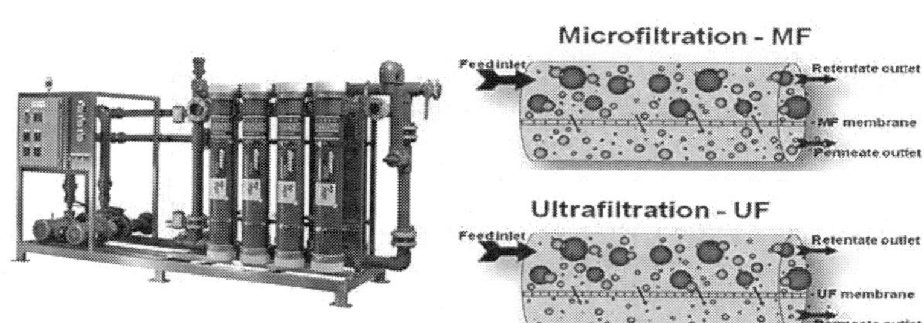

Figure 3. Schematic representation of membrane-based separation process.

http://dx.doi.org/10.5772/64821

In their study, Galanakis et al. [47] evaluated ultrafiltration processes in application of recovery, concentration, and fraction of polyphenolic compounds extracted from winery sludge. Ultrafiltration removes high molecular-weight substances, colloidal materials, and organic and inorganic polymeric molecules (i.e., soluble dietary fibers or polysaccharides) from low molecular-weight organics and ions (i.e., phenols, simple sugars) in a non-destructive way. In the mentioned work were tested three membrane types (100- and 20-kDapolysulfone, 1-kDa fluoropolymer), and the results indicated that solute retention was affected mainly by severe fouling phenomena due to polar solutes adsorption on membrane surface instead of size exclusion. Finally, it was separated successfully hydroxycinnamic acid derivatives from anthocyanins and flavonols. Giacobbo et al. [48] investigated the aqueous extraction associated with microfiltration for the recovery of phenolic compounds present in the effluent of wine lees. They proposed that effluents are rich in polyphenols and can be potential sources. Therefore, authors worked first in reducing the charge of the suspended solids and then used this permeate in an ultrafiltration process (V0.2 and MFP5 membranes) with dilutions combined with vacuum filtration. At the optimal conditions, a solution diluted 10 (v/v) followed by microfiltration led to the achievement of a limpid permeate, rich in phenolics obtaining a recovery rate of 21% of the total content of phenolic compounds. The results demonstrated that this technology is up to 6 times more efficient than others. On the other hand, Fernández et al. [49] studied the maximization of the permeate flux in the purification by ultrafiltration of a grape seed extract, by evaluating the effect of operating variables: transmembrane pressure and tangential velocity on permeate flux and on the extracts chemical characteristics. The authors concluded that the UF process (10-kDa membrane to 5 bar and 1.3 m/s) reduced the mean degree of polymerization of the extracts from 7.15 up to 1–3 units of flavan-3-ols, corresponding to dimmers and trimmers in the permeate. Those evidences stand out membrane separation as an attractive alternative for recovery specific phenolic compounds from grape fruits and wastes.

7. Supercritical fluid extraction (SFE)

Supercritical fluid extraction (SFE) is a technique that uses supercritical fluids (systems formed by one or more compounds at conditions over their critical values of pressure and temperature) as an extraction solvent in the separating one component (the extractant) from another (the matrix). In this process, the mobile phase is subjected to pressures and temperatures near or above the critical point for the purpose of enhancing the mobile phase-solvating power [50]. The supercritical fluid is used as an alternative to traditional organic liquid solvents. The most widely used supercritical fluids are CO_2 (Tc = 31°C, Pc = 74 bar) and water (as above described) (Tc = 374°C, Pc = 221 bar), but some processes involve the use of supercritical methanol, ethanol, propane, ethane [51, 52]. A basic SFE system consists of the following parts: the delivery system of supercritical fluid is very important because a high purity is required. The pumps employed in supercritical fluid extraction must be able to drive carbon dioxide at high pressures required, maintaining a constant flow. A heater capable of controlling the temperature in the furnace, and a cell or stop able to withstand the pressures generated by the pump is required. The most important part is the restrictor which controls the flow of the supercritical fluid flowing through the cell and, moreover, is responsible for depressurizing the fluid by passing existing supercritical conditions in the cell extraction atmospheric conditions. Finally, the collection system of solute is responsible for increasing the fluid density and hence its solvent power decreases, achieving the separation of the solute and fluid (usually is achieved by depressurizing the fluid) [53]. A symmetric diagram of typical SFE instrumentation is given in **Figure 4**.

Most of the studies evaluating the potential of SFE to recover valuable compounds from grape by-products have been focused on seed oil and proanthocyanidins recovery. In this line, Oliveira et al. [54] proposed an increase in aggregated value of the huge amount of residues generated by wineries. In their study, it was evaluated the global extraction yield, the antimicrobial activity and the composition profile of Merlot and Syrah grape pomace during the application of supercritical fluids supplemented with a co-solvent (300 bar at 50 and 60°C). Even though the extraction yields were remarkably low, the supercritical fluid extracts presented the highest antimicrobial effectiveness (against four strains of bacteria) compared to the other grape pomace extracts due to the presence of bioactive compounds. Da Porto et al. [7] evaluated supercritical carbon dioxide (SC–CO_2) extraction of grape marc using water (W) and ethanol (EtOH) as co-solvent at 15% (w/w), 100, and 200 MPa, and 313.15, 323.15, and 333.15°K to analyze their influence upon total phenols of the extracts. Supercritical extraction obtained the highest phenolic yield (68.0 g/kg of extract), phenol content (733.6 mg GAE/100 g DM), proanthocyanidins concentration (572.8 mg catechin/100 g DM), and antioxidant activity (2649.6 mg—tocopherol/100 g DM) in comparison with conventional extraction.

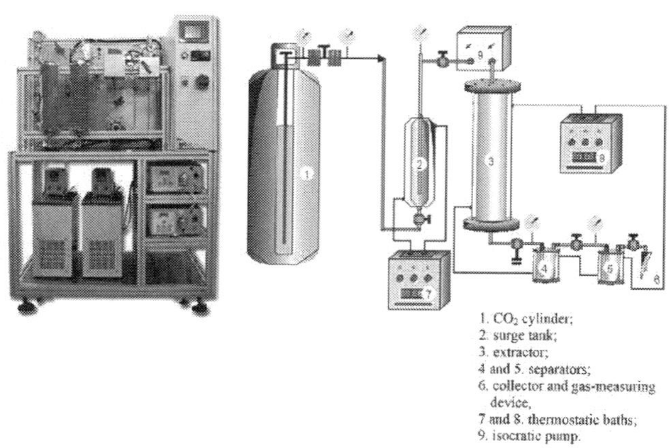

Figure 4. Schematic representation of a supercritical fluid extraction (SFE) system.

Rombaut et al. [55] compared three seed oil extraction methods (screw pressing, extraction by supercritical CO_2 percolation, and the combination of these two processes) and evaluated their efficiency for producing oil rich in phenolic compounds. The results suggested that the processes using supercritical CO_2 permit an increase in the co-extraction of phenolic with oil.

By combining a uniaxial compression with supercritical CO_2, oil yield is enhanced from 0 (hydraulic pressing, without supercritical CO_2) to 35%. On the other hand, Farías-Campomanes et al. [56] evaluated the economic feasibility of large-scale operations of supercritical fluid extraction (supercritical CO_2 containing 10% ethanol (w/w) at 313°K and 20–35 MPa) for the recovery of phenolics using grape bagasse. The supercritical CO_2/ethanol extraction process produced extracts with higher concentrations of phenolics (23 g/kg) than that extracts produced using conventional techniques with an economic evaluation of the process that estimated a cost of manufacturing of US$ 133.16/kg. However, more investigation about the effect of this extraction technique on functionality and change of the extracted-phenolics and it application in biotechnological processes is needed.

8. Pulsed-electric field extraction (PEF)

Exposing a plant cell to a high-intensity electric field (kV/cm) in the form of very short pulses (μs to ms) induces the formation of temporary or permanent pores on the cell membrane. This phenomenon, named electroporation, causes the permeabilization of cell membrane and increases its permeability and if the intensity of the treatment is sufficiently high, cell membrane disintegration occurs. During a pulsed-electric field extraction, the material is placed between two electrodes forming a treatment chamber and high-voltage-repetitive pulses are applied across the system in order to achieve membrane breakdown. Pulse amplitude in PEF equipment is ranging from 100 to 300 V/cm to 20–80 kV/cm [57]. Normally, PEF treatment is conducted at ambient temperature or slightly higher than the ambient temperature and for a treatment time less than 1 s (ms or μs) [58]. A PEF system, in general, consists of three basic components: a high voltage pulse generator, a treatment chamber, and a control system for monitoring the process parameters. In recent year, PEF technology has been mostly investigated in the recuperation of winery wastes and grape skin polyphenols. The recovered antioxidative compounds depend on the nature of raw materials, and in particular, the tissue structure of the source and the PEF treatment conditions applied. El Darra et al. [59] evaluated the influence of PEF (0.8–5 kV/cm, 1–100 ms, 42–53 kJ/kg) on the recuperation of phenolic compounds from Cabernet Franc grapes and its relationship with the process fermentation compared with a conventional treatment (50°C for 15 min). The study showed a significant improvement in phenolics extraction (anthocyanin and tannin contents), color intensity, and scavenging activity of the samples during red wine fermentation after applying PEF (51–62%) and thermal treatments (20%). In addition, Delsart et al. [60] determined the presence of phenolic compounds from Merlot grapes and the effect of PEF on the fermentation process of this grape variety and the related wine characteristics. The experiments focused in the application of PEF treatments (500–700 V/cm) with times of incidence of 40–100 ms where the measured responses were color intensity, anthocyanins, and phenolic content during the alcoholic fermentation and seven months during storage. The results suggest that pulsed electric field treatment has the advantage of nonthermal-selective extraction (<5°C) involving no loss of product quality in with respect to the classical process.

Other authors investigated the application of PEF treatment combined with densification to recover phenolic compounds from grape pomace of low moisture content, without any addition of conductive liquid [61]. Moreover, they studied the influence of a supplementary hydro-alcoholic extraction under various temperatures. The results indicate that PEF treatment (1.2 kV/cm, 18 kJ/kg) in grape pomace (1 g/cm^3) allows greater recovery of polyphenolic compounds for this matrix. Also, it was determined that this technology allows more selective recovery of anthocyanins after applying the treatment and finding compounds like anthocyanins/total flavan-3-ols at 20°C of 7.1 and 9.0, for control and PEF-treated samples, respectively. With the above aspects in view and the improvement of later biotechnological stages on winemaking, PEF could be considered as a good technique of enhancement of phenolic compounds on wine industry products.

9. Biotechnology applied on phenolic releasing from grape waste

Biotechnological releases of phenolic compounds from plant materials are associated to the degradation of cell-wall polysaccharides for microbial enzymes by use crude enzymatic extracts or purified commercial enzymes, which are able to eliminate this physical barrier and opens up the cell. Biotechnology techniques such as enzyme technology and solid-state fermentation have been successfully applied for phenolic extraction from grape samples with important environmental advantages. In addition, the enzymatic extraction method excludes the use of xenobiotics or toxic reagents, something that must also be taken into account, as it is more environmentally friendly [62].

In this sense, it has been reported that the use of cell-wall hydrolyzing enzymes can significantly increase the release of phenolic compounds from grape skins and seeds in a very short time. In a compressively study, Xu et al. [63] reported that β-glucosidase and pectinase can increase the releasing of total phenolic compounds from grape skins at 12 and 72%, respectively, when compared to the control. Therefore, both enzyme types could be considered to achieve an effective enzyme method for releasing phenolics from grape skins. Those findings are according to Fernández et al. [64], who observed an increment on phenolic compounds of 1.26-, 1.32-, and 1.34-fold when pectinase, cellulose, and tannase were used for describing the enzymatic effect of such enzymes on grape skins and seeds. In addition, due to the complexity of plant cell structure, combinations of those enzymes were evaluated; however, significant a significant effect on phenolic releasing was not observed. According to these studies, pectinase could be the enzyme, which allows the major amounts of phenolic compounds may be due to the high pectin content on this material. More recently, endoprotease mixtures have studied in order to obtain not only phenolic compounds, but also other functional biomolecules with anti-inflammatory and antioxidant capacities such as peptides. From this, flavonoids (flavanols and flavonols) and phenolic acids were observed as the main phenolic compounds present in grape pomace [62]. **Table 3** summarizes some contribution on biotechnological releasing of phenolic compounds from grape wastes. Nevertheless, it is important to observe that just one study regarding to the application on the solid-state fermentation of phenolic recovery from grape by-products was found, which indicates that the application of this process is still a matter of research for production of bioactive compounds from such agroindustrial material. However, biotechnological tools are very important environmental advantages since it reduces the use of xenobiotics or toxic reagents used in the recovery of phenolics in other techniques.

Biotechnology technique	Grape tissue	Enzyme/ microorganism	Phenolic compounds	Reference
Enzymatic	Whole grape pomace	Trypsin and chymotrypsin mixture	C, EC, Q, PB1, PB2, K, R	[71]
	Whole grape waste	Commercial enzymatic preparation (Novoferm)	GA, Rs, OCA	[72]
	Seeds	Cellulase	Do not identified for this enzyme	[63]
		β-Glucosidase	EA, EAH, ECG, EC and GA	
		Pectinase	Do not identified for this enzyme	
	Skin	Pectinase	EGC-P, EC-P, ECG-P and C	[64]
	Seed		C-P, EC-P, C, ECG-P, EC and ECG	
	Skin	Cellulase	EGC-P, C-P, EC-P, C and ECG-P	
	Seed		C-P, EC-P, C, ECG-P, EC and ECG	
	Skin	Tanasse	EGC-P, C-P, EC-P, C and ECG-P	
	Seed		EC-P, C, ECG-P, EC and ECG	
	Whole grape pomace	Mixture of proteases	GA, CA, CfA, PA, C, EC, PB1 and some glucosides	[62]
Solid-state fermentation	Whole grape waste	*Aspergillus niger GH1*	GA (as a main phenolic)	[73]

C, catechin; EC, epicatechin; EGC, epigallocatechin; C-P, catechin-phloroglucinol; EC-P, epicatechin-phloroglucinol; ECG-P, epicatechin gallate-phloroglucinol; EGC-P, epigallocatechin-phloroglucinol; EA, ellagic acid; EAH, ellagic acid hexoside; ECG, epicatechin gallate; GA, gallic acid; CA, caffeic acid; CfA, caftaric acid; PA, protocatechuic acid; PB1, procyanidin B1; PB2, procyanidin B2; Q, quercetin; K, kaempferol; R, resveratrol; Rs, resorcinol; and OCA, O-coumaric acid.

Table 3. Biotechnology techniques applied for phenolic compounds recovery from grape by-products.

10. Concluding remarks

As shown, grape fruit and by-products have demonstrated be an excellent source for obtaining phenolic compounds with the use of several conventional and/or emerging technologies. The growing demand to extract high-quality and high-activity extracts rich in phenolic compounds from plant materials encourages researches for found convenient extraction methods to this proposal. Since all the methods above described are based on different mechanism and extraction processes, the possible arrangement and development of hybrid procedures must be investigated to select the appropriate extracting-technique considering the target phenolic compounds, the physicochemical characteristic of the source, and the economic and environmental advantages of those methodologies.

Acknowledgements

Authors wish to thank the Mexican Council for Science and Technology (CONACYT) for the INFRA 2015-254178 Project.

Author details

Cecilia Castro-López[2], Romeo Rojas[1], Ernesto J. Sánchez-Alejo[1], Guillermo Niño-Medina[1] and Guillermo C.G. Martínez-Ávila[1*]

*Address all correspondence to: guillermo.martinezavl@uanl.edu.mx; cristian_mtza@hotmail.com

1 Laboratory of Chemistry and Biochemistry, Autonomous University of Nuevo León, General Escobedo, Nuevo León, Mexico

2 Department of Food Science and Technology, School of Chemistry, Autonomous University of Coahuila, Saltillo, Coahuila, Mexico

References

[1] Martins S, Mussatto SI, Martínez-Avila G, Montañez-Sáenz J, Aguilar CN & Teixeira JA. Bioactive phenolic compounds: production and extraction by solid-state fermentation. A review. Biotechnology Advances. 2011;29:365–373. doi:10.1016/j.biotechadv.2011.01.008

[2] Jara-Palacios M, Hernanz D, Cifuentes-Gómez T, Escudero-Gilete M, Heredia FJ & Spencer JPE. Assessment of white grape pomace form winemaking as source of

bioactive compounds, and its antiproliferative activity. Food Chemistry. 2015;183:78–82. doi:10.1016/j.foodchem.2015.03.022

[3] Food and Agriculture Organization. http://faostat3.fao.org/browse/Q/QC/S [Internet]. 2015 [Updated: 09/11/2015]. doi: 10.1016/j.foodchem.2015.03.022

[4] Tournour HH, Segundo MA, Magalhaes LM, Berreiros L, Queiroz J & Cunha L. Valorization of grape pomace: extraction of bioactive phenolics with antioxidant properties. Industrial Crops and Products. 2015;74:397–406. doi:10.1016/j.indcrop.2015.05.055

[5] Cicero MA, Pietrantonio E, Romanelli G, & Di Muccio A. Comparison of Soxhlet, shaking, and microwave assisted extraction techniques for determination of PCB congeners in a marine sediment. Bulletin of Environmental Contamination and Toxicology. 2007;65:307–313. doi:10.1007/s001280000129

[6] Heleno SA, Diz P, Prieto MA, Barros L, Rodrigues A, Barreiro MF & Ferreira ICFR. Optimization of ultrasound-assisted extraction to obtain mycosterols from *Agaricus bisporus* L. by response surface methodology and comparison with conventional Soxhlet extraction. Food Chemistry. 2016;197:1054–63. doi:10.1016/j.foodchem.2015.11.108

[7] Da Porto C, Decorti D, & Natolino A. Water and ethanol as co-solvent in supercritical fluid extraction of proanthocyanidins from grape marc: a comparison and a proposal. The Journal of Supercritical Fluids. 2014a;87:1–8. doi:10.1016/j.supflu.2013.12.019

[8] García-Becerra L, Mitjans M, Rivas-Morales C, Verde-Star J, Oranday-Cárdenas A, & Vinardell María P. Antioxidant comparative effects of two grape pomace Mexican extracts from vineyards on erythrocytes. Food Chemistry. 2016;194:1081–1088. doi:10.1016/j.foodchem.2015.08.131

[9] Drosou C, Kyriakopoulou K, Bimpilas A, Tsimogiannis D & Krokida M. A comparative study on different extraction techniques to recover red grape pomace polyphenols from vinification byproducts. Industrial Crops and Products. 2015a;75:141–149. doi:10.1016/j.indcrop.2015.05.063

[10] Manna L, Bugnone CA, & Banchero M. Valorization of hazelnut, coffee and grape wastes through supercritical fluid extraction of triglycerides and polyphenols. The Journal of Supercritical Fluids. 2015;104:204–211. doi:10.1016/j.supflu.2015.06.012

[11] Da Porto C, Porretto E & Decorti D. Comparison of ultrasound-assisted extraction with conventional extraction methods of oil and polyphenols from grape (*Vitis vinifera* L.) seeds. Ultrasonics Sonochemistry. 2013;20:1076–1080. doi:10.1016/j.ultsonch.2012.12.002

[12] Danlami JM, Arsad A & Zaini MAA. Characterization and process optimization of castor oil (*Ricinus communis* L.) extracted by the Soxhlet method using polar and non-

polar solvents. Journal of the Taiwan Institute of Chemical Engineers. 2015;47:99–104. doi:10.1016/j.jtice.2014.10.012

[13] Dutta R, Sarkar U & Mukherjee A. Extraction of oil from Crotalaria Juncea seeds in a modified Soxhlet apparatus: physical and chemical characterization of a prospective bio-fuel. Fuel. 2014;116:794–802. doi:10.1016/j.fuel.2013.08.056

[14] Ramluckan K, Moodley KG & Bux F. An evaluation of the efficacy of using selected solvents for the extraction of lipids from algal biomass by the soxhlet extraction method. Fuel. 2014;116:103–108. doi:10.1016/j.fuel.2013.07.118

[15] Delaunay-Bertoncini N, van der Wielen FWM, de Voogt P, Erlandsson B & Schoenmakers PJ. Analysis of low-molar-mass materials in commercial rubber samples by Soxhlet and headspace extractions followed by GC-MS analysis. Journal of Pharmaceutical and Biomedical Analysis. 2004;35:1059–73. doi:10.1016/j.jpba.2004.03.025

[16] Barba FJ, Brianceau S, Turk M, Boussetta N & Vorobiev E. Effect of alternative physical treatments (ultrasounds, pulsed electric fields, and high-voltage electrical discharges) on selective recovery of bio-compounds from Fermented Grape Pomace. Food and Bioprocess Technology. 2015;8:1139–1148. doi:10.1007/s11947-015-1482-3

[17] Barba FJ, Zhu Z, Koubaa M, Sant'Ana AS & Orlien V. Green alternative methods for the extraction of antioxidant bioactive compounds from winery wastes and by-products: a review. Trends in Food Science and Technology. 2016;49:96–109. doi:10.1016/j.tifs.2016.01.006

[18] Vergara-Salinas JR, Vergara M, Altamirano C, Gonzalez Á, & Pérez-Correa JR. Characterization of pressurized hot water extracts of grape pomace: chemical and biological antioxidant activity. Food Chemistry. 2015;171:62–69. doi:10.1016/j.foodchem.2014.08.094

[19] Hawthorne SB, Miller DJ, Lagadec AJ-M, Hammond PJ & Clifford AA. Method of manipulating the chemical properties of water to improve the effectiveness of a desired process. United States Patent. 2002.

[20] Plaza M & Turner C. Pressurized hot water extraction of bioactives. TrAC Trends in Analytical Chemistry. 2015;71:39–54. doi:10.1016/j.trac.2015.02.022

[21] Chérif R, Guillaume A, Léon C, Christine L, Jérome B, Joanna WB, & Ouazzani J. Antiangiogenic-like properties of fermented extracts of ayurvedic medicinal plants. Journal of Medicinal Food. 2015;18:1065–1072. doi:10.1089/jmf.2014.0128

[22] Morata A, Loira I Vejarano R Bañuelos MA, Sanz PD, Otero L & Suárez-Lepe JA. Grape processing by high hydrostatic pressure: effect on microbial populations, phenol extraction and wine quality. Food Bioprocess Technology. 2015;8:277–286. doi:10.1007/s11947-014-1405-8

[23] Virot M, Tomao V, Le-Bourvellec C, Renard CMCG & Chemat F. Towards the industrial production of antioxidants from food processing byproducts with ultrasound-assisted

extraction. Ultrasonics Sonochemistry. 2010;17:1066–1074. doi:10.1016/j.ultsonch.2009.10.015

[24] Chemat F, Vian MA & Cravotto G. Green extraction of natural products: concept and principles. International Journal of Molecular Sciences. 2012;13:8615–8627. doi:10.3390/ijms13078615

[25] Ghafoor K, Choi YH, Jeon JY & Jo IH. Optimization of ultrasound-assisted extraction of phenolic compounds, antioxidants, and anthocyanins from grape (*Vitis vinifera*) seeds. Journal of Agricultural and Food Chemistry. 2009;57:4988–4994. doi:10.1021/jf9001439

[26] Zu G, Zhang R, Yang L, Ma C, Zu Y, Wang W & Zhao C. Ultrasound assisted extraction of carnosic acid and rosmarinic acid using ionic liquid solution from *Rosmarinus officinalis*. International Journal of Molecular Science. 2012;13:11027–11043. doi:10.3390/ijms130911027

[27] Vinatoru M, Toma M, Radu O, Filip PI, Lazurca D & Mason TJ. The use of ultrasound for the extraction of bioactive principles from plant materials. Ultrasonics Sonochemistry. 1997;4:135–140. doi:10.1016/S1350-4177(97)83207-5

[28] American Institute of Engineers. Understand the basics of membrane filtration [Internet]. 05/04/2013. Available from: http://research.che.tamu.edu/groups/seminario/numerical-topics/membrane%20filtration.pdf [Accessed: 10/03/2016]

[29] Luque de Castro MD & Priego-Capote F. Ultrasound assistance to liquid-liquid extraction: a debatable analytical tool. Analytica Chimica Acta. 2007;583:2–9. doi:10.1016/j.aca.2006.09.039

[30] Ghassempour A, Heydari R, Talebpou Z, Fakhari AR, Rassouli A, Davies N & Aboul-Enein H. Study of new extraction methods for separation of anthocyanins from red grape skins: analysis by HPLC and LC-MS/MS. Journal of Liquid Chromatography and Related Technologies. 2008;31:2686–2703. doi:10.1080/10826070802353247

[31] González-Centeno MR, Comas-Serra F, Femenia A, Rosselló C & Simal S. Effect of power ultrasound application on aqueous extraction of phenolic compounds and antioxidant capacity from grape pomace (*Vitis vinifera* L.): experimental kinetics and modeling. Ultrasonics Sonochemistry. 2015;22:506–514. doi:10.1016/j.ultsonch.2014.05.027

[32] Tao Y, Zhang Z & Sun D. Kinetic modeling of ultrasound-assisted extraction of phenolic compounds from grape marc: influence of acoustic energy density and temperature. Ultrasonics Sonochemistry. 2014;21:1461–1469. doi:10.1016/j.ultsonch.2014.01.029

[33] Chemant F, Tomao V & Virot M. Ultrasound-assisted extraction in food analysis. In: Semih Otles, editor. Handbook of Food Analysis Instruments. Boca Raton, FL, USA: CRC Press, Taylor & Francis Group; 2008. p. 85–94.

[34] Chemat F, Abert-Vian M & Zille-Huma YJ. Microwave assisted separations: green chemistry in action. In: Pearlman JT, editor. Green Chemistry Research Tends. New York: Nova Science Publishers; 2009. p. 33–62.

[35] Jain T. Microwave assisted extraction for phytoconstituents—an overview. Asian Journal of Research in Chemistry. 2009;2:19–25.

[36] Alupului A, Calinescu I & Lavric V. Microwave extraction of active principles from medicinal plants. UPB Science Bulletin Series B. 2012;74:129–142.

[37] Eskilsson CS & Björklund E. Analytical-scale microwave-assisted extraction. Journal of Chromatography A. 2000;902:227–250. doi:10.1016/S0021-9673(00)00921-3

[38] Routray W & Orsat V. Microwave-assisted extraction of flavonoids: a review. Food Bioprocess Technology. 2011;5:11–68. doi:10.1007/s11947-011-0573-z

[39] Liazid A, Guerrero RF, Cantos E, Palma M & Barroso CG. Microwave assisted extraction of anthocyanins from grape skins. Food Chemistry. 2011;124:1238–1243. doi:10.1016/j.foodchem.2010.07.053

[40] Bittar SA, Perino-Issartier S, Dangles O & Chemat F. An innovative grape juice enriched in polyphenols by microwave-assisted extraction. Food Chemistry. 2013;141:3268–3272.

[41] Li Y, Skouroumounis GK, Elsey GM & Taylor DK. Microwave-assistance provides very rapid and efficient extraction of grape seed polyphenols. Food Chemistry. 2011;129:570–576. doi:10.1016/j.foodchem.2011.04.068

[42] Ahmad S & Ahmed M. Application of membrane technology in food processing. Food Processing: Strategies for Quality Assessment, Food Engineering Series. 2014; 1: 379–394 doi:10.1007/978-1-4939-1378-7_15

[43] Moura-Bernardes A. General aspects of membrane separation processes. Electrodialysis and Water Reuse, Topics in Mining, Metallurgy and Materials Engineering. 2014; doi:10.1007/978-3-642-40249-4_2

[44] Giacobbo M, Oliveira EC, Duarte NF, Mira HMC, Bernardes AM & de Pinho MN. Ultrafiltration based process for the recovery of polysaccharides and polyphenols from winery effluents. Separation Science and Technology. 2013;48:438–444. doi:10.1080/01496395.2012.725793

[45] Pinto PCR, Mota IF, Loureiro JM & Rodrigues AE. Membrane performance and application of ultrafiltration and nanofiltration to ethanol/water extract of Eucalyptus bark. Separation and Purification Technology. 2014;132:234–243. doi:10.1016/j.seppur.2014.04.04

[46] Mohanty K & Purkait MK, editor. Membrane Technologies and Applications. Boca Raton, FL, USA: CRC Press, Taylor & Francis Group; 2012. 505 p.

[47] Galanakis C. M., Markouli E. & Gekas V. Fractionation and recovery of different phenolic classes from winery sludge via membrane filtration. Separation and Purification Technology. 2013; 107:245–251. doi:10.1016/j.seppur.2013.01.034

[48] Giacobbo A, Matos do Prado J, Meneguzzi A, Moura-Bernardes A, De Pinho MN. Microfiltration for the recovery of polyphenols from winery effluents. Separation and Purification Technology. 2015;143:12–18. doi:10.1016/j.seppur.2015.01.019

[49] Fernández K, Paiva R & Aspé E. Purification of grape proanthocyanidins by membrane ultrafiltration. Journal of Medical and Bioengineering. 2015; doi:10.12720/jomb.4.3.178-183

[50] Sairam P, Ghosh S, Jena S, Rao KNV & Banji D. Supercritical fluid extraction (SFE)—an overview. Asian Journal of Pharmaceutical Science. 2012;2:112–120.

[51] Sapkale GN, Patil SM, Surwase US & Bhatbhage PK. Supercritical fluid extraction—a review. International Journal of Chemistry Science. 2010;8:729–743.

[52] Ibáñez E, Herrero M, Mendiola JA & Castro-Puyana M. Extraction and characterization of bioactive compounds with health benefits from marine resources: macro and micro algae, cyanobacteria, and invertebrates. In: Hayes M, editors. Marine Bioactive Compounds: Sources, Characterization and Applications. New York: Springer; 2012. p. 55–98. doi:10.1007/978-1-4614-1247-2

[53] Azmir J, Zaidul ISM, Rahman MM, Sharif KM, Mohamed A, Sahena F, Jahurul MHA, Ghafoor K, Norulaini NAN & Omar AKM. Techniques for extraction of bioactive compounds from plant materials: a review. Journal of Food Engineering. 2013;117:426–436. doi:10.1016/j.jfoodeng.2013.01.014

[54] Oliveira DA, Salvador AA, Smania A, Smania EFA, Maraschin M & Ferreira SRS. Antimicrobial activity and composition profile of grape (*Vitis vinifera*) pomace extracts obtained by supercritical fluids. Journal of Biotechnology. 2013;164:423–432. doi:10.1016/j.jbiotec.2012.09.014

[55] Rombaut N, Savoire R, Thomasset B, Belliard T, Castello J, Van Hecke E & Lanoiselle J. Grape seed oil extraction: interest of supercritical fluid extraction and gas-assisted mechanical extraction for enhancing polyphenol co-extraction in oil. Comptes rendus Chimie. 2014;17:284–292. doi:10.1016/j.crci.2013.11.014

[56] Farías-Campomanes AM, Rostagno MA & Meireles MA. Production of polyphenol extracts from grape bagasse using supercritical fluids: yield, extract composition and economic evaluation. Journal of Supercritical Fluids. 2013;77:70–78. doi:10.1016/j.supflu.2013.02.006

[57] Koubaa M, Rosello-Soto E, Sic-Zlabur J, Rezek-Jambrak A, Bencic M & Grimi N. Current and new insights in the sustainable and green recovery of nutritionally valuable compounds from Stevia rebaudiana Bertoni. Journal of Agricultural and Food Chemistry. 2015;63:6835–6846. doi:10.1021/acs.jafc.5b01994

[58] Barbosa-Canovas GV, Pierson MD, Zhang QH & Schaffner DW. Pulsed electric fields. Journal of Food Science. 2000; 65: 65–79. doi:10.1111/j.1750-3841.2000.tb00619.x

[59] El Darra N, Grimi N, Maroun RG, Louka N & Vorobiev E. Pulsed electric field, ultrasound, and thermal pretreatments for better phenolic extraction during red fermentation. European Food Research and Technology. 2013;236:47–56. doi:10.1007/s00217-012-1858-9

[60] Delsart C, Ghidossi R, Poupot C, Cholet C, Grimi N & Vorobiev E. Enhanced extraction of phenolic compounds from Merlot grapes by pulsed electric field treatment. American Journal of Enology and Viticulture. 2012;63:205–211. doi:10.5344/ajev.2012.11088

[61] Brianceau S, Turk M, Vitrac X & Vorobiev E. Combined densification and pulsed electric field treatment for selective polyphenols recovery from fermented grape pomace. Innovative Food Science and Emerging Technologies. 2015;29:2–8. doi:10.1016/j.ifset.2014.07.010

[62] Rodríguez-Morgado B, Candiracci M, Santa-María C, Revilla E, Gordillo B, Parrado P & Castaño A. Obtaining form grape pomace an enzymatic extract with anti-inflammatory properties. Plant Foods Human Nutrition. 2015;70(1):42–9. doi:10.1007/s11130-014-0459-0

[63] Xu C, Yagiz Y, Borejsza-Wysocki W, Lu J, Gu L, Ramírez-Rodrigues MM, Marshall MR. Enzyme release of phenolics form muscadine grape (Vitis rotundifolia Michx). Food Chemistry. 2104;157:20–29. doi:10.1016/j.foodchem.2014.01.12

[64] Fernández K, Vaga M & Aspé E. An enzymatic extraction of proanthocyanidins from País grape seeds and skins. Food Chemistry. 2015;168:7–13. doi:10.1016/j.foodchem.2014.07.021

[65] Delgado Adámez J, Gamero Samino E, Valdés Sánchez E & González-Gómez D. In vitro estimation of the antibacterial activity and antioxidant capacity of aqueous extracts from grape-seeds (*Vitis vinifera* L.). Food Control. 2012;24:136–141. doi:10.1016/j.foodcont.2011.09.016

[66] Sagdic O, Ozturk I, Ozkan G, Yetim H, Ekici L & Yilmaz M T. RP-HPLC–DAD analysis of phenolic compounds in pomace extracts from five grape cultivars: evaluation of their antioxidant, antiradical and antifungal activities in orange and apple juices. Food Chemistry. 2011;126:1749–1758. doi:10.1016/j.foodchem.2010.12.075

[67] Rajha HN, Ziegler W, Louka N, Hobaika Z, Vorobiev E, Boechzelt HG & Maroun RG. Effect of the drying process on the intensification of phenolic compounds recovery from grape pomace using accelerated solvent extraction. International Journal of Molecular Sciences. 2014;15:18640–58. doi:10.3390/ijms151018640

[68] Vergara-Salinas JR, Bulnes P, Zúñiga MC, Pérez-Jiménez J, Torres JL, Mateos-Martín ML, Pérez-Correa J. Effect of pressurized hot water extraction on antioxidants from

grape pomace before and after enological fermentation. Journal of Agricultural and Food Chemistry. 2013;61:6929–6936. doi:10.1021/jf4010143

[69] Monrad JK, Srinivas K, Howard LR & King JW. Design and optimization of a semicontinuous hot–cold extraction of polyphenols from grape pomace. Journal of Agricultural and Food Chemistry. 2012;60:5571–5582. doi:10.1021/jf300569w

[70] Srinivas K, King JW, Monrad JK, Howard LR & Zhang D. Pressurized solvent extraction of flavonoids from grape pomace utilizing organic acid additives. Italian Journal of Food Science. 2011;23:90–105. Retrieved from http://www.scopus.com/inward/record.url?eid=2-s2.0-79953902522&partnerID=40&md5=18a5c7d3fd44219d988dbae56dc4c342.

[71] Rodriguez-Rodriguez R, Justo MA, Claro CM, Vila E, Parrado J, Herrera MD & de Sotomayor MA. Endothelium-dependent vasodilator and antioxidant properties of a novel extract of grape pomace from wine industrial waste. Food Chemistry. 2012;135:1044–1051. doi:10.1016/j.foodchem.2012.05.089

[72] Gómez-García R, Martínez-Avila GCG & Aguilar CN. Enzyme-assisted extraction of antioxidative phenolics from grape (*Vitis vinifera* L.) residues. 3 Biotech. 2012;4(2):297–300. doi:10.1007/s13205-012-0055-7

[73] Martínez-Ávila GCG, Aguilera-Carbó AF, Rodríguez-Herrera R & Aguilar CN. Fungal enhancement of the antioxidant properties of grape wastes. Annals of Microbiology. 2012;62:923–930. doi:10.1007/s13213-011-0329-z

Chapter 6

Benefits of Vine Leaf on Different Biological Systems

Denise S. Lacerda, Pedro C. Costa, Cláudia Funchal,
Caroline Dani and Rosane Gomez

Additional information is available at the end of the chapter

http://dx.doi.org/10.5772/64930

Abstract

For centuries, the therapeutic benefits of grapes and other byproducts have been empirically used for medical purposes such as bleeding, pain, inflammation, nausea, diarrhea, gastroenteritis, or skin diseases. Moderated intake of the red wine improves parameters as blood lipids, endothelial dysfunction, platelet aggregation, and other risk factors for cardiovascular disease. However, few studies have been explored the potential benefits from vine byproducts. Vine leaves, a waste product from the vine, are also rich source of polyphenols and other therapeutic compounds. In this chapter, we explored the therapeutic properties from vine leaf in different biological systems.

Keywords: polyphenols, organic viticulture, grapevine, natural products, live, heart, kidney, brain

1. Introduction

The production of grapes is considered an economically important activity in many countries, mainly related to the wine production [1]. Beyond their lucrative potential, grapes and their byproducts show nutritional and functional properties [2–4]. Since centuries ago, grapes have been used for medical purposes, preventing or treating diseases as nausea, diarrhea, gastroenteritis, or skin disorders [5]. More recently, the therapeutic effect of red wine has been reported, and moderate intake has been related to improved blood lipid parameters, endothelial dysfunction, platelet aggregation, and other risk factors for cardiovascular disease [6, 7]. Apart from the grape or wine, studies have been shown that grape byproducts such as juice, or extracts from the skin, seed, or leaf also present therapeutic proprieties [8–12]. Grape leaves, for example, have been popularly used to stop bleeding, relieve pain, inflammation, and diarrhea (**Figure 1**) [13, 14].

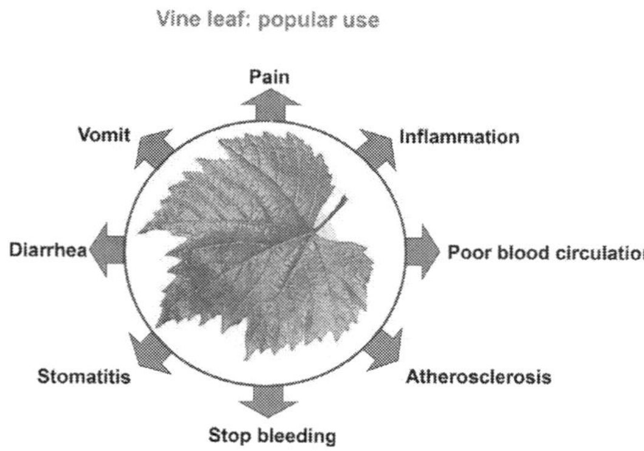

Figure 1. Popular use of vine leaf for health purpose.

Therapeutic proprieties by grapes, wine, or byproducts are mainly related to the polyphenolic compounds [7, 15]. Leaves, which are a waste product from the grapevine, usually discarded by grape farmers, are also rich source of polyphenols and other therapeutic compounds [16]. More recently, their therapeutic properties have been explored, mainly because grape juices are rich in carbohydrates and wine is an alcoholic beverage, nonrecommended to diabetic or alcoholics individuals, respectively.

2. Bioactive polyphenols in vine leaves

The grapevine (*Vitis* spp.) is cultivated across the world in different regions, mainly in temperate climate with adequate rain, warm and dry summers, and mild winters [17]. Climate, soil, conventional or organic cultivation method, and different cultivars are determinant to phytochemical constitution of grapevines [18]. These phytochemical compounds include a variety of bioactive organic acids (e.g., malic, oxalic, fumaric, ascorbic, citric, linoleic, and tartaric acids), vitamin E, terpenes, tannins, carotenoids, and polyphenols that have been highlighted for their beneficial effect on human health [19, 20]. The most important grape polyphenols as flavanols (e.g., epicatechin and gallocatechin), flavonols (e.g., quercetin and myricetin), anthocyanins (e.g., pelargonidin and cyanidin), and resveratrol are secondary metabolites synthesized by plants and associated with growth, pigmentation, pollination, environmental stress, and resistance against pathogens and predators [13, 17].

Polyphenols present biological activities, such as antioxidant, anti-inflammatory, anticancer, antimicrobial, cardioprotective, and antiaging effects [9, 14]. Polyphenols therapeutic properties have been related to their chemical structure and ability to act as radical scavengers of the lipid peroxidation chain reactions, donating electrons, and neutralizing free radicals [21]. Moreover, they are chelators of metals as iron (Fe^{2+}) and copper (Cu^{2+}), preventing oxidation caused by highly reactive hydroxyl radicals [21, 22]. They also inhibit the immune cell recruitment (T lymphocytes and natural killer cells) and decrease the nuclear factor kappa B (NFκB) expression [23, 24].

Species	Viticulture method	Preparation	Total phenolic mg/g gallic acid	Phytochemicals detected	Reference
Vitis vinifera	NI	Ethanolic extract	216.0 ± 5.1	Total flavonoids	[26]
Vitis vinifera	NI	Aqueous extract	149.93 ± 0.35	Total proanthocyanidin; total flavonoid	[27]
Vitis vinifera	NI	Aqueous extract	146.3 ± 4.2	Anthocyanin: cyanidin-3-*O*-glucoside > peonidin-3-*O*-glucoside Flavonols: quercetin-3-*O*-glucuronide Caffeic acid derivatives: caftaric acid	[28]
Vitis vinifera	NI	Ethanolic extract	98.84 ± 9.26	NI	[29]
Vitis labrusca	Organic	Aqueous extract	81.79 ± 2.68	Catechin; resveratrol	[30]
Vitis vinifera	NI	Ethanolic extract	60.4 ± 0.4	Flavonols; quercetin-3-*O*-glucuronide > kaempherol-3-*O*-glucoside Anthocyanin: peonidin-3-glucoside > cyanidin-3-glucoside Hydroxycinnamic acid: *trans*-caftaric acid	[10]
Vitis labrusca	Organic	Ethanolic extract	20.2 ± 1.8	Catechin; resveratrol; quercetin; rutin; kaempherol	[9]
Vitis labrusca	Conventional	Aqueous extract	19.83 ± 0.76	Catechin; resveratrol	[30]
Vitis labrusca	Conventional	Ethanolic extract	19.0 ± 1.8	Catechin; resveratrol; quercetin; rutin; kaempherol, naringin	[9]
Vitis vinifera	NI	Acetone/methanol extract		Anthocyanins: peonidin 3-glucoside > malvidin > cyaniding 3-glucoside; flavonols: quercetin 3-*O*-β-D-glucuronide > isoquercitrine quercetin 3-*O*-β-D-glucoside phenolic acids	[16]

NI: not informed.

Table 1. Phenolic compounds from different vine leaf extracts.

A study comparing 10 grape cultivars grown in southern Georgia, USA, showed that the total concentration of phenolic compounds was higher in seed (2178.8 mg/g gallic acid equivalent), followed by skin (374.6 mg/g), and leaf (351.6 mg/g) [25], evidencing that the leaf is also an important source of phenolic compounds. Although gallic acid was a dominant phenolic acid in the vine leaf, other constituents may contribute to the beneficial properties of its extract. **Table 1** shows the phenolic contend in different extracts from *V. vinifera* and *V. labrusca* leaves.

It reveals that ethanolic extracts show the highest extraction rate and the *V. labrusca* varietal shows the highest total phenolic concentration. Optimal or prolonged low-temperature exposure decreases the phenolic contends in *Vitis vinifera* leaves from 526 g/g to 458 mg/g of extract [31]. Antioxidant index, measured by trolox equivalent antioxidant capacity (TEAC) assay, showed that, similarly to grape seeds, leaves present 10 times higher antioxidant activity than grape juice or pulp [7]. Moreover, total phenolic levels in leaf are not affected by brining, a method assumed to preserve vine leaves for future use in the Turkish cuisine [32]. Resveratrol, a compound with therapeutic properties, accumulates in the surface of leaves at range of 40–400 μg/g fresh weight in according to environmental conditions [33].

In addition to environmental influences, farming practices, as organic or conventional viticulture, also interfere with the production of polyphenols [34]. In the organic viticulture, grapevine grown in the absence of pesticides, chemicals, or genetic engineering modification, and it is more vulnerable to external attacks from insets or microorganisms, which may contribute to the higher production of phytochemicals, responsible for plant defenses [35]. A study showed that organic vine leaf extract presents higher concentrations of resveratrol than conventional vine extract, although total polyphenols were similar and catechin and quercetin were lower [9, 30] (**Table 1**). Given the variability in the phenolic composition of the vine leaf, the quantification of the phenolic constituents may estimate the quality and therapeutic potential in vine leaves [16].

3. Effect of vine leaf extract on hepatic and gastrointestinal systems

Alcoholic and nonalcoholic liver diseases have been related to chronic exposition to risk factors as alcohol, tobacco smoking, drugs, environmental pollutants, and irradiation. It is well known that these risk factors promote excessive formation of oxygen and nitrogen reactive species and may lead to oxidative damage in the liver [36]. Although clinical studies are scarce, preclinical studies show that natural antioxidants from products as vine leaves prevent or attenuate the severity of liver diseases induced by oxidative mechanisms. Animal studies have explored some morphological and biochemistry changes by hepatotoxic substances and the protective effect of vine leaf extracts. Aspartate aminotransferase (AST), alanine aminotransferase (ALT), γ-glutamyl transferase (GGT), and alkaline phosphatase (ALP) are some biomarkers that predict liver function and explored in these studies. Aqueous extract from *Vitis coignetiae Pulliat* leaves shows hepatoprotective effect after chronic oral administration in an animal model of nonalcoholic steatohepatitis (NASH), evidenced by decreasing on AST and ALP activity, confirmed by increasing in plasma antioxidants and delaying in the progression of liver fibrosis [37] (**Table 2**). Similarly, alcoholic or butanolic extract of vine leaves (*Vitis vinifera*) decreased AST and ALT activity after acute hepatotoxicity induced by carbon tetrachloride (CCl_4) in rats [26]. Vine leaves extract also decreased AST, ALT, ALP, and GGT activity after chronic alcohol administration [29]. For both, CCl_4 and alcohol-induced hepatotoxicity models, vine leaves extract decreased biomarkers of serum oxidative stress as malondialdehyde (MDA), superoxide dismutase (SOD), catalase (CAT), glutathione peroxidase

(GPx) enzyme, as well decreased histopathological lesions [26, 29]. Preincubation with organic and conventional vine leaf (*Vitis labrusca*) extracts also prevents both lipid and protein oxidative damage in the rat liver after oxidative stress induced by hydrogen peroxide [39]. Moreover, the organic vine leaves extract restored SOD and the conventional vine leaves extract restored CAT activity, both decreased by hydrogen peroxide-induced stress and related to different phenolic contend in each extract [39] (**Table 2**). The liver of diabetic individual is also subject to damage due to exposure to self-oxidation of free glucose and deficiency of antioxidant system [41]. Indeed, chronic oral administration of aqueous extract of organic grape leaves (*Vitis labrusca*) reduced the AST activity in an experimental model of diabetes in rats [38]. The synergistic effects of different polyphenols in the vine leaf extract reduced the oxidative stress, preventing lipid and protein damage and increasing enzymatic and nonenzymatic antioxidant defenses in the liver of diabetic rats, suggesting a promising therapeutic approach to hepatic complications induced by diabetes [38].

Species	Culture method	Treatment	Condition	Results	Reference
Vitis labrusca	Organic	Aqueous extract Orally	Diabetes (rats)	↓ Lipid peroxidation ↓ Protein damage ↑ Nonenzymatic antioxidant defenses ↑ SOD activity ↓ CAT activity ↓ AST	[38]
Vitis labrusca *Vitis labrusca*	Organic Conventional	Aqueous extract preincubation	H_2O_2-induced stress (*in vitro*)	↓ Lipid peroxidation ↓ Protein damage ↑ SOD activity ↓ Lipid peroxidation ↓ Protein damage ↑ CAT activity	[39]
Vitis vinifera	NI	*n*-BuOH extract Orally	Cirrhosis (rats)	↓ Lipid peroxidation ↑ GSH content ↓ Histopathological injury ↓ AST, ALT	[26]
Vitis vinifera	NI	Ethanolic extract Orally	Alcohol induced oxicity (rats)	↓ AST, ALT, ALP, GGT ↓ Lipid peroxidation ↓ Hydroperoxides ↑ Vitamin E ↑ Vitamin C ↑ GSH ↑ SOD activity ↑ CAT activity ↑ GPx activity ↑ GST activity	[29]
Vitis coignetiae	NI	Aqueous extract Orally	Nonalcoholic steatohepatitis (rats)	↓ ALT ↓ Fibrosis area ↓ MPO activity ↓ Mitochondrial ROS ↓ NFκB expression	[40]
Vitis coignetiae	NI	Aqueous extract Orally	Nonalcoholic steatohepatitis (rats)	↓ AST and ALP ↓ CYP2E1 induction ↓ Fibrosis	[37]

NI: not informed; AST: aspartate aminotransferase; ALT: alanine aminotransferase; GGT: γ-glutamyl transferase; ALP: alkaline phosphatase; SOD: superoxide dismutase, CAT: catalase; GPx: glutathione peroxidase; GSH: reduced glutathione; GST: glutathione-S-transferase; ROS: reactive oxygen species; MPO: myeloperoxidase; NFκB; factor nuclear kappa B.

Table 2. Hepatoprotective effects of vine leaf.

Vine leaves extract (*Vitis coignetiae Pulliat*) decreases the leakage of biliary enzymes and attenuates liver fibrosis after 3 weeks of treatment in a model of nonalcoholic steatohepatitis in rats [40]. Improving on hepatic fibroses or suppression of its progression by the extract was associated to increasing on plasma antioxidant activity, decreasing on reactive species and NFκB activity, a key pathway linking oxidative stress and inflammation [40]. Vine leaves extract from *Vitis vinifera* preserved the integrity of the membrane of hepatocytes in CCl_4-intoxicated rats, evidenced by the reduction in plasma levels of AST, ALT, and GGT [27]. Additionally, the extract reduced the concentration of bilirubin, lipoproteins, lipid oxidation and, in parallel, preserved histological injuries of the liver [27].

Among the gastrointestinal diseases, the prevalence and incidence of gastritis, peptic ulcers, and inflammatory bowel disease have increased in recent years, associated to the consumption of processed foods and lifestyle [42]. The activation of inflammatory pathways is the common pathological mechanism of these diseases and initiates by the activation of NFκB, which is related with transcriptional control of multiple proinflammatory mediators as IL-1β, TNF-α, and IL-8 in the gastrointestinal tissue [43].

In this context, the biological activity of the aqueous extract of vine leaves (*Vitis vinifera*) was assessed *in vitro* in a model of gastric inflammation (human gastric and intestinal epithelial cell) [28]. Vine leaf extracts impaired the NFκB pathway and, consequently, reduced the TNF-α and IL-8 secretion and expression by gastric epithelial cells. The anti-inflammatory effect of the extract decreased significantly after simulation of intestinal digestion, explained by the poor stability and high rate of degradation of anthocyanins and flavonoids present in the aqueous extract of vine leaves in an alkaline pH [28].

4. Effect of vine leaf extract on the cardiovascular system

Cardiovascular diseases are the most common causes of morbidity and mortality worldwide, currently responsible for over 17 million deaths, with growth forecast to 23.6 million per year to 2030 [44]. Hypertension, dyslipidemia, obesity, and smoking are considered the main cardiovascular risk factors [45]. These factors adversely affect the vascular endothelium, reducing the availability of nitric oxide, facilitating the deposition of oxidized LDL cholesterol by activating oxidative and inflammatory cascades leading to atherosclerosis, endothelial dysfunction, and cardiovascular damage [46].

Studies suggest that consumption of grape polyphenols and its derivatives is associated with reduction in cardiovascular risk related to their antioxidant, anti-inflammatory, and antithrombotic properties [6]. It is well known that there is a correlation between moderate consumption of red wine and the lowest risk of death associated with heart disease [47, 48]. Indeed, the daily consumption of low to moderate doses of wine reduces by half the risk of death compared to individuals who did not drink wine [49]. Aqueous extract of grape leaves has been tested in rodents and evidenced also an antioxidant effect, decreasing lipid and protein damage, as well increasing SOD and CAT activity in a heart homogenates injured by H_2O_2 in

http://dx.doi.org/10.5772/64930

rats [39]. These antioxidant effects were more significant compared to those extracts prepared from organic grape leaves [39] (**Table 3**).

Specie	Viticulture method	Treatment	Condition	Target tissue	Results	Reference
Vitis labrusca	Organic and conventional	Aqueous extract Preincubation	H_2O_2-induced stress (*in vitro*)	Heart	↓ Lipid peroxidation ↓ Damage protein ↑ CAT activity	[39]
Vitis vinifera	NI	Aqueous extract Orally	Diabetic (rats)	Heart	↑ GSH content	[50]
Vitis vinifera	NI	Ethanolic extract Orally	Alcohol-induced toxicity (rats)	Kidney	↓ TBARS ↓ Hydroperoxides ↑ Vitamin E and vitamin C ↑ GSH ↑ SOD activity ↑ CAT activity ↑ GPx activity ↑ GST activity	[29]
Vitis labrusca	Organic	Aqueous extract preincubation	H_2O_2-induced stress (*in vitro*)	Kidney	↓ Lipid peroxidation ↓ Damage protein ↑ SOD activity	[39]
Vitis vinifera	NI	Aqueous extract Orally	Toxicity-induced by CCl_4 (rats)	Kidney	↓ Creatinine, uric acid, and calcium levels ↓ MDA ↑ NP-SH	[27]
Vitis labrusca	Organic and conventional	Aqueous extract preincubation	H_2O_2-induced stress (*in vitro*)	Cerebral cortex, cerebellum and hippocampus	↓ Lipid peroxidation: cerebellum, hippocampus ↓ Damage protein: cerebral cortex, cerebellum, hippocampus ↓ Lipid peroxidation: cerebellum ↓ Damage protein: cerebral cortex	[9]
Vitis labrusca	Organic	Aqueous extract pretreatment (intraperitoneally)	CCl_4-induced stress (in rats)	Cerebral cortex, cerebellum and hippocampus	↓ Damage protein: cerebral cortex, cerebellum, hippocampus ↓ SOD activity: cerebral cortex, hippocampus ↑ SOD activity: cerebellum ↑ SOD/CAT ratio: cerebral cortex, cerebellum	[30]

NI: not informed; CCl_4: carbon tetrachloride; MDA: malondialdehyde; NP-SH: nonprotein sulfhydryl; SOD: superoxide dismutase; CAT: catalase; GPx: glutathione peroxidase; GSH: reduced glutathione; GST: glutathione-S-transferase.

Table 3. Therapeutic effects of vine leaf in different tissues.

Aqueous extract of grape leaves also presents an *in vivo* antioxidant effect, increasing the GHS levels in the heart of streptozotocin-induced diabetic rats, at doses of 500 mg/kg [50], although it did not change MDA levels (**Table 3**).

Vine leaves are rich in polyphenols such as flavonoids and anthocyanins (**Table 1**). Beside antioxidant activities, polyphenols inhibit pro-oxidant enzymes (e.g., xanthine oxidase,

NADPH oxidase, lipoxygenases), chelate transient metals, interact with some ion channels, reduce platelet aggregation and leukocyte adhesion, and promote vasodilatation, decreasing the resistance to blood flow [51, 52]. Anthocyanins are also responsible for increasing in the strength and vascular permeability, as well as the inhibition of platelet aggregation [53]. Studies suggest that they promote vasorelaxation by increasing nitric oxide levels and by inhibiting the action of phosphodiesterase-5 enzyme, which metabolizes the cyclic guanosine monophosphate (cGMP), an important vasodilator, reducing the risk of cardiovascular disease [54].

Anti-inflammatory properties from flavonoids and other grapevine constituents also contribute to the cardioprotective mechanism against injury caused by ischemia-reperfusion [51, 52]. Flavonoids inhibit phospholipase A2 and cyclooxygenase enzymes, decreasing prostaglandins synthesis and, indirectly, all inflammatory cascade [55]. Studies show that flavonoids inhibit the TNF-α, IL1-β, and interferon-γ synthesis [51]. All these mechanisms contribute to LDL cholesterol reduction and increasing on HDL cholesterol, useful to protect against cardiovascular disease [56].

A commercial standardized red vine leaf aqueous extract (Antistax®, Boehringer Ingelheim Pharma GmbH & Co, Ingelheim am Rhein, Germany) from *Vitis vinifera Folium* is available for chronic venous insufficiency, improving the cutaneous microcirculation and oxygen supply in humans [57]. A randomized, double blind study showed that this vine leaf extract decreased the lower leg edema and circumference in chronic venous insufficiency patients [58]. Additionally, the vine leaf extract was investigated in women in long-term hormone replacement therapy with phlebopathy of the lower limbs [59]. After 3 months, leaf extract treatment decreased the calf and ankle circumference, besides the diameter of the great saphenous vein (GSV), relieving venous symptoms, and improving the quality of life of users [59]. Regulation of blood flow by vine left extract has been positively associated to NO (nitric oxide) synthesis by endothelial and red blood cells, adding to its antioxidant properties [60].

5. Effect of vine leaf extract on the renal system

Diseases that affect the renal system are related to progressive and irreversible loss of kidney function, and inability of the kidney to adequately clean waste products from the blood. This condition is characterized by a reduction in glomerular filtration rate, decreased urine output, proteinuria and microalbunuria, common in diabetes, and hypertensive patients [61, 62].

Oxidative stress is considered an important pathogenic mechanism in renal diseases [61]. In diabetic individuals, particularly, high levels of final advanced glycation end products (AGEs), reactive species, and oxidative stress promote protein oxidation, DNA damage, and apoptosis [62, 63]. Glomerular hypertrophy and tubulointerstitial fibrosis in the kidney in diabetic individuals may progress to nephropathy [63]. Buffering the generation of oxidative pathway may represent a nephroprotective effect against oxidative damage by diabetes [62].

In this context, unpublished results from our group (**Figure 2**) showed the beneficial effects of an organic aqueous vine leaves extract on the kidney of diabetes rats, agreeing with the results from others [62]. In our experimental protocol, nondiabetic (C) and streptozotocin-induced diabetic (D) rats were daily administered with 50, 100, and 200 mg/kg of an organic vine leaf extract, by oral gavage, for 30 days (design details showed at [38]). The kidney was collected for analysis of oxidative stress parameters and the blood, for urea and creatinine determination. A two-away ANOVA showed that diabetes significantly increased protein oxidation (carbonyl), and SOD activity ($P < 0.05$) and decreased the total sulfhydryl levels ($P < 0.001$) in the kidney of diabetic rats. All three doses prevented the protein carbonylation ($P < 0.05$) increased by diabetes, but only the dose of 50 mg/kg restored sulfhydryl levels ($P < 0.05$) and decreased the SOD activity ($P < 0.05$) (**Figure 2**).

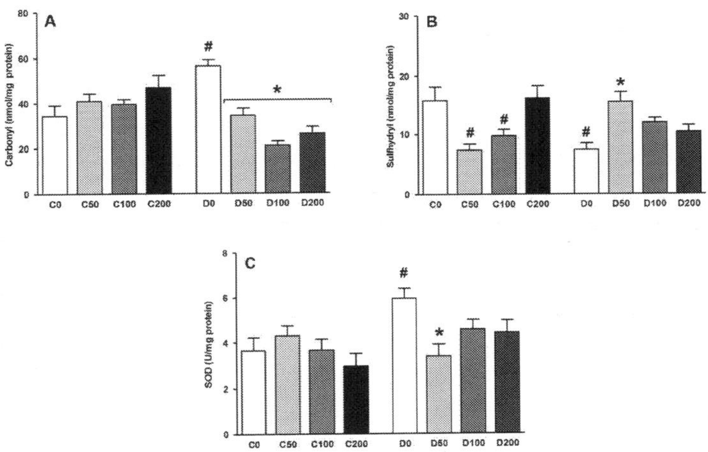

Figure 2. Effect of different doses (50, 100, and 200 mg/kg) of an organic aqueous vine leaf extract in the (A) carbonyl levels, (B) total sulfhydryl levels, and (C) SOD activity in the kidney of nondiabetic (C) and diabetic (D) rats. Values were represented as mean ± standard error; $n = 10$/group; ANOVA two-way + Bonferroni. (#) Different from C0 group, $P < 0.05$; (*) different from D0 group, $P < 0.001$.

We also showed that diabetes increased the relative kidney weight ($P < 0.001$) and urea ($P < 0.05$) and decreased creatinine levels (**Table 4**). The organic vine leaf extract did not change kidney weight, but the dose of 50 mg/kg significantly decreased urea levels in diabetic rats. Moreover, the organic extract decreased creatinine at doses of 50 and 100 mg/kg in diabetic rats and at dose of 50 mg/kg in nondiabetic rats.

The nephroprotective effect of our vine leaf extract is related to its ability to inhibit *in vivo* oxidative stress. Our results replayed *in vitro* assays that showed that the organic and conventional vine leaf extracts prevent both lipids and proteins oxidative damages in the kidney after hydrogen peroxide or alcohol-induced stress [29, 39]. Polyphenols are the main antioxidants from vine leaf, since these compounds undergo redox reactions and hydrogen atoms transfer from the phenolic hydroxyl group to the free radicals, stabilizing them [29, 64]. Bioactive phytochemicals in our extract showed a remarkable antioxidant activity as evidenced by the reduction of protein oxidation and increase in nonenzymatic antioxidants in the renal tissue of diabetic rats. Resveratrol, one of these bioactive compounds, restores the nonenzymatic levels of antioxidants in the kidney of diabetic rats by reducing the availability of reactive species and improving antioxidant status in this tissue [65]. Indeed, flavonoids increase the expression of enzyme γ-glutamylcysteine synthetase, a rate-limiting enzyme in the synthesis of glutathione, a potent antioxidant [66].

Groups	Kidney weight (g)	Urea (mg/dl)	Creatinine (mg/dl)
C0	0.32 ± 0.03	31.01 ± 12.58	0.29 ± 0.04
C50	0.31 ± 0.03	29.20 ± 3.70	0.25 ± 0.03**
C100	0.31 ± 0.03	28.05 ± 7.59	0.28 ± 0.03
C200	0.33 ± 0.08	27.80 ± 4.08	0.28 ± 0.07
D0	0.53 ± 0.08*	77.03 ± 25.52*	0.29 ± 0.08
D50	0.52 ± 0.07*	40.71 ± 16.78#	0.20 ± 0.03*#
D100	0.52 ± 0.04*	56.02 ± 16.39	0.25 ± 0.07#
D200	0.50 ± 0.06*	59.80 ± 22.32	0.31 ± 0.09
P	<0.001	<0.05	<0.05

Values are represented as mean ± standard deviation, *n* = 10/group; ANOVA two-way + Bonferroni; (*) different from C groups; (**) different from C0 group, (#) different from D0 group.

Table 4. Relative weights of kidney (g/% body weight), as well as urea and creatinine levels, after 30 days of daily oral administration from an organic aqueous vine (*Vitis labrusca*, L.) leaf extract, in different doses (50, 100, or 200 mg/kg), in diabetic (D) and nondiabetic (C) rats.

Regarding the antioxidant enzymes, we found that only the dose of 50 mg/kg prevented the increasing on SOD activity in the kidney by the chronic hyperglycemia. Because SOD catalyzes the dismutation of superoxide to H_2O_2 and water, we may infer that this was the main reactive species produced in this tissue in our diabetic rats, prevented by polyphenols present in the organic vine leaf extract [38]. We do not discard that diverse effect would be found after chronic treatment with conventional vine leaf extract, since a study showed that only the organic extract from vine leaf (*Vitis labrusca*) restored SOD activity after *in vitro* alcohol-induced stress in the kidney of rats [39].

Lower urea and creatinine in diabetic rats treated with vine leaf extract at the dose of 50 and 100 mg/kg suggested a dose-related nephroprotective effect and consequently, improving on renal function. These results agree with another study that showed that polyphenols extracts from *Hibiscus sabdariffa* improved renal function in an experimental model of diabetes [67]. Resveratrol also decreased creatinine and urea levels, protecting against kidney damage caused by chronic hyperglycemia [65]. In nondiabetic rats, our extract decreased creatinine levels at dose of 50 mg/kg, suggesting an improvement on renal function by hemodynamic mechanisms, already evidenced by regulation of blood flow from poliphenols [57]. Indeed, acute aqueous extracts of *Vitis indica* leaves increased the urine volume, sodium and potassium chloride excretion in rats [68]. Moreover, the pretreatment with epicatechin in rats exposed to an animal model of nephrolithiasis increased creatinine excretion and urine volume, reducing renal calcium and preventing papillary renal tissue from subepithelial calcification [69]. Pretreatment with vine leaf (*Vitis vinifera*) extract also restored the renal function, evidenced by decreasing on creatinine, urea, uric acid, and calcium plasma levels, associate to lower histopathologic injuries (**Table 3**) [27]. In addition, all these parameters were related to lower lipid oxidation and restoring on nonenzymatic antioxidant defenses in the kidney. Such nefroprotetive effects were attributed to the antioxidant properties of proanthocyanidins and other flavonoids present in vine leaf [27].

6. Effect of vine leaf extracts on the central nervous system

The brain is susceptible to the oxidative damage and shows high oxygen consumption rate and abundant lipid content. Indeed, evidence shows that oxidative stress and inflammation are associated with Parkinson, Alzheimer, and other neurodegenerative diseases [70–72].

Bioactive compounds as flavonols, flavan-3-ols, anthocyanins, phenolic acids, or resveratrol, in red wine and other grapevine byproducts have been extensively studied by their central effect. Conventional and organic vines leaf extracts decrease lipid and protein oxidative damage induced by hydrogen peroxide (H_2O_2) in the rat brain, reestablishing the SOD and CAT activity [9]. The same neuroprotective effect was found after treatment with both conventional and organic vines leaf extracts in the cortex, hippocampus, and cerebellum after carbon tetrachloride-induced stress in rats [30].

Although poor central bioavailability, resveratrol is effective for the treatment of aging-related learning and memory deficits [73]. A recent study showed that oral resveratrol (20 and 40 mg/kg) ameliorated learning and memory impairment and prevented memory extinction in mice in an *in vivo* animal model of Alzheimer disease [74]. Agreeing with these results, resveratrol also improve learning and memory in old mice, related to increasing on CREB (cAMP response element-binding) and BDNF (brain-derived neurotrophic factor) proteins in the hippocampus [75]. Adding to these neurochemical mechanisms, the beneficial effect of resveratrol have been related to its anti-inflammatory and antioxidant properties in different brain areas as in the hippocampus, and frontal cortex of diabetic and nondiabetic rats [76]. Natural products, rich in anthocyanin, as purple sweet potato extracts also exhibit antioxidant properties and memory enhancing effects in rats [77].

Methanolic *Vitis amurensis* leaf extract (25–100 mg/kg, oral gavage) prevented oxidative stress after cerebral ischemic in rats indicated by increasing on GSH and decreasing on lipid peroxidation, beyond inhibition of cyclooxygenase-2, and phosphorylated mitogen-activated protein kinases (MAPKs) [78]. Moreover, that extract inhibited the glutamate-induced neuronal death *in vitro*, and changed the apoptosis-related proteins, suggesting that the neuroprotective effect of this extract is related to its antioxidant, anti-inflammatory, and anti-excitotoxic properties, preventing the neurodegeneration in stroke [78]. Glutamate-induced neural cytotoxicity *in vitro* was prevented by a *V. vinifera* seed extract [79]. Grape seed extract also inhibited DNA damage in the CA1 region of gerbil hippocampus after transient forebrain ischemia, evidencing a neuroprotective effect [80].

7. Conclusion

For centuries, the therapeutic benefits of grapevines and other byproducts have been empirically explored. Recently, it has grown the interest in the health benefits from vine leaves. Leaves remain a waste product from many vine farming, although they show 10 times higher antioxidant activity than grape juice or pulp. Vine leaf extracts, for medical use, or freshly/cooked, for eating as a supplement, are devoid of alcohol (as wine) or sugar (as juice) providing an additional advantage from other vine byproducts. Here, we showed the effect of vine leaf extract in different tissues and point the needed of increase the researchers in the area to explore clinical use of this natural product.

Author details

Denise S. Lacerda[1], Pedro C. Costa[2], Cláudia Funchal[2], Caroline Dani[2] and Rosane Gomez[1*]

*Address all correspondence to: rosane.gomez@ufrgs.br

1 Post-Graduation Program in Biological Science: Pharmacology and Therapeutics, Federal University of Rio Grande do Sul (UFRGS), Porto Alegre, Brazil

2 Methodist University Center - IPA, Porto Alegre, Brazil

References

[1] Bisson LF, Waterhouse AL, Ebeler SE, Walker MA, Lapsley JT. The present and future of the international wine industry. Nature 2002; 418:696–9. doi:10.1038/nature01018.

[2] Bernal J, Mendiola JA, Ibáñez E, Cifuentes A. Advanced analysis of nutraceuticals. J Pharm Biomed Anal 2011; 55:758–74. doi:10.1016/j.jpba.2010.11.033.

[3] Georgiev V, Ananga A, Tsolova V. Recent advances and uses of grape flavonoids as nutraceuticals. Nutrients 2014; 6:391–415. doi:10.3390/nu6010391.

[4] Giovinazzo G, Grieco F. Functional properties of grape and wine polyphenols. Plant Foods Hum Nutr Dordr Neth 2015; 70:454–62. doi:10.1007/s11130-015-0518-1.

[5] Yang J, Xiao Y-Y. Grape phytochemicals and associated health benefits. Crit Rev Food Sci Nutr 2013; 53:1202–25. doi:10.1080/10408398.2012.692408.

[6] Dohadwala MM, Vita JA. Grapes and cardiovascular disease. J Nutr 2009; 139:1788S–93S. doi:10.3945/jn.109.107474.

[7] Xia E-Q, Deng G-F, Guo Y-J, Li H-B. Biological activities of polyphenols from grapes. Int J Mol Sci 2010; 11:622–46. doi:10.3390/ijms11020622.

[8] Dani C, Oliboni LS, Pasquali MAB, Oliveira MR, Umezu FM, Salvador M, et al. Intake of purple grape juice as a hepatoprotective agent in Wistar rats. J Med Food 2008; 11:127–32. doi:10.1089/jmf.2007.558.

[9] Dani C, Oliboni LS, Agostini F, Funchal C, Serafini L, Henriques JA, et al. Phenolic content of grapevine leaves (*Vitis labrusca* var. Bordo) and its neuroprotective effect against peroxide damage. Toxicol Vitro Int J Publ Assoc BIBRA 2010; 24:148–53. doi:10.1016/j.tiv.2009.08.006.

[10] Monagas M, Hernández-Ledesma B, Gómez-Cordovés C, Bartolomé B. Commercial dietary ingredients from *Vitis vinifera* L. leaves and grape skins: antioxidant and chemical characterization. J Agric Food Chem 2006; 54:319–27. doi:10.1021/jf051807j.

[11] Park E-J, Bae JH, Kim S-Y, Lim J-G, Baek W-K, Kwon TK, et al. Inhibition of ATP-sensitive K^+ channels by taurine through a benzamido-binding site on sulfonylurea receptor 1. Biochem Pharmacol 2004; 67:1089–96. doi:10.1016/j.bcp.2003.11.003.

[12] Suwannaphet W, Meeprom A, Yibchok-Anun S, Adisakwattana S. Preventive effect of grape seed extract against high-fructose diet-induced insulin resistance and oxidative stress in rats. Food Chem Toxicol Int J Publ Br Ind Biol Res Assoc 2010; 48:1853–7. doi:10.1016/j.fct.2010.04.021.

[13] Duthie GG, Gardner PT, Kyle JAM. Plant polyphenols: are they the new magic bullet? Proc Nutr Soc 2003; 62:599–603. doi:10.1079/PNS2003275.

[14] Nassiri-Asl M, Hosseinzadeh H. Review of the pharmacological effects of *Vitis vinifera* (grape) and its bioactive compounds. Phytother Res 2009;23(9):1197–204. doi: 10.1002/ptr.2761.

[15] Fraga CG, Galleano M, Verstraeten SV, Oteiza PI. Basic biochemical mechanisms behind the health benefits of polyphenols. Mol Aspects Med 2010; 31:435–45. doi:10.1016/j.mam.2010.09.006.

[16] Schneider E, von der Heydt H, Esperester A. Evaluation of polyphenol composition in red leaves from different varieties of *Vitis vinifera*. Planta Med 2008; 74:565–72. doi: 10.1055/s-2008-1034370.

[17] Ali K, Maltese F, Choi YH, Verpoorte R. Metabolic constituents of grapevine and grape-derived products. Phytochem Rev Proc Phytochem Soc Eur 2010; 9:357–78. doi:10.1007/s11101-009-9158-0.

[18] Vivier MA, Pretorius IS. Genetic improvement of grapevine: tailoring grape varieties for the third millennium—a review. Importance Biotechnol South Afr Vitic Wine Ind New Millenium 2000; 21:5–26.

[19] Quiñones M, Miguel M, Aleixandre A. Beneficial effects of polyphenols on cardiovascular disease. Pharmacol Res 2013; 68:125–31. doi:10.1016/j.phrs.2012.10.018.

[20] Ross JA, Kasum CM. Dietary flavonoids: bioavailability, metabolic effects, and safety. Annu Rev Nutr 2002; 22:19–34. doi:10.1146/annurev.nutr.22.111401.144957.

[21] Corcoran MP, McKay DL, Blumberg JB. Flavonoid basics: chemistry, sources, mechanisms of action, and safety. J Nutr Gerontol Geriatr 2012; 31:176–89. doi:10.1080/21551197.2012.698219.

[22] Perron NR, Brumaghim JL. A review of the antioxidant mechanisms of polyphenol compounds related to iron binding. Cell Biochem Biophys 2009; 53:75–100. doi:10.1007/s12013-009-9043-x.

[23] Canali R, Comitato R, Ambra R, Virgili F. Red wine metabolites modulate NF-kappaB, activator protein-1 and cAMP response element-binding proteins in human endothelial cells. Br J Nutr 2010; 103:807–14. doi:10.1017/S0007114509992479.

[24] Janega P, Klimentová J, Barta A, Kovácsová M, Vranková S, Cebová M, et al. Red wine extract decreases pro-inflammatory markers, nuclear factor-kB and inducible NOS, in experimental metabolic syndrome. Food Funct 2014; 5:2202–7. doi:10.1039/c4fo00097h.

[25] Pastrana-Bonilla E, Akoh CC, Sellappan S, Krewer G. Phenolic content and antioxidant capacity of muscadine grapes. J Agric Food Chem 2003; 51:5497–503. doi:10.1021/jf030113c.

[26] Orhan DD, Orhan N, Ergun E, Ergun F. Hepatoprotective effect of *Vitis vinifera* L. leaves on carbon tetrachloride-induced acute liver damage in rats. J Ethnopharmacol 2007; 112:145–51. doi:10.1016/j.jep.2007.02.013.

[27] Ahmed AF, Al-Yousef HM, Al-Qahtani JH, Al-Said MS, Ashour AE, Al-Sohaibani M, et al. Hepatorenal protective effect of Antistax(®) against chemically-induced toxicity. Pharmacogn Mag 2015; 11:S173–81. doi:10.4103/0973-1296.157726.

[28] Sangiovanni E, Di Lorenzo C, Colombo E, Colombo F, Fumagalli M, Frigerio G, et al. The effect of in vitro gastrointestinal digestion on the anti-inflammatory activity of *Vitis vinifera* L. leaves. Food Funct 2015; 6:2453–63. doi:10.1039/c5fo00410a.

[29] Pari L, Suresh A. Effect of grape (*Vitis vinifera* L.) leaf extract on alcohol induced oxidative stress in rats. Food Chem Toxicol Int J Publ Br Ind Biol Res Assoc 2008; 46:1627–34. doi:10.1016/j.fct.2008.01.003.

[30] Wohlenberg M, Almeida D, Bokowski L, Medeiros N, Agostini F, Funchal C, et al. Antioxidant activity of grapevine leaf extracts against oxidative stress induced by

carbon tetrachloride in cerebral cortex, hippocampus and cerebellum of rats. Antioxid Basel Switz 2014; 3:200–11. doi:10.3390/antiox3020200.

[31] Król A, Amarowicz R, Weidner S. The effects of cold stress on the phenolic compounds and antioxidant capacity of grapevine (*Vitis vinifera* L.) leaves. J Plant Physiol 2015; 189:97–104. doi:10.1016/j.jplph.2015.10.002.

[32] Kosar M, Küpeli E, Malyer H, Uylaser V, Türkben C, Baser KHC. Effect of brining on biological activity of leaves of *Vitis vinifera* L. (Cv. Sultani Cekirdeksiz) from Turkey. J Agric Food Chem 2007; 55:4596–603. doi:10.1021/jf070130s.

[33] Jeandet P, Bessis R, Gautheron B. The production of resveratrol (3,5,4′-trihydroxystilbene) by grape berries in different developmental stages. Am J Enol Vitic 1991; 42:41–6.

[34] Soyer Y, Koca N, Karadeniz F. Organic acid profile of Turkish white grapes and grape juices. J Food Comp Anal 2003; 16(5):629–36.

[35] Herrmann K. Flavonols and flavones in food plants: a review. Int J Food Sci Technol 1976; 11:433–48. doi:10.1111/j.1365-2621.1976.tb00743.x.

[36] Li S, Tan H-Y, Wang N, Zhang Z-J, Lao L, Wong C-W, et al. The role of oxidative stress and antioxidants in liver diseases. Int J Mol Sci 2015; 16:26087–124. doi:10.3390/ijms161125942.

[37] Takayama F, Nakamoto K, Kawasaki H, Mankura M, Egashira T, Ueki K, et al. Beneficial effects of *Vitis coignetiae* Pulliat leaves on nonalcoholic steatohepatitis in a rat model. Acta Med Okayama 2009; 63:105–11.

[38] Lacerda DdS, Santos CF, Oliveira AS, Zimmermann R, Schneider R, Agostini F, et al. Antioxidant and hepatoprotective effects of an organic grapevine leaf (*Vitis labrusca* L.) extract in diabetic rats. RSC Adv 2014; 4:52611–9. doi:10.1039/C4RA08396B.

[39] Oliboni LS, Dani C, Funchal C, Henriques JA, Salvador M. Hepatoprotective, cardioprotective, and renal-protective effects of organic and conventional grapevine leaf extracts (*Vitis labrusca* var. Bordo) on Wistar rat tissues. An Acad Bras Ciênc 2011; 83:1403–11. doi:10.1590/S0001-37652011000400027.

[40] Pak W, Takayama F, Hasegawa A, Mankura M, Egashira T, Ueki K, et al. Water extract of *Vitis coignetiae* Pulliat leaves attenuates oxidative stress and inflammation in progressive NASH rats. Acta Med Okayama 2012; 66:317–27.

[41] Dey A, Lakshmanan J. The role of antioxidants and other agents in alleviating hyperglycemia mediated oxidative stress and injury in liver. Food Funct 2013; 4:1148–84. doi: 10.1039/c3fo30317a.

[42] Marco Gasparetto and Graziella Guariso, "Highlights in IBD Epidemiology and Its Natural History in the Paediatric Age," Gastroenterology Research and Practice, vol. 2013, Article ID 829040, 12 pages, 2013. doi:10.1155/2013/829040.

[43] Lawrence T. The nuclear factor NF-kappaB pathway in inflammation. Cold Spring Harb Perspect Biol 2009; 1:a001651. doi:10.1101/cshperspect.a001651.

[44] Wong ND. Epidemiological studies of CHD and the evolution of preventive cardiology. Nat Rev Cardiol 2014; 11:276–89. doi:10.1038/nrcardio.2014.26.

[45] Laslett LJ, Alagona P, Clark BA, Drozda JP, Saldivar F, Wilson SR, et al. The worldwide environment of cardiovascular disease: prevalence, diagnosis, therapy, and policy issues: a report from the American College of Cardiology. J Am Coll Cardiol 2012; 60:S1–49. doi:10.1016/j.jacc.2012.11.002.

[46] Widmer RJ, Lerman A. Endothelial dysfunction and cardiovascular disease. Glob Cardiol Sci Pract 2014; 2014:291–308. doi:10.5339/gcsp.2014.43.

[47] Hansen AS, Marckmann P, Dragsted LO, Finné Nielsen I-L, Nielsen SE, Grønbaek M. Effect of red wine and red grape extract on blood lipids, haemostatic factors, and other risk factors for cardiovascular disease. Eur J Clin Nutr 2005; 59:449–55. doi:10.1038/sj.ejcn.1602107.

[48] Lippi G, Franchini M, Favaloro EJ, Targher G. Moderate red wine consumption and cardiovascular disease risk: beyond the "French paradox". Semin Thromb Hemost 2010; 36:59–70. doi:10.1055/s-0030-1248725.

[49] Grønbaek M, Deis A, Sørensen TI, Becker U, Schnohr P, Jensen G. Mortality associated with moderate intakes of wine, beer, or spirits. BMJ 1995; 10:1165–9.

[50] Orhan N, Aslan M, Orhan DD, Ergun F, Yesilada E. In-vivo assessment of antidiabetic and antioxidant activities of grapevine leaves (*Vitis vinifera*) in diabetic rats. J Ethnopharmacol 2006; 108:280–6. doi:10.1016/j.jep.2006.05.010.

[51] Akhlaghi M, Bandy B. Mechanisms of flavonoid protection against myocardial ischemia-reperfusion injury. J Mol Cell Cardiol 2009; 46:309–17. doi:10.1016/j.yjmcc.2008.12.003.

[52] Mladenka P, Zatloukalová L, Filipský T, Hrdina R. Cardiovascular effects of flavonoids are not caused only by direct antioxidant activity. Free Radic Biol Med 2010; 49:963–75. doi:10.1016/j.freeradbiomed.2010.06.010.

[53] Lila MA. Anthocyanins and human health: an in vitro investigative approach. J Biomed Biotechnol 2004; 2004(5):306–13. doi:10.1155/S111072430440401X.

[54] Dell'Agli M, Galli GV, Vrhovsek U, Mattivi F, Bosisio E. In vitro inhibition of human cGMP-specific phosphodiesterase-5 by polyphenols from red grapes. J Agric Food Chem 2005; 53:1960–5. doi:10.1021/jf048497+.

[55] Kim HP, Son KH, Chang HW, Kang SS. Anti-inflammatory plant flavonoids and cellular action mechanisms. J Pharmacol Sci 2004; 96:229–45.

[56] Castilla P, Echarri R, Dávalos A, Cerrato F, Ortega H, Teruel JL, et al. Concentrated red grape juice exerts antioxidant, hypolipidemic, and antiinflammatory effects in both hemodialysis patients and healthy subjects. Am J Clin Nutr 2006; 84:252–62.

[57] Kalus U, Koscielny J, Grigorov A, Schaefer E, Peil H, Kiesewetter H. Improvement of cutaneous microcirculation and oxygen supply in patients with chronic venous insufficiency by orally administered extract of red vine leaves AS 195: a randomised, double-blind, placebo-controlled, crossover study. Drugs RD 2004; 5:63–71.

[58] Kiesewetter H, Koscielny J, Kalus U, Vix JM, Peil H, Petrini O, et al. Efficacy of orally administered extract of red vine leaf AS 195 (folia *Vitis viniferae*) in chronic venous insufficiency (stages I-II). A randomized, double-blind, placebo-controlled trial. Arzneimittelforschung 2000; 50:109–17. doi:10.1055/s-0031-1300174.

[59] Tsukanov IT, Tsukanov AI. Experience with six-month administration of red vine leaf extract in hormone-induced phlebopathy. Angiol Sosud Khirurgiia Angiol Vasc Surg 2014; 20:102–7.

[60] Grau M, Bölck B, Bizjak DA, Stabenow CJA, Bloch W. The red-vine-leaf extract AS195 increases nitric oxide synthase-dependent nitric oxide generation and decreases oxidative stress in endothelial and red blood cells. Pharmacol Res Perspect 2016; 4:e00213. doi:10.1002/prp2.213.

[61] Small DM, Coombes JS, Bennett N, Johnson DW, Gobe GC. Oxidative stress, anti-oxidant therapies and chronic kidney disease. Nephrol Carlton Vic 2012; 17:311–21. doi: 10.1111/j.1440-1797.2012.01572.x.

[62] Ha H, Hwang I-A, Park JH, Lee HB. Role of reactive oxygen species in the pathogenesis of diabetic nephropathy. Diabetes Res Clin Pract 2008; 82(Suppl. 1):S42–5. doi:10.1016/j.diabres.2008.09.017.

[63] Kashihara N, Haruna Y, Kondeti VK, Kanwar YS. Oxidative stress in diabetic nephropathy. Curr Med Chem 2010; 17:4256–69.

[64] Goupy P, Dufour C, Loonis M, Dangles O. Quantitative kinetic analysis of hydrogen transfer reactions from dietary polyphenols to the DPPH radical. J Agric Food Chem 2003; 51:615–22. doi:10.1021/jf025938l.

[65] Schmatz R, Perreira LB, Stefanello N, Mazzanti C, Spanevello R, Gutierres J, et al. Effects of resveratrol on biomarkers of oxidative stress and on the activity of delta aminolevulinic acid dehydratase in liver and kidney of

streptozotocin-induced diabetic rats. Biochimie 2012; 94:374–83. doi:10.1016/j.biochi.2011.08.005.

[66] Moskaug JØ, Carlsen H, Myhrstad MCW, Blomhoff R. Polyphenols and glutathione synthesis regulation. Am J Clin Nutr 2005; 81:277S–283S.

[67] Lee W-C, Wang C-J, Chen Y-H, Hsu J-D, Cheng S-Y, Chen H-C, et al. Polyphenol extracts from Hibiscus sabdariffa Linnaeus attenuate nephropathy in experimental type 1 diabetes. J Agric Food Chem 2009; 57:2206–10. doi:10.1021/jf802993s.

[68] Shastry CS, Ashok K, Aravind MB, Joshl SD. Diuretic activity of the extracts of *Vitis vinifera* leaves. Indian Drugs 2002; 39:497–9.

[69] Grases F, Prieto RM, Fernandez-Cabot RA, Costa-Bauzá A, Tur F, Torres JJ. Effects of polyphenols from grape seeds on renal lithiasis. Oxid Med Cell Longev 2015; 2015:813737. doi:10.1155/2015/813737.

[70] Sarrafchi A, Bahmani M, Shirzad H, Rafieian-Kopaei M. Oxidative stress and Parkinson's disease: new hopes in treatment with herbal antioxidants. Curr Pharm Des 2015; 22:238–46.

[71] Chen Z, Zhong C. Oxidative stress in Alzheimer's disease. Neurosci Bull 2014; 30:271–81. doi:10.1007/s12264-013-1423-y.

[72] Chen G, Shi J, Hu Z, Hang C. Inhibitory effect on cerebral inflammatory response following traumatic brain injury in rats: a potential neuroprotective mechanism of N-acetylcysteine. Mediators Inflamm 2008; 2008:716458. doi: 10.1155/2008/716458.

[73] Ahmed T, Javed S, Javed S, Tariq A, Šamec D, Tejada S, et al. Resveratrol and Alzheimer's disease: mechanistic insights. Mol Neurobiol. 2016. [Epub ahead of print] doi:10.1007/s12035-016-9839-9.

[74] Wang G, Chen L, Pan X, Chen J, Wang L, Wang W, et al. The effect of resveratrol on beta amyloid-induced memory impairment involves inhibition of phosphodiesterase-4 related signaling. Oncotarget 2016; 5. doi:10.18632/oncotarget.8041.

[75] Zhao Y-N, Li W-F, Li F, Zhang Z, Dai Y-D, Xu A-L, et al. Resveratrol improves learning and memory in normally aged mice through microRNA-CREB pathway. Biochem Biophys Res Commun 2013; 435:597–602. doi:10.1016/j.bbrc.2013.05.025.

[76] Venturini CD, Merlo S, Souto AA, Fernandes MDC, Gomez R, Rhoden CR. Resveratrol and red wine function as antioxidants in the nervous system without cellular proliferative effects during experimental diabetes. Oxid Med Cell Longev 2010; 3:434–41.

[77] Cho J, Kang JS, Long PH, Jing J, Back Y, Chung K-S. Antioxidant and memory enhancing effects of purple sweet potato anthocyanin and cordyceps mushroom extract. Arch Pharm Res 2003; 26:821–5.

[78] Kim JY, Jeong HY, Lee HK, Kim S, Hwang BY, Bae K, et al. Neuroprotection of the leaf and stem of *Vitis amurensis* and their active compounds against ischemic brain damage in rats and excitotoxicity in cultured neurons. Phytomedicine Int J Phytother Phytopharm 2012; 19:150–9. doi:10.1016/j.phymed.2011.06.015.

[79] Narita K, Hisamoto M, Okuda T, Takeda S. Differential neuroprotective activity of two different grape seed extracts. PLoS One 2011; 6(1). doi:10.1371/journal.pone.0014575.

[80] Hwang IK, Yoo K-Y, Kim DS, Jeong Y-K, Kim JD, Shin H-K, et al. Neuroprotective effects of grape seed extract on neuronal injury by inhibiting DNA damage in the gerbil hippocampus after transient forebrain ischemia. Life Sci 2004; 75:1989–2001. doi:10.1016/j.lfs.2004.05.013.

Chapter 7

Grape Drying: Current Status and Future Trends

Jun Wang, Arun S. Mujumdar, Weisong Mu,
Jianying Feng, Xiaoshuan Zhang, Qian Zhang,
Xiao-Ming Fang, Zhen-Jiang Gao and
Hong-Wei Xiao

Additional information is available at the end of the chapter

http://dx.doi.org/10.5772/64662

Abstract

With high moisture and sugar content, fresh grapes respire and transpire actively after harvest, which contribute to quality loss. Drying can process grapes into raisins for longer shelf-life as well as dehydrated grapes, which can be used for wines or juice production. The pre-treatments, drying method and drying conditions, can significantly influence the quality of final products. In this chapter, firstly, different pre-treatments as a necessary operation previous to the drying of grapes into raisins is introduced. These pre-treatments include chemical pre-treatment, physical pre-treatment, and blanching. In addition, the quality and drying characteristics of different pre-treatments is summarized too. Secondly, the current status of different technologies for grape drying and their effects on drying kinetics and quality attributes of seedless grapes are described to highlight the advantages and disadvantages of each drying method. These drying methods include the traditional open sun drying, shade drying, hot-air drying, freezing drying, microwave drying, as well as the vacuum impulsed drying. Thirdly, influences of drying on bioactive substances (flavonoids, phenolics, anthocyanin, and resveratrol) and antioxidant capacity of grape by-products including seed, skin, stem, and stalk are also examined. Finally, the future research trends of grape and its by-product drying are indentified and discussed.

Keywords: grape drying, pre-trements, drying methods, quality attributes, by-products drying

1. Introduction

Grape is one of the most popular and largest fruit corps and is cultivated in more than 100 countries around the world. Grape production all over the world was about 7.7×10^9 tons according to Food and Agriculture Organization (FAO) data for 2013 [1]. The top five grape production countries are China (about 1.16×10^9 t), Italy (about 8.01×10^8 t), United States of America (about 7.74×10^8 t), Spain (about 7.48×10^8 t), and France (about 5.52×10^8 t).

As one of the most popular fruits, grape can be consumed directly or processed into various products, such as raisin, grape juice, and wine, as illustrated in **Figure 1**. Fresh grapes with relatively high moisture and sugar contents respire and transpire actively after harvest and are very sensitive to microbial spoilage during storage, even at refrigerated conditions [2, 3]. As one of the most frequently used methods for food and bioproducts preservation, drying can remove moisture content to a very low content and drastically reduce microbial, enzymatic degradation or any moisture-mediated deteriorative reactions [4–7]. In addition, drying can bring some benefits such as substantial reduction in weight and volume, minimizing packing, storage, and transportation costs [8–12]. Drying is one of the most frequently used methods for grape processing. It can process grapes into raisins for longer shelf-life as well as dehydrated grapes, which can be used for wines or juice production. Such as many world-renowned wines, e.g. Passito wines, Sauternes, Tokaj, Porto, Pedro Ximénez and Amarone are produced using dehydrated grapes [13, 14]. Additionally, the main by-products during juice and wine production are grape seed, skin, stem and stalks, which are usually treated as waste [15, 16]. Recently, how to improve the utilization value of grape by-products become more and more popular as they are good source of phytochemicals including flavonoids, phenolics, anthocyanin, and resveratrol [17]. However, raw grape by-products also with high moisture contents are very sensitive to microbial spoilage and component degradation, dehydration become an essential processing prior to effective constituent extracting from them. Additionally, the drying methods and drying conditions also have great effect on effective extraction of the constituent.

For raisin processing, pre-treatments including chemical pre-treatment, physical pre-treatment and blanching have been investigated and applied to remove the wax layer on grape surface and enhance drying rate. Drying has a great effect on the quality of the grape raisins product, such as its texture and nutrients [18]. However, presently the natural sun drying and shade drying are still the most common drying methods performed in many countries for grape drying [19]. Although the investments and operation of natural sun drying is small and simplicity, it has several drawbacks. Such as long drying time usually taking more than two or three weeks, the rewetting products caused by bad weather, contamination by dust and insects, tedious and laborious to make the product more uniform, nutrients deterioration caused by long exposure to solar radiation. Therefore, application of a suitable drying technology and selection of appropriate drying conditions are therefore of great important in the production of good raisins products.

For grape by-products drying, different drying methods and drying conditions have great influences on their antioxidant capacity [17]. Therefore, preservation of the constituent and the bioactive contents throughout the drying process is necessary.

In this chapter, the background of different pre-treatment methods to enhance grape drying were outlined since a thin-layer wax covers on grape surface and forms the main resistance hindering moisture transfer during dehydration process [20, 21]. Then, different drying technologies for grape drying were presented, such as natural open sun drying, shade drying, solar drying, hot air drying, microwave drying, vacuum pulsed drying etc. After that, the drying of grape by-products and their influences on bioactive and antioxidant capacity were also discussed. Finally, the future research trends of grape and its by-product drying are also identified and discussed. It is hoped that the information provided in the current review would not only contribute to a better understanding of the research status of grape and its by-products drying, but also trigger new research opportunities to develop innovative drying technologies for grape drying.

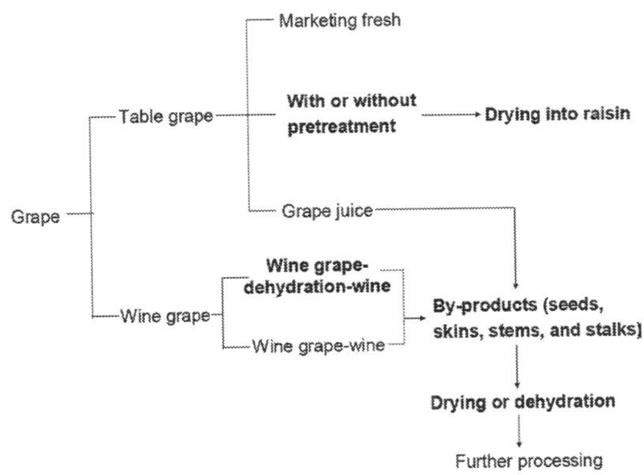

Figure 1. Flow diagram of grape processing.

2. Drying of grape into raisin

2.1. Pre-treatments of grapes pre-drying

Low moisture diffusion rate has become the basic problem during grape dehydration process. This can be attributed to the peculiar structure of a thin-layer of wax covered on grape surface which prevents the rate of moisture diffusion [21, 22]. The skin of the grape consists of an epidermis and six to ten layers of small thick-walled cells. The outer epidermis is covered by non-living layers, namely cuticle, lenticels, wax, and collenchymatous hypodermal cells [23]. Wax on grape skin serves as a protective barrier against fungal pathogens and protects the grape from UV light and physical injuries. However, the presence of waxes in the skin cuticle is an obstacle to drying. Therefore, it is necessary to remove the wax layer before drying [24]. Currently, various pre-treatments including chemical, physical, and blanching treatments have been carried out to remove the wax layer prior to the drying process. All the pre-treated results showed an increase in drying rate with reduction in drying time for the grapes to reach a safe moisture content required for storage. The different pre-treatments and main conclusions are summarized in **Table 1**.

The main constitutes of chemical pre-treatment usually contain two or three solution such as NaOH, K_2CO_3, $NaHCO_3$, olive oil, and ethyl oleate solution with a certain proportion. Chemical dipping pre-treatments could dissolve the grape skins and increase their permeability to water, by thus to improve the drying rate [31, 32]. The chemical dipping pre-treatment methods have been widely applied in commercial production [39]. However, there are some disadvantages of chemical pre-treatments, such as the residual chemical additives in the raisins, which are harmful for our health and may cause food safety problems; larger quantities of corrosive chemicals, which could pollute surroundings and their disposal is a high cost operation. As the food safety issues have attracted much more attention and the natural food consumption is becoming more and more popular, using of chemical additives in foods is being discouraged.

In order to avoid chemical residues during pre-treatment, some physical pre-treatments have been developed to remove the wax layer on grape surface. Di Matteo et al. [34] and Adiletta et al. [40] pre-treated the grape samples with some abrasion of the peel before drying. The results showed that the drying rate was significantly increased compared to untreated samples (**Table 1**). As the same treatment, Adiletta et al. [40] and Senadeera et al. [41] used a shaker with abrasive sheets created by Prof. Marisa Di Matteo, Department of Industrial Engineering, University of Salerno. The results also found that the pre-treatment affected the drying kinetics of grape samples, reduced drying times and rehydration time, and the surface structures of the pre-treated samples were detected by SEM. However, the physical pre-treated grape, which the final dried products occurred serious browning and the feasibility of this practice on larger scale has not been considered. Microwave-assisted pre-treatment [36] and ohmic pre-treatment [37] have also been explored. It was found both of them could enhance drying rate significantly.

Besides, pulsed electric fields (PEF) and ultrasounds are two other physical approaches to increase agricultural products drying rate by pre-treatment [42, 43]. Due to the advantages of short processing time, little heating of the medium, and low energy-consume, PEF is used for many material pre-treatments previous to drying and the drying rate was increased in various degrees [44]. For examples, compared to untreated samples, a 20, 34.7, and 12% drying rate increasing were obtained for PEF-pre-treated carrots [45], red pepper [46], and apple tissue [47], respectively. To get rid off the use of chemicals in raisin processing, Dev et al. [38] employed PEF pre-treatment to improve drying rate of grape, and a 20% decrease of drying time was obtained compared to untreated samples, and the highest drying rate was chemically treated samples (40% less). Ultrasound as one of pre-treatment methods also has been widely applied in extraction and prior to drying of grape and by-products [48].

Pre-treatment methods	Materials	Treatment	Control	Drying methods characteristics	Detecting indexes	Main conclusions	References
Chemical	Grapes (var. *sultanas*)	(1) 5% K_2CO_3 + 1.5% olive oil; (2) 4% K_2CO_3 + 2% ethyl oleate; (3) 5% K_2CO_3 + 2% ethyl oleate; (4) 6% K_2CO_3 + 2% ethyl oleate; (5) 7% K_2CO_3 + 2% ethyl oleate; All treatments with 20–25 s	Natural (untreated)	(1) Solar drying; (2) Sun drying on concrete ground; (3) Sun drying on wooden racks, or polypropylene canvas sheets	Drying rate, colour, and colour changes during storage	(1) Drying rate: solar drying> sun drying on concrete ground> sun drying on wooden racks, or polypropylene canvas sheets; (2) The drying rate increased with the increasing K_2CO_3 concentration from 4 to 7%; (3) The moisture content and colour intensities of the sun-dried grapes were found to be non-uniform	[25]
	Sultana (Thompson seedless) Initial moisture content 78% (w.b.)	2% $KHCO_3$ + 0.2% olive oil of 2 min	Natural (without treatment after collection from the farms)	Sun drying Temperature: 23–35°C Relative humidity: >72% Forced air drying Temperature: 60°C Air velocity: 0.5–1.5 m/s	Drying rate	(1) The drying time: sun drying took 179 h, forced air drying took 56 h (2)Different drying period with different temperature and air velocity would be reasonable	[26]
	Thompson seedless grapes Average diameter: 18± 1 mm Average brix: 23	D1: 0.5% NaOH solution of 5 s at 93°C±1.0°C; D2: 2.0% commercial dipping oil + 2.5% K_2CO_3 solution of 3 min at ambient temperature; D3: 2.0% ethyl oleate + 2.5% K_2CO_3 solution of 3 min at ambient temperature; D4: 0.4% olive oil + 7.0% K_2CO_3 solution of 3 min at temperature	D5: untreated grapes	The dryer was laboratory setup. Temperature: 60°C Air velocity: 0.5 m/s	Drying rate and organoleptic quality	(1) The drying time of different treatment of D1: 8 h, D2: 26 h, D3: 27 h, D4: 30 h, and D5: 46 h; (2) Hot dipping pre-treatment, though reduced the drying time, the quality of products found to be poor; (3) Page's model is accurate enough to predict the drying behaviour of pre-treated grapes.	[27]
	Sultana seedless grapes (*Vitis vinifera L.*) The initial moisture content: 77.3%-80.5% (w.b.)	*POTAS*: 0.5 kg K_2CO_3 + 10 L water + 0.05 kg olive oil; *AEEO*: ethyl oleate: 0.5 kg K_2CO_3 + 10 L water + 0.2 kg ethyl oleate Both *POTAS* and *AEEO* pre-treated 1 min at ambient temperature	NAT: untreated grapes	Cabinet drier: produced by APV&PASILAC firm (England) Temperature: 50, 55, 60, and 70°C; Air velocity: 1.2 m/s	Colour, drying rate	(1) Pre-treatment with the AEEO solution is effective in increasing the drying rate; (2) The use of ethyl oleate as pre-treatment solution for the drying of grapes leads to a better colour; (3) Exponential equations agree satisfactorily with the drying	[28]
	Grapes, initial moisture content of 80.20% (w.b.)	Immersed for 2 min in emulsion of 5% K_2CO_3 + 0.5% olive oil	Some other materials: apricots, peaches, figs, and plums	Open-air sun drying Temperature 31–43°C Solar radiation: 1.10–2.93 MJ/m^{2}h	Drying rate, mathematical modelling of drying curves, and uncertainty analysis	(1) Water removal from the selected fruits in the drying process occurs in the falling rate period (2) The drying time of grape samples over 5 days (7000 min) (3) Verma et al. model could adequately describe grape open-air sun drying behaviour ($MR = a\exp(-kt) + (1-a)\exp(-gt)$)	[29]

Pre-treatment methods	Materials	Treatment	Control	Drying methods characteristics	Detecting indexes	Main conclusions	References
	Seedless Sultanine grapes Initial moisture: 83.33–86.11% (w.b.)	1% NaOH solution at 90°C of 2–3 times for 2–3 s	(1) In the open air; (2) Under greenhouse; (3) In the drier.	Solar tunnel greenhouse drying Temperature: 18–60°C Relative humidity: 18–96% Solar radiation: 0–600 W/m^2	Drying rate	(1) The drying time of solar drier was 77 h (about 4 days), greenhouse was 119 h (about 5 days), open sun was 250 h (more than 11 days) (2) The solar greenhouse drying is advantageous in regard to solar dried, greenhouse have a big drying capacity and don't require a large initial investment or additional running cost.	[30]
	Black grapes (*var. Muscat*) Average radius, length and weight 1.83 cm, 2.78 cm and 5.85 g, respectively; The initial moisture content: 79.3% ± 0.2 (w/w)	POTAS: 5% K_2CO_3 + 0.5% olive oil; EO1: 2% ethyl oleate + 2.5% K_2CO_3; EO2: 2% ethyl oleate + 2.5% KOH; EO3: 2% ethyl oleate + 2.5% Na_2CO_3; All treatments immersed for 1 min	NAT: untreated samples	Cabinet dryer: installed in the Chemical Engineering Department of Yildiz Technical University, Istanbul, Turkey Temperature: 60°C Air velocity: 1.1 m/s	Drying rate	(1) EO1 (2% ethyl oleate + 2.5% K_2CO_3) obtained the shortest drying time (about 25 h) among all treated and untreated samples; (2) No constant-rate period was found of black grapes drying. (3) Page model showed a better fit to the experimental data; The effective moisture diffusivity: $3.82{\times}10^{-10}$~$1.28{\times}10^{-10}$ m^2/s	[31]
	Seedless grapes (*Vitis vinifera L.*) Dry matter 23.62%±1.38 Total sugar: 19.97% ±1.06	D1: 2% ethyl oleate + 5% K_2CO_3 solution of 60 s at ambient temperature; D2: 4% PAKSAN oil (contains free oleic acid and chiefly ethyl esters of fatty acids; C_{14}-C_{18}) + K_2CO_3 solution of 60 s at ambient temperature	Hot water (HW) of 15 s at 95°C	Laboratory -scale tray dryer Drying temperature: 40, 50, 60, 70°C; Humidity ranged from 10% to 15%; Air velocity: 1 m/s	Thermal diffusivity, moisture diffusivity, and heat and mass transfer coefficients	(1) Effective moisture diffusivity strongly depends on dipping as well as moisture content and the temperature of the product; (2) Thermal diffusivity of the grapes varies with the moisture content of the grapes (3) Pre-treatments affect thermal gradients in the early stage of the drying process, but had no significant effect on the thermal diffusivities.	[32]
	Thompson seedless grapes (*Vitis vinifera*) Average diameter: 17.5–18.5 mm The initial moisture content: 80.3–82.6% (w.b.)	5% (w/v) K_2CO_3 + 2% (v/v) ethyl oleate at 30, 40, 50, and 60°C for 1, 2, and 3 min	Not pre-treated with dipping solution	A tray dehydrator (Excalibur, Sacramento, CA) Temperature: 60°C Air velocity: 0.6 m/s	Effective diffusivity and colour	(1) Dipping time of 2- and 3-min played an important role at 30 and 40°C (2) Browning occurred at all dipping times and temperatures (3) Midilli model best described the drying kinetics of grapes pre-treated with dipping solutions (MR = $a\exp(-kt^n) + bt$)	[33]
Physical	Seedless white grapes (var. Nevado) The initial moisture content: 84.0% ± 1.6	The abrasion of the grape peel was carried out in a shaker the walls of which were covered by coating with abrasive sheets Shaker for 10 min (Abr) EtOl: 2% (v/v) ethyl oleate + 2.5% (v/v) K_2CO_3 at 40°C for 3 min	Untreated samples (UT)	Convection oven Temperature: 50°C Air speed: 0.5 m/s	Drying rate, colour parameters, and microstructure	(1) The physical treatment found to be as effective as the chemical dipping method, mass transport coefficient was about four times greater than untreated samples (drying time about 35 h) (2) Physical treated samples gives rise to a more coloured final product than the chemical one	[34]

Pre-treatment methods	Materials	Treatment	Control	Drying methods characteristics	Detecting indexes	Main conclusions	References
	Red grapes (*Vitis Vinifera*) Initial moisture content: 6.43±0.02 kg/kg (d.b.) Average diameter: 24.4±1.95 mm	Abraded grape (TR-Abr): the abrasion of the grape peel was carried out in a motorized rotating drum (*D*=240 mm, *L*=250 mm) made of plexiglass, lined inside with sandpaper. Rotation speed of drum: 10 rpm; Pre-treatment time: 15 min; Mass of grapes: 4 kg.	Untreated grape (UTR) Chemical solution (TR-EtOl): 2% (v/v) ethyl oleate + 2.5% (v/v) Na_2CO_3 at 40°C for 3 min	Convective dryer (*Zanussi FCV/E6L3*) Temperature: 40, 50, 60, and 70°C Air velocity: 2.3 m/s	Drying rate, colour, total phenolic content, antioxidant activity, shrinkage, microstructure, rehydration.	(1) The highest drying rate was found for abraded grapes at 50 and 60°C, about 1/3 drying time of untreated grapes; (2)The colour of abraded grapes was darkest; (3) Based on total phenolic content, the best drying temperature was 50°C for both untreated and pre-treated samples (4) Abraded grape drying: The logarithmic model was the best fitting for all temperatures except at 70°C; Page model gave the highest correlation factor; Quadratic model showed an acceptable fit to experimental data for all the samples and temperatures investigated.	[35]
	Sultana seedless grape (with length 15–18 mm and diameter 12–14 mm). Average weight was 1.28 g.	Microwave pre-treatment: fresh, dipped (2.5% K_2CO_3 + 0.5% olive oil for 1 min) or blanched (boiling water for 0.5 min) for 0.5–2 min at 215 W, 325 W or 420W.	Untreated samples	Sun-drying: average daylight temperature was 22°C.	Colour and water activity	(1) Microwave pre-treated grapes dried nearly two times faster than the control; (2) The same drying rate be found of blanching and microwaves; (3) Colour and appearance of treated grapes were comparable to commercial products.	[36]
	Seedless red grape, the size and weight of each of the individual grape berries were relatively uniform to minimize their effects.	The treated bulk samples were ohmically heated in a solution containing 2% citric acid to a final medium temperature of 60°C using a field strength of 15 V/cm. The ohmic pre-treatment was conducted at 30 Hz, 60 Hz, and 7.5 kHz.	Untreated samples	Food dehydrator (Excalibur Products, Sacramento, CA), drying temperature was maintained 57°C.	Drying rate & adsorption isotherm	(1) Ohmically pre-treatment increased grape drying rate significantly; (2) The highest extent of the drying rate increased at the 30 Hz frequency of alternating; (3) Ohmic pre-treatment caused a shift in the sorption isotherm.	[37]
	Grapes (raisins variety)	pulsed electric fields (PEF), chemicals, microwave	No treatment	Convective drier at 65°C.	Colour (L, a, and b), total soluble solids (TSS), bulk density, appearance and market quality.	(1) Chemical treated grape obtained the highest drying rate; (2) PEF and microwave-treated samples had a significantly high TSS, appearance and market quality.	[38]
Blanching	Thompson seedless grapes Average length, width, and weight are 18.4 mm, 12.3 mm, and 3.34 g, respectively. The initial moisture content: 3.95 kg/kg (d.b.)	High-humidity hot air impingement blanching (HHAIB) Blanching time: 30, 60, 90, 120 s; Blanching temperature: 90, 100, 110, and 120°C Relative humidity: 40–45%	Fresh grapes	Air impingement dryer was installed in the College of Engineering of China Agricultural University, Beijing, China. Drying temperature: 55, 60, 65, 70°C.	Drying rate, polyphenol oxidase (PPO) activity, moisture diffusivity, and colour	(1) The PPO residual activity decreased with the increase of blanching time and temperature; (2) Fick's second law used to describe the drying kinetics of samples; (3) The colour analysis of the dried grape products showed that the Thompson seedless grapes pre-treated by HHAIB result in desirable green-yellow to green raisins.	[21]

Table 1. Comparison of different pre-treatments for grape drying.

As an essential step before processing of agricultural products blanching has been widely applied to inactivate enzymes, preserve colour, improve drying rate, or even to soften tissue, etc. Hot water blanching is the most popular and commercially used blanching method due to the advantages of low-cost, simplicity and convenient, and the small capital investments. However, there are several disadvantages of hot water blanching, including excessive loss of nutritional substances and how to deal with the hot water after blanching which contented large quantity nutrients [49]. Based on the disadvantages of hot water blanching, Bai et al. [21] used high-humidity hot air impingement blanching (HHAIB) pre-treatment for seedless grapes drying, which combines the advantages of steam blanching and impingement technologies, and they found that the drying rate in this case remarkably enhanced and the dried grape products obtained desirable green-yellow to green raisins. Xiao et al. [49] reviewed the application of superheated steam impingement blanching (SSIB) in agricultural products processing especially the fruits with a thin-layer of wax on their surface.

2.2. Different drying methods and their effects on grape drying

Grape drying is one of the most important methods to prolong its shelf-life and reduce economic losses. Therefore, how to improve the drying rate as well as obtain desirable products are the main objectives of grape drying. A larger number of studies focused on different drying methods and quality change kinetics during drying process. Currently, there are three frequently used drying methods for grape drying: natural sun drying or solar drying, shade drying, and mechanical drying.

2.2.1. Natural sun drying

Natural drying of grapes includes the open sun drying (with or without cover) and shade drying [19]. As a traditional method (**Figures 2** and **3**), natural drying of grape can be dated to 1490 BC in Greece and even today it is still widely applied, especially in developing countries due to its low initial and running costs [22, 24].

As the oldest drying method, natural open sun-drying is widely used method for thousands of years by human beings even nowadays. When open sun drying is performed the grapes are spread over the grape bunches either the ground or on a platform in a thin layer directly exposed to the sun or on a plastic sheet. During sun drying process, part of the solar radiation may penetrate the material and be absorbed within the grape itself, thus generating heat in the interior of the material as well as at its surface, therefore, increasing the heat transfer and enhancing moisture evaporation. This method is cheapest and is successfully employed in grapes producing countries [25]. Practically, no capital outlay for equipment is required, although considerable labour may be involved, which is seldom costly. However, the drying time is nearly 8–10 days, even much longer if the weather is sunny-less. Insect attacks, dust and potential rain resulting in a risk of grape deterioration. In addition, direct exposure to intense sun radiation and various temperature ranges would cause colour, appearance, and aroma deterioration and difference [19, 50]. The sensory quality of dehydrated grape especially colour and aroma is closely related to its' acceptability and wine-making. Ruiz et al. [50] found

http://dx.doi.org/10.5772/64662

that different temperatures have a significant effect on aroma profile of musts from the dried grapes, a less loss of raisiny aroma for a lower temperature was found.

Figure 2. The open sun drying of grape into raisin.

Figure 3. Shade drying of grape into raisin and the structure of shade-room.

Above all, a series of disadvantages limited the application of natural sun drying, such as lack of ability to control the drying operation properly, the length of the drying time, weather uncertainties, high labour costs, large area requirement, insect infestation, mixing with dust and other foreign materials and so on [29].

2.2.2. Solar drying

With rich solar energy radiation and available free of cost in many countries, solar energy has been widely used for heat production or power generation. Solar drying is the most commonly used for drying agricultural products. There are several types for grape solar drying, such as direct type [51], indirect type [52], and mixed type [53, 54]. For each type the solar energy is used as either the sole source of the required or as a supplement source. For grape drying, direct solar radiation causing poor quality formed due to light-sensitive of ascorbic acid and polyphenol, especially undesirable discolouration and aroma loss. Therefore, the indirect and mixed type solar dryer are more suitable for raisin [55].

2.2.3. Shade drying

Shade drying is also a kind of natural method and extensively used for grape drying in China (**Figure 3**), Australia, and India. Shade drying is also known as natural rack dryer, the ambient air is the principal source of heat required for drying [19]. Raisin of shade drying obtained better colour than sun drying, avoid the directly contact with sundries. However, there are some disadvantages of shade drying, such as long drying time, high labour require, and poor sanitary conditions.

2.2.4. Mechanical drying

With the rapid development of mechanization in agricultural production, mechanical drying has been widely used in raisin production due to its rapid, controllable, low labour, and high quality of products. Using solar energy as the heating generator, combined with some thermal-energy supplied dryer have been widely developed. Besides, microwave drying [56–58], vacuum pulsed drying, as well as combination of different drying methods also used for grape drying [59]. Heat pump dryer is also been developed due to its' improved efficiency, accurate control of drying conditions, wide range of drying conditions, better product quality, and increased throughput [60–62]. However, there are some limitations for the use of heat pump dryer, such as high maintenance cost, refrigerant leak causing environment pollution, and the initial capital cost [63, 64].

3. Drying of grape by-products from wine production and effects of their quality

Grape by-products from wine and juice production include grape seeds, skins, stems and stalks. Many researches have demonstrated that those sub-products are source of phenolic compounds, flavonoids, and anthocyanin pigments, which are natural antioxidants and of interest for food, cosmetic and pharmaceutical industries [65–67]. Wet grape residues with an approximate moisture content of 70% (wet basis) are generated as final residues, which are very sensitive to microbial spoilage and degradation of its effective components [68]. Traditionally, grape by-products are mainly used to obtain rectified alcohol, livestock feed production, and usually they are regarded as fertilizer, and even as waste into the environment [69]. Furthermore, the grape seed can be used extraction of oil, which is an alternative option for industrial application. For all grape residues from wine-making, about 15% is seed, and the extraction of oil from grape seed would be an excellent case [70]. However, such process is quite limited nowadays, and grape seed oil is only available at specialised dietetic shops. More and more researchers focus on the high valuable functional components extraction and their contribution for human beings [71–76].

Drying is a necessary step before antioxidants extraction, which may affect not only drying kinetics and energy efficiency but also product quality. However, drying could provoke a change in the physical, chemical and biological properties of the treated biomaterials [77]. The phenolic content degradation has been linked to the drying temperature-time combination [48, 78]. Different drying methods have been studied to obtain high effective ingredients reservations, and their influences and main results are summarized in **Table 2**.

Drying methods	Type of by-products	Drying condition	Main results	References
Freeze-drying & oven-drying	Skin from Carmenere and Cabernet Sauvignon, respectively.	Freeze-drying: samples were frozen at −78°C for 12 h and then freeze-dried in a vacuum ($2.4{\times}10^{-2}$ mB) for 24 h; Oven-drying: 60°C for 24 h.	(1) many volatile compounds decreased significantly with the oven-drying method, in contrast to the freeze-drying method; [2] Both phenolic compounds, anthocyanins and flavonols, were identified in fresh and dehydrated samples, thus resulting in the freeze-drying method being less aggressive than oven-drying methods.	[79]
Air-circulating oven	Red grape pomace (*Vitis vinifera* var. Cencidel)	Flow rate of 2.3 m^3/min; Temperature of 60, 100, and 140°C	(1) The total extractable polyphenols, condensed tannins, and antioxidant activity decreased significantly of 18.6.	[65]
Convective hot air drying	Seeds of Riesling, Concord, and Cab Franc.	Temperatures: 40, 50, 60°C; Air velocity: 1.5 m/s.	(1) Effective moisture diffusivity: Riesling seeds of 1.57–3.96 × 10^{-10} m^2/s, Concord seeds of 2.93–5.91×10^{-10} m^2/s, and Cab Franc seeds of 3.89–8.03× 10^{-10} m^2/s; (2) The activation energies of Riesling seeds was 40.14 kJ/mol, Concord seeds was	[80]

Drying methods	Type of by-products	Drying condition	Main results	References
			30.45 kJ/mol, and Cab Franc seeds was 31.47 kJ/mol; (3) Lewis model was shown to be an excellent model for predicting all three grape seed varieties.	
Convective drying with air-borne	Skins	Temperature: 40, 50, 60, and 70°C with (21.7 kHz, 45 W) and without power ultrasound application.	(1) Drying kinetics, total phenolic content and antioxidant capacity are influenced by both temperature and ultrasound; (2) Ultrasound application reduced the antioxidant potential, and increased as a consequence activation temperature drying.	[78]
Freeze-drying & oven drying	Muscat skin	Freeze-drying: -49±2°C under vacuum (2.4×10^{-2} mB) for 24 h; Oven drying: 30 and 45°C	(1) Freeze-drying is a good technique to preserve characteristic volatiles loss and phenolic compounds decrease of grape skins; (2) Freeze-dried grape skin could apply to enhance the flavour of white wines and other fields.	[79]
Hot air drying	Grape seed	Temperature: 40, 50, 60, 70°C; Velocity: 1.0, 1.5, 2.0, 3.0 m/s; With or without ultrasound application.	(1) Peleg's model could well describe grape seed drying; (2) Air velocity no significant influence on the dehydration process according to experimental result; (3) Ultrasound application had no influence on the dehydration kinetics of grape seeds.	[48]
Freeze-drying & hot-air drying	Grape stalk (*Vitis vinifera* var Bobal)	Hot-air drying: temperatures of 40, 55, 70, 85, 100 and 115°C; Freeze-drying: initial temperature -48±2 C, pressure 10^{-3} mbar	(1) The drying method have an significant effect on antioxidant of grape stalk, and hot air drying has a lower antioxidant and a slower extraction process; (2) The minimum antioxidant diffusivity and concentration was found for grape stalks dried at temperature ranging between 60 and 80°C; (3) Reduction of the mass transfer coefficient in hot air samples suggested the formation of a crust or shell during drying caused by the higher drying rate.	[81]

Drying methods	Type of by-products	Drying condition	Main results	References
Convective drying + ultrasound	Grape stalk from *Vitis vinifera* var. Bobal	Temperature of 40 and 60°C with or without ultrasound (45 and 90 W), velocity of 1 m/s.	(1) Ultrasound power enhance the diffusion and heat transfer coefficient during grape stalk drying; (2) The use of ultrasound increased the energy efficiency during the drying of grape stalk.	[82]
Infrared drying	Wet grape residues	Temperature: 100, 120, 140, 160°C	(1) Midilli model can well decrease the change of moisture ratio with drying time in the temperature range from 100 to 160°C; (2) The values of effective diffusivity and activation energy for moisture diffusion were determined.	[69]
Infrared, Convective, and Sequential infrared + convective	Wine grape pomace	Convective Drying (CD): 60, 70, 80, and 90°C; Infrared Drying (IR): the distance from the infrared emitter to the pomace was about 20 cm, far infrared range of 12,250 W; Sequential infrared and convective drying (SIRCD): IR7 min-CD, IR14 min-CD, IR21 min-CD, IR28 min-CD	(1) IR drying had the highest drying rate, which reduced the drying time by more than 47.3% compared with other methods; (2) SIRCD had a faster drying rate than CD; (3) Midilli et al. model had the highest R^2 and lowest *RMSE* and χ^2 for experimental data.	[83]

Table 2. Effects of different drying methods on grape by-products.

4. Summary and future research opportunities

1. For raisin production, pre-treatment is an important step to enhance drying rate. Chemical dipping pre-treatment is the most commonly used method in practical production. However, chemical residues in products has become a serious problem as the residual chemicals are bad for human being's health and can trigger food safety problems. While, different pre-treatments have a quite influence on grape quality, especially colour, bioactive component, and texture. Therefore, novel pre-treatment method should be developed to improve the permeability of the grape skin without damaging the product attributes. Microscopic analysis as the tool of evaluating pre-treatments should been taken into consideration in the future research.

2. Different varieties have different requirements for quality of raisin, proper drying method, drying condition, and processing should be classified. High quality products are the target of grape drying, therefore, except texture, aroma, colour, and rehydration, the change of bioactive components should accounted during drying processing. To explore the mechanism of quality changes, the form of moisture in grape and diffusion mechanism should be studied.

3. Grape by-products during juice or wine-making, has attracted more and more attention because of their rich content of bioactive component and high natural antioxidant capacity. Dehydration is the necessary processing prior to further operation, such as extract of phenolic compounds, flavonoids, and anthocyanin pigments. Temperature is the key influence parameter for maintaining bioactive components. So, lyophilisation has been widely studied and indicated ideal drying conditions. However, large-scale commercial production and high cost of lyophilisation should be considered.

Acknowledgements

This research is supported by joint project of China Agricultural University and Xinjiang Agricultural University Program, the Chinese Agricultural Research System (CARS-30), the National Natural Science Foundation of China (nos. 31360399, 31501548), the Project in the National Science & Technology Pillar Program during the Twelfth Five-year Plan Period (2015BAD19B010201), and the Chinese Transformation Fund of Agricultural Scientific and Technological Achievements (no. 2014GB2G410112).

Author details

Jun Wang[1], Arun S. Mujumdar[2], Weisong Mu[3], Jianying Feng[3], Xiaoshuan Zhang[1], Qian Zhang[4], Xiao-Ming Fang[5], Zhen-Jiang Gao[1] and Hong-Wei Xiao[1*]

*Address all correspondence to: xhwcaugxy@163.com

1 College of Engineering, China Agricultural University, Beijing, China

2 Department of Bioresource Engineering, McGill University, Quebec, Canada

3 College of Information and Electrical Engineering, China Agricultural University, Beijing, China

4 College of Mechanical and Electrical Engineering, Shihezi University, Shihezi, China

5 Institute of Bee Research, Chinese Academy of Agricultural Sciences, Beijing, China

References

[1] FAO. (2013). FaoStat database. Available from http://faostat.fao.org.

[2] Xiao HW, Pang CL, Wang LH, Bai JW, Yang WX, Gao ZJ. (2010). Drying kinetics and quality of Monukka Seedless grapes dried in an air-impingement jet dryer. *Biosystems Engineering, 105*(2), 233–240.

[3] Xiao HW, Mujumdar AS. Chapter 12: Impingement drying: applications and future trends. In Drying Technologies for Foods: Fundamentals & Applications (ISBN978-938-14-5074-1) edited by Prabhat K. Nema, Barjiinder Pal Kaur, and Arun S. Mujumdar. Printed in December 2014 by New India Publishing Agency, New Delhi, India. (pp. 279–299).

[4] Xiao HW. (2015). Guest Editorial: Some mitigation strategies for climate change. *Drying Technology, 33*(14), 1679–1680.

[5] Mujumdar AS (Ed.). Handbook of Industrial Drying, Fourth Edition (ISBN: 978-1-4665-9665-8). Printed in 2014 by CRC Press, Boca Raton, USA.

[6] Xiao HW, Lin H, Yao XD, Du ZL, Lou Z, Gao ZJ. (2009). Effects of different pretreatments on drying kinetics and quality of sweet potato bars undergoing air impingement drying. *International Journal of Food Engineering, 5*(5), Article 5.

[7] Xiao HW, Law CL, Sun DW, Gao ZJ. (2014). Color change kinetics of American ginseng (*Panax quinquefolium*) slices during air impingement drying. *Drying Technology, 32*(4), 418–427.

[8] Xiao HW, Bai JW, Xie L, Sun DW, Gao ZJ. (2015). Thin-layer air impingement drying enhances drying rate of American ginseng (*Panax quinquefolium* L.) slices with quality attributes considered. *Food and Bioproducts Processing, 94*, 581–591.

[9] Dai JW, Rao JQ, Wang D, Xie L, Xiao HW, Liu YH, Gao Z.J. (2015). Process-based drying temperature and humidity integration control enhances drying kinetics of apricot halves. *Drying Technology, 33*(3), 365–376.

[10] Wang D, Dai JW, Ju HY, Xie L, Xiao HW, Liu YH, Gao ZJ. (2015). Drying kinetics of American ginseng slices in thin-layer air impingement dryer. *International Journal of Food Engineering, 11*(5), 701–711.

[11] Ju HY, El-Mashad H, Fang XM, Pan Z, Xiao HW, Liu YH, Gao ZJ. (2016). Drying characteristics and modeling of yam slices under different relative humidity conditions. *Drying Technology, 34*(3), 296–306.

[12] Gao ZJ, Lin H, Xiao HW. (2008). Air-impingement de-shelling of chestnuts (*C. mollisima*): process parameter optimization. *International Journal of Food Engineering, 4*(2), Article 14.

[13] Mencarelli F, Tonutti P. (2013). Sweet, reinforced and fortified wines: Grape biochemistry, technology and vinification, Wiley-Blackwell, A John Wiley & Sons, Ltd., publication.

[14] Corona O, Torchio F, Giacosa S, Segade SR, Planeta D, Gerbi V, Squadrito M, Mencarelli F, Rolle L. (2016). Assessment of postharvest dehydration kinetics and skin mechanical properties of "Muscat of Alexandria" grapes by response surface methodology. *Food and Bioprocess Technology, 9*(6), 1060–1069.

[15] Spanos GA, Wrolstad RE. (1990). Influence of processing and storage on the phenolic composition of Thompson seedless grape juice. *Journal of agricultural and food chemistry, 38*(7), 1565–1571.

[16] Yang J, Martinson TE, Liu RH. (2009). Phytochemical profiles and antioxidant activities of wine grapes. *Food Chemistry, 116*(1), 332–339.

[17] De Torres C, Díaz-Maroto MC, Hermosín-Gutiérrez I, Pérez-Coello MS. (2010). Effect of freeze-drying and oven-drying on volatiles and phenolics composition of grape skin. *Analytica Chimica Acta, 660*(1), 177–182.

[18] Yang W, Gao ZJ, Tan H, Yang Y, Chen Z, Xiao HW. (2009). Drying Monukka grapes with air-impingement jet technique and quality analysis. *Transactions of the Chinese Society of Agricultural Engineering, 25*(4), 237–242.

[19] Pangavhane DR, Sawhney RL. (2002). Review of research and development work on solar dryers for grape drying. *Energy Conversion and Management, 43*(1), 45–61.

[20] Xiao HW, Wang J, Bai JW, Gao ZJ. (2016). Chapter 5: Novel high-humidity hot air impingement blanching (HHAIB) in agricultural products processing. In Amit K. Jaiswal edited Food Processing Technologies: Impact on Product Attributes (ISBN978-1-482-25744-0). Printed in 2016, By CRC Press, Boca Raton, Fl., USA (pp. 79–91).

[21] Bai JW, Sun DW, Xiao HW, Mujumdar AS, Gao ZJ. (2013). Novel high-humidity hot air impingement blanching (HHAIB) pretreatment enhances drying kinetics and color attributes of seedless grapes. *Innovative Food Science & Emerging Technologies, 20*, 230–237.

[22] Jairaj KS, Singh SP, Srikant K. (2009). A review of solar dryers developed for grape drying. *Solar Energy, 83*(9), 1698–1712.

[23] Kök D, Çelik S. (2004). Determination of characteristics of grape berry skin in some table grape cultivars (*V. vinifera* L.). *Journal of Agronomy, 3*(2), 141–146.

[24] Esmaiili M, Sotudeh-Gharebagh R, Cronin K, Mousavi MAE, Rezazadeh G. (2007). Grape drying: a review. *Food Reviews International, 23*(3), 257–280.

[25] Mahmutoğlu T, Emír F, Saygi YB. (1996). Sun/solar drying of differently treated grapes and storage stability of dried grapes. *Journal of Food Engineering, 29*(3), 289–300.

[26] Karathanos VT, Belessiotis V G. (1997). Sun and artificial air drying kinetics of some agricultural products. *Journal of Food Engineering, 31*(1), 35–46.

[27] Pangavhane DR, Sawhney RL, Sarsavadia PN. (1999). Effect of various dipping pretreatment on drying kinetics of Thompson seedless grapes. *Journal of Food Engineering, 39*(2), 211–216.

[28] Doymaz I, Pala M. (2002). The effects of dipping pretreatments on air-drying rates of the seedless grapes. *Journal of Food Engineering, 52*(4), 413–417.

[29] Toğrul İT, Pehlivan D. (2004). Modelling of thin layer drying kinetics of some fruits under open-air sun drying process. *Journal of Food Engineering, 65*(3), 413–425.

[30] Fadhel A, Kooli S, Farhat A, Bellghith A. (2005). Study of the solar drying of grapes by three different processes. *Desalination, 185*(1), 535–541.

[31] Doymaz I. (2006). Drying kinetics of black grapes treated with different solutions. *Journal of Food Engineering, 76*(2), 212–217.

[32] Esmaiili M, Sotudeh-Gharebagh R, Mousavi MA, Rezazadeh G. (2007). Influence of dipping on thin-layer drying characteristics of seedless grapes. *Biosystems engineering, 98*(4), 411–421.

[33] Bingol G, Roberts JS, Balaban MO, Devres YO. (2012). Effect of dipping temperature and dipping time on drying rate and color change of grapes. *Drying Technology, 30*(6), 597–606.

[34] Di Matteo M, Cinquanta L, Galiero G, Crescitelli S. (2000). Effect of a novel physical pretreatment process on the drying kinetics of seedless grapes. *Journal of Food Engineering, 46*(2), 83–89.

[35] Adiletta G, Russo P, Senadeera W, Di Matteo M. (2016). Drying characteristics and quality of grape under physical pretreatment. *Journal of Food Engineering, 172*, 9–18.

[36] Kostaropoulos AE, Saravacos GD. (1995). Microwave pre-treatment for sun-dried Raisins. *Journal of Food Science, 60*(2), 344–347.

[37] Salengke S, Sastry SK. (2005). Effect of ohmic pretreatment on the drying rate of grapes and adsorption isotherm of raisins. *Drying Technology, 23*(3), 551–564.

[38] Dev SRS, Padmini T, Adedeji A, Gariépy Y, Raghavan GSV. (2008). A comparative study on the effect of chemical, microwave, and pulsed electric pretreatments on convective drying and quality of raisins. *Drying Technology, 26*(10), 1238–1243.

[39] Zhao H, Fang XM, Wang J, Ju HY, Zhang Y, Chen X, Zhang Q, Liu Y, Gao ZJ, Xiao HW. (2016). Effect of ethanol dipping pretreatment on drying characteristics and quality of

eggplant slices. *Transactions of the Chinese Society of Agricultural Engineering*, 2016, *32*(9), 233–340.

[40] Adiletta G, Senadeera W, Liguori L, Crescitelli A, Albanese D, Russo P. (2015). The influence of abrasive pretreatment on hot air drying of grape. *Food and Nutrition Sciences*, *6*(3), 355–364.

[41] Senadeera W, Adilettta G, Di Matteo M, Russo P. (2014). Drying kinetics, quality changes and shrinkage of two grape varieties of Italy. *Applied Mechanics and Materials*, *553*, 362–366.

[42] Dai C, Zhou X, Zhang S, Zhou N. (2016). Influence of ultrasound assisted nucleation on freeze drying of carrots. *Drying Technology*, published online with doi: 10.1080/07373937.2015.1100203

[43] Tao Y, Sun DW. (2015). Enhancement of food processes by ultrasound: a review. *Critical Reviews in Food Science and Nutrition*, 55, 570–594.

[44] Nahimana H, Zhang M, Mujumdar AS, Ding Z. (2011). Mass transfer modeling and shrinkage consideration during osmotic dehydration of fruits and vegetables. *Food Reviews International*, *27*(4), 331–356.

[45] Amami E, Khezami L, Vorobiev E, Kechaou N. (2008). Effect of pulsed electric field and osmotic dehydration pretreatment on the convective drying of carrot tissue. *Drying technology*, *26*(2), 231–238.

[46] Won YC, Min SC, Lee DU. (2015). Accelerated drying and improved color properties of red pepper by pretreatment of pulsed electric fields. *Drying Technology*, *33*(8), 926–932.

[47] Wiktor A, Iwaniuk M, Śledź M, Nowacka M, Chudoba T, Witrowa-Rajchert D. (2013). Drying kinetics of apple tissue treated by pulsed electric field. *Drying Technology*, *31*(1), 112–119.

[48] Clemente G, Sanjuán N, Cárcel JA, Mulet A. (2014). Influence of temperature, air velocity, and ultrasound application on drying kinetics of grape seeds. *Drying Technology*, *32*(1), 68–76.

[49] Xiao HW, Bai JW, Sun DW, Gao ZJ. (2014). The application of superheated steam impingement blanching (SSIB) in agricultural products processing—a review. *Journal of Food Engineering*, *132*, 39–47.

[50] Ruiz MJ, Moyano L, Zea L. (2014). Changes in aroma profile of musts from grapes cv. Pedro Ximenez chamber-dried at controlled conditions destined to the production of sweet Sherry wine. *LWT-Food Science and Technology*, *59*(1), 560–565.

[51] Sharma VK, Sharma S, Ray RA, Garg HP. (1986). Design and performance studies of a solar dryer suitable for rural applications. *Energy Conversion and Management*, *26*(1), 111–119.

[52] Diamante LM, Munro PA. (1993). Mathematical modelling of the thin layer solar drying of sweet potato slices. *Solar Energy, 51*(4), 271–276.

[53] Tsamparlis M. (1990). Solar drying for real applications. *Drying Technology, 8*(2), 261–285.

[54] Barnwal P, Tiwari GN. (2008). Grape drying by using hybrid photovoltaic-thermal (PV/T) greenhouse dryer: An experimental study. *Solar Energy, 82*(12), 1131–1144.

[55] Gee DL, Webb RL (1980). Forced convection heat transfer in helically rib-roughened tubes. *International Journal of Heat and Mass Transfer, 23*(8), 1127–1136.

[56] Tulasidas TN, Raghavan GSV, Norris ER. (1993). Microwave and convective drying of grapes. *Transactions of the ASAE, 36*(6), 1861–1865.

[57] Vega-Mercado H, Góngora-Nieto MM, Barbosa-Cánovas GV. (2001). Advances in dehydration of foods. *Journal of Food Engineering, 49*(4), 271–289.

[58] Kassem AS, Shokr AZ, EI-Mahdy AR, Aboukarima AM, Hamed EY. (2011). Comparison of drying characteristics of Thompson seedless grapes using combined microwave oven and hot air drying. *Journal of the Saudi Society of Agricultural Sciences, 10*(1), 33–40

[59] Tulasidas T. (1995). Combined convective and microwave drying of grapes. *Drying Technology, 13*(4), 1029–1031.

[60] Minea V. (2015). Overview of heat-pump-assisted drying systems, Part I: integration, control complexity, and applicability of new innovative concepts. *Drying Technology, 33*(5), 515–526.

[61] Zhu Z, Yang Z, Wang F. (2016). Experimental research on intermittent heat pump drying with constant and time-variant intermittency ratio. *Drying Technology*, published online with doi: 10.1080/07373937.2016.1138966

[62] Gan SH, Ong SP, Chin NL, Law CL. (2016). A comparative quality study and energy saving on intermittent heat pump drying of Malaysian edible bird's nest. *Drying Technology*, published online with doi: 10.1080/07373937.2016.1155053

[63] Chua KJ, Chou SK, Ho JC, Hawlader MNA. (2002). Heat pump drying: recent developments and future trends. *Drying technology, 20*(8), 1579–1610.

[64] Vazquez G, Chenlo F, Moreira R, Cruz E. (1997). Grape drying in a pilot plant with a heat pump. *Drying Technology, 15*(3–4), 899–920.

[65] Larrauri JA, Rupérez P, Saura-Calixto F. (1997). Effect of drying temperature on the stability of polyphenols and antioxidant activity of red grape pomace peels. *Journal of Agricultural and Food Chemistry, 45*(4), 1390–1393.

[66] Da Silva DI, Nogueira GD, Duzzioni AG, Barrozo MA. (2013). Changes of antioxidant constituents in pineapple (Ananas comosus) residue during drying process. *Industrial Crops and Products, 50*, 557–562.

[67] Rockenbach II, Gonzaga LV, Rizelio VM, Gonçalves AEDSS, Genovese MI, Fett R. (2011). Phenolic compounds and antioxidant activity of seed and skin extracts of red grape (*Vitis vinifera* and *Vitis labrusca*) pomace from Brazilian wine making. *Food Research International, 44*(4), 897–901.

[68] Celma AR, Rojas S, López-Rodríguez F. (2007). Waste-to-energy possibilities for industrial olive and grape by-products in Extremadura. *Biomass and Bioenergy, 31*(7), 522–534.

[69] Celma AR, López-Rodríguez F, Blázquez FC. (2009). Experimental modelling of infrared drying of industrial grape by-products. *Food and Bioproducts Processing, 87*(4), 247–253.

[70] Luque-Rodríguez JM, De Castro ML, Pérez-Juan P. (2005). Extraction of fatty acids from grape seed by superheated hexane. *Talanta, 68*(1), 126–130.

[71] Shi J, Yu J, Pohorly J, Young JC, Bryan M, Wu Y. (2003). Optimization of the extraction of polyphenols from grape seed meal by aqueous ethanol solution. *Food, Agriculture & Environment, 1*(2), 42–47.

[72] Spigno G, Tramelli L, De Faveri DM. (2007). Effects of extraction time, temperature and solvent on concentration and antioxidant activity of grape marc phenolics. *Journal of Food Engineering, 81*(1), 200–208.

[73] Corrales M, Toepfl S, Butz P, Knorr D, Tauscher B. (2008). Extraction of anthocyanins from grape by-products assisted by ultrasonics, high hydrostatic pressure or pulsed electric fields: a comparison. *Innovative Food Science & Emerging Technologies, 9*(1), 85–91.

[74] Pedroza MA, Amendola D, Maggi L, Zalacain A, De Faveri DM, Spigno G. (2015). Microwave-assisted extraction of phenolic compounds from dried waste grape skins. *International Journal of Food Engineering, 11*(3), 359–370.

[75] Wenzel J, Samaniego CS, Wang L, Nelson LS, Ketchum K, Ammerman M, Zand A. (2015). Superheated liquid and supercritical denatured ethanol extraction of antioxidants from Crimson red grape stems. *Food Science & Nutrition, 3*(6), 569–576.

[76] Spigno G, De Faveri DM. (2007). Antioxidants from grape stalks and marc: influence of extraction procedure on yield, purity and antioxidant power of the extracts. *Journal of Food Engineering, 78*(3), 793–801.

[77] Nowicka P, Wojdyło A, Lech K, Figiel A. (2015). Chemical Composition, Antioxidant Capacity, and Sensory Quality of Dried Sour Cherry Fruits pre-Dehydrated in Fruit Concentrates. *Food and Bioprocess Technology, 8*(10), 2076–2095.

[78] Cruz L, Clemente G, Mulet A, Ahmad-Qasem MH, Barrajón-Catalán E, García-Pérez JV. (2016). Air-borne ultrasonic application in the drying of grape skin: kinetic and quality considerations. *Journal of Food Engineering, 168*, 251–258.

http://dx.doi.org/10.5772/64662

[79] De Torres C, Schumacher R, Alañón ME, Pérez-Coello MS, Díaz-Maroto MC (2015). Freeze-dried grape skins by-products to enhance the quality of white wines from neutral grape varieties. *Food Research International, 69*, 97–105.

[80] Roberts JS, Kidd DR, Padilla-Zakour O. (2008). Drying kinetics of grape seeds. *Journal of Food Engineering, 89*(4), 460–465.

[81] Garcia-Perez JV, García-Alvarado MA, Carcel JA, Mulet A. (2010). Extraction kinetics modeling of antioxidants from grape stalk (Vitis vinifera var. Bobal): influence of drying conditions. *Journal of Food Engineering, 101*(1), 49–58.

[82] Garcia-Perez JV, Carcel JA, Simal S, Garcia-Alvarado MA, Mulet A. (2013). Ultrasonic intensification of grape stalk convective drying: kinetic and energy efficiency. *Drying Technology, 31*(8), 942–950.

[83] Sui, Y., Yang, J., Ye, Q., Li, H., & Wang, H. (2014). Infrared, convective, and sequential infrared and convective drying of wine grape pomace. *Drying Technology, 32*(6), 686–694.

Chapter 8

Earthworms and Grape Marc: Simultaneous Production of a High-Quality Biofertilizer and Bioactive-Rich Seeds

Jorge Domínguez, Hugo Martínez-Cordeiro and Marta Lores

Additional information is available at the end of the chapter

http://dx.doi.org/10.5772/64751

Abstract

Winemaking produces annually millions of tons of grape marc as a byproduct, which is a revaluable resource having many potential uses, including a nutrient-rich organic soil amendment. However, its application as untreated raw material can damage crops owing to the release of phytotoxic polyphenols. This agronomic problems can be minimized by vermicomposting, as earthworms can partly digest polyphenols. This chapter reports the results obtained in the processing of grape marc derived from white wine through vermicomposting on an industrial scale to yield both a high quality organic, polyphenol-free fertilizer and grape seeds as a source of bioactive compounds. Vermicomposting reduced substantially the residue biomass. In a very short-term, the process yielded a nutrient-rich, microbiologically active and stabilized peat-like material that can be easily separated from the seeds by sieving. The isolation of the seeds eliminates the polyphenol-associated phytotoxicity from the vermicompost and left those seeds prepared to be easily processed to get different bioactive compounds, mainly rich-polyphenols extracts but also rich-fatty acids seed oil. The procedure described is effective, simple, environmental-friendly and economical, and can easily be scaled up for industrial application yielding a variety of added-value products from the initial grape marc.

Keywords: wine and winemaking residues, earthworms, grape polyphenols, vermicomposting, vermicompost, earthworm humus

1. Introduction

The annual worldwide production of grapes in the world keeps increasing and accounts to nearly 78 million tonnes [1], and most of these grapes (up to 80%) are utilized to make wine. The main residue of winemaking is grape marc, also known as grape bagasse or grape pomace, which consists of the stalks, skin, pulp and seeds that remain after pressing the grapes. After the pressing process of the grapes to obtain the stum or grape juice, the grape marc is nearby 20% of the grapes weight. The overall composition of grape marc depends on the pre-treatment process in the winery and consists of 40% seeds and 60% skin and pulp grape, when stems are removed before pressing. When grapes are directly processed including their stalks, the composition of grape mars is 30% stems, 30% seeds and 40% skin and pulp grape.

Traditionally, grape marc has been used to produce pomace brandy spirits (orujo, grappa, zivania, törkölypálinka, ...). Nowadays, a relative small fraction of the grape marc produced during the winemaking process in the wine industry is utilized for the production of ethanol, to extract organic acids and to produce grape seed oil and other food ingredients [2–4]. Due to its high acidity, it is easy to make silage and thus grape marc has also been used as fodder to feed livestock animals, although its high lignin content makes it rather indigestible [5].

This byproduct or subproduct is potentially a very valuable resource that could be used as a nutrient-rich organic soil amendment; however, overproduction in small geographic areas has led to inappropriate disposal of the material on agricultural land. Moreover, the application of the untreated raw material can damage crops owing to the release of excessive amounts of phytotoxic polyphenols to soils [6]. These phenolic compounds are responsible for the potential phytotoxic and anti-microbial activity of the grape marc, including potential negative effects on the physical, chemical and biological properties of the soil, potential phytotoxic effects on crops and potential groundwater pollution [7].

Since earthworms can digest polyphenols, at least partly [8], the agronomic problems associated with the application of the grape marc to soil can be minimized or eliminated by vermicomposting technologies [9].

From another point of view, polyphenols have well-known human health-promoting effects and other properties in different biological and food systems [3, 10]. Over the past 20 years, the level of scientific and public interest in grapevine polyphenols has increased greatly. Over this period, increasing numbers of potential human health applications for polyphenols have been suggested and experimental data supporting various uses have been accumulated, including anti-cancer and cardioprotective effects and also anti-inflammatory, anti-obesity, anti-ageing, anti-diabetic and neuroprotective properties. Polyphenols are potent antioxidants and can neutralize free radicals, thus halting lipid oxidation and other oxidative side effects [11, 12]. This characteristic makes polyphenols of interest for many different applications, such as the treatment of inflammation [13] and cancer [14, 15]. Moreover, these natural phenolic compounds have many industrial applications, for example, they may be used as natural colourants and have been reported to have excellent properties as food preservatives [11, 12], for anti-ageing purposes in cosmetics [16, 17] and for nutraceutical purposes [18].

Polyphenols also have nootropic properties, i.e. they can enhance several brain functions, such as learning, memory, attention and motivation [19].

A large proportion of the polyphenols (ca. 60%) in grape marc is contained in the seeds [20]. Consequently, another interesting approach is to recover these polyphenols as functional compounds for the pharmaceutical, cosmetic and food industries [2, 3, 11, 21, 22].

A possible alternative to optimize the extraction of these polyphenols is to use the vermicomposting process as a pre-treatment technique of the grape marc. In this way, the potential agronomic problems associated with the application of grape marc to soil can be simultaneously minimized or eliminated [9, 23].

2. The process of vermicomposting

Vermicomposting is a bio-oxidative process in which detritivorous earthworms intensively interact with microorganisms, thus strongly affecting decomposition processes, accelerating the stabilization of organic matter and greatly modifying its physical and biochemical properties [24, 25]. Although microbiota produces the enzymes for the biochemical decomposition of organic wastes, earthworms are the crucial drivers of the vermicomposting process. Thus, they are responsible for the activation and acceleration of microbial activity through the processes of ingestion and breaking up of fresh organic matter, which result in a larger surface area accessible for microbial attack, altering significantly biological activity. Furthermore, the passage through the earthworm's gut and the associated process, as well as their interactions with other organisms in those detritivorous networks, changes the structure and function of the microbial communities [26, 27].

Vermicomposting includes two different phases in relation to earthworm activity (**Figure 1**):

i. An active phase during which earthworms ingest, process and digest the dead organic matter; thereby modifying its physical and chemical properties and the structure and function of the microbial communities [27–29]; and

ii. A maturation phase characterized by the shift of the earthworms towards raw layers of unprocessed organic waste, during which microorganisms alone take over control of the decomposition of the earthworm's casts [9, 24].

The extent of the active phase is variable and depends on the species and density of earthworms (the main drivers of the process), and the rates at which they ingest and process the organic waste [30].

In the first instance, the effect of earthworms on the decomposition of organic waste during vermicomposting is due to gut-associated processes (GAPs). GAPs include all those modifications that the dead organic matter and the microorganisms undergo during the transit though the earthworm's gut. These alterations include the addition of carbohydrates, enzymes and other metabolites, change diversity and activity of the microbial and microfaunal populations and communities, physical homogenization of the ingested material and the inherent processes of digestion, assimilation and production of mucus and excretory substances such as urea and ammonia, which constitute an easily assimilable pool of nutrients for microorganisms (**Figure 1**). In addition, endosymbiotic microorganisms that live in the earthworms' guts and produce extracellular enzymes that degrade cellulose, polyphenols and other macromolecules [9] boost decomposition of the organic waste. The continuous burrowing activities of earthworms aerate and homogenize the substrate, producing important physical modifications of the substrate, accelerating microbial activity and further increasing the breakdown of organic wastes [24].

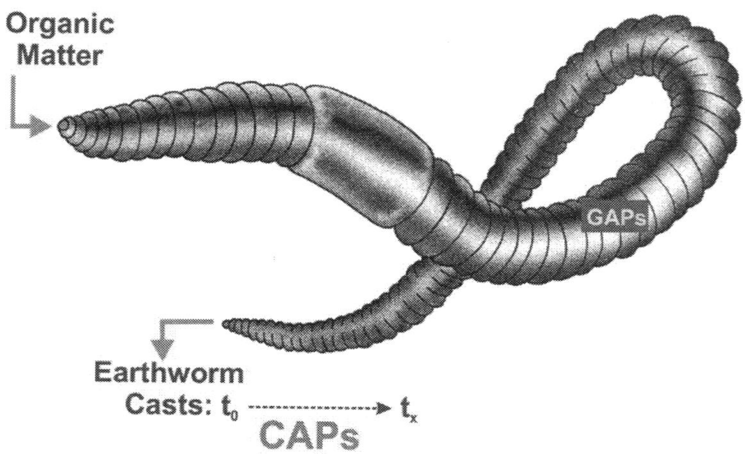

Figure 1. Earthworms influence the vermicomposting of organic matter primarily through gut-associated processes (GAPs, including ingestion, digestion and assimilation in the gut) and secondary through cast-associated processes (CAPs, ageing and microbial modifications of their excreta or casts).

Once GAPs are completed, the earthworm casts undertake cast-associated processes (CAPs), which are more related with ageing and maturation stages, with the physical and chemical changes of the egested casts and with the modifications that the microflora and microfauna present in the vermicompost exert over those cast materials [31]. During cast-associated processes, earthworms exert secondary transformations resulting from the GAPs (**Figure 1**). In vermicomposting systems, earthworm casts are mixed with other materials not ingested and/or digested by the earthworms, and depending on the heterogeneity of the organic wastes, the resulting vermicompost consists of a mixture of the two different portions. During this ageing process, vermicompost will reach an optimum stage in terms of its biological properties promoting plant-growth enhancement and suppression of plant diseases [32].

Vermicompost, the end product of vermicomposting, is a finely divided and porous peat-like material with a high water-holding capacity; it also contains many nutrients in forms that are readily taken up by plants [24]. At the end of the vermicomposting process, the vermicompost can be separated easily from the more recalcitrant fractions of the waste material.

3. Vermicomposting of grape marc

Grape marc derived from white grapes (*Vitis vinifera*, v. Albariño) was collected in a cellar of the Rías Baixas DO (Terras Gauda, O Rosal, Galicia, NW Spain). As a simple pre-treatment, the grape marc was moisturized and revolved before vermicomposting. The main chemical features of the grape marc are detailed in **Table 1**. The system used for the large-scale vermicomposting process was a pilot-scale vermireactor housed in a greenhouse. To prevent desiccation, the vermireactor was watered daily with an automatic system. At the beginning of the trial, the initial earthworm (*Eisenia andrei*) density was 214 ± 26 individuals m^2. The grape marc was then placed on top of a plastic mesh in the vermireactor to facilitate the removal of grape marc after processing by the earthworms [33].

	Grape marc	Vermicompost
pH	4.36 ± 0.04^a	7.1 ± 0.003^b
Electrical conductivity (mS cm^{-2})	1.34 ± 0.15^a	0.27 ± 0.009^b
Organic matter (%)	91.21 ± 0.30^a	74.98 ± 0.34^b
Total carbon (g kg^{-1} dw)	484.23 ± 1.60^a	375.96 ± 1.47^b
Total nitrogen (g kg^{-1} dw)	20.19 ± 0.62^a	29.63 ±0.13^b
C/N ratio	24.02 ± 0.72^a	12.68 ± 0.07^b
Total phosphorus (g kg^{-1} dw)	4.03 ± 0.08^a	8.36 ± 0.32^b
Total potassium (g kg^{-1} dw)	30.46 ± 0.56^a	11.40 ± 0.65^b
Basal respiration (mg O_2 kg OM^{-1} h^{-1})	312.39 ± 40.57^a	68.40 ± 27.11^b
Lignin (g kg^{-1} dw)	516.32 ± 9.56^a	323.54 ± 2.36^b
Cellulose (g kg^{-1} dw)	225.3 ± 10.39^a	58.26 ± 10.48^b
Hemicellulose (g kg^{-1} dw)	100.6 ± 1.39^a	30.56 ± 0.54^b

Values are means ± SE (n = 5). Different letters indicate significant differences between the values, based on post hoc tests (Tukey HSD).

Table 1. Chemical properties and microbial activity of the fresh grape marc and the final vermicompost.

The density and biomass of the earthworm population were determined periodically by collecting 10 samples with a core sampler (five from above and five from below the plastic mesh) of the material in the vermireactor every 14 days during the whole trial (112 days). For the analysis of polyphenols and the biological and physicochemical properties, five samples (10 g) were collected every 7 days during the trial. The samples were stored in plastic bags at 20°C until analysis [33].

Samples of the material were dried at 105°C for 24 h, for the determination of the moisture content, and combusted at 550°C for 4 h, for the determination of the organic matter content. Electrical conductivity and pH were measured in aqueous extracts (1:10 w/v). The total C and N and total P and K contents were analysed in oven-dried (60°C) samples, using a C/N analyser and optical emission spectrometry with inductively coupled plasma (ICP-OES), respectively.

Microbial activity of the grape marc during vermicomposting was determined according to DIN ISO 16072, by measuring the oxygen consumption with the OxiTop® Control System. The total contents of cellulose, hemicellulose and lignin in the grape marc and vermicompost samples were determined by detergent fibre methods. Values of neutral detergent fibre (NDF), acid detergent fibre (ADF) and acid detergent lignin (ADL) were determined, as described by Aira et al. [34], using the FibreBag System® [32].

To determine the total and individual polyphenols, samples were extracted by means of pressurized liquid extraction (PLE), as described by Alvarez-Casas et al. [2]. The concentration of total polyphenols (TP) in grape marc extracts was determined according to the Folin-Ciocalteu colorimetric method, and the absorbance values were measured at 760 nm. TP were quantified from a calibration curve prepared with gallic acid standard solutions and expressed as mg gallic acid per g of dry weight (mg gallic g^{-1} dw). A 5 mL aliquot of each PLE grape marc extract was concentrated to a final volume of 0.5 mL under an N_2 stream at 40°C. Finally, the concentrated extract was filtered through a 0.22 μm PVDF filter and analysed in a high-performance liquid chromatography (Varian Prostar HPLC system with a diode array detector). The determination chromatographic method is described in detail elsewhere [33].

Data were statistically analysed by repeated measures analysis of variance (rANOVA) with sampling time as the within-subject factor. Mauchly's test confirmed that all variables satisfied the assumption of sphericity, and significant differences in the main effects were further analysed by paired comparisons, with the Tukey HSD test.

4. Evolution of the earthworm population during vermicomposting of grape marc

Before adding the grape marc to the vermireactor, the population density of earthworms (in this study belonging to the species *Eisenia andrei*) in the vermireactor was around 300 individuals m^2, including 19 ± 3 adult and mature earthworms m^2, 215 ± 37 juveniles m^2 and 63 ± 18 cocoons m^2. The total earthworm biomass in the vermireactor was 58.4 ± 15 g live weight m^2 (**Figure 2**).

The total population density of earthworms and the population density of adult earthworms, juveniles and cocoons augmented considerably until day 70, when the population density reached its maximum. Then, since no more grape marc was added to the vermireactor, the earthworm population density started decreasing thereafter until reaching its minimum value at the end of the trial (day 112). Earthworm biomass increased in the same way, with maximum values after 70 days of vermicomposting, decreasing then until day 112 (**Figure 2**). The earthworm population density in the vermireactor before adding grape marc was quite small. As a consequence of the input of earthworm food from the grape marc, it increased rapidly and noticeably, but it reached values far from its maximum capacity. Detritivorous earthworms as *Eisenia andrei* live in pure organic matter environments where the availability of food increases earthworm growth, development and reproduction, leading to very large earthworm populations. Thus, when large amounts of food are available, the population density of this type of earthworms can reach very high values, as for example up to 8000 earthworms m^2 in cow manure and 14,600 individuals m^2 in pig manure [35].

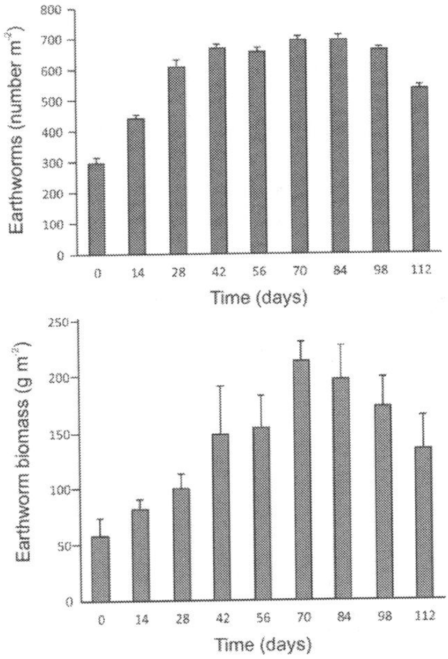

Figure 2. Earthworm density and earthworm biomass during vermicomposting of white grape marc.

Although microorganisms are the main agents responsible for organic matter decomposition, earthworms affect the rates of decomposition directly by feeding and fragmenting activities and indirectly through interactions with microorganisms [9, 24] (Domínguez 2004; Domínguez et al. 2010). Thus, the decomposition rates are directly related to the earthworm population density [31].

5. Vermicompost of grape marc

The pH of the fresh grape marc was quite acid and increased rapidly due to the action of the earthworms, reaching neutrality after seven weeks and remaining neutral in the final vermicompost (**Table 1**).

The rapid mineralization of the organic C of the grape marc leads to a significant reduction of the waste mass and volume. Thus, the mass of grape marc was reduced in 60% as a consequence of the vermicomposting process. On the other hand, this important reduction in mass implies increments in the concentration and availability of mineral nutrients.

Electrical conductivity (EC) of the grape marc was relatively high and decreased significantly during vermicomposting reaching quite low values in the final vermicompost (**Table 1**). The organic matter content of the fresh grape marc is very high and decreased rapidly over time reducing its values to almost 20%. During vermicomposting, the total carbon content of the grape marc depleted rapidly and reduced considerably in the vermicompost (**Table 1**). The nitrogen content of the grape marc was quite high (2%) and increased significantly during vermicomposting until reaching values of 3% in the vermicompost. The C to N ratio decreased gradually and quickly during the process until values around 12 (**Table 1**). While the total K content decreased significantly, the total P content increased significantly during vermicomposting. The microbial activity, measured as basal respiration, decreased very significantly over time reaching much lower values in the vermicompost (**Table 1**). Vermicomposting of grape marc drastically reduced the contents of cellulose, hemicellulose and lignin (**Table 1**).

Some other studies have demonstrated that vermicomposting can be an interesting and efficient alternative for the treatment of grape marc derived from the elaboration of red wine [36, 37] and white wine [38–40]. In the case study presented here, the positive and high dynamics of the earthworm population density together with the correct evolution of the chemical and biological properties indicate that vermicomposting was optimal and produced excellent quality vermicompost. Vermicompost is a mineral-rich, microbiologically active organic amendment that results from the interactions between earthworms and microorganisms during the breakdown of organic matter. It is a stabilized, finely divided peat-like material with a low C:N ratio, high porosity and high water-holding capacity, and it functions as a concentrated source of mineral nutrients that are released slowly and gradually, through mineralization, when plants require them [24]. The speeded breakdown and mineralization of organic wastes, the changes in the structure and function of the microbial communities and the high humification rates achieved during vermicomposting explain the quality and quantity of the nutrients in the vermicompost [40]. At the same time, the organic compounds, extracellular enzymes and other biological characteristics of vermicomposts make these outstanding biological fertilizers. Consequently, when added to the soil or to plant growing substrates, this complex mixture of earthworm faeces, humified organic matter and microorganisms also known as earthworm humus increases germination, growth, flowering and fruit production and accelerates the development of a wide range of plant species. The boosted plant growth may be indorsed to different direct and indirect effects, including biologically mediated mechanisms such as the supply of plant growth regulating substances and improvements in a vast array of soil biological functions [41].

6. Evolution of the polyphenol content of the grape marc and the grape seeds during vermicomposting

The polyphenol content of the initial white grape marc was 58 ± 10 mg GAE g^{-1} dw and decreased significantly throughout the vermicomposting process; the amount of polyphenols was reduced by almost one half in a period of only 14 days. At the end of the trial, the decrease was about 98% of the initial amount, with very low levels maintained during the past weeks, compared with the pre-vermicomposting levels (**Figure 3**), reaching a residual concentration in the final vermicompost.

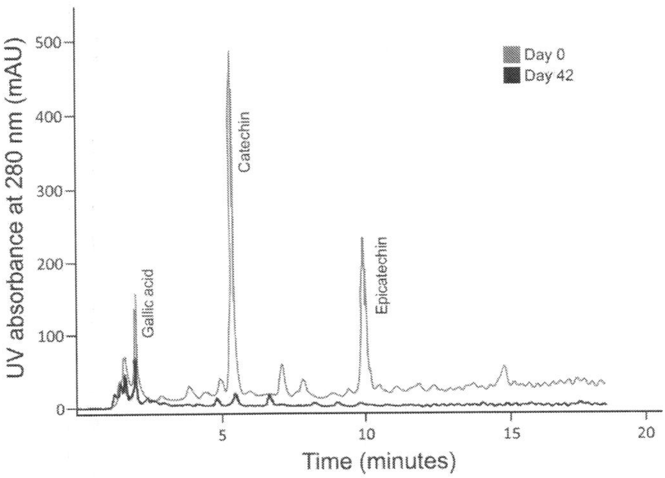

Figure 3. HPLC chromatograms of the fresh grape marc (green) and the vermicompost after 42 days (black), showing the dramatic reduction on the polyphenolic content during the vermicomposting process.

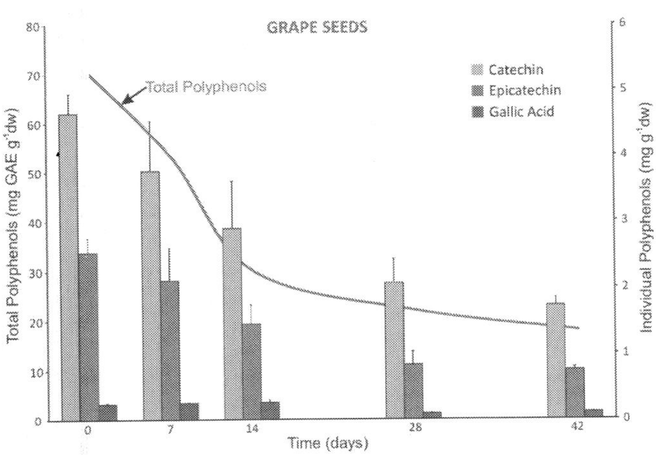

Figure 4. Evolution of the total polyphenol index (blue line) and of the concentration of gallic acid, catechin and epicatechin in white grape seeds during vermicomposting.

The initial concentration of polyphenols in the grape seeds was 70 ± 5 mg GAE g^{-1} dw. The polyphenol content also decreased gradually throughout the vermicomposting process (**Figure 4**). Grape seeds contain large quantities of polyphenolic compounds, being an interesting source for exploiting the biological properties of these natural substances on an industrial scale. After 2 weeks of vermicomposting, it starts to be possible the separation of the grape seeds and the earthworm by sieving. The optimal separation of the seeds is between weeks 4 and 6; latter, although the separation process is even easier the polyphenolic content of the seeds is lower. From the sixth week, the sieved biofertilizer does not contain polyphenols, due to the biodegradation of the 98% of the initial concentration, with residual values in the final vermihumus (**Figure 4**).

Gallic acid, catechin and epicatechin were the main polyphenols identified in the grape seeds (**Figure 4**). Their concentration was determined until the sixth week, corresponding to the end of the optimum time to collect the grape seeds, and because later the concentrations are much lower (**Figure 3**). The concentration of gallic acid and the flavanols, catechin and epicatechin, in the grape seeds decreased gradually and significantly during the first stages of vermicomposting (**Figure 4**).

The polyphenol content of the vermicomposted grape marc and seeds decreased gradually over time. The earthworm activity and the effects on decomposition are enhanced by the action of endosymbiotic microorganisms that produce extracellular enzymes that degrade phenolic compounds [32]. Seeds have greater resistance to the combined biodegradation action of earthworms and microorganisms than the remained vermicomposted grape marc, and this explains the higher concentration of polyphenols in the grape seeds during vermicomposting.

Earthworm activity during vermicomposting speeded the mechanical separation of vermihumus and grape seeds. They break down the grape marc acting as mixing machines, expanding the superficial area for microbial attack, and translocating materials and microbial-rich casts throughout the waste, thus homogenizing it. The most readily assimilable parts are rapidly decomposed to fine particles by the combined action of earthworms and microorganisms, whereas grape seeds stay almost entire. Polyphenols are mainly included in these more recalcitrant parts of the grape marc [20]. Grape seeds can easily be separated from the vermicompost after 2 weeks of vermicomposting. The total polyphenolic content of the seeds after these 2 weeks is lower than in the fresh grape marc, but the separation of seeds and vermihumus is easy, whereas this separation is much more difficult in the grape marc. The removal of the grape seeds also eliminates the phytotoxicity caused by the polyphenols in the vermihumus. Likewise, the lack of phytotoxic compounds is an indication of maturity in organic amendments [42]. This is important because the application of immature vermicompost can negatively affect crop development [43, 44].

Several studies have shown that grapes are a major source of polyphenolic compounds, especially benzoic acids, cinnamic acids, anthocyanins, flavonols, catechins and tannins, which are largely preserved in the grape marc [2]. The concentrations of gallic acid, catechin and epicatechin decreased in the same way as the total polyphenols, as they were degraded by earthworms and microorganisms. Nevertheless, the seeds obtained on day 14 still contain useful amounts of the three major polyphenols (**Figure 4**). These polyphenols have many beneficial properties mainly attributed to their anti-oxidant properties and anti-bacterial activities [3], making them interesting substances for use in the cosmetic and food industries. The three main polyphenols contained in the grape seeds act particularly well as hydrogen atom donors, the main mechanism by which these compounds express their anti-oxidant action [22]. Specifically, flavonols are used as natural anti-oxidants preventing degradation of lipids, as anti-microbial agents and functional supplements in foodstuffs to improve animal health and to preserve animal products, and as bioactive components in nutritional and dietary supplements [20].

7. Overview of the vermicomposting process of grape marc

Vermicomposting of grape marc has proven to be a very useful procedure that yields simultaneously an organic fertilizer and grape seeds. During the vermicomposting process, the activity of earthworms favours the mechanical separation of the different fractions of grape marc. The earthworms act as mechanical mixers, thus decomposing the organic material and increasing the surface area exposed to microorganisms; and moving the fragments and excreta rich in bacteria through the residue profile and thus, homogenizing the organic material.

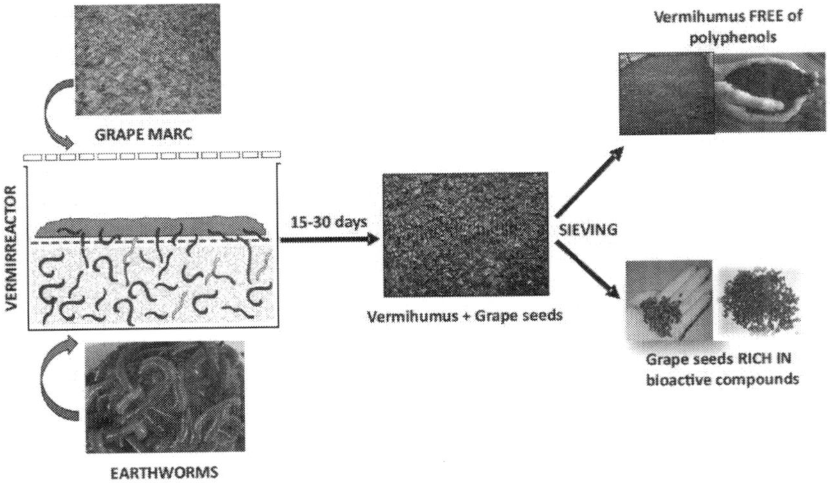

Figure 5. Schematic representation of the potential of earthworms during the vermicomposting process to obtain simultaneously a high-quality biofertilizer free of polyphenols and grape seeds rich in bioactive compounds.

Earthworms reduce the more digestible parts to a finer particle size, while seeds remain almost unaffected. These most recalcitrant parts of grape marc contain the highest amounts of polyphenols and, as mentioned previously, seeds can be easily separated from the vermicompost after 2 weeks of vermicomposting. Sieving the material at the earlier stages of the process led to the separation of the organic fertiliser (vermicompost) from the remaining residual material that mainly consists in grape seeds (**Figure 5**) (Patent no. ES2533501 [45]). The seeds maintain a high proportion of the initial polyphenol content, and separation of the material facilitates extraction of the polyphenols, which have several potential industrial applications.

The separation of the seeds also eliminates the phytotoxicity in the final vermicompost. The degradation of the phytotoxic compounds is a good indicator of the maturity of the vermicompost, an important fact because the immature earthworm humus can affect adversely the development of crops. Thus, a mature and stable product with great potential for use in agriculture is obtained.

Interestingly, the polyphenol content of a wine depends on how grapes have been processed in the winery. Consequently, the polyphenol content of the grape marc also depends on the winemaking process. During red wine vinification, skins and seeds remain for several days in contact with the fermentation broth, giving the red wine a high polyphenol concentration. However, in the white winemaking, the grape juice ferments without being in contact with the grape marc, which remains as a final residue of the process, retaining much of the initial polyphenolic load of the grapes.

8. Conclusions

In recent years, the wastes derived from the wine industry have been object of a growing interest due to several environmental and industrial issues. The excessive accumulation of this waste and the problems associated with its agricultural use led to the search for new techniques for its valorization. In the present study, the application of a vermicomposting process represents an interesting method for the treatment of grape marc, environmentally friendly and rendering a new resource with industrial and commercial interest.

The overall conclusion of this study is the patent viability of vermicomposting to:

- Transform rapidly the most labile parts of the grape marc into a high-quality, polyphenol-free organic fertilizer.

- Facilitate the mechanical separation of grape seeds with a high proportion of their initial polyphenol content.

- Increase promptly earthworm populations, susceptible to be used as fish bait, animal protein and source of bioactive compounds.

As well as yielding these beneficial added-value products, the process is inexpensive and environmentally friendly.

http://dx.doi.org/10.5772/64751

Acknowledgements

This study was supported by grants from the Ministerio de Economía y Competitividad (CTM2013-42540-R) and the Xunta de Galicia (CN2012/305 and GPC2014/035).

Author details

Jorge Domínguez[1*], Hugo Martínez-Cordeiro[1] and Marta Lores[2]

*Address all correspondence to: jdguez@uvigo.es

1 Department of Ecology and Animal Biology, University of Vigo, Vigo, Pontevedra, Spain

2 Department of Analytical Chemistry, Nutrition and Food Sciences, University of Santiago de Compostela, Santiago de Compostela, Galicia, Spain

References

[1] FAO. FAO STAT 2015 [Internet]. Available from: http://faostat.fao.org. [Accessed: 2016-04-01].

[2] Alvarez-Casas M, Garcia-Jares C, Llompart M, Lores M. Effect of experimental parameters in the pressurized solvent extraction of polyphenolic compounds from white grape marc. Food Chemistry.2014;157:524–532. DOI: 10.1007/s00217-015-2573-0

[3] Fontana AR, Antoniolli A, Bottini R. Grape pomace as a sustainable source of bioactive compounds: extraction, characterization, and biotechnological applications of phenolics. Journal of Agricultural and Food Chemistry.2013;61:8987–9003. DOI: 10.1021/jf402586f

[4] Negro C, Tommasi L, Miceli A. Phenolic compounds and antioxidant activity from red grape marc extracts. Bioresource Technology. 2003;87:41–44. DOI: 10.1016/S0960-8524(02)00202-X

[5] Kammerer D, Schieber A, Carle R. Characterization and recovery of phenolic compounds from grape pomace: a review. Journal of Applied Botany and Food Quality. 2005;79:189–196.

[6] Inderjit. Plant phenolics in allelopathy. The Botanical Review. 1996;62:186–202.

[7] Barbera AC, Maucieri C, Cavallaro V, Ioppolo A, Spagna G. Effects of spreading olive mill wastewater on soil properties and crops, a review. Agricultural Water Management. 2013;19:43–53. DOI: 10.1016/j.agwat.2012.12.009

[8] Hättenschwiler S, Vitousek PM. The role of polyphenols in terrestrial ecosystem nutrient cycling. Trends in Ecology & Evolution. 2000;15:238–243. DOI: 10.1016/S0169-5347(00)01861-9

[9] Domínguez J, Aira M, Gómez-Brandón M. Vermicomposting: earthworms enhance the work of microbes. In Insam H, Franke-Whittle I, Goberna M, editors. Microbes at Work. Berlin: Springer; 2010. pp. 93–114. DOI: 10.1007/978-3-642-04043-6_5

[10] Aizpurua-Olaizola O, Ormazabal M, Vallejo A, Olivares M, Navarro P, Etxebarria N, Usobiaga A. Optimization of supercritical fluid consecutive extractions of fatty acids and polyphenols from *Vitis vinifera* grape wastes. Journal of Food Science. 2015;80:101–107. DOI: 10.1111/1750-3841.12715

[11] Makris DP, Boskou G, Andrikopoulos NK. Polyphenolic content and in vitro antioxidant characteristics of wine industry and other agri-food solid waste extracts. Journal of Food Composition and Analysis. 2007;20:125–132. DOI: 10.1016/j.jfca.2006.04.010

[12] Leopoldini M, Russo N, Toscano M. The molecular basis of working mechanism of natural polyphenolic antioxidants. Food Chemistry. 2011;125:288–306. DOI: 10.1016/j.foodchem.2010.08.012

[13] Kim J, Lee KW, Lee HJ. Polyphenols suppress and modulate inflammation: possible roles in health and disease. In: Watson RR, Preedy VR, Zibadi S, editors. Polyphenols in Human Health and Disease. MA, USA: Academic Press; 2014. pp 393–408.

[14] Hertog MGL. Flavonols and flavones in foods and their relation with cancer and coronary heart disease risk [thesis]. Wageningen, The Netherlands: Agricultural University; 1994.

[15] Cordova FM, Watson RR. Food and supplement polyphenol action in cancer recurrence. In: Watson RR, Preedy VR, Zibadi S, editors. Polyphenols in Human Health and Disease. MA, USA: Academic Press; 2014. pp. 191–195. DOI: 10.1016/B978-0-12-398456-2.00016-5

[16] Jean-Gilles D, Li I, Vaidyanathan VG, King R, Cho B, Worthen DR, Chichester CO, Seeram NP. Inhibitory effects of polyphenol punicalagin on type-II collagen degradation in vitro and inflammation in vivo. Chemico-Biological Interactions. 2013;205:90–99. DOI: 10.1016/j.cbi.2013.06.018. Accessed 2014 April 15

[17] Menaa F, Menaa A. Skin photoprotection by polyphenols in animal models and humans. In: Watson RR, Preedy VR, Zibadi S, editors. Polyphenols in Human Health and Disease. MA, USA: Academic Press; 2014. pp. 831–838.

[18] Gollücke AP, Peres RC, Odair A, Ribeiro DA. Polyphenols: a nutraceutical approach against diseases. Recent Patents on Food, Nutrition & Agriculture. 2013;5:214–219. DOI: 10.2174/2212798405666131129153239.

[19] Papandreou MA, Dimakopoulou A, Linardaki ZI, Cordopatis P, Klimis-Zacas D, Margarity M, and others. Effect of a polyphenol-rich wild blueberry extract on cognitive

performance of mice, brain antioxidant markers and acetylcholinesterase activity. Behavioural Brain Research. 2009;198:352–358. DOI: 10.1016/j.bbr.2008.11.013

[20] Yilmaz Y, Toledo RT. Major flavonoids in grape seeds and skins: antioxidant capacity of catechin, epicatechin, and gallic acid. Journal of Agricultural and Food Chemistry. 2004;52:255–260. DOI: 10.1021/jf030117h

[21] El Gharras H. Polyphenols: food sources, properties and applications—a review. International Journal of Food Science & Technology. 2009;44:2512–2518. DOI: 10.1111/j.1365-2621.2009.02077.x

[22] Quideau S, Deffieux D, Douat-Casassus C, Pouységu L. Plant Polyphenols: Chemical Properties, Biological Activities, and Synthesis. Angewandte Chemie International Edition. 2011;50:586–621. DOI: 10.1002/anie.201000044

[23] Gómez-Brandón M, Aira M, Lores M, Domínguez J. Changes in microbial community structure and function during vermicomposting of pig slurry. Bioresource Technology. 2011;102:4171–4178. DOI: 10.1016/j.biortech.2010.12.057

[24] Domínguez J. State of the art and new perspectives on vermicomposting research. In: Edwards CA, editor. Earthworm Ecology. Boca Raton, FL: CRC Press; 2004. pp. 401–424.

[25] Domínguez J, Gómez-Brandón M. Vermicomposting: composting with earthworms to recycle organic wastes. In: Kumar S, Bharti A, editors. Management of Organic Waste. Rijeka, Croacia: Intech Open Science; 2012. pp. 29–48. DOI: 10.5772/33874

[26] Domínguez J, Parmelee RW, Edwards CA. Interactions between *Eisenia andrei* (Oligochaeta) and nematode populations during vermicomposting. Pedobiologia. 2003;47:53–60. DOI: 10.1078/0031-4056-00169

[27] Lores M, Gómez-Brandón M, Pérez-Díaz D, Domínguez J. Using FAME profiles for the characterization of animal wastes and vermicomposts. Soil Biology and Biochemistry. 2006;38:2993–2996. DOI: 10.1016/j.soilbio.2006.05.001

[28] Aira M, Bybee S, Pérez-Losada M, Domínguez J. Feeding on microbiomes: effects of detritivory on the taxonomic and phylogenetic bacterial composition of animal manures. FEMS Microbiology Ecology. 2015;91:117. DOI: 10.1093/femsec/fiv117

[29] Aira M, Olcina J, Pérez-Losada M, Domínguez J. Characterization of the bacterial communities of casts from *Eisenia andrei* fed with different substrates. Applied Soil Ecology. 2016;98:103–111. DOI: 10.1016/j.apsoil.2015.10.002

[30] Aira M, Domínguez J. Optimizing vermicomposting of animal wastes: effects of dose of manure application on carbon loss and microbial stabilization. Journal of Environmental Management. 2008;88: 1525–1529. DOI: 10.1016/j.jenvman.2007.07.030

[31] Aira M, Sampedro L, Monroy F, Domínguez J. Detritivorous earthworms directly modify the structure, thus altering the functioning of a microdecomposer

food web. Soil Biology & Biochemistry. 2008;40:2511–2516. DOI: 10.1016/j.soilbio.2008.06.010

[32] Domínguez J, Edwards CA. Biology and ecology of earthworm species used for vermicomposting. In: Edwards CA, Arancon NQ, Sherman RL, editors. Vermiculture Technology: Earthworms, Organic Waste and Environmental Management. Boca Raton, FL: CRC Press; 2011. pp. 25–37.

[33] Domínguez J, Martínez-Cordeiro H, Álvarez Casas M, Lores M. Vermicomposting grape marc yields high quality organic biofertilizer and bioactive polyphenols. Waste Management & Research. 2014;32:1235–1240. DOI: 10.1177/0734242X14555805

[34] Aira M, Monroy F, Domínguez J. *Eiseniafetida* (Oligochaeta, Lumbricidae) activates fungal growth, triggering cellulose decomposition during vermicomposting. Microbial Ecology. 2006;52:738–747. DOI: 10.1007/s00248-006-9109-x

[35] Monroy F, Aira M, Domínguez J, Velando A. Seasonal population dynamics of *Eiseniafetida* (Savigny, 1826) (Oligochaeta, Lumbricidae) in the field. Comptes Rendus Biologies. 2006;329:912–915. DOI: 10.1016/j.crvi.2006.08.001

[36] Nogales R, Cifuentes C, Benítez E. Vermicomposting of winery wastes: a laboratory study. Journal of Environmental Science and Health, Part B. 2005;40:659–673. DOI: 10.1081/PFC-200061595

[37] Romero E, Plaza C, Senesi N, Nogales R, Polo A. Humic acid-like fractions in raw and vermicomposted winery and distillery wastes. Geoderma. 2007;139: 397–406. DOI: 10.1016/j.geoderma.2007.03.009

[38] Gómez-Brandón M, Lazcano C, Lores M, Domínguez J. Papel de las lombrices de tierra en la degradación del bagazo de uva: efectos sobre las características químicas y la microflora en las primeras etapas del proceso. Acta Zoologica Mexicana. 2010;26:397–408. DOI: 0065-1737

[39] Gómez-Brandón M, Lazcano C, Lores M, Domínguez J. Short-term stabilization of grape marc through earthworms. Journal of Hazardous Materials. 2011;187:291–295. DOI: 10.1016/j.jhazmat.2011.01.011

[40] Domínguez J, Gómez-Brandón M. The influence of earthworms on nutrient dynamics during the process of vermicomposting. Waste Management & Research. 2013;31:859–868. DOI: 10.1177/0734242X13497079

[41] Lazcano C., Domínguez J. The use of vermicompost in sustainable agriculture: impact on plant growth and soil fertility. In: Miransari M, editor. Soil Nutrients. New York: Nova Science Publishers; 2011. pp.230–254. DOI: 978-1-61324-785-3

[42] Wu L, Ma LQ, Martinez GA. Comparison of methods for evaluating stability and maturity of biosolids compost. Journal of Environmental Quality. 2000;29:424–429. DOI: 10.2134/jeq2000.00472425002900020008

[43] Hirai MF, Chanyasak V, Kubota M. A standard measurement for maturity. Biocycle. 1983;24:54–56.

[44] He XT, Logan TJ, Traina SJ. Physical and chemical characteristics of selected U.S. Municipal solid waste compost. Journal of Environmental Quality. 1995;24:543–552. DOI: 10.2134/jeq1995.00472425002400030022x

[45] Domínguez J, Lores M, Álvarez Casas M, Martínez-Cordeiro H. Procedimientopara la obtención y aislamiento de un fertilizanteorgánico y de semillas de uva a partir de residuos de uva. 2015. Patent number: ES2533501. 30/11/2015. Universidade de Vigo y Universidade de Santiago de Compostela. Spain.

Section 2

Fermentation and Wine Biotechnolgy

Chapter 9

Grape and Wine Metabolites: Biotechnological Approaches to Improve Wine Quality

Fernanda Cosme, Berta Gonçalves, António Inês, António M. Jordão and Alice Vilela

Additional information is available at the end of the chapter

http://dx.doi.org/10.5772/64822

Abstract

Grape metabolites can be affected by many extrinsic and intrinsic factors, such as grape variety, ripening stage, growing regions, vineyard management practices, and edaphoclimatic conditions. However, there is still much about the *in vivo* formation of grape metabolites that need to be investigated. The winemaking process also can create distinct wines. Nowadays, wine fermentations are driven mostly by single-strain inoculations, allowing greater control of fermentation. Pure cultures of selected yeast strains, mostly *Saccharomyces cerevisiae,* are added to grape must, leading to more predictable outcomes and decreasing the risk of spoilage. Besides yeasts, lactic acid bacteria also play an important role, in the final wine quality. Thus, this chapter attempts to present an overview of grape berry physiology and metabolome to provide a deep understanding of the primary and secondary metabolites accumulated in the grape berries and their potential impact in wine quality. In addition, biotechnological approaches for wine quality practiced during wine alcoholic and malolactic fermentation will also be discussed.

Keywords: grape physiology, grape metabolites, wine biotechnology, alcoholic fermentation, malolactic fermentation, microbial metabolites

1. Introduction

Grape berry chemical composition is complex, containing hundreds of compounds. Water (75–85%) is the main component followed by sugars and then organic acids. Other important compounds include amino acids, proteins, and phenolic compounds. Berry sugar composition has a key role in wine quality, since it determines alcohol content in wines [1]. Grape sugar, acidity, pH, and color are considered to mark harvest. *Bouquet* and flavor are related to the winemaker's expertise, stabilization, and storage processes, but primarily they are related to grape varietal character and its particular expression in a given *terroir*.

Nowadays, wine fermentations are driven mostly by single-strain inoculations, allowing greater fermentation control, leading to more predictable outcomes and decreasing the risk of spoilage by other yeasts [2]. During must fermentation, *Saccharomyces cerevisiae* produces a plethora of active-aroma secondary metabolites and releases many aroma compounds from inactive precursors present in grape juice, which significantly affect the sensory quality of the final wine [3, 4]. Besides yeasts, lactic acid bacteria (LAB) are members of the normal microbiota that appears in all type of wines (white and red), and, therefore, they also play an important role in their final quality. Malolactic fermentation (MLF), a long-standing process of deacidification in winemaking carried by LAB, is a reaction of L-malic acid decarboxylation to L-lactic acid. Complex metabolic activities also occur, thus suggesting that MLF can positively or negatively affect the final wine quality [5, 6].

2. Grape berry physiology and metabolome

2.1. Morphology and anatomy of grape berries

After successful pollination and fertilization of ovules within a flower berry development initiates [7]. The formation and growth of grape (*Vitis vinifera*) berries follows a double sigmoid pattern with three distinct phases [8]: I, rapid cell division and expansion in green berries; II or lag phase, in which cell expansion ceases; and III, in which growth is reinitiated and the fruit matures. The berry fruit comprises up to four seeds surrounded by the inner endocarp, the middle mesocarp, pulp or flesh, and the outer exocarp or skin [8, 9] (**Figure 1**).

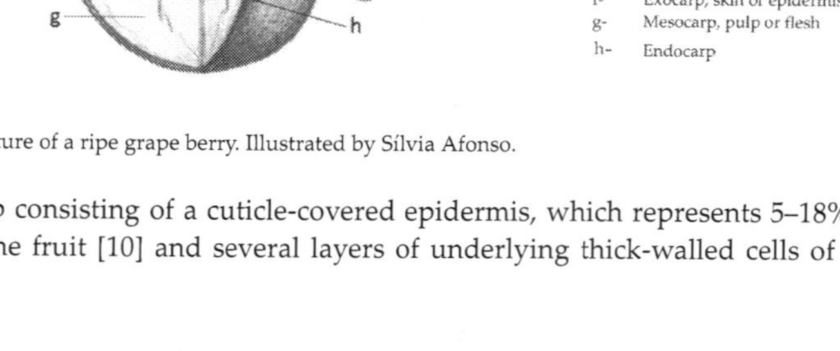

Figure 1. Structure of a ripe grape berry. Illustrated by Sílvia Afonso.

The exocarp consisting of a cuticle-covered epidermis, which represents 5–18% of the fresh weight of the fruit [10] and several layers of underlying thick-walled cells of hypodermis, contains most of the skin flavonoids [11], notably anthocyanins in the outermost layers of the red grape varieties [8], interspersed with cells rich in needle-like crystals (raphides) [12]. Epidermis has non-photosynthetic cells with vacuoles containing large oil droplets [8]. Small berries have greater color, tannins, and flavor compounds than large berries because skin has a higher percentage of the total mass of small berries [7]. Scanning electron microscopy showed very few but functional stomata on young berries and wax-filled stomata on older berries [13], which accumulate polyphenolics and abnormally high concentrations of silicon and calcium in the peristomatal protuberances of up to 200 μm diameter [14].

At harvest, the cuticle of grape berry had an amorphous outer region and a mainly reticulate inner region [15]. During fruit development, the composition of the cuticular waxes changed, being oleanolic acid the main constituent, representing 50–80% of the total weight [16]. The soft wax was a mixture of long chain fatty acids (C_{16} and C_{18} fatty acid esters [17]), alcohols, aldehydes, esters, and hydrocarbons [18].

The mesocarp consists of thin-walled parenchyma [12]. The cells are round to ovoid and contain large vacuoles, which are the primary sites for the accumulation of sugars and phenolics [8], water, and organic acids [9] during grape berry ripening. According to Coombe [19], the translucent and hydrated mesocarp composes 85–87% of the berry's spherical volume. Altogether these make up 99.5% of the juice mass and hence are the major determinants of berry size and quality [9, 20]. The remaining 0.5% of berry components are phenolics, terpenoids, lipids, cellulose, and pectin [20]. The endocarp consists of crystal-containing cells (druses) and an inner epidermis [12].

Grape seeds are contained in locules (**Figure 1**), and are composed of an outer seed coat, the endosperm, and the embryo [9]. As with most seeds, the endosperm comprises the bulk of the grape seed and serves to nourish the embryo during early growth. The normal or perfect number of seeds in the grape is four [9], but lack of ovule fertilization or ovule abortion reduces the number of developing seeds, generally resulting in smaller berry size [7]. Based upon recent molecular evidence, auxin is synthesized in the ovule and transported to the pericarp upon fertilization, where it induces gibberellin (GA) biosynthesis. The GA then degrades DELLA proteins that repress ovary growth and fruit initiation [21]. The size of mature berries at harvest is also a function of the number of cells divisions before and after flowering, extent of growth of these cells [22], and the extent of preharvest shrinkage [23].

High level of tannins is observed in the seed coat [9, 11]. Similar to the tannins and phenols found in the flesh, these tannins also decline greatly on a per-berry basis after *véraison* [24].

Berry vascular tissue develops directly from that of the ovary. It consists primarily of a series of peripheral bundles that ramify throughout the outer circumference of the berry and axial bundles that extend directly up through the stem [8]. Grape berry is provided through the berry stem or pedicel by a vascular system composed of xylem and phloem vessels [25]. Water, minerals, hormones, and nutrients are transported from the root system throughout the vine by the xylem tissue [25]. Present evidence indicates that in the final stages of grape development, water movement through the xylem vessels decreases markedly [25]. But, it seems that the fruit is not hydraulically isolated from the parent grapevine by xylem occlusion then, rather, is "hydraulically buffered" by water delivered via the phloem [9]. Berry is also supplied by the phloem, which is the vasculature involved in photosynthate (sucrose) transport from the canopy to the vine [25].

2.2. Grape primary and secondary metabolites

2.2.1. *Sugars*

One of the main features of the grape-ripening process is the accumulation of sugars in the form of glucose and fructose within the cellular medium, specific in vacuole. In addition, sugar content is an important indicator often used to assess ripeness and to mark grape harvest. But, it is also possible to quantify small traces of sucrose in *V. rotundifolia* and hybrids between *V. labrusca* and *V. vinifera* grapevines [26]. Liu et al. [27] analyzed sugar concentration of 98 different grape cultivars and concluded that glucose (45.86–122.89 mg/mL) and fructose (47.64–131.04 mg/mL) were the predominant sugars in grape berries. During grape berry maturation, sucrose is produced in leaves by photosynthetic carbon assimilation and is transported to the berry in the phloem [24]. Sucrose is loaded into the phloem by either a symplastic or apoplastic mechanism [28]. However, it is at *véraison* that begins the sugar accumulation and the imported sucrose is converted into hexoses as a result of the activity of invertases [29].

Grape berries accumulate glucose and fructose in equal amounts at a relatively constant rate during ripening [29]. In addition, after *véraison* there is a considerable accumulation of glucose and fructose in the vacuoles of mesocarp cells, while 20 days after this period, the hexose content of the grape berry is close to 1 M, with a glucose/fructose ratio of 1 [19, 30]. Grape sugar concentration and composition is mainly determined by several factors, such as genotype [26, 31], vineyard management [32, 33], and climatic conditions [34, 35]. Moreover, in last years, as a result of climate change, there is a tendency for a sugar increase in grapes [36]. But, according to Mira de Orduña [35], the extremely high sugar levels reached at harvest today, especially in warm climates, may be rather associated with the desire to optimize technical or polyphenolic and/or aromatic maturity.

2.2.2. *Organic acids and nitrogenous compounds*

L-Tartaric and L-malic acids contribute to around 90% of the organic acid content in mature grapes [37, 38]. Minor amounts of citric, succinic, lactic, and acetic acids are also present in ripened grapes [39]. Despite L-tartaric and L-malic acids having similar chemical structures, they are synthesized and degraded by evidently different metabolic pathways in the grape berries. L-Tartaric acid synthesis in grape berries occurs during the period of grape growth [19, 40]. Tartaric acid pathway using L-ascorbic acid (vitamin C) is considered to be responsible for >95% of grape L-tartaric acid production [41]. L-Malic acid synthesis indicates that-carboxylation of pyruvate or of phosphoenol pyruvate is the most important pathway [42]. Accumulation of acids usually occurs at the beginning of berry development. The organic acid content increases up to *véraison* and then declines. The content of organic acids is determined by a balance between their synthesis and degradation. L-Tartaric acid was the most prominent acid from *véraison* until the fruits were fully mature. L-Malic acid content increased gradually until *véraison*, after which it decreased with fruit ripening [37]. Grape acid composition is influenced by many factors such as grape variety, environmental conditions, and cultural practices [43]. High malate-producing grape varieties have been identified, such as Carignane, Chardonnay, Grenache, Malbec, and Pinot Noir, as well as high tartrate-producing grape varieties such as Merlot, Semillon, Riesling, and Thompson Seedless [44]. Temperature is a key factor in the rate of L-malic acid degradation during the berries ripening; with low temperatures, higher concentration of L-malic acid was observed [43]. L-Tartaric acid is presumed to be more stable when exposed to higher temperature, being the slight decreases during ripening due to dilution from berry expansion [45, 46].

Grapes nitrogenous compounds include ammonium cations and organic nitrogenous compounds such as amino acids, hexose amines, peptides, nucleic acids, and proteins. As maturation happens, organic nitrogen progressively increases while ammonia slightly declines. The synthesis of amino acids, peptides, and protein occurs during the last 6–8 weeks of berry ripening [47]. In grapes, the main free amino acids include proline (up to 2 g/L), arginine (up to 1.6 g/L), and to a lesser extent, alanine, aspartic acid, and glutamic acid [48]. However, compositional differences in amino acids were observed by Stines et al. [49] among grape varieties, proline and arginine always being the major grape amino acids. In all grape varieties, most of the proline accumulation happened late in ripening, nearby 4 weeks of post-*véraison*. In opposite, arginine accumulation started before *véraison* and continued to maturity, excluding grape varieties in which a great level of proline accumulated [49]. The variation of amino acid profile and their concentration in grapes depends on grape variety, but also on viticultural management and environmental conditions [43, 50, 51].

According to Hsu and Heatherbell [52], grapes contain naturally a wide range of different proteins, up to 41 protein fractions with molecular mass ranging from 11.2 to 190 kDa and isoelectric point from 2.5 to 8.7 [53, 54]. Soluble proteins in grape are globular proteins, mainly albumins [55, 56]. There is a significant increase in grape total protein content after *véraison* being a small content of proteins synthesized significantly during grape ripening [55, 57]. The most abundant grape proteins synthesized during ripening are pathogenesis-related proteins, including chitinases (32 kDa) and thaumatin-like proteins (24 kDa) [29, 57, 58].

2.2.3. *Aroma and flavor compounds*

Free and bound terpene grape content has been used to measure berry flavorant development and potential. Numerous types of flavorants existed in the form of glycosidic precursors. Analysis of the total precursor content by assessment of the glycoside glucose (GG) content of the grapes may yield a more complete depiction of the grape flavorant potential [59]. During grape maturity, changes in the concentration and diversity of aroma precursors and volatile compounds occurred [60, 61]. Lacey et al. [60] observed that grapes grown under cool temperatures showed higher grape methoxypyrazine concentration than grapes grown under hot temperatures. Grape methoxypyrazine levels were relatively high at *véraison* but decreased markedly with grape ripening. However, since grape maturation is genetically controlled, it is considerably influenced by environmental conditions [60].

2.2.4. *Phenolic compounds*

Phenolic compounds are very important for wine quality because they are responsible for most of the wine sensory characteristics, particularly color and astringency. These groups of compounds constitute a diverse group of secondary metabolites that exist in grapes, mainly in the grape berries' skins and seeds [62] and also in grape stems [63]. The phenolic compounds in *V. vinifera* grapes include two classes of phenolic compounds: non-flavonoids and flavonoids. The non-flavonoid compounds include phenolic acids divided into hydroxybenzoic acids and hydroxycinnamic acids, but also other phenol derivatives such as stilbenes (**Figure 2**). Non-flavonoids incorporate C_6-C_3 hydroxycinnmates acids, C_6-C_1 hydroxybenzoic acids, and C_6-C_3-C_6 stilbenes *trans*-resveratrol, *cis*-resveratrol, and *trans*-resveratrol glucoside.

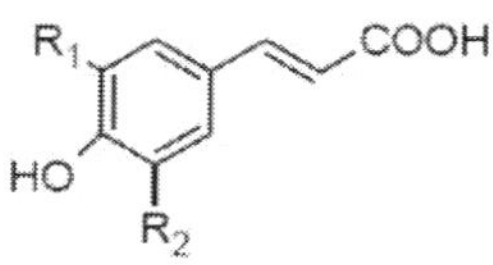

Hydroxycinnamic acids	*R₁*	*R₂*
Caffeic acid	OH	H
Ferulic acid	OCH_3	H
p-Coumaric acid	H	H
Sinapic acid	OCH_3	OCH_3

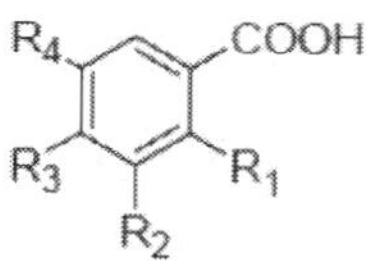

Hydroxybenzoic acids	*R₁*	*R₂*	*R₃*	*R₄*
Gallic acid	H	OH	OH	OH
p- Hydroxybenzoic acid	H	H	OH	H
Vanillic acid	H	OCH_3	OH	H
Syringic acid	H	OCH_3	OH	OCH_3

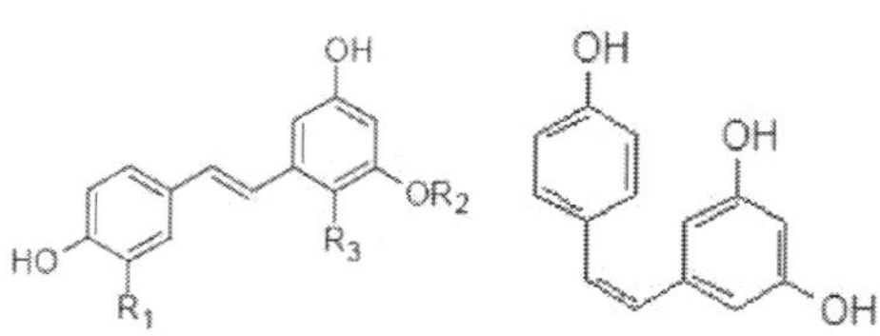

Trans-resveratrol *Cis*-resveratrol

Stilbenes	*R₁*	*R₂*	*R₃*
Trans-resveratrol	H	H	H
Piceid	H	OCH_3	OH

Figure 2. Main non-flavonoid compounds found in *V. vinifera* grapes.

For flavonoid compounds, there are a large number of subclasses, such as flavonols, flavanols, and anthocyanins [64]. Flavonols are the most abundant phenolic compounds in grape skins [65], while grape seeds are rich in flavan-3-ols [66]. Flavonoids are characterized by a basic structure of 15 carbon atoms comprising two aromatic rings bound through a three carbon chain (C_6-C_3-C_6). The major C_6-C_3-C_6 flavonoids in grapes include conjugates of flavonols quercetin, and myricetin; flavan-3-ols (+)-catechin and (−)-epicatechin; and malvidin-3-*O*-glucoside and other anthocyanins (**Figure 3a–c**).

According to Pastrana-Bonilla et al. [67], the average concentration of the total phenolic compounds in different grape fractions varied from 2178.8 mg/g gallic acid equivalent in seeds to 374.6 mg/g gallic acid equivalent in skins. In addition, it is also possible to found low concentrations of phenolic compounds in pulps (23.8 mg/g gallic acid equivalent).

(a)

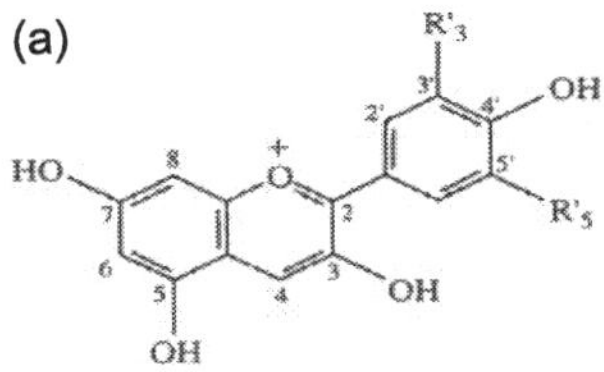

Anthocyanidins	R'_3	R'_5
Delphinidin	OH	OH
Cyanidin	OH	H
Petunidin	OH	OCH_3
Peonidin	OH	OCH_3
Malvidin	OCH_3	OCH_3

(b)

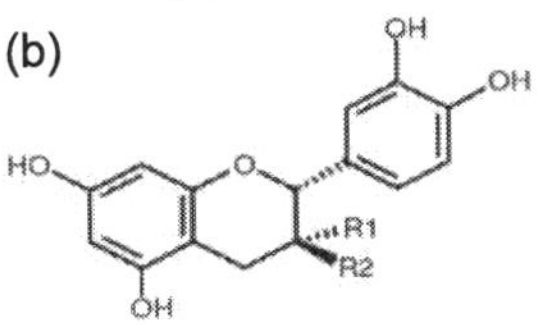

Flavan-3-ols	R_1	R_2
(+) - catechin	H	OH
(+) - epicatechin	OH	H

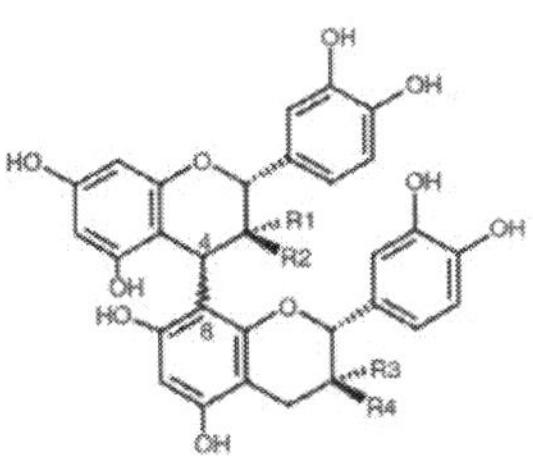

Dimers (C_4-C_8)	R_1	R_2	R_3	R_4
Procyanidin B1	OH	H	H	OH
Procyanidin B2	OH	H	OH	H
Procyanidin B3	H	OH	H	OH
Procyanidin B4	H	OH	OH	H

(c)

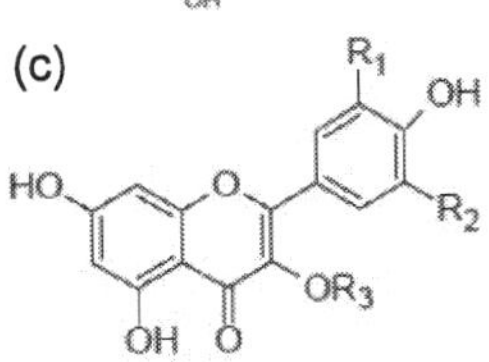

Flavonols	R_1	R_2	R_3
Myricetin	OH	OH	H
Myricetin-3-O-glucoside	OH	OH	gluc
Quercetin	OH	H	H
Quercetin-3-O-glucoside	OH	H	gluc
Kaempferol	H	H	H
Kaempferol-3-O-glucoside	H	H	gluc

Figure 3. (a) Main flavonoid compounds (anthocyanidins) found in *V. vinifera* grape varieties. (b) Main flavonoid compounds (flavan-3-ols and procyanidins) found in *V. vinifera* grape varieties. (c) Main flavonoid compounds (flavonols) found in *V. vinifera* grape varieties.

In general, the phenolic composition of grapes is influenced by different factors, such as grape variety [68, 69], sunlight exposition [70], solar radiation [71] altitude [72], soil composition [73], climate [70, 74–76], cultivation practices [43, 74], exposure to diseases [77], and the degree of grape ripeness [63, 69].

The quantification of phenolic acids, stilbenes, monomeric anthocyanins, flavan-3-ols, and proanthocyanidins in red grape varieties is summarized in **Tables 1–3** and the quantification of phenolic acids, stilbenes, flavan-3-ols, and proanthocyanidins in white grape varieties is summarized in **Table 4**.

Phenolic compounds	Grape variety	Concentration	References
Phenolic acids	Negroamaro	7.3[a]	
Gallic acid	Susumaniello	45.0[a]	Nicoletti et al. [78]
	Malvasia Nera	77.3[a]	
	Aglianico	151.9[a]	
	Merlot	66.6[a]	
	Carménère	2.8[b]	Obreque-Slier et al. [79]
	Cabernet Sauvignon	3.5[b]	
	Merlot	9.8[c]	Montealegre et al. [80]
	Cencibel	7.3[c]	
	Cabernet Sauvignon	9.0[c]	
	Shiraz	6.8[c]	
Protocatechuic acid	Negroamaro	42.0[a]	Nicoletti et al. [78]
	Susumaniello	8.5[a]	
	Malvasia Nera	46.0[a]	
	Aglianico	37.4[a]	
	Cesanese	31.1[a]	
	Merlot	328.7[a]	
	Cencibel	1.5[b]	Montealegre et al. [80]
	Cabernet Sauvignon	2.4[b]	
	Merlot	1.7[b]	
	Shiraz	2.4[b]	
	Merlot	8.7[c]	Montealegre et al. [80]
	Cencibel	3.3[c]	
	Cabernet Sauvignon	7.1[c]	
	Shiraz	6.2[c]	
Caftaric acid	Primitivo	1.89[a]	Nicoletti et al. [78]
	Negroamaro	8.5[a]	
	Susumaniello	171.7[a]	
	Malvasia Nera	171.9[a]	
	Aglianico	320.4[a]	
	Cesanese	28.8[a]	
	Alphonse	645.0[a]	
	Merlot	746.3[a]	
	Carménère	0.6[b]	Obreque-Slier et al. [79]

Phenolic compounds	Grape variety	Concentration	References
	Cabernet Sauvignon	0.7[b]	
Stilbenes	Primitivo	30.7[a]	Nicoletti et al. [78]
Trans-piceid	Negroamaro	4.14[a]	
	Susumaniello	150.3[a]	
	Uva di Troia	15.3[a]	
	Malvasia Nera	98.0[a]	
	Aglianico	75.7[a]	
	Cesanese	12.0[a]	
	Merlot	26.3[a]	
	Alphonse Lavallée	24.1[a]	
	Castelão	67.24[c]	Sun et al. [81]
	Syrah	10.43[c]	
	Tinta Roriz	11.57[c]	
Trans-resveratrol	Primitivo	13.9[a]	Nicoletti et al. [78]
	Negroamaro	3.6[a]	
	Susumaniello	63.0[a]	
	Uva di Troia	4.6[a]	
	Malvasia Nera	48.5[a]	
	Aglianico	61.1[a]	
	Cesanese	8.1[a]	
	Merlot	9.2[a]	
	Alphonse Lavallée	40.0[a]	
	Blauer Burgunder	0.5[d]	Mikeš et al. [82]
	Lemberger	0.3[d]	
	Saint Laurent	1.0[d]	
	Saint Laurent	2.3[d]	Balík et al. [83]
	Blauer Portugieser	0.4[d]	
	Andre	0.4[d]	
	Castelão	6.8[d]	Sun et al. [81]

[a]mg/kg of berry dry weight.
[b]mg/kg of fresh grape skin.
[c]mg/kg of fresh grape seed.
[d]mg/kg dry skin.

Table 1. Quantification of phenolic acids and stilbenes in red grape varieties.

Monomeric anthocyanins	Grape variety	Concentration	References
Delphinidin 3-*O*-glucoside	Cabernet-Sauvignon	431.6[a]	Ortega-Regules et al. [84]
	Merlot	231.7[a]	
	Syrah	258.0[a]	
	Cabernet Sauvignon	4.67[b]	Revilla et al. [85]
	Garnacha	2.26[b]	
	Graciano	6.81[b]	
	Mencia	5.13[b]	
	Merlot	7.53[b]	
	Tempranillo	10.9[b]	
	Castelão Francês	6.2[c]	Jordão et al. [86]
	Touriga Francesa	0.9[c]	
Cyanidin 3-*O*-glucoside	Cabernet-Sauvignon	53.1[a]	Ortega-Regules et al. [84]
	Merlot	48.2[a]	
	Syrah	27.9[a]	
	Cabernet Sauvignon	0.90[b]	Revilla et al. [85]
	Garnacha	1.02[b]	
	Graciano	1.28[b]	
	Mencia	2.15[b]	
	Merlot	5.52[b]	
	Tempranillo	3.26[b]	
	Castelão Francês	2.6[c]	Jordão et al. [86]
	Touriga Francesa	0.1[c]	
Petunidin-3-*O*-glucoside	Cabernet-Sauvignon	337.4[c]	Ortega-Regules et al. [84]
	Merlot	270.9[a]	
	Syrah	385.2[a]	
	Cabernet Sauvignon	4.21[b]	Revilla et al. [85]
	Garnacha	3.73[b]	
	Graciano	7.21[b]	
	Mencia	6.68[b]	
	Merlot	7.0[b]	
	Tempranillo	11.11[b]	
	Castelão Francês	8.5[c]	Jordão et al. [86]
	Touriga Francesa	2.5[c]	
Peonidin 3-*O*-glucoside	Cabernet-Sauvignon	259.5[a]	Ortega-Regules et al. [84]

Monomeric anthocyanins	Grape variety	Concentration	References
	Merlot	381.9[a]	
	Syrah	299.2[a]	
	Cabernet Sauvignon	4.87[b]	Revilla et al. [85]
	Garnacha	12.69[b]	
	Graciano	12.79[b]	
	Mencia	14.85[b]	
	Merlot	14.27[b]	
	Tempranillo	7.81[b]	
	Castelão Francês	11.7[c]	Jordão et al. [86]
	Touriga Francesa	3.6[c]	
	Cabernet-Sauvignon	2506.3[a]	Ortega-Regules et al. [84]
Malvidin 3-*O*-glucoside	Merlot	1834.7[a]	
	Syrah	2889.7[a]	
	Cabernet Sauvignon	41.45[b]	Revilla et al. [85]
	Garnacha	64.69[b]	
	Graciano	53.69[b]	
	Mencia	47.40[b]	
	Merlot	35.54[b]	
	Tempranillo	46.35[b]	
	Castelão Francês	59.2[c]	Jordão et al. [86]
	Touriga Francesa	46.3[c]	

[a]μg/g grape skin.
[b]Relative amount of anthocyanidins (%).
[c]% weight of anthocyanins/weight grape.

Table 2. Quantification of monomeric anthocyanins in red grape varieties.

Phenolic compounds	Grape variety	Concentration	References
Flavan-3-ols	Baboso Negro	51.61[a]	
(+)-Catechin	Listán Negro	54.25[a]	Pérez-Trujillo et al. [87]
	Negramoll	51.31[a]	
	Tintilla	50.10[a]	
	Vijariego Negro	49.09[a]	
	Touriga Nacional	0.012–0.021[b]	Mateus et al. [88]
	Touriga Francesa	0.012[b]	

Phenolic compounds	Grape variety	Concentration	References
	Merlot	240.0[c]	Montealegre et al. [80]
	Cencibel	82.0[c]	
	Cabernet Sauvignon	270.0[c]	
	Shiraz	120.0[c]	
(−)-Epicatechin	Baboso Negro	16.50[a]	Pérez-Trujillo et al. [87]
	Listán Negro	13.77[a]	
	Negramoll	15.07[a]	
	Tintilla	20.55[a]	
	Vijariego Negro	16.13[a]	
	Touriga Francesa	0.010[b]	Mateus et al. [88]
	Merlot	210.0[c]	Montealegre et al. [80]
	Cencibel	60.0[c]	
	Cabernet Sauvignon	130.0[c]	
	Shiraz	130.0[c]	
Proanthocyanidins	Touriga Nacional	0.013[b]	Mateus et al. [88]
Procyanidin B3	Merlot	64.0[c]	Montealegre et al. [80]
	Cencibel	43.0[c]	
	Cabernet Sauvignon	50.0[c]	
	Shiraz	55.0[c]	
Procyanidin B1	Baboso Negro	15.95[a]	Pérez-Trujillo et al. [87]
	Listán Negro	15.00[a]	
	Negramoll	14.69[a]	
	Tintilla	13.64[a]	
	Vijariego Negro	13.39[a]	
	Touriga Nacional	0.184–0.260[b]	Mateus et al. [88]
	Touriga Francesa	0.090–0.138[b]	
	Merlot	170.0[c]	Montealegre et al. [80]
	Cencibel	74.0[c]	
	Cabernet Sauvignon	150.0[c]	
	Shiraz	100.0[c]	
Procyanidin B4	Merlot	80.0[c]	Montealegre et al. [80]
	Cencibel	39.0[c]	
	Cabernet Sauvignon	57.0[c]	
	Shiraz	33.0[c]	

Phenolic compounds	Grape variety	Concentration	References
Procyanidin B2	Baboso Negro	10.39[a]	Pérez-Trujillo et al. [87]
	Listán Negro	5.74[a]	
	Negramoll	7.55[a]	
	Tintilla	9.92[a]	
	Vijariego Negro	7.44[a]	
	Touriga Nacional	0.020[b]	Mateus et al. [88]
	Touriga Francesa	0.011–0.015[b]	
	Merlot	37[c]	Montealegre et al. [80]
	Cencibel	21.0[c]	
	Cabernet Sauvignon	41.0[c]	
	Shiraz	23.0[c]	

[a]Molar percentages.
[b]mg/g dry weight.
[c]mg/kg of fresh grape seed.

Table 3. Quantification of flavan-3-ols and proanthocyanidins in red grape varieties.

Phenolic compounds	Grape variety	Concentration	References
Phenolic acids	Grüner Veltliner	3.9[a]	
Gallic acid	Hibernal	4.0[a]	Mikeš et al. [82]
	Malverina	3.5[a]	
	Müller Thurgau	2.6[a]	
	Rheinriesling	2.1[a]	
	Welschriesling	1.8[a]	
	Neuburger	3.9[a]	
Protocatechuic acid	Chardonnay	4.8[b]	Montealegre et al. [80]
	Sauvignon Blanc	4.4[b]	
	Moscatel	3.6[b]	
	Gewürztraminer	6.0[b]	
Caftaric acid	Moscato	48.4[c]	Nicoletti et al. [78]
Stilbenes	Chardonnay	1.1[a]	Balík et al. [83]
Trans-piceid	Welschriesling	0.4[a]	
	Pinot Gris	0.6[a]	
Trans-resveratrol	Moscato	3.89[c]	Nicoletti et al. [78]
	Grüner Veltliner	0.1[a]	Mikeš et al. [82]
	Hibernal	0.3[a]	

Phenolic compounds	Grape variety	Concentration	References
	Malverina	0.3[a]	
	Müller Thurgau	0.3[a]	
	Rheinriesling	0.2[a]	
	Welschriesling	0.5[a]	
	Neuburger	1.5[a]	
	Chardonnay	0.3[b]	
	Welschriesling	1.6[b]	Balík et al. [83]
	Pinot Gris	1.1[b]	
Flavan-3-ols	Chardonnay	123[a]	
(+)-Catechin	Welschriesling	61.0[a]	Balík et al. [83]
	Pinot Gris	481[a]	
	Ugni blanc	2.6–222.0[d]	De Freitas and Glories [89]
	Sémillon	12–35.2[d]	
	Chardonnay	390.0[c]	Montealegre et al. [80]
	Sauvignon Blanc	200.1[c]	
	Moscatel	350.0[c]	
	Gewürztraminer	500.0[c]	
	Riesling	400.0[c]	
	Viogner	120.0[c]	
(−)-Epicatechin	Chardonnay	144[a]	Balík et al. [83]
	Welschriesling	84.3[a]	
	Pinot Gris	251[a]	
	Ugni blanc	0.04–3.0[d]	De Freitas and Glories [89]
	Sémillon	0.03–1.6[d]	
	Chardonnay	310.0[c]	Montealegre et al. [80]
	Sauvignon Blanc	130.0[c]	
	Moscatel	120.0[c]	
	Gewürztraminer	150.0[c]	
	Riesling	160.0[c]	
	Viogner	110.0[c]	
Proanthocyanidins	Ugni blanc	0.2–0.3[d]	De Freitas and Glories [89]
Procyanidin B3	Sémillon	0.01–0.2[d]	
	Chardonnay	52.0[c]	Montealegre et al. [80]
	Sauvignon Blanc	52.0[c]	
	Moscatel	39.0[c]	
	Gewürztraminer	56.0[c]	

Phenolic compounds	Grape variety	Concentration	References
	Riesling	43.0[c]	
	Viogner	51.0[c]	
Procyanidin B1	Ugni blanc	1.1–1.9[d]	De Freitas and Glories [89]
	Sémillon	0.02–0.4[d]	
	Chardonnay	380.0[c]	Montealegre et al. [80]
	Sauvignon Blanc	250.0[c]	
	Moscatel	330.1[c]	
	Gewürztraminer	460.0[c]	
	Riesling	620.0[c]	
	Viogner	200.0[c]	
Procyanidin B4	Ugni blanc	0.04[d]	De Freitas and Glories [89]
	Chardonnay	71.5[c]	Montealegre et al. [80]
	Sauvignon Blanc	54.0[c]	
	Moscatel	40.0[c]	
	Gewürztraminer	70.0[c]	
	Riesling	95.0[c]	
	Viogner	53.0[c]	
Procyanidin B2	Ugni blanc	0.06–0.2[d]	De Freitas and Glories [89]
	Chardonnay	33.0[c]	Montealegre et al. [80]
	Sauvignon Blanc	19.0[c]	
	Moscatel	15.0[c]	
	Gewürztraminer	22.0[c]	
	Riesling	33.0[c]	
	Viogner	19.0[c]	

[a]mg/kg fresh grape weight.
[b]mg/kg of fresh grape seed.
[c]mg/kg of berry dry weight.
[d]mg/g dry weight.

Table 4. Quantification of phenolic acids, stilbenes, flavan-3-ols, and proanthocyanidins in white grape varieties.

3. Biotechnological approaches for wine quality

More than 800 volatile compounds have been identified in wines, with a concentration range from hundreds of mg/L to the μg/L or ng/L [90]. The wine bouquet is formed by secondary metabolites synthesized by an extensive range of microbial species (yeasts and bacteria). Wine alcoholic fermentation (AF) is the key for innovation or creation of biotechnology that will change the expanding market [91] (**Figure 4**).

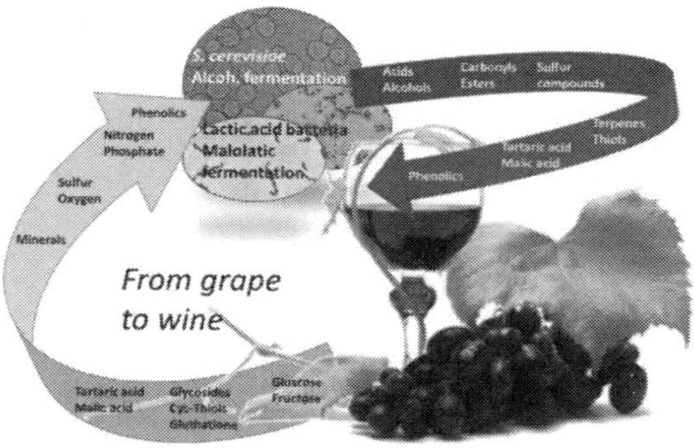

Figure 4. Grape juice is converted into wine by the action of wine yeast and bacteria during alcoholic and malolactic fermentations. Some wine components are wholly generated by these microorganisms as part of metabolism, while others are essentially synthesized by the grapevine. Wine quality and style is determined by the quality and quantity of compounds produced or modified by must/wine microflora.

In addition to yeasts, LAB also appears in all type of wines, being responsible for MLF that normally occurs after AF but may also occur simultaneously [92]. During the winemaking process, indigenous populations of LAB vary quantitatively and qualitatively [93], through a succession of species and strains before, during and after the AF [94]. After a phase of latency, the surviving cells begin to multiply and entering the exponential growth phase, reaching populations from 10^6 to 10^8 cfu/mL, almost exclusively, constituted by strains of *Oenococcus oeni,* species that dominate this stage and performs the MLF. Normally, a great diversity of strains of *Oenococcus oeni* at the beginning of the MLF is detected, while at the end only one or two predominate [95].

3.1. Yeasts metabolites: the imperceptible search of perfection

Wine yeasts contribute to wine aroma by a number of mechanisms: (i) they utilize grape juice constituents and transform them into flavor-impacting components, then (ii) they produce enzymes capable to transform neutral grape compounds into flavor-active compounds, and finally (iii) they can synthesize many flavor-active compounds such as primary and secondary metabolites [96].

Esters, in wine, are mainly originated from yeast metabolism during AF. But, some esters are also found in grape berry [97], where they occur in small amounts, contributing to the aroma of *V. vinifera* varieties [98]. Esters are formed via an intracellular process, catalyzed by an acyl transferase or ester synthase [99]. The concentration of esters usually found in wine is mostly well above their sensory threshold levels. Fruity and floral terms in Chardonnay wines were related to 2-phenylethyl acetate, as a rose-like/honey aroma [100] (**Table 5**). In red wines, ethyl butyrate (pineapple aroma), ethyl 2-methylbutyrate (sweet, floral, fruity, and apple), ethyl 3-methylbutyrate (strawberry, ethereal, buttery, and ripe), isoamyl acetate (banana-like aroma), ethyl hexanoate (anise seed, apple, or pineapple aroma), and ethyl octanoate (sweet, cognac, and apricot aroma) made a main contribution to the fruity character of wines [101] (**Table 5**).

These esters also appear in higher levels in wines after bio-reduction (deacidification) of wine's volatile acidity [102]. A study of overexpression *S. cerevisiae* alcohol acetyltransferases genes, ATF1p, ATF2p, and Lg-ATF1p, was performed by Verstrepen et al. [103]. Analysis of the fermentation products confirmed that the expression levels of ATF1 and ATF2 greatly affected the production of ethyl acetate and isoamyl acetate. But, factors such as oxygen and temperature that allow ester and higher alcohol synthesis must be monitored during AF [104].

Compounds	**Odor description**	**Det. Threshold (μg/L)**	**References**
Isoamyl acetate	Banana	30	Guth [115]
2-Phenylethylacetate	Roses, honey	250	Guth [115]
Ethylpropionate	Ethereal, fruity, rum-like	1800	Etievant [116]
Ethylisobutyrate	Strawberry, ethereal, buttery, ripe	15	Etievant [116]; Ong and Acree [117]
Ethyl butyrate	Pineapple	20	Guth [115]
Ethyl 2-methylbutyrate	Sweet, floral, fruity, apple	1–18	Guth [115]; Ferreira et al. [118]
Ethylisovalerate	Fruity	3	Ferreira et al. [118]
Ethyl hexanoate	Anise seed, apple, pineapple	5–14	Guth [115]; Ferreira et al. [118]
Ethyl octanoate	Sweet, cognac, apricot	2–5	Guth [115]; Ferreira et al. [118]
Diethylsuccinate	Fruity, melon	1200	Peinado et al. [119]
Acetaldehyde	Grass, green, apple, sherry	100,000	Carlton et al. [120]
Benzaldehyde	Almond	3500	Delfini et al. [121]
Linalool	Rose, lavender	25	Ferreira et al. [118]
α-Terpineol	Lily of the valley	300	Mateo and Jiménez [122]
Citronellol	Citronella	100	Guth [115]
Geraniol	Rose-like; geranium flowers	~75	Pardo et al. [109]
2-phenylethanol	Roses	10,000	Guth [115]
Isoamyl alcohol	Marzipan, burnt, whisky -like	30,000	Guth [115]
Butyric acid	Rancid, cheese	173	Ferreira et al. [118]
Isovaleric acid	Rancid, sweaty	33.4	Ferreira et al. [118]
Hexanoic acid	Sweaty, cheesenotes	420–3000	Guth [115]; Ferreira et al. [118]
Octanoic acid	Grass acid- like	500–8800	Etievant [116]; Ferreira et al. [118]
Decanoic acid	Soapy	1000–15,000	Guth [115]; Ferreira et al. [118]

Table 5. Major wine-yeast aromatic compounds, odor description, and detection thresholds in white and red wines.

Ethanol and glycerol are quantitatively the largest group of alcohols found in wine. Both contribute to the textural aspects of wines [1]. The search of yeast that can impart specific desirable characteristics to wines led to investigations such as the production of optimal levels of glycerol (the overexpression of GPD1, GPD2, and FPS1, together with the deletion of the ALD6 acetaldehyde dehydrogenase gene) [105].

Medium-chain fatty acids and their ethyl esters are natural components of alcoholic beverages. Fatty acids (butyric, isovaleric, hexanoic, octanoic, and decanoic acids, among others; **Table 5**) are produced by yeasts as intermediates in the biosynthesis of long-chain fatty acids, important components of yeast membrane [106]. Their aroma goes from vinegar to pungent, rancid, and soapy, sweetie, fruit and butter [106] (**Table 5**). One of the major problematic volatile acids is acetic acid. It can be formed as a by-product of AF, MLF, or as a product of the metabolism of acetic bacteria. Acetic acid affects the quality of certain types of wine when it is present above a given concentration [107] due to its unpleasant vinegar aroma.

Terpenes are one of the major grape components that contribute to wine aroma. This is especially valid to wines of Gewürztraminer and Muscat varieties, but these flavor compounds are also present in other grape varieties, where they supplement other varietal flavors and aromas. They are present in two forms: a free volatile and a non-volatile sugar-conjugated [108]. Geraniol (geranium flowers aroma) and linalool (rose or lavender-like aroma) are considered to be the most important of the monoterpene alcohols as they are present in higher levels and have lower perception thresholds than other major wine monoterpenes [109]. Monoterpenes can be released from their glycosides either by acid or by enzymatic hydrolysis. Hydrolysis during winemaking is caused by grape [110] or microorganisms enzymes taking part in the process [111]. In the yeasts that were selected in the past years, glycosidase activities have been used for the hydrolysis of glycoconjugated aromatic precursors in order to enhance wine sensorial quality [112]. Fungi are considered a promising genetic source for commercial production of recombinant β-glucosidase [113]. In a work by Zietsman et al. [114], an yeast strain (*S. cerevisiae* VIN13) was built to express and secrete the *Aspergillus awamori* encoding a B-type α-l-arabinofuranosidase (AwAbfB) in combination with either the β-glucosidases BGL2 from *Saccharomycopsis fibuligera* or the BGLA from *Aspergillus kawachii*. Coexpression of AwAbfB and BGL2 in VIN13 increased free monoterpenes in wines. Panelists confirmed wine aroma profile improvement, mainly in floral character [114]. Recently, Pardo et al. [109] found that the expression of *Ocimum basilicum* (sweet basil) geraniol synthase (GES) gene in an *S. cerevisiae* wine strain greatly changed terpene profile of wine made from a non-aromatic grape variety.

3.2. Lactic acid bacteria metabolites: beyond malolactic fermentation

The complexity and diversity of LAB metabolic activities in wine illustrates that MLF is more than a mere decarboxylation of L-malic acid into L-lactic acid, and it may affect positively and/or negatively the quality of wine [123] (**Table 6**). Besides to the decrease in acidity, MLF also improves sensorial characteristics and increases wines microbiological stability that undergone this important second fermentation [124, 125].

Aromatic modifications are due to L-lactic acid, less aggressive, and due to the increase of a number of other compounds such as diacetyl, acetoin, 2,3-butanediol, esters mainly ethyl lactate and diethyl succinate, and some higher alcohols and aromatic aglycones released by the action of β-glucosidases [126–128]. Sumby et al. [129] have verified the impact that different strains of *O. oeni* had on wine aroma and related that to their ester hydrolysis and synthesis abilities. For the aromatic complexity of wines, the production of volatile sulfur compounds, particularly 3-methylsulfanyl-propionic acid with chocolate and toasted odors [130], and the activity of taninoacil hydrolase enzyme, commonly termed tannase, reducing wine astringency and turbidity [131], also contribute.

Compounds	Odor description	Det. threshold (μg/L)	References
4-Ethylguaiacol	Bacon, spice, clove, or smoky aromas	33	Dai et al. [26]; Bartowsky [123]
4-Ethylphenol	Horse and barnyard odor	440	Barthelmebs et al. [147], [148]
Tetrahydropyridines	Mousy off-odor	60	Swiegers et al. [149]; Harrison and Dake [150]
3- Methylsulfanyl-propionic acid	Chocolate and toasted odors	244	Pripis-Nicolau et al. [151]
Ethyl lactate	Lactic, raspberry	154–636	Ferreira et al. [118]; Bartowsky [152]
Diethyl succinate	Fruity, melon	1200	Peinado et al. [119]; Bartowsky [152]
Diacetyl	Butter	200–2800	Martineau and Henick-Kling [153]; Bartowsky and Henschke [154]
Acetoin	No negative organoleptic influence. Unpleasant buttery flavor at concentrations higher than threshold	150	Swiegers et al. [155]; Ehsani et al. [156]
2,3-Butanediol	Neutral sensory qualities	150	Swiegers et al. [155]; Romano and Suzzi [157]

Table 6. Major LAB aromatic compounds, odor description, and detection thresholds in wine.

Concerning to negative effects on wine quality, LAB may be responsible for the formation of ethyl carbamate by the degradation of arginine [124] and for the formation of biogenic amines such as histamine, tyramine, and putrescine by the degradation of precursor amino acids [132, 133]. Also, although less frequent nowadays, bitterness by acrolein formation from glycerol degradation [134], butter aroma due to excessive production of diacetyl [135], flocculent growth [136], mannitol taint [137], ropiness [138], tartaric acid degradation [137], mousy off-odor by acetamide production of tetrahydropyridines [139], the geranium off-odor [140], and the formation of 4-ethilguaiacol and 4-ethylphenol volatile phenols [141, 142] are spoilage phenomenons that may occur after malolactic fermentation. Nevertheless, it is thought that the time between the completion of alcoholic fermentation and the start of malolactic fermentation is the most likely time that *Brettanomyces* multiplies and produces "Brett character," 4-ethylphenol of flavor, in wine [143].

As what happens to other food products, some researchers defend the use of autochthones LAB strains, more adapted and efficient to regional vinification conditions, for keeping the typicity of wines, instead of using universal ones that may impart similar characteristics and thus leading to final products that are too similar and also for preserving the local microbial biodiversity [144, 145]. According to Marcobal and Mills [146], the knowledge of some wine LAB whole genome, including the PSU1 *O. oeni* strain, allows deeper phylogenetic analyses and their relation with key pathways involved in carbon and nitrogen metabolism, which will foster modeling of *O. oeni* growth and metabolism in order to predict optimum strategies for efficiently performing the MLF with a desired flavor outcome.

4. Composition of grapes and wines: new analytical techniques

Several different analytical approaches are increasingly used to profile the volatile, non-volatile, and elemental composition of grapes and wines (see recent reviews, e.g., [158, 160]).

According to a review made by Ebeler [159], we can group these analytical approaches in (i) targeted analysis of compounds, (ii) non-targeted analysis and profiling of metabolites, (iii) elemental analysis, and (iv) relating chemical composition and sensory attributes (**Table 7**).

Therefore, wine composition and hence wine origin are possible by combining several analytical techniques (**Table 7**) that offer significant advantages for trace quantification of important aroma-active volatiles [174], [175] and taint compounds [163]. It is also possible to comprehensively profile metals [178], including those that affect chemical stability and oxidative reactions, and to characterize aroma qualities of complex mixtures [182]. Each of these tools, alone and in combination, is providing significant new insights into variables influencing grape and wine composition and flavor. Moreover, concerning to specific grape compounds, in past years, several methodologies were also developed focused on the identification, quantification, and also in extraction techniques. For example for phenolic compounds, substantial developments for individual phenolic analysis, such as benzoic and cinnamic acid, coumarins, tannins, lignins, lignans, and flavonoids, have occurred over the last 25 years. Thus, several extraction techniques have been employed namely for grape phenolic compounds, such as ultrasounds and microwaves [183], supercritical fluid extraction [184], subcritical water extraction [185], high hydrostatic pressure extraction [186], pulsed electric fields [187], and enzymatic treatment [188].

Analytical approaches	Analytical techniques	Examples and references
Targeted analysis of compounds (i)	Selected ion monitoring and tandem mass spectrometric, MS/MS or MS^n	Analysis of trace analytes, with important sensory properties—Ebeler [160] and Robinson et al. [161, 162]—such as 2,4,6-trichloroanisole (TCA)—Hjelmeland et al. [163]
	Combination of liquid chromatography, LC with mass spectrometry, MS. MS/MS is the combination of two mass analyzers in one mass spectrometry instrument, LC-MS/MS/LC-MS/MS. Supercritical fluid chromatography (SFC)	Smoke-derived volatile phenols—guaiacol and their glycoside precursors, and anthocyanins from grapes and wines—Kennison et al. [164–166], Hayasaka et al. [167], and Pati et al. [168]. Polyphenols from grape seed extracts—Kamangerpour et al. [169]
Non-targeted analysis and profiling of metabolites (ii)	Ultra-high performance liquid chromatography, UHPLC wish operates in the 20,000 psi range, combined with quadrupole time-of-flight mass spectrometry, qTOF and UHPLC-qTOF-MS	Varietal classification of wines—Vaclavik et al. [170] and Flamini [171]
	Ion cyclotron resonance mass spectrometry, ICR-MS	Characterization of Pinot Noir grapes and wines and chemodiversity comparison of different appellations: Vintage vs terroir effects—Roullier-Gall et al. [172, 173]
	Gas chromatography combined with time-of-flight mass spectrometry, GC GC-TOF-MS	Identification of over 350 volatile compounds in Australian Cabernet Sauvignon wines—Robinson et al. [174, 175]
	Nuclear Magnetic Resonance, NMR	^1H NMR metabolite profiling to relate chemical composition to sensory perception of body and mouthfeel of white wines—Kogerson et al. [176]
Elemental analysis (iii)	Inductively coupled plasma mass Spectrometry, ICP-MS	Relating elemental composition of wines to the vineyard that the grapes were grown or in wish winery they were made—Hopfer et al. [177]. Leaching of metals from stainless steel containers and from closures—Hopfer et al. [178]
Relating chemical composition and sensory attributes (iv)	Categorical principal components analysis, CATPCA; principal components analysis, PCA and partial least squares analysis, PLS	One or more compounds that correlate with specific aroma or flavor attributes—Polaskova et al. [179] and development of a flavor lexicon using new statistical nonparametric approaches—Vilela et al. [180] and Monteiro et al. [181]
	In-instrument gas chromatography recomposition-olfactometry, GC-RO	Perceptual characterization and analysis of aroma mixtures—Johnson et al. [182]

Table 7. Analytical approaches, analytical techniques used to profile the volatile, non-volatile and elemental composition of grapes and wines.

5. Final remarks

The study of the grape berry physiology and metabolome will provide a deep understanding of the primary metabolites including sugars, organic acids and amino acids, and some secondary metabolites accumulated in the grape berries such as phenolic compounds. This issue is of particular importance for viticulturists and oenologists in order to know how grape composition could affect wine quality. In addition, biotechnological approaches for wine quality, practiced during wine AF and MLF, are also a promising tool available for oenologists that improve wine quality, namely, their sensorial value.

Author details

Fernanda Cosme[1], Berta Gonçalves[2], António Inês[1], António M. Jordão[3] and Alice Vilela[1*]

*Address all correspondence to: avimoura@utad.pt

1 CQ-VR, Chemistry Research Centre, Department of Biology and Environment, School of Life Sciences and Environment, University of Trás-os-Montes and Alto Douro (UTAD), Vila Real, Portugal

2 CITAB, Centre for the Research and Technology of Agro-Environmental and Biological Sciences, Department of Biology and Environment, School of Life Sciences and Environment, University of Trás-os-Montes and Alto Douro, Vila Real, Portugal

3 Department of Food Industries, Agrarian Higher School, Polytechnic Institute of Viseu (CI&DETS), Viseu, Portugal

References

[1] Jordão AM, Vilela A, Cosme F: From sugar of grape to alcohol of wine: sensorial impact of alcohol in wine. Beverages. 2015;1(4):292–310. Doi: 10.3390/beverages1040292.

[2] Bisson LF: The biotechnology of wine yeast. Food Biotechnology. 2004;18(1):63–96. Doi: 10.1081/FBT-120030385.

[3] Swiegers JH, Pretorius IS: Modulation of volatile sulfur compounds by wine yeast. Applied Microbiology and Biotechnology. 2007;74:954–960. Doi: 10.1007/s00253-006-0828-1.

[4] Chambers PJ, Pretorius IS: Fermenting knowledge: the history of winemaking, science and yeast research. EMBO Reports. 2010;11(12):914–920. Doi: 10.1038/embor.2010.179.

[5] Inês A, Tenreiro T, Tenreiro R, Mendes-Faia A: Review: The lactic acid bacteria of wine- Part I. Ciência e Técnica Vitivinícola. 2008;23:81–96.

[6] Inês A, Tenreiro T, Tenreiro R, Mendes-Faia A: The lactic acid bacteria of wine- Part II. Ciência e Técnica Vitivinícola. 2009;24:1–23.

[7] Hellman EW: Grapevine structure and function. Oregon Viticulture. Hellman, EW (ed.). Oregon State University Press, Corvallis, 2003; pp. 5–19.

[8] Jackson RS: Grapevine Structure and Function, Chapter 5 in Wine Science, 2nd ed. Academic Press, San Diego, 2000; pp. 66–71.

[9] Keller M: The Science of Grapevines: Anatomy and Physiology. 2nd ed. Academic Press, Oxford, 2015; pp. 522.

[10] Wilson B, Strauss CR, Williams PJ: The distribution of free and glycosidically-bound monoterpenes among skin, juice and pulp fractions of some white grape varieties. American Journal of Enology and Viticulture. 1986;37:107–111.

[11] Teixeira A, Eiras-Dias J, Castellarin SD, Gerós H: Berry phenolics of grapevine under challenging environments. International Journal of Molecular Sciences. 2013;14(9): 18711–18739. Doi: 10.3390/ijms140918711.

[12] Hardie WJ, O'Brien TP, Jaudzems VG: Morphology, anatomy and development of the pericarp after anthesis. Australian Journal of Grape and Wine Research. 1996;2(2):97–141.

[13] Rogiers SY, Hatfield JM, Jaudzems VG, White RG, Keller M: Grape berry cv. Shiraz epicuticular wax and transpiration during ripening and pre harvest weight loss. American Journal of Enology and Viticulture. 2004;55:121–127.

[14] Blanke MM, Pring RJ, Bake, EA: Structure and elemental composition of grape berry stomata. Journal of Plant Physiology. 1999;154:477–481. Doi: 10.1016/S0176-1617(99)80286-7.

[15] Comménil P, Brunet L, Audran J-C: The development of the grape berry cuticle in relation to susceptibility to bunch rot disease. Journal of Experimental Botany. 1997;48:1599–1607.

[16] Yamamura H, Naito R: The surface wax of several grapes in Japan. Journal Japan Society of Horticultural Science. 1983;52:266–272.

[17] Walton TJ, Kolattukudy PE: Determination of the structures of cutin monomers by a novel depolymerization procedure and combined gas chromatography and mass spectrometry. Biochemistry. 1972;11:1885–1897.

[18] Kolattukudy PE: Plant waxes. Lipids. 1969;5:259–275.

[19] Coombe B: Research on development and ripening of the grape berry. American Journal of Enology and Viticulture. 1992;43:101–110.

[20] Tilbrook J, Tyerman SD: In: Finishing the job—Optimal ripening of Cabernet Sauvignon and Shiraz, Water, sugar and acid: How and where they come and go during berry ripening. Oag D., DeGaris K., Partridge S., Dundon C., Francis M., Johnstone R., Hamilton R (Eds.). Austral. Soc. Viticult. Oenol. Adelaide, Australia, 2006; pp. 4–12.

[21] Sundberg E, Ostergaard L: Distinct and dynamic auxin activities during reproductive development. Cold Spring Harbor Perspective Biology. 2009;1:a001628. Doi: 10.1101/cshperspect.a001628.

[22] Coombe BG: The development of fleshy fruits. Annual Review of Plant Physiology. 1976,27: 507–528.

[23] Gerling C: Environmentally Sustainable Viticulture: Practices and Practicality. Apple Academic Press, Canada, 2015; 424 pp.

[24] Conde BC, Silva P, Fontes N, Dias ACP, Tavares RM, Sousa MJ, Agasse A, Delrot S, Geros H: Biochemical changes throughout grape berry development and fruit and wine quality. Food. 2007;1:1–22. http://hdl.handle.net/1822/6820.

[25] Kalliopi A, Roubelakis-Angelakis KA (eds.). Grapevine Molecular Physiology and Biotechnology, 2nd ed. Springer Dordrecht Heidelberg London New York 2009; 599 pp. Doi: 10.1007/978-90-481-2305-6.

[26] Dai ZW, Ollat N, Gomès E, Decroocq S; Tandonnet J-P, Bordenave L, Pieri P, Hilbert G, Kappel C, Van Leeuwen C, Vivin P, Delrot S: Ecophysiological, genetic, and molecular causes of variation in grape berry weight and composition: a review. American Journal of Enology and Viticulture. 2011;62:413–425.

[27] Liu HF, Wu BH, Fan PG, Li SH, Li LS: Sugar and acid concentrations in 98 grape cultivars analyzed by principal component analysis. Journal of the Science of Food and Agriculture. 2006;86:1526–1536. Doi: 10.1002/jsfa.2541.

[28] Boss PK, Davies C: Molecular biology of sugar and anthocyanin accumulation in grape berries. In: Molecular Biology and Biotechnology of the Grapevine, Roubelakis-Angelakis KA (ed.). Kluwer Academic Publishers, Dordrecht, The Netherlands, 2001; pp. 1–33.

[29] Robinson SP, Davies C: Molecular biology of grape berry ripening. Australian Journal of Grape and Wine Research. 2000;6:175–188. Doi: 10.1111/j.1755-0238.2000.tb00177.x

[30] Fillion L, Ageorges A, Picaud S, Coutos-Thevenot P, Lemoine R, Romieu C, Delrot S: Cloning and expression of a hexose transporter gene expressed during the ripening of grape berry. Plant Physiology. 1999;120:1083–1093. Doi: http://dx.doi.org/10.1104/pp.120.4.1083.

[31] Esteban MA, Villanueva MJ, Lissarrague JR: Relationships between different berry components in Tempranillo (*Vitis vinifera* L.) grapes from irrigated and non-irrigated vines during ripening. Journal of the Science of Food and Agriculture. 2002;82:1136–1146. Doi: 10.1002/jsfa.1149

[32] Jordão AM, Ricardo-da-Silva JM, Laureano O: Influence of irrigation on phenolic composition of the French Touriga red wine-grapes variety (*Vitis vinifera* L.). Ciência e Tecnologia dos Alimentos. 1998;2:60–73.

[33] Van Leeuwen C, Tregoat O, Choné X, Bois B, Pernet D, Gaudillère J-P: Vine water status is a key factor in grape ripening and vintage quality for red Bordeaux wine. How can it be assessed for vineyard management purposes? Journal International des Sciences de la Vigne et du Vin. 2009;43:121–134. http://prodinra.inra.fr/record/33055.

[34] Hawker JS: Effect of temperature on lipid, starch and enzymes of starch metabolism in grape, tomato and broad bean-leaves. Phytochemistry. 1982;21:33–36. Doi: 10.1016/0031-9422(82)80009-5.

[35] Mira de Orduña R: Climate change associated effects on grape and wine quality and production. Food Research International. 2010;43:1844–1855. Doi: 10.1016/j.foodres. 2010.05.001.

[36] García-Martín N, Perez-Magariño S, Ortega-Heras M, González-Huerta C, Mihnea M, González-Sanjosé ML, Palacio L, Prádanos P, Hernández A: Sugar reduction in musts with nanofiltration membranes to obtain low alcohol-content wines. Separation and Purification Technology. 2010;76:158–170. Doi: 10.1016/j.seppur.2010.10.002.

[37] Kanellis AK, Roubelakis-Angelakis KA: Grape. In: Biochemistry of Fruit Ripening, Seymour G, Taylor J, Tucker G (Eds.). London: Chapman &Hall, 1993; pp. 189–234.

[38] Lamikanra O, Inyang ID, Leong S: Distribution and Effect of Grape Maturity on Organic Acid Content of Red Muscadine Grapes. Journal of Agricultural and Food Chemistry. 1995;43(12):3026–3028. Doi: http://dx.doi.org/10.1021/jf00060a007.

[39] Kliewer WM: Sugars and Organic Acids of *Vitis vinifera*. Plant Physiology. 1966;41:923–931.

[40] Rüffner HP: Metabolism of tartaric and malic acid in Vitis: a review—Part B. Vitis. 1982;21:346–358.

[41] DeBolt S, Cook DR, Ford CM: L-Tartaric acid synthesis from vitamin C in higher plants. Proceedings of the National Academy of Sciences USA. 2006;103:5608–5613.

[42] Ruffner HP, Possner D, Brem S, Rast DM: The physiological role of malic enzyme in grape ripening. Planta. 1984;160:444–448.

[43] Jackson DI, Lombard PB: Environmental and management practices affecting grape composition and wine quality—a review. American Journal of Enology and Viticulture. 1993;44:409–430.

[44] Kliewer WM, Howarth L, Omori M: Concentrations of tartaric acid and malic acids and their salts in *Vitis vinifera* grapes. American Journal of Enology and Viticulture. 1967;18:42–54.

[45] Hale CR: Relation between potassium and the malate and tartrate contents of grape berries. Vitis. 1977;16: 9–19.

[46] Saito K, Kasai Z: Conversion of labeled substrates to sugars, cell wall polysaccharides, and tartaric acid in grape berries. Plant Physiology. 1978;62(2):215–219.

[47] Wermelinger B: Nitrogen Dynamics in Grapevine. Physiology and Modeling. Proc. Int. Symp. Nitrogen in grape and wines, Seatle WA, USA (Ed J. M. Rantz) 1999; pp. 23–31.

[48] Huang Z, Ough CS: Amino acid profiles of commercial grape juices and wines. American Journal of Enology and Viticulture. 1991;45:261–267.

[49] Stines AP, Grubb J, Gockowiak H, Henschke PA, Høj PB, van Heeswijck R: Proline and arginine accumulation in developing berries of *Vitis vinifera* L. in Australian vineyards: Influence of vine cultivar, berry maturity and tissue type. Australian Journal of Grape and Wine Research. 2000;6:150–158. Doi: 10.1111/j.1755-0238.2000.tb00174.x

[50] Feuillat M: The nitrogenous constituents of grapes and wines. Le Vigneron Champenois. 1974;5:201–210.

[51] Spayd SE, Andersen-Bagge J: Free amino acid composition of grape juice from 12 *Vitis vinifera* cultivars in Washington. American Journal of Enology and Viticulture. 1996;47:389–402.

[52] Hsu JC, Heatherbell DA: Isolation and characterization of soluble proteins in grapes, grape juice, and wine. American Journal of Enology and Viticulture. 1987;38:6–10.

[53] Anelli G: The proteins of must. American Journal of Enology and Viticulture. 1977;28:200–203.

[54] Yokotsuka K, Yoshii M, Aihara T, Kushida T: Isolation and characterization of soluble glycoproteins in red wine. Journal of Fermentation Technology. 1977;55:510–515.

[55] Murphey JM, Spayd SE, Powers JR: Effect of grape maturation on soluble protein characteristics of Gewürztraminer and White Riesling juice and wine. American Journal of Enology and Viticulture. 1989;40:199–207.

[56] Yokotsuka K, Ebihara T, Sato T: Comparison of soluble proteins in juice and wine from Koshu juice. Journal of Fermentation Technology. 1991;71:248–253. Doi: 10.1016/0922-338X(91)90276-M

[57] Tattersall DB, Heeswijck R, Hoj PB: Identification and characterization of a fruit-specific, thaumatin-like protein that accumulates at very high levels in conjunction with the onset of sugar accumulation and berry softening in grapes. Plant Physiology. 1997;114:759–769. Doi: http://dx.doi.org/10.1104/pp.114.3.759

[58] Waters EJ, Hayasaka Y, Tattersall DB, Adams KS, Williams PJ: Sequence analysis of grape (*Vitis vinifera*) berry chitinases that cause haze formation in wines. Journal of Agricultural and Food Chemistry. 1998;46:4950–4957. Doi: 10.1021/jf980421o

[59] Williams PJ, Francis IL: Wine flavor research: Experiences from the past offer a guide to the future. Proceedings of the ASEV 50th Anniversary Annual Meeting, Seattle, American Society for Enology & Viticulture, Davis, CA,2000;pp.191–195.

[60] Lacey MJ, Allen MS, Harris RLN, Brown WV: Methoxypyrazines in Sauvignon Blanc grapes and wines. American Journal of Enology and Viticulture. 1991;42:103–108.

[61] Dunlevy JD, Kalua CM, Keyzers RA, Boss PK: The production of flavour and aroma compounds in grape berries. In Grapevine Molecular Physiology and Biotechnology. K.A. Roubelakis-Angelakis (ed.), Springer, Dordrecht, Netherlands, 2009; pp. 293–340.

[62] Lorrain B, Chira K, Teissedre P-L: Phenolic composition of Merlot and Cabernet–Sauvignon grapes from Bordeaux vineyard for the 2009-vintage: Comparison to 2006, 2007 and 2008 vintages. Food Chemistry. 2011;126:1991–1999. Doi: 10.1016/j.foodchem.2010.12.062

[63] Jordão AM, Ricardo-da-Silva JM, Laureano O: Evolution of proanthocyanidins in bunch stems during berry development (*Vitis vinifera* L.). Vitis. 2001;40:17–22.

[64] Hernandez-Jimenez A, Gomez-Plaza E, Martinez-Cutillas A, Kennedy JA: Grape skin and seed proanthocyanidins from Monastrell x Syrah grapes Journal of Agricultural and Food Chemistry. 2009; 57:10798–10803. Doi: 10.1021/jf903465p

[65] Cheynier V, Rigaud J: HPLC separation and characterization of flavonols in the skins of Vitis *Vinifera* var. Cinsault. American Journal of Enology and Viticulture. 1986;37:248–252.

[66] Jordão AM, Ricardo-da-Silva JM, Laureano O: Evolution of catechins and oligomeric procyanidins during grape maturation of Castelão Francês and Touriga Francesa. American Journal of Enology and Viticulture. 2001;53:231–234.

[67] Pastrana-Bonilla E, Akoh, CC, Sellappan S, Krewer G: Phenolic content and antioxidant capacity of Muscadine grapes. Journal of Agriculture and Food Chemistry. 2003;51:5497–5503. Doi: 10.1021/jf030113c

[68] Monagas M, Gómez-Cordovés C, Bartolomé B, Laureano O, Ricardo-da-Silva JM: Monomeric, oligomeric, and polymeric flavan-3-ol composition of wines and grapes from *Vitis vinifera* L. Cv. Graciano, Tempranillo, and Cabernet Sauvignon Journal of Agricultural and Food Chemistry. 2003;51:6475–6481. Doi: 10.1021/jf030325

[69] Obreque-Slier E, Peña-Neira A, Lopez-Solís R, Zamora-Marín F, Ricardo-da-Silva JM, Laureano O: Comparative study of the phenolic composition of seeds and skins from Carménère and Cabernet Sauvignon grape varieties (*Vitis vinifera* L.) during ripening. Journal of Agricultural and Food Chemistry. 2010;58:3591–3599. Doi: 10.1021/jf904314u

[70] Bergqvist J, Dokoozlian N, Ebisuda N: Sunlight exposure and temperature effects on berry growth and composition of Cabernet Sauvignon and Grenache in the Central San Joaquin Valley of California. American Journal of Enology and Viticulture. 2001;52:1–7.

[71] Madeira FC: Effect of ultraviolet radiation b on the behavior of the wine variety (*Vitis Vinífera* L.) Cabernet Sauvignon in São Joaquim, Santa Catarina. Master Thesis (Mestrado em Recursos Genéticos Vegetais), University Federal of Santa Catarina, Florianópolis, Brasil. 2011.

[72] Mateus N, Marques S, Gonçalves AC, Machado JM, De Freitas V: Proanthocyanidin composition of red *Vitis vinifera* varieties from the Douro Valley during ripening: Influence of cultivation altitude. American Journal of Enology and Viticulture. 2001;52:115–121.

[73] Ubalde JM, Sort X, Alicia Zayas A, Poch RM: Effects of soil and climatic conditions on grape ripening and wine quality of Cabernet Sauvignon. Journal of Wine Research 2010;21:1–17. Doi: 10.1080/09571264.2010.495851.

[74] Downey MO, Dokoozlian NK, Krstic MP: Cultural practice and environmental impacts on the flavonoid composition of grapes and wine: a review of recent research. American Journal of Enology and Viticulture. 2006;57:257–268.

[75] Costa E, Cosme F, Jordão AM, Mendes-Faia A: Anthocyanin profile and antioxidante activity from 24 grape varieties cultivated in two Portuguese wine regions. Journal International des Sciences de la Vigne et du Vin. 2014;48:51–62.

[76] Costa E, Da Silva JF, Cosme F, Jordão AM: Adaptability of some French red grape varieties cultivated at two different Portuguese terroirs: comparative analysis with two Portuguese red grape varieties using physicochemical and phenolic parameters. Food Research International. 2015;78:302–312. Doi: 10.1016/j.foodres.2015.09.029

[77] Bruno G, Sparapano L: Effects of three esca-associated fungi on *Vitis vinifera* L. Changes in the chemical and biological profile of xylem sap from diseased cv. Sangiovese vines. Physiological and Molecular Plant Pathology. 2007;71:210–229. Doi: 10.1016/j.pmpp.2008.02.005

[78] Nicoletti I, Bello C, De Rossi A, Corradini D: Identification and quantification of phenolic compounds in grapes by HPLC-PDA-ESI-MS on a semimicro separation scale. Journal of Agricultural and Food Chemistry. 2008;56:8801–8808. Doi: 10.1021/jf801411m.

[79] Obreque-Slier E, Peña-Neira A, Lopez-Solís R, Zamora-Marín F, Ricardo-da-Silva JM, Laureano O: Comparative study of the phenolic composition of seeds and skins from Carménère and Cabernet Sauvignon grape varieties (*Vitis vinifera* L.) during ripening. Journal of Agricultural and Food Chemistry. 2010;58:3591–3599. Doi: 10.1021/jf904314u.

[80] Montealegre RR, Peces R., Vozmediano JLC, Gascueña JM, Romero EG: Phenolic compounds in skins and seeds of ten grape *Vitis vinifera* varieties grown in a warm climate. Journal of Food Composition and Analysis. 2006;19:687–693. Doi: 10.1016/j.jfca.2005.05.003.

[81] Sun B, Ribes AM, Conceição ML, Belchior AP, Spranger MI: Stilbenes: quantitative extraction from grape skins, contribution of grape solids to wine and variation during

wine maturation. Analytica Chimica Acta. 2006;563:382–390. Doi: 10.1016/j.aca.2005.12.002.

[82] Mikeš O, Vrchotová N, Tříska J, Kyseláková M, Šmidrkal J: Distribution of major polyphenolic compounds in vine grapes of different cultivars growing in South Moravian vineyards. Czech Journal of Food Sciences. 2008;26:182–189.

[83] Balík J, Kyseláková M, Vrchotová N, Tříska J, Kumšta M, Veverka J, Híc P, Totušek J, Lefnerová D: Relations between polyphenols content and antioxidant activity in vine grapes and leaves. Czech Journal of Food Sciences. 2008;26(Special Issue):S25-S32.

[84] Ortega-Regules A. Romero-Cascales I, López-Roca JM, Ros-García JM, Gómez-Plaza E: Anthocyanin fingerprint of grapes: environmental and genetic variations. Journal of Science of Food and Agriculture. 2006;86:1460–1467. Doi: 10.1002/jsfa.2511.

[85] Revilla E, García-Beneytez E, Cabello F, Martín-Ortega G, Ryan JM: Value of high-performance liquid chromatographic analysis of anthocyanins in the differentiation of red grape cultivars and red wines made from them. Journal of Chromatography A. 2001;915:53–60. Doi: 10.1016/S0021-9673(01)00635-5.

[86] Jordão AM, Ricardo-da-Silva JM, Laureano O: Evolution of anthocyanins during grape maturation of two varieties (*Vitis vinífera* L.), Castelão Francês and Touriga Francesa. Vitis. 1998;37:93–94.

[87] Pérez-Trujillo JP, Hernández Z, López-Bellido FJ, Hermosín-Gutiérrez I: Characteristic phenolic composition of single-cultivar red wines of the Canary Islands (Spain). Journal of Agricultural and Food Chemistry. 2011;59:6150–6164. Doi: 10.1021/jf200881s.

[88] Mateus N, Marques S, Gonçalves AC, Machado JM. De Freitas V: Proanthocyanidin composition of red *Vitis vinifera* varieties from the Douro Valley during ripening: Influence of cultivation altitude. American Journal of Enology and Viticulture. 2001;52:115–121.

[89] De Freitas VAP, Glories Y: Concentration and compositional changes of procyanidins in grape seeds and skin of white *Vitis vinifera* varieties. Journal of the Science of Food and Agriculture. 1999;79:1601–1606. Doi: 10.1002/(SICI)1097-0010(199909)79:12<1601::AID-JSFA407>3.0.CO;2-1.

[90] Li H: Wine tasting. China Science Press, Beijing, China. 2006.

[91] Fleet GH: Wine Yeasts for the Future. FEMS Yeast Research. 2008;8:979–995. Doi: http://dx.doi.org/10.1111/j.1567-1364.2008.00427.x

[92] Lonvaud-Funel A: Lactic acid bacteria in the quality improvement and depreciation of wine. Antonie van Leeuwenhoek. 1999;76:317–331. Doi: 10.1023/A:1002088931106.

[93] Wibowo D, Eschenbruch R, Davis C, Fleet G, Lee T: Occurrence and growth of lactic acid bacteria in wine: a Review. American Journal of Enology and Viticulture. 1985;36:302–313.

[94] Fleet GH, Lafon-Lafourcade S, Ribéreau-Gayon P: Evolution of yeasts and Lactic Acid Bacteria during fermentation and storage of Bordeaux Wines. Applied and Environmental Microbiology. 1984;48:1034–1038.

[95] Reguant C, Bordons A: Typification of *Oenococcus oeni* strains by multiplex RAPD-PCR and study of population dynamics during malolactic fermentation. Journal of Applied Microbiology. 2003;95:344–353. Doi: 10.1046/j.1365-2672.2003.01985.x.

[96] Styger G, Prior B, Bauer FF: Wine favour and aroma—review. Journal of Industrial Microbiology & Biotechnology. 2011;38:1145–1159. Doi: 10.1007/s10295-011-1018-4.

[97] Perestrelo R, Fernandes A, Alburquerque FF, Marques JC, Camara JS: Analytical characterization of the aroma of Tinta Negra Mole red wine: identification of the main odorants compounds. Analytica Chimica Acta. 2006;563:154–164. Doi: 10.1016/j.aca.2005.10.023.

[98] Swiegers JH, Pretorius IS: Yeast modulation of wine flavour. Advances in Applied Microbiology. 2005;57:131–175. Doi: 10.1016/S0065-2164(05)57005-9.

[99] Saerens SMG, Delvaux FR, Verstrepen KJ, Thevelein JM: Production and biological function of volatile esters in *Saccharomyces cerevisiae*—minireview. Microbial Biotechnology. 2010;3(2):165–177. Doi: 10.1111/j.1751-7915.2009.00106.x.

[100] Lee SJ, Noble AC: Characterization of odor-active compounds in Californian Chardonnay wines using GC–olfactometry and GC–mass spectrometry. Journal of Agricultural and Food Chemistry. 2003;51:8036–8044. Doi: 10.1021/jf034747v.

[101] Li H, Tao Y, Wang H, Zhang L: Impact odorants of Chardonnay dry white wine from Changli County (China). European Food Research and Technology. 2008;227:287–292. Doi: 10.1007/s00217-007-0722-9.

[102] Vilela-Moura A., Schuller D, Mendes-Faia A, Silva RF, Chaves SR, Sousa MJ, Côrte-Real M: The impact of acetate metabolism on yeast fermentative performance and wine quality: reduction of volatile acidity of grape-musts and wines—minireview. Applied Microbiology and Biotechnology. 2011;89:271–280. Doi: 10.1007/s00253-010-2898-3.

[103] Verstrepen KJ1, Van Laere SD, Vanderhaegen BM, Derdelinckx G, Dufour JP, Pretorius IS, Winderickx J, Thevelein JM, Delvaux FR: Expression levels of the yeast alcohol acetyltransferase genes ATF1, Lg-ATF1, and ATF2 control the formation of a broad range of volatile esters. Applied Environmental Microbiology. 2003;69(9):5228–37. Doi: 10.1128/AEM.69.9.5228-5237.2003.

[104] Procopio S, Qian F, Becker T: Function and regulation of yeast genes involved in higher alcohol and ester metabolism during beverage fermentation. European Food Research and Technology. 2011;233:721–729. Doi: 10.1007/s00217-011-1567-9.

[105] Pretorius IS: Tailoring wine yeast for the new millennium: novel approaches to the ancient art of winemaking. Yeast. 2000;16:675–729. Doi: 10.1002/1097-0061(20000615)16:8<675::AID-YEA585>3.0.CO;2-B.

[106] Lambrechts MG, Pretorius IS: Yeast and its importance to wine aroma. A review. South African Journal of Enology and Viticulture. 2000;21:97–129.

[107] Vilela-Moura A, Schuller D, Mendes-Faia A, Côrte-Real M: Effects of acetic acid, ethanol and SO_2 on the removal of volatile acidity from acidic wines by two *Saccharomyces cerevisiae* commercial strains. Applied Microbiology and Biotechnology. 2010;87(4):1317–1326. Doi: 10.1007/s00253-010-2558-7.

[108] Gunata Z, Bayonove C, Baumes RL, Cordonnier R: The aroma of grapes. Extraction and determination of free and glycosidically bound fractions of some grape aroma components. Journal of Chromatography. 1985;331:83–90.

[109] Pardo E, Rico J, Gil, JV, Orejas M: De novo production of six key grape aroma monoterpenes by a geraniol synthase-engineered *S. cerevisiae* wine strain. Microbial Cell Factories. 2015;14:136. Doi: 10.1186/s12934-015-0306-5.

[110] Biron C, Cordonnier R, Glory O, Gunata Z, Sapis JC: Study, in grapes, of the beta-glucosidase activity. Connaissance de la Vigne et du Vin. 1988;22:125–134.

[111] Delcroix A, Gunata, Z, Sapis JC, Salmon JM, Bayonove C: Glycosidase activities of three enological yeast strains during winemaking: effect on the terpenol content of Muscat wine. American Journal of Enology and Viticulture. 1994;45:291–296.

[112] Palmeri R, Spagna G: β-Glucosidase in cellular and acellular form for winemaking application. Enzyme and Microbial Technology. 2007;40:382–389. Doi: 10.1016/j.enzmictec.2006.07.007.

[113] Van Rensburg P, Stidwell T, Lambrechts MG, Otero RC, Pretorius IS: Development and assessment of a recombinant *Saccharomyces cerevisiae* wine yeast producing two aroma-enhancing β-glucosidases encoded by the *Saccharomycopsis fibuligera* BGL1 and BGL2 genes. Annals of Microbiology. 2005;55(1):33–42. Doi: http://hdl.handle.net/10019.1/11876.

[114] Zietsman A, Klerk JJD, van Rensburg P. Coexpression of α-l-arabinofuranosidase and β-glucosidase in *Saccharomyces cerevisiae*. FEMS Yeast Research. 2011;11:88–103. Doi: 10.1111/j.1567-1364.2010.00694.x.

[115] Guth H: Identification of character impact odorants of diferente white wine varieties. Journal of Agricultural and Food Chemistry.1997;45:3027–3032. Doi: 10.1021/jf9608433

[116] Etievant PX: Volatile compounds in foods and beverages. In: Wine, Maarse H (ed.). 2nd ed. Marcel Dekker, New York, 1991; pp 483–546

[117] Ong PKC, Acree TE: Similarities in the aroma chemistry of Gewürztraminer variety wines and lychee (Litchi chinesis Sonn.) fruit. Journal of Agricultural and Food Chemistry. 1999;47:665–670. Doi: 10.1021/jf980452j

[118] Ferreira V, Lopez R, Cacho JF: Quantitative determination of the odorants of young red wines from different grape varieties. Journal of the Science of Food and Agriculture.

2000;80(11):1659–1667. Doi: 10.1002/1097-0010(20000901)80:11<1659::AID-JSFA693>3.0.CO;2–6.

[119] Peinado RA, Moreno J, Bueno JE, Moreno JA, Mauricio JC: Comparative study of aromatic compounds in two young white wines subjected to pre-fermentative cryomaceration. Food Chemistry. 2004;84:585–590. Doi: 10.1016/S0308-8146(03)00282-6.

[120] Carlton WK, Gump B, Fugelsang K, Hasson AS: Monitoring acetaldehyde concentrations during micro-oxygenation of red wine by headspace solid-phase microextraction with on-fiber derivatization. Journal of Agricultural and Food Chemistry. 2007;55:5620–5625. Doi: 10.1021/jf070243b.

[121] Delfini C, Cocito C, Bonino M: A review: biochemical and molecular mechanisms in *Saccharomyces cerevisiae* that are involved in the formation of some volatile compounds in wines. Journal International des Sciences de la Vigne et du Vin. 1999;33(4):195–211.

[122] Mateo JJ, Jiménez M: Monoterpenes in grape juice and wines. Journal of Chromatography A. 2000;881:557–567. Doi: 10.1016/S0021-9673(99)01342-4.

[123] Bartowsky E: *Oenococcus oeni* and malolactic fermentation—moving into the molecular arena. Australian Journal Grape and Wine Research. 2005;11:174–187. Doi: 10.1111/j.1755-0238.2005.tb00286.x.

[124] Liu S: A review: malolactic fermentation in wine – beyond deacidification. Journal of Applied Microbiology. 2002;92:589–601. Doi: 10.1046/j.1365-2672.2002.01589.x.

[125] Arnink K, Henick-Kling T: Influence of *Saccharomyces cerevisiae* and *Oenococcus oeni* strains on successful malolactic conversion in wine. American Journal of Enology and Viticulture. 2005;56:228–237.

[126] Bartowsky E, Burvill T, Henschke P: Diacetyl in wine: Role of malolactic bacteria and citrate. The Australian Grapegrower and Winemaker. 1997; (25th Technical Issue 402a): 130–135.

[127] Bartowsky E, Henschke P: Management of malolactic fermentation for the 'buttery' diacetyl flavour in wine. The Australian Grapegrower and Winemaker. 2000;(28th Technical Issue 438a)58-67.

[128] Bartowsky E., Henschke P: The 'buttery' attribute of wine–diacetyl-desirability, spoilage and beyond. International Journal of Food Microbiology. 2004;96:235–252. Doi: 10.1016/j.ijfoodmicro.2004.05.013.

[129] Sumby KM, Jiranek V, Grbin PR: Ester synthesis and hydrolysis in an aqueous environment, and strain specific changes during malolactic fermentation in wine with *Oenococcus oeni*. Food Chemistry. 2013;141:1673–1680. Doi: 10.1016/j.foodchem.2013.03.087.

[130] Pripis-Nicolau L, De Revel G, Bertrand A, Lonvaud-Funel A: 2004. Methionine catabolism and production of volatile sulphur compounds by *Oenococcus oeni*. Journal of Applied Microbiology 96:1176–1184. Doi: 10.1111/j.1365-2672.2004.02257.x.

[131] Vaquero I, Marcobal A, Munoz R: Tannase activity by lactic acid bacteria isolated from grape must and wine. International Journal of Food Microbiology. 2004;96: 199–204. Doi: 10.1016/j.ijfoodmicro.2004.04.004.

[132] Arena M, Manca de Nadra M: Biogenic amine production by Lactobacillus. Journal of Applied Microbiology. 2001; 90:158–162. Doi: 10.1046/j.1365-2672.2001.01223.x.

[133] Lonvaud-Funel A: Biogenic amines in wines: role of lactic acid bacteria. FEMS Microbiololy Letters. 2001;199:9–13. Doi: 10.1111/j.1574-6968.2001.tb10643.x.

[134] Claisse O, Lonvaud-Funel A: Detection of lactic acid bacteria producers of 3-hydroxypropionaldehyde (acrolein precursor) from glycerol by molecular tests. Lait. 2001;81:173–181. Doi: <10.1051/lait:2001121>.<hal-00895470>.

[135] Martineau B, Henick-Kling T: Performance and diacetyl production of commercial strains of malolactic bacteria in wine. Journal of Applied Bacteriology. 1995;78:526–536. Doi: 10.1111/j.1365-2672.1995.tb03095.x.

[136] Amerine M, Kunker RE: Microbiology of winemaking. Annual Review of Microbiology. 1968;22:323–358.

[137] Sponholz WR: Wine spoilage by microorganisms. In: Wine Microbiology and Biotechnology. Editor Fleet, G.H., Harwood Academic Publishers, Chur, Switzerland,1993; pp. 395–420.

[138] Walling E, Gindreau E, Lonvaud-Funel A: The biosynthesis of exopolysaccharide by strains of *Pediococcus damnosus* isolated from wine: development of molecular detection tools. Lait. 2001;81,289–300. Doi: http://dx.doi.org/10.1051/lait:2001132.

[139] Costello P, Henschke P: Mousy off-flavor of wine: precursors and biosynthesis of the causative N-heterocycles 2-ethyltetrahydropyridine, 2-acetyltetrahydropyridine, and 2-acetyl-1-pyrroline by Lactobacillus hildargii DSM 20176. Journal of Agricultural and Food Chemistry; 2002:50:7079–7087. Doi: 10.1021/jf020341r.

[140] Crowell EA, Guymon JF: Wine constituents arising from sorbic acid addition, and identification of 2-ethoxyhexa-3,5-diene as source of geranium-like off-odor. American Journal of Enology and Viticulture. 1975; 26:97–102.

[141] Chatonnet P, Viala C, Dubourdieu D: Influence of polyphenolic components of red wines on the microbial synthesis of volatile phenols. American Journal of Enology and Viticulture. 1997;48:443–448.

[142] Barthelmebs L, Divies C, Cavin J: Molecular characterization of the phenolic acid metabolism in the lactic acid bacteria *Lactobacillus plantarum*. Lait. 2001;81:161–171.

[143] Lin W, You W, Feng T: Review of *Brettanomyces* of grape wine and its inhibition methods. In: Advanced Engineering and Technology, Xie and Huang (eds.), CRC Press, London, 2014; pp 635–640. Doi: 10.1201/b16699-99.

[144] Zapparoli G, Moser M, Dellaglio F, Tourdot-Marechal R, Guzzo J: Typical metabolic traits of two *Oenococcus oeni* strains isolated from Valpolicella wines. Letters in Applied Microbiology. 2004;39:48–54. Doi: 10.1111/j.1472-765X.2004.01541.x

[145] Pramateftaki P, Metafa M, Kallithraka S, Lanaridis P: Evolution of malolactic bacteria and biogenic amines during spontaneous malolactic fermentations in a Greek winery. Letters in Applied Microbiology. 2006;43:155–160. Doi: 10.1111/j.1472-765X.2006.01937.x

[146] Marcobal A, Mills DA: Genomics of *Oenococcus oeni* and other lactic acid bacteria. In: Biology of Microorganisms on Grapes, in Musts and Wine. Editor Koning, H., Uden, G., Frohlich, J., Springer, New York, 2008; pp. 351–360. Doi: 10.1007/978-3-540-85463-0_19

[147] Barthelmebs L, Divies C, Cavin J: Molecular characterization of the phenolic acid metabolism in the lactic acid bacteria Lactobacillus plantarum. Lait. 2001;81:161–171.

[148] Barthelmebs L, Lecomte B, Divies C, Cavin J: Inducible metabolism of phenolic acids in Pediococcus pentosaceus is encoded by an autoregulated operon which involves a new class of negative transcriptional regulator. Journal of Bacteriology. 2000;182:6724–6731.

[149] Swiegers J., Bartowsky E., Henschke P., Pretorius I: Yeast and bacterial modulation of wine aroma and flavour. Australian Journal of Grape and Wine Research. 2005;11:139–173. Doi: 10.1111/j.1755-0238.2005.tb00285.x.

[150] Harrison T.J., Dake G.R: An expeditious, high-yielding construction of the food aroma compounds 6-acetyl-1,2,3,4-tetrahydropyridine and 2-acetyl-1-pyrroline. The Journal of Organic Chemistry. 2005;70:10872–10874. Doi: 10.1021/jo051940a.

[151] Pripis-Nicolau L., De Revel G., Bertrand A., Lonvaud-Funel A: Methionine catabolism and production of volatile sulphur compounds by *Oenococcus oeni*. Journal of Applied Microbiology. 2004;96:1176–1184.

[152] Bartowsky E: 2005. *Oenococcus oeni* and malolactic fermentation—moving into the molecular arena. Australian Journal of Grape and Wine Research. 2005;11:174–187.

[153] Martineau B., Henick-Kling T: Performance and diacetyl production of commercial strains of malolactic bacteria in wine. Journal of Applied Microbiology. 1995;78:526–536. Doi: 10.1111/j.1365-2672.1995.tb03095.x.

[154] Bartowsky E., Henschke P: The 'buttery' attribute of wine–diacetyl-desirability, spoilage and beyond. The International Journal of Food Microbiology. 2004;96:235–252.

[155] Swiegers J., Bartowsky E., Henschke P., Pretorius I: Yeast and bacterial modulation of wine aroma and flavour. Australian Journal of Grape and Wine Research. 2005;11:139–173.

[156] Ehsani M, Fernandez MR, Biosca JA, Julien A, Dequin S: Engineering of 2,3-butanediol dehydrogenase to reduce acetoin formation by glycerol-overproducing, low-alcohol *Saccharomyces cerevisiae*. Applied and Environmental Microbiology. 2009;75:3196–3205. Doi: 10.1128/AEM.02157-08.

[157] Romano P., Suzzi G: Origin and production of acetoin during wine yeast fermentation —a review. Applied and Environmental Microbiology. 1996;62:309–315.

[158] Lambert M, Meudec E, Verbaere A, Mazerolles G, Wirth J, Masson G, Cheynier V, Sommerer N: A high-throughput UHPLC-QqQ-MS method for polyphenol profiling in rosé wines. Molecules. 2015;20:7890–7914. Doi: 10.3390/molecules20057890.

[159] Ebeler SE: Analysis of Grapes and Wines: An Overview of New Approaches and Analytical Tools. In: Advances in Wine Research, ACS Symposium Series; American Chemical Society, Washington, DC, 2015; pp. 3–12. Doi: 10.1021/bk-2015-1203.ch001.

[160] Ebeler SE. Chapter 30 – Gas Chromatographic Analysis of Wines: Current Applications and Future Trends. In: Gas Chromatography, Poole, C. F., Elsevier, New York, 2012; pp 689–710. Doi: 10.1016/B978-0-12-385540-4.00030-4.

[161] Robinson AL, Boss PK, Solomon PS, Trengove RD, Heymann H, Ebeler SE: Origins of grape and wine aroma. Part 1. Chemical components and viticultural impacts. American Journal of Enology and Viticulture. 2014;65:1–24. Doi: 10.5344/ajev.2013.12070.

[162] Robinson AL, Boss PK, Solomon PS, Trengove RD, Heymann H, Ebeler SE: Origins of grape and wine aroma. Part 2. Chemical and sensory analysis. American Journal of Enology and Viticulture. 2014;65:25–42. Doi: 10.5344/ajev.2013.13106.

[163] Hjelmeland AK, Collins TS, Miles JL, Wylie PL, Mitchell AE, Ebeler SE: High-throughput, sub ng/L analysis of haloanisoles in wines using HS-SPME with GC-triple quadrupole MS. American Journal of Enology and Viticulture. 2012;63:494–499. Doi: 10.5344/ajev.2012.12043.

[164] Kennison KR, Wilkinson KL, Williams HG, Smith JH, Gibberd MR: Smoke-derived taint in wine: effect of postharvest smoke exposure of grapes on the chemical composition and sensory characteristics of wine. Journal of Agricultural and Food Chemistry. 2007;55:10897–10901. Doi: 10.1021/jf072509k.

[165] Kennison KR, Gibberd MR, Pollnitz AP, Wilkinson KL: Smoke-derived taint in wine: The release of smoke-derived volatile phenols during fermentation of merlot juice following grapevine exposure to smoke. Journal of Agricultural and Food Chemistry. 2008;56:7379–7383. Doi: 10.1021/jf800927e.

[166] Kennison KR, Wilkinson KL, Pollnitz AP, Williams HG, Gibberd MR: Effect of timing and duration of grapevine exposure to smoke on the composition and sensory prop-

erties of wine. Australian Journal of Grape and Wine Research. 2009;15:228–237. Doi: 10.1111/j.1755-0238.2009.00056.x.

[167] Hayasaka Y, Dungey KA, Baldock GA, Kennison KR, Wilkinson KL: Identification of a β-d-glucopyranoside precursor to guaiacol in grape juice following grapevine exposure to smoke. Analytica Chimica Acta. 2010;660:143–148. Doi: 10.1016/j.aca.2009.10.039.

[168] Pati S, Liberatore MT, Gambacorta G, Antonacci D, La Notte E: Rapid screening for anthocyanins and anthocyanin dimers in crude grape extracts by high performance liquid chromatography coupled with diode array detection and tandem mass spectrometry. Journal of Chromatography A. 2009;1216:3864–3868. Doi: 10.1016/j.chroma.2009.02.068.

[169] Kamangerpour A, Ashraf-Khorassani M, Taylor LT, McNair HM, Chorida L: Supercritical fluid chromatography of polyphenolic compounds in grape seed extract. Chromatographia. 2002;55:417–421. Doi: 10.1007/BF02492270.

[170] Vaclavik L, Lacina O, Hajslova J, Zweigenbaum J: The use of high performance liquid chromatography–quadrupole time-of-flight mass spectrometry coupled to advanced data mining and chemometric tools for discrimination and classification of red wines according to their variety Analytica Chimica Acta. 2011;685:45–51. Doi: 10.1016/j.aca.2010.11.018.

[171] Flamini R: Recent Applications of Mass Spectrometry in the Study of Grape and Wine Polyphenols. ISRN Spectroscopy Article. 2013;2013:Article ID 813563, 45 pages. http://dx.doi.org/10.1155/2013/813563.

[172] Roullier-Gall C, Lucio M, Noret L, Schmitt-Kopplin P, Gougeon RD: How Subtle Is the "Terroir" Effect? Chemistry-Related Signatures of Two "Climats de Bourgogne". PLoS One. 2014;9(5):e97615. Doi: 10.1371/journal.pone.0097615.

[173] Roullier-Gall C, Boutegrabet L, Gougeon RD, Schmitt-Kopplin P: A grape and wine chemodiversity comparison of different appellations in Burgundy: Vintage vs terroir effects. Food Chemistry. 2014;152:100–107. Doi: 10.1016/j.foodchem.2013.11.056.

[174] Robinson AL, Adams DO, Boss PK, Heymann H, Solomon PS, Trengove RD: The relationship between sensory attributes and wine composition for Australian Cabernet Sauvignon wines. Australian Journal of Grape and Wine Research. 2011;17:327–340. Doi: 10.1111/j.1755-0238.2011.00155.x.

[175] Robinson AL, Boss PK, Heymann H, Solomon PS, Trengove RD: Influence of Yeast Strain, Canopy Management, and Site on the Volatile Composition and Sensory Attributes of Cabernet Sauvignon Wines from Western Australia Journal of Agricultural and Food Chemistry.201;59:3273–3284. Doi: 10.1021/jf104324d.

[176] Kogerson K, Runnebaum R, Wohlgemuth G, de Ropp J, Heymann H, Fiehn O: Comparison of gas chromatography-coupled time-of-flight mass spectrometry and 1H nuclear magnetic resonance spectroscopy metabolite

identification in white wines from a sensory study investigating wine body. Journal of Agricultural and Food Chemistry. 2009;57:6899–6907. Doi: 10.1021/jf9019322.

[177] Hopfer H, Nelson J, Collins TS, Heymann H, Ebeler SE: The combined impact of vineyard origin and processing winery on the elemental profile of red wines. Food Chemistry. 2015;172:486–496. Doi: 10.1016/j.foodchem.2014.09.113.

[178] Hopfer H, Nelson J, Mitchell AE, Heymann H, Ebeler SE: Profiling the trace metal composition of wine as a function of storage temperature and packaging type. Journal of Analytical Atomic Spectrometry. 2013;28:1288–1291. Doi: 10.1039/c3ja50098e.

[179] Polaskova P, Herszage J, Ebeler SE: Wine flavor: chemistry in a glass. Chemical Society Reviews. 2008;37:2478–2489. Doi: 10.1039/B714455P.

[180] Vilela A, Monteiro B, Correia E: Sensory profile of port wines: categorical principal component analysis, an approach for sensory data treatment. Ciência e Técnica Vitivinícola. 2015;30(1):1–8. http://dx.doi.org/10.1051/ctv/20153001001.

[181] Monteiro B, Vilela A, Correia E: Sensory profile of pink port wines: development of a flavour lexicon. Flavour and Fragrance Journal. 2014;29:50–58. Doi: 10.1002/ffj.3178.

[182] Johnson AJ, Hirson GD, Ebeler SE: Perceptual characterization and analysis of aroma mixtures using gas chromatography recomposition-olfactometry. PLoS One. 2012;7(8):e42693. Doi: 10.1371/journal.pone.0042693.

[183] Casazza AA, Aliakbarian B, Mantegna S, Cravotto G, Perego P: Extraction of phenolics from *Vitis vinifera* wastes using non-conventional techniques. Journal of Food Engineering. 2010;100:50–55. Doi: 10.1016/j.jfoodeng.2010.03.026.

[184] Vatai T, Skerget M, Knez, Z: Extraction of phenolic compounds from elder berry and different grape marc varieties using organic solvents and/or supercritical carbon dioxide. Journal of Food Engineering. 2009;90:246–254. Doi: 10.1016/j.jfoodeng.2008.06.028.

[185] Luque-Rodríguez JM, Luque de Castro MD, Pérez-Juan P: Dynamic superheated liquid extraction of anthocyanins and other phenolics from red grape skins of wine making residues. Bioresource Technology. 2007;98:2705–2713. Doi: 10.1016/j.biortech.2006.09.019.

[186] Corrales M, Toepfl S, Butz P, Knorr D, Tauscher B: Extraction of anthocyanins from grape by-products assisted by ultrasonics, high hydrostatic pressure or pulsed electric fields: A comparison. Innovative Food Science and Emerging Technologies. 2008;9:85–91. Doi: 10.1016/j.ifset.2007.06.002.

[187] López N, Puértolas E, Condón S, Álvarez I, Raso J: Effects of pulsed electric fields on the extraction of phenolic compounds during the fermentation of must of Tempranillo grapes. Innovative Food Science and Emerging Technologies. 2008;9:477–482. Doi: 10.1016/j.ifset.2007.11.001.

[188] Maier T, Goppert A, Kammerer DR, Schieber A, Carle R: Optimization of a process for enzyme-assisted pigment extraction from grape (*Vitis vinifera* L.) pomace. European Food Research and Technology. 2008;227:267–275. Doi: 10.1007/s00217-007-0720-y.

Chapter 10

Grape Microbiome: Potential and Opportunities as a Source of Starter Cultures

Despina Bozoudi and Dimitris Tsaltas

Additional information is available at the end of the chapter

http://dx.doi.org/10.5772/64806

Abstract

Grape microbiome is the source of a vastly diverse pool of filamentous fungi, yeast and bacteria that will play a coordinated role for the quality of the produced wines. In recent times, the significance of this pool of microorganisms with a long list of studies of the microbial ecology of grape berries of different geographical origin, cultural practices, grape varieties and climatic conditions has been acknowledged. Similarly, the ongoing microbial evolution of must fermentations has been fully uncovered. All these ecology studies, along with detailed metabolic studies and sensorial characterisations of the produced wines, led to the suggestion of the microbial *terroir*. These new concepts are today leading worldwide research efforts to the production of unique wines, preserving their historical identity and verifying their quality and geographical origin. This chapter is a quick but thorough and up-to-date review of how autochthonous microbiota highlight the *terroir* in wines, a comparison of commercial and wild yeast strains and how this biodiversity has been explored. Moreover, technological, physiological and oenological selection criteria will be under consideration. At the end, the positive and negative aspects of wild vinifications, the technological problems of wild strains and some suggestions for the future in starter cultures will be presented.

Keywords: grape microbiome, autochthonous yeasts, microbial *terroir*, *Saccharomyces*, starter culture

1. Introduction

Traditionally, wine making process relies on spontaneous fermentation without the addition of any chemical compounds or externally added microbes to begin the fermentation. Under these conditions, the biodiversity of the fermenting microorganisms, mainly yeasts and lactic acid bacteria (LAB), as well as the final quality of the resulting wine, is considered to be quite unpredictable. However, several works have shown the positive effects of spontaneous fermentations on the organoleptic complexity of wine as a consequence of the growth of different species and/or strains together, while commercial starter culture driven fermentations show "universally flatten" characteristics [1, 2].

On the other hand, modern oenological practices commonly use commercial starter cultures in order to ensure a controlled fermentation. Although starter cultures are subjected to strict selection for their technological properties for fermentation, they may not be able to compete with the indigenous microbiota of a certain must and for this reason, they cannot dominate the vinification process. The addition of sulphites is usually beneficial in that direction. Recently, the request for wines with unique style is on the increase as well as the demand of special wines, such as Marsala, Madeira, Sherry and Commandaria [3–6].

Several kinds of microorganisms, i.e., yeasts, bacteria and filamentous fungi, are responsible for turning the grape juice into wine, throughout the fermentation. During this process, some species are replaced by others, mostly due to antagonistic actions in order to gain access to nutrients and as a result, eventually, dominate. The substitution of species normally takes place because of the changes that occur in the must matrix turning into wine. Yeasts, such as *Hanseniaspora* (*Kloeckera*), *Torulaspora*, *Candida* and *Zygosaccharomyces*, are commonly present on the surface of grapes. Although grape cultivar and cultivation provide the foundations of wine flavour, microorganisms, and especially yeasts, impact on the subtlety and individuality of the flavour response. Generally, species of *Hanseniaspora* (*Kloeckera*), *Candida* and *Metschnikowia*, initiate the fermentation. Sometimes, species of *Pichia*, *Issatchenkia* and *Kluyveromyces* may also grow at this stage. The survival of non-*Saccharomyces* species during the fermentation process is regulated by ethanol production as the main *Saccharomyces cerevisiae* metabolic product. Specific species of *Hanseniaspora*, *Candida*, *Pichia*, *Kluyveromyces*, *Metschnikowia* and *Issatchenkia* isolated from grapes and must are sensitive in high ethanol concentrations (more than 5–7%), and that is probably the reason for their decline and finally their death, as the fermentation progresses [7]. Surviving indigenous microorganisms seem to be better adapted to the environmental conditions of a given wine producing area, as well as to the cellar where the winemaking process takes place. Function-targeted ecology studies, also referred as metagenomics, are at the time among the most reliable approaches to analyse the microbiota of fermented products (i.e. wine) and is expected to reveal astonishing results helping to understand the undergoing functions of many times unknown microorganisms in certain substrates [8].

This chapter aims to review in a thorough but concise way all latest literature in the scope of helping connect basic research with application. From the race to define originality of different types of wine worldwide, to basic scientific questions and technological obstacles of oenology, we are reviewing the most current literature in an effort to offer to the reader a conclusive opinion on modern wine microbiology.

2. Autochthonous microbiota in order to highlight *terroir*

Terroir is defined as high complex ecosystem in which the vine interacts with the environmental factors (i.e. soil, climate, humans, etc.) affecting the quality and typicality of the wine produced in a particular location. The biogeography model presented by the uniqueness of the wine grapes, including the microbial heterogeneity at the different viticultural areas, is important in order to preserve and sustain this biodiversity. Additionally, the product quality is enhanced, and as a result the consumer acceptance, as well. Therefore, a financial benefit for both the consumer and the producer is being established. On their journey from the vineyard to the wine bottle, grapes are transformed to wine through microbial activity, which determines a wide range of the wine quality parameters. Wine grapes harbour a wide range of microorganisms originating from the vine, the soil, the fauna, and the humans, many of which are recognised for their role in vine and grapes health and therefore, the wine quality. Nevertheless, the factors affecting the specific region wine characteristics have not been acknowledged, but are frequently assumed to originate from viticultural practices. It has been shown that these microbial aggregations are correlated to specific regional factors, suggesting a link between vineyard environmental conditions and microbial distribution. Bukolich et al. [9] reported that these factors taken together shape a unique microbial fingerprint to regional wines, setting the existence of non-random "microbial *terroir*" as a determining factor in regional variation among wine grapes [9].

Currently, there is a continuously rising interest for autochthonous yeast starters, which are potentially adapted to a specific grape must and reflect the biodiversity of a particular area, which support the idea that indigenous yeast strains can be associated with a "*terroir*" [9–11]. The composition of yeast communities on grapes had been shown to be dependent on several factors, including the geographical location of the vineyard, the type of soil, the age of the vineyard, the grape variety, the harvesting technique, the degree of grape maturation and the grape sanity [12]. Furthermore, it has been demonstrated that certain yeast strains are fully adapted to a specific climatic environment and/or substrate [13]. Some good results have been obtained when selected yeast starters from the micro-area where wines are produced were used for must fermentation [11]. It is quite obvious that in a given wild fermented wine, most of the yeasts derive from the vineyard environment. Further studies are needed in order to better understand the factors influencing yeast diversity in vineyards towards facilitating the selection process.

So far, studies on grape microbiota biogeography are mostly focused on the distribution of yeasts, where *S. cerevisiae* populations vary in respect of their presence or absence at the different regions, often affected by climate and vineyard age and size [14]. Setati et al. [15] interpreted their findings of higher yeast heterogeneity on grape samples collected at different sites inside individual vineyards due to the many microclimates existing even because of differential shading by leaves and grape bunch structure. In this study, fewer differences in the spatial distribution of fungal microbial communities were found between vineyards with very contrasting farming strategies. Bokulich et al. [9] proved that differentiation between regions increases dramatically at the biogeography within a grape variety of vintage. These findings suggest that factors such as host genotype and of course the vintage also play a significant role.

Introducing microbial ecology into agriculture, observations by farmers and viticulturists can be now better understood. This practice can be used to help improve the wine *terroir* or even reproduce those *terroirs* in sites a priori unsuitable for generating a wine with such characteristics. Upon till recently, the contribution of, and link between, microbes and differential geographic phenotypes (or *terroirs*), of agricultural products has not been objectively verified. It was the work of Knight et al. [16] that performed the first empirical test for whether there is a microbial aspect to *terroir*. The researchers conducted a crucial next step experimentation testing whether the genetic variance in microbial populations correlates with altered crop phenotypes. Their results show a quantifiable microbial aspect to *terroir*.

Generally, only few native *S. cerevisiae* strains are able to dominate the final phases of the process. Some predominant *S. cerevisiae* strains, recovered from spontaneous fermentation in the same winery, could occur over year, assuming that might be some correlation between strain and winery environment. Additionally, some *S. cerevisiae* strains isolated from different wineries located in the same region could be very similar, highlighting a correlation between strains and oenological region [17]. Studies based on genetic and microbiological analyses suggest that a significant part of the mechanisms that generate this genetic polymorphism in this yeast, occur during the vegetative phase of its growth cycle, where meiosis is an infrequent event [18]. This means that, once yeasts reproduce clonally and they are constantly adapting to a specific habitat, there must be a link between the genetic similarity of the strains and their ecological/geographic origin. Geographic or ecological isolation is one of the mechanisms involved in the species differentiation, as it is an obstacle for genetic flow. Thus, strains originating from the same microenvironment will be more alike to each other than with those from other geographic origins [19].

The selection and the employment of autochthonous microorganisms could be a powerful instrument to improve the organoleptic and sensory characteristics of wine produced from indigenous grape cultivars. In fact, autochthonous yeasts are the microorganisms better adapted to a specific must, which detain characteristics determined by the variety of the grapes and the *terroir* and therefore, they are able to exalt the peculiarities (aromas, structure, and colour) of the wine.

3. Commercial versus wild yeast strains

The importance of molecularly determining the autochthonous character of strains collected in strain selection programs for fermentation, is shown by the detection of commercial yeasts from the isolation of wild-type strains [19]. This is in spite of the studies which suggest that the continual use of commercial yeasts on the autochthonous yeast populations has a limited influence [20]. Therefore, there is a possibility that commercial strains used disseminate in the wine cellar and the vineyard of the same or neighbouring vineyards. This is due to oenological practices that facilitate the dispersion of these yeasts, allowing commercial strains to be erroneously recollected and selected as native strains.

Two main practices are usually used by oenologists. The first is to inoculate the must with commercial dry yeasts according to the manufacturer's instructions. The second one is to let the must ferments spontaneously. This last practice gives quite questionable results, since annual variations on quality and quantity of the dominant autochthonous microbiota, have been observed. However, there might be pitfalls in production also when using commercial dry yeasts, since commercial starter culture driven fermentations show "universally flatten" characteristics. Recently, Orlic et al. [21] used indigenous *Saccharomyces paradoxus* strains in order to study their influence in the aromatic profile of regional wine [21]. The inoculation of musts with *S. cerevisiae* strains selected from indigenous populations, at concentrations allowing the development of wild yeasts, can control the alcoholic fermentation better than commercial yeasts, as well as contribute to the production of more balanced wines [22].

Although there are plenty industrial yeast strains on the market promising to give special sensorial features to the produced wine, they do not possess the necessary metabolic pathways to enhance the typicity of local wines, as the indigenous yeasts has been proved to do so [23]. In a recent study by Borneman et al. [24], the results suggest that many commercial strains from multiple suppliers are nearly genetically identical, suggesting that the limits of effective gene variation within this genetically narrow group may be approaching saturation. They propose that, future strain development efforts should be introgressing new variation from outside of the wine yeast clade into these commercial yeasts, in order to enhance their genetic diversity and as a result their phenotypic one. Obviously, this work also reinforces the point that genetic homogeneity equals to genotypic homogeneity and therefore, to wine homogeneity.

4. Wine yeasts diversity, phylogeny and genomics

As the different strains of *S. cerevisiae* encompass different fermentation properties, their identification is a fundamental process which includes phenotypic, genotypic and karyotypic characterisation and can be applied with several molecular methods such as Polymerase Chain Reaction (PCR) amplification, capillary electrophoresis and fluorescence-based techniques. As more than one *Saccharomyces* strain is involved and interacts with the other strains during fermentation, their identification is of great importance. Despite the fact that non-*Saccharomyces* species that grow during fermentation have a low fermentative capacity, they play a key role in wine flavour as they produce flavour compounds such as esters, higher alcohols, acetic acid and acetaldehyde. Therefore, the identification and differentiation between *S. cerevisiae* and non-*Saccharomyces* species allows the creation of specific mixtures which are used to improve the sensory quality of the wine [7]. The discrimination of different strains is also of ecological interest. New studies aim to exploit the interactions between *S. cerevisiae* wild strains and their environment which highly affects the fermentation products [25]. Characterisation of the strains at a molecular level with high discrimination power techniques, such as multilocus

understanding that a plethora of microorganisms that colonise the grape skin and internal tissues, take place. These microorganisms are primarily yeasts and bacteria and modern studies show that their presence is significantly influenced by the grape varieties, the agronomical practices and the microenvironment [14]. Verginer et al. [57] had shown the role of the autochthonous microbiota on volatile organic compounds. The researchers showed that three single grape associated isolates of *Paenibacillus* sp., *Sporobolomycesroseus*, and *Aureobasidium pullulans* emit typical, well-known flavour components of red wine (i.e. 2-methylbutanoic acid, 3-methyl-1-butanol and ethyl octanoate). It is not yet reported if endophytic microorganisms have a role on grape aromatic compounds but grapevine endophyte studies have progressed [58] and it is very likely to identify such interplay in the near future.

The microbiome consists of yeasts from the basidiomycetous species, that are not able to ferment the juice sugars and are therefore, of non-importance to winemaking and ascomycetous fungi, such as *Aurobasidium pullulans* (also technologically not useful) and the fermentative *Candida* spp., *Hanseniaspora uvarum*/*Klockera apiculata*, *Metschnikowia* spp. and *Pichia* spp. [59]. More fermentative yeasts will follow their dominant presence when alcohol levels exceed 4–5%. These are of the *Saccharomyces* genus, with *S. cerevisiae* as the most prominent, followed by *Saccharomyces bayanus*, *Saccharomyces pastorianus* and *S. paradoxus*. Unfortunately, fermentative yeasts including the species *Brettanomyces bruxellensis* and *Zygosaccharomyces bailii* can spoil wine with off flavours and sediment/cloudiness formation, respectively.

Bacterial species common in grape microbiota are acetic acid bacteria (AAB) and LAB although their control is relatively easy to be succeeded by good manufacturing practices. In addition, *Oenococcus oeni* is of high interest in recent years due to its worldwide appreciation for malolactic fermentation.

Populations of yeasts on grapes surface are 10^2–10^4 cells/g although higher counts have been observed while bacteria are usually lower at maximum 10^2 cfu/g. These numbers though vary significantly depending on sampling methods and more importantly on berries conditions. In an exhaustive review by Barata et al. [14] the authors propose a very simple but applicable systematic for the wine microbial consortia on grape berries. The following three main groups characterised by similar behaviour on the berries are particularly dependent on nutrient availability on berry skins:

1. Oligotrophic, oxidative basidiomycetous yeasts, the yeast-like fungi *A. pullulans*, and LAB (*Lactobacillus* spp., *O. oeni*). Species favoured on poor environments-intact berries.

2. Copiotrophic, oxidative ascomycetes (several *Candida* spp.); weakly fermentative apiculate (*Hanseniaspora* spp.), film-forming (*Pichia* spp.), fermentative (*Candida zemplinina*, *Metschnikowia* spp.) yeasts. The emergence of these species is likely a result of juice and volatile organic compounds release as berries initiate their ripening process and cuticle is softened releasing these compounds.

3. Copiotrophic strongly fermentative yeasts (*Saccharomyces* spp., *Torulaspora* spp., *Zygosaccharomyces* spp., *Lachancea* spp. and *Pichia* spp.) and the obligate aerobic acetic acid bacteria (*Gluconobacter* spp., *Gluconacetobacter* spp., *Acetobacter* spp.). This group may be explained by the high nutrient availability as a result of berries damage.

Species	France	Italy	Spain	Greece	Portugal	Slovenia	Austria	Canada	Brazil	Argentina	Japan	Australia	India	China	South Africa	Hungary	Chile-Peru-Uruguay	New Zealand
Basidiomycetesa	+++	++	+++		+++	+++		+++	+	+	+++	++		+++				
Aspergillus tubingensis															+			
Aureobasidium pullulans	++	+++	+++	+		++		+++	+			+++			+++			++
Botryotinia fuckelian															+			
Botrytis elliptica															+			
Brettanomyces spp.		+																
Bulleromyces albus	++																	
Candida spp.	+++	+	++		+		+			+	++		++	++				
C. stellata/zemplinina	+++	+		++	+		+			+				+				+
C. diversa					+													
Cryptococcus mangus	++																	+
Debaryomyces spp.	+					+							+					
Hanseniaspora spp.	++		+++	++							+		+++	+				
H. guilliermindii					+++													
H. opuntiae					+													
H. uvarum	+++	+++	+++	+++	++	++	+++	++	+++	+++	+++		+	+++				+
Issatchenkia spp.	+		++	+					+++	+			++	+				
I. occidentalis							+											
I. terricola		++																
Kabatiella microsticta															+++			
*Kluyveromyces*spp./*Lachancea* spp.	+		++															

Species	France	Italy	Spain	Greece	Portugal	Slovenia	Austria	Canada	Brazil	Argentina	Japan	Australia	India	China	South Africa	Hungary	Chile-Peru-Uruguay	New Zealand
Kregervanrija fluxuum							+											
Lachancea thermotolerans							+											
Metschnikowia spp.	+	++	+			+	+++			+++				+				
M. fructicola		++																
Metschnikowia pulcherrima	++														+++			
Metschnikowia viticola							+											
Penicillium brevicompactum															+			
P. corylophilum															+			
P. glabrum															+			
Pleospora herbarum															+			
Pichia spp.	+	+	+		+	+				++			++	++				
Pichia terricola					+													
Pichia fermentans		+																
Pichia kluyveri							++											
Pichia kudriavzevii					+													
Rhodotorula glutinis		+																+++
Saccharomyces spp.	+	+																
S. uvarum																+		
S. bayanus							++									+		
S. cerevisiae	++	++	++				+++						++			+++	+++	

http://dx.doi.org/10.5772/64806

Species	France	Italy	Spain	Greece	Portugal	Slovenia	Austria	Canada	Brazil	Argentina	Japan	Australia	India	China	South Africa	Hungary	Chile-Peru-Uruguay	New Zealand
Saccharomycopsis spp.		++																
Saccharomycodes ludwigii										+								
Sporobolomyces roseus	++																	
Starmerella bacillaris					+++										+			
Pichia manshurica					+													
Torulaspora spp.			+		+													
T. delbrueckii	+	+																
Zygoascus hellenicus					+		+											
Zygosaccharomyces spp.		+												+				
Z. bisporus					+													
Z. bailii			+		+									+				
Sampling	Be	Be	Be, Bu	Bu	Be	Bu	Bu	Bu	Be, Bu	Bu	Bu	Bu	Bu	Bu	Be	Bu	Be	Be
Culture media	G	G	G, A	G,S	G,S	G	A, S	G	G	G	–	G	G	G,S	NGS	G, S	G	G
References	[60–62]	[63, 64]	[65, 66]	[67]	[38, 68, 69]	[70]	[42]	[71]	[72]	[73]	[74]	[12]	[75]	[76]	[77]	[78]	[19]	[63]

Yeast and yeast like species isolated from sound grapes or berries at harvest. The data are collected from published surveys (see references) and "+" in dicates relative proportion of the detected species

Be: berry, Bu: bunch

G: general purpose, A: autoenrichment, S: selective media

NGS: Next Generation Sequencing (metagenomic approach)

Table 1. Yeast and yeast like species isolated from sound grapes or berries at harvest.

Worldwide surveys seem to indicate that apparently sound grapes are colonised by a wide variety of yeast species without any obvious explanation.

Table 1 summarises the most indicative surveys on yeast species found in the respective countries.

7. Vinification examples with autochthonous starter cultures: pros and cons

The last four decades, wine industries worldwide try to exploit new indigenous strains of *S. cerevisiae* in order to produce wines with specific characteristics resulting from the biodiversity of each different area. Studies done on Debina must, a white-wine producing variety, in Zitsa (Epirus, Greece) have shown that a specific indigenous strain was the most predominant and responsible for a variety of aromas in the produced wine [25]. Another interesting example of application of indigenous *S. cerevisiae* strains in winemaking is that of Negroamaro wines, where selected strains are used to produce Negroamaro wines in Salento (Apulia, Italy) and share interesting volatile profiles that are associated with their geographical origin [79]. The application of combined mixtures of *S. cerevisiae* and non-*Saccharomyces* strains has widely been used, in cases such as the production of Italian passito wines, where studies have shown that the combination of *Botrytis cinerea* strains (non-*Saccharomyces* species with great esterase, glucosidase and protease activities) with *S. cerevisiae* strains can result to the production of highly improved passito varieties [1]. Moreover, studies on Italian Amarone wine have shown that mixtures of *S. cerevisiae* and *S. bayanus* strains, which are used during fermentation in different wineries in Valpolicella area (Italy), all produce specific amounts of isobutanol and amylic alcohols and therefore contribute to the production of traditional varieties with desired aromatic and flavour features [2]. An indigenous *S. cerevisiae* strain can be used in both primary and secondary fermentations which are needed for the production of Champenoise sparkling wine, as it responds perfectly to the stressful conditions presented in both fermentations such as low nitrogen content and increased accumulation of toxic by-products [80]. Moreover, Aponte and Blaiotta [81], used a selected *S. cerevisiae* autochthonous strain as starter culture in the production of *"Moscato di Saracena"*, a southern Italy passito wine, and suggested that the physicochemical traits obtained, showed better characteristics compared to those obtained by spontaneous fermentation. Finally, various studies were undertaken in order to develop region-specific starter cultures, such as wines in 'El Penedes' area of Spain [13] and sweet white wine in Tokaj area of Hungary [82]. They demonstrated that native selected strains may be better adapted to fermentation conditions than commercial strains, and selected inoculated strains were found to play an important role in the resulting wine.

As the importance of *S. cerevisiae* role in winemaking has long been established, the use of the commercial strains of these yeast cultures in fermentation is an ordinary practice in order to ensure a reproducible product and to reduce the risk of wine spoilage. However, this approach can cause a progressive substitution of local microflora and a consequent reduction of microbial biodiversity. Indeed, knowledge of the autochthonous yeast strains will help to preserve and employ the most representative strains which will enhance the quality characteristics and retain the product typicity. The selection and the employment of autochthonous microorganisms could be a powerful tool in order to improve the organoleptic and sensory characteristics of wine produced from indigenous grape cultivars.

8. Problems with wild strains

Wine obtained with pure culture fermentation of non-*Saccharomyces* yeast may show several problems, due to their fermentative behaviour or metabolite compounds production. Non-*Saccharomyces* yeasts can produce several secondary compounds, such as acetic acid, acetaldehyde, acetoin and ethyl acetate, compounds which are undesirable even at low concentrations. They also cause the presence of off-odours, such as ethyl and vinyl phenols, generally produced by *Brettanomyces* spp. and/or *Dekkera* spp. [83]. In addition, the majority of the non-*Saccharomyces* strains lack of good fermentative parameters, i.e., poor SO_2 resistance, low power and rate of fermentation. Nevertheless, some negative traits of non-*Saccharomyces* yeasts may not be expressed or could be modified during multi-starters fermentations in the presence of *S. cerevisiae* strains [31].

Similarly, spoilage species of LAB, AAB and, occasionally, *Bacillus* and *Clostridium* species may grow in wines during storage in the cellar and after bottling [59]. Their growth is probably encouraged by nutrients released by autolysis of wine Yeasts, as well as *O. oeni* strains [84]. Fornachon [85] reported that the spoilage yeasts, *Pichia* spp., *Saccharomycodes ludwigii* and *Candida pulcherrima*, showed an inhibitory activity towards spoilage LAB (i.e. *Lactobacillus hilgardii, Lactobacillus brevis, Leuconostoc mesenteroides*) which could be caused by the toxic concentrations of sulphur dioxide produced by the above mentioned yeasts.

Moreover, studies concluded that besides LAB, some yeasts such as *S. cerevisiae* and *Brettanomyces bruxellensis* are also responsible for biogenic amine formation. Various histaminogenic abilities of the yeasts have been confirmed in fermentation tests [86]. However, the relation between the concentrations of the biogenic amines and their precursor amino acids during fermentation depend on the yeast strain involved in the fermentation. Together with the decarboxylating aptitude of the starter cultures, the presence and relative activity of amino-oxidases (or amino-acid oxidases) should be considered as an important factor in the selection of starter cultures for wine production. Inoculation with species and strains of LAB with none or low forming capacities of biogenic amines reduces their occurrence in wine [87].

9. Future perspectives

Several studies undertaken in different countries attributed an important contribution of non-*Saccharomyces* species to yeast growth dynamics during wine fermentations [88, 89]. Hence, non-*Saccharomyces* yeast species supply a factor of diversity that requires specific studies to avoid any negative consequences, and to exploit their beneficial contributions [88]. Yeasts populations on grapes and in must, the effect of winemaking practices on these yeasts, as well as how their metabolites interact with each other and with LAB, must be known [48]. In addition, during the last years, the improvement and application of molecular approaches for the analysis of yeast populations have shown that, together with species variability, spontaneous fermentation is characterised by a significant intraspecific biodiversity [34], as well as by a high genetic polymorphism observed in the population of *S. cerevisiae* present during spontaneous fermentation. That is to say, the population of yeasts correlated to wild wine fermentation is composed of genotypically different strains with most likely different phenotype and therefore, potentially capable of influencing, in proportion to their relative abundance, the flavour profile of the resulting wine [90].

As the demand for high quality wines is emerging worldwide, the need for discovering new strains and new innovative techniques for their application in wine production is increasing. An example of the effort given by wine industry to implement new techniques is the management of nutrient availability and uptake before and during fermentation which has the potential to increase the biomass production by *S. cerevisiae* [91]. The same nutrient demands should be explored for the non-*Saccharomyces* species as well. Another aspect of interest is the understanding of the kinetic and metabolic behaviour developed by mixtures of *Saccharomyces* and non-*Saccharomyces* strains, as it can contribute to the production of wine yeasts with improved technological characteristics which can be used for the production of improved quality wines [8, 92]. In addition, studies done on experimental hybrids of different *Saccharomyces* species like *S. cerevisiae* and *S. bayanus* exploit the production of new yeasts through a variety of evolutionary programmes [29]. Moreover, as *S. cerevisiae* is a stable microorganism that can survive under the unfavourable conditions during the winemaking process, studies on recombinant yeast strains aim to the creation of yeasts with excellent fermentative behaviour and improved oenological characteristics [93]. Also an interesting case is the one of the application of auxotrophic strains of *S. cerevisiae* which have the ability to produce large quantities of high quality fermentation products at very low growth rate [94].

Saccharomyces cerevisiae is by far the most widely used yeast in oenology. However, many studies of wine fermentation ecology have shown that several other yeast species participate in the phenomenon and can positively impact wine quality. *Torulaspora delbrueckii, Metschnikowia pulcherrima, Pichia kluyveri, Lachancea thermotolerans, Hanseniaspora uvarum, Starmerella bacillaris* are now proposed as starter cultures in mixed fermentations with *S. cerevisiae.* The knowledge of these non-conventional yeasts is increasing because of the advancement in genomic and proteomic analysis tools. The next step lies on the development of selection programs and/or genetic improvement of these non-conventional species. In addition, next generation sequencing is for seeing to help the efforts in wine differentiation based on the biological/genetic fingerprint [95].

The scientific community should enhance its efforts studying microbial genetic fingerprint and metabolic footprints, resulting from biodiversity and microbial activity, respectively, in order to preserve food heritage and support the typicality and authenticity of traditional fermented products.

Author details

Despina Bozoudi and Dimitris Tsaltas*

*Address all correspondence to: dimitris.tsaltas@cut.ac.cy

Department of Agricultural Sciences, Biotechnology and Food Science, Cyprus University of Technology, Limassol, Cyprus

References

[1] Azzolini M, Tosi E, Faccio S, Lorenzini M, Torriani S, Zapparoli G. Selection of *Botrytis cinerea* and *Saccharomyces cerevisiae* strains for the improvement and valorisation of Italian passito style wines. FEMS Yeast Research. 2013; 13: 540–552.

[2] Torriani S, Zapparoli G, Suzzi G. Genetic and phenotypic diversity of *Saccharomyces sensu stricto* strains isolated from Amarone wine. Antonie van Leeuwenhoek. 1999; 75: 207–215.

[3] Settanni L, Sannino C, Francesca N, Guarcello R, Moschetti G. Yeast ecology of vineyards within Marsala wine area (western Sicily) in two consecutive vintages and selection of autochthonous *Saccharomyces cerevisiae* strains. Journal of Bioscience and Bioengineering. 2012; 114: 606–614.

[4] Alves RF, Nascimento AMD, Nogueir JMF. Characterisation of the aroma profile of Madeira wine by sorptive extraction techniques. Analytica Chimica Acta. 2005; 546: 11–21.

[5] Martínez P, Pérez Rodríguez L, Benítez T. Evolution of Flor yeast population during the biological aging of fino sherry wine. American Journal of Enology and Viticulture. 1997; 48: 160–168.

[6] Ioannou-Papayianni E, Kokkinofta RI, Theocharis CR. Authenticity of Cypriot sweet wine commandaria using FT-IR and chemometrics. Journal of Food Science. 2011; 76: C420–C427.

[7] Maturano YP, Assof M, Fabani MP, Nally MC, Jofre V, Rodriguez ALA, Toro ME, Catelanos FLI, Vazquez F. Enzymatic activities produced by mixed *Saccharomyces* and non-*Saccharomyces* cultures: relationship with wine volatile composition. Antonie van Leeuwenhoek. 2015; 108: 1239–1256.

[8] Ercolini D. High-throughput sequencing and metagenomics: moving forward in the culture-independent analysis of food microbial ecology. Applied and Environmental Microbiology. 2013; 79: 3148–3155.

[9] Bokulich NA, Thorngate JH, Richardson PM, Mills DA. Microbial biogeography of wine grapes is conditioned by cultivar, vintage, and climate. PNAS. 2014; 111: E139–E148.

[10] Gilbert JA, van der Lelie D, Zarraonaindia I. Microbial terroir for wine grapes. PNAS. 2014; 111: 5–6.

[11] Mas A, Padilla B, Esteve-Zarzoso B, Beltran G, Reguant C, Bordons A. Taking advantage of natural biodiversity for wine making: The WILDWINE Project. Agriculture and Agricultural Science Procedia. 2016; 8: 4–9.

[12] Prakitchaiwattana CJ, Fleet GH, Heard GM. Application and evaluation of denaturing gradient gel electrophoresis to analyse the yeast ecology of wine grapes. Fems Yeast Research. 2004; 4: 865–877.

[13] Esteve-Zarzoso B, Gostincar A, Bobet R, Uruburu F, Querol A. Selection and molecular characterization of wine yeasts isolated from "El Penedes" area (Spain). Food Microbiology. 2000; 17: 553–562.

[14] Barata A, Malfeito-Ferreira M, Loureiro V. The microbial ecology of wine grape berries. International Journal of Food Microbiology. 2012; 153: 243–259.

[15] Setati ME, Jacobson D, Andong U-C, Bauer F. The vineyard yeast microbiome, a mixed model microbial map. PLoS One. 2012; 7: e52609.

[16] Knight S, Klaere S, Fedrizzi B, Goddard MR. Regional microbial signatures positively correlate with differential wine phenotypes: evidence for a microbial aspect to terroir. Nature Scientific Reports. 2015; 5: 14233.

[17] Vigentini I, De Lorenzis G, Fabrizio V, Valdetara F, Faccincani M, Panont, AC, Picozzi C, Imazio S, Failla O, Foschino R. The vintage effect overcomes the terroir effect: a three-year survey on the wine yeast biodiversity in Franciacorta and Oltrepo Pavese, two northern Italian vine-growing areas. Microbiology. 2015; 161: 362–373.

[18] Longo E, Vezinhet F. Chromosomal rearrangements during vegetative growth of a wild strain of *Saccharomyces cerevisiae*. Applied and Environmental Microbiology. 1993; 59: 322–326.

[19] Martínez C, Cosgaya P, Vásquez C, Gac S, Ganga A. High degree of correlation between molecular polymorphism and geographic origin of wine yeast strains. Journal of Applied Microbiology. 2007; 103: 2185–2195.

[20] Valero E, Schuller D, Cambon B, Casal M, Dequin S. Dissemination and survival of commercial wine yeast in the vineyard: a large-scale, three-years study. FEMS Yeast Research. 2005; 5: 959–969.

[21] Orlic S, Redzepovic S, Jeromel A, Herjavec S, Iacumin L. Influence of indigenous *Saccharomyces paradoxus* strains on Chardonnay wine fermentation aroma. International Journal of Food Science and Technology. 2007; 42: 95–101.

[22] Fleet GH. Growth of yeasts during wine fermentations. Journal of Wine Research. 1990; 1: 211–223.

[23] Rainieri S, Pretorius IS. Selection and improvement of wine yeasts. Annals of Microbiology. 2000; 50: 15–31.

[24] Borneman AR, Forgan AH, Kolouchova R, Fraser JA, Schmidt SA. Whole genome comparison reveals high levels of inbreeding and strain redundancy across the spectrum of commercial wine strains of *Saccharomyces cerevisiae*. Genes, Genomes, Genetics. 2016; 6: 957–971.

[25] Parapouli M, Hatziloukas E, Drainas C, Perisynakis A. The effect of Debina grapevine indigenous yeast strains of *Metschnikowia* and *Saccharomyces* on wine flavour. Journal of Industrial Microbiology and Biotechnology. 2009; 37: 85–93.

[26] Muñoz R, Gómez A, Robles V, Rodríguez P, Cebollero E, Tabera L, Carrascosa AV, Gonzalez R. Multilocus sequence typing of oenological *Saccharomyces cerevisiae* strains. Food Microbiology. 2009; 26: 841–846.

[27] Goffeau A, Barrell BG, Bussey H, Davis RW, Dujon B, Feldmann H, Galibert F, Hoheisel JD, Jacq C, Johnston M, Louis EJ, Mewes HW, Murakami Y, Philippsen P, Tettelin HS, Oliver G. Life with 6000 genes. Science. 1996; 274: 546–567.

[28] Borneman AR, Pretorius IS, Chambers PJ. Comparative genomics: a revolutionary tool for wine yeast strain development. Current Opinion in Biotechnology. 2013; 24: 192–199.

[29] Naumov GI, Naumova ES, Martynenko NN, Pomarede MI. Taxonomy, ecology, and genetics of the yeast *Saccharomyces bayanus*: a new object for science and practice. Microbiology. 2011; 80: 735–742.

[30] Rantsiou K, Campolongo S, Alessandria V, Rolle L, Torchio F, Cocolin L. Yeast populations associated with grapes during withering and their fate during alcoholic fermentation of high-sugar must. Australian Journal of Grape and Wine Research. 2013; 19: 40–46.

[31] Fleet, GH. Wine yeasts for the future. FEMS Yeast Research. 2008; 8: 979–995.

[32] Englezos V, Rantsiou K, Torchio F, Rolle L, Gerbi V, Cocolin L. Exploitation of the non-*Saccharomyces* yeast *Starmerella bacillaris* (synonym *Candida zemplinina*) in wine fermentation: Physiological and molecular characterizations. International Journal of Food Microbiology. 2015; 199: 3–40.

[33] Cocolin L, Campolongo S, Alessandria V, Dolci P, Rantsiou K. Culture independent analyses and wine fermentation: an overview of achievements 10 years after first application. Annals of Microbiology. 2011; 61: 17–23.

[34] Cocolin L, Bisson LF, Mills DA. Direct profiling of the yeasts dynamics in wine fermentations. FEMS Microbiology Letters. 2000; 189:81–87.

[35] Alessandria V, Marengo F, Englezos V, Gerbi V, Rantsiou K, Cocolin L. Mycobiota of Barbera grapes from the Piedmont region from a single vintage year. American Journal of Enology and Viticulture. 2015; 66: 244–250.

[36] Tofalo R, Chaves-López C, Di Fabio F, Schirone M, Felis GE, Torriani S, Paparella A, Suzzi G. Molecular identification and osmotolerant profile of wine yeasts that ferment a high sugar grape must. International Journal of Food Microbiology. 2009; 130: 179–187.

[37] Tristezza M, Fantastico L, Vetrano C, Bleve G, Corallo D, Grieco F, Mita G, Grieco F. Molecular and technological characterization of *Saccharomyces cerevisiae* strains isolated from natural fermentation of susumaniello grape must in Apulia, southern Italy. International Journal of Microbiology. 2014; 897428: 1–11.

[38] Teixeira A, Caldeira I, Luz Duarte F. Molecular and oenological characterization of Touriga Nacional non-*Saccharomyces* yeasts. Journal of Applied Microbiology. 2015; 118: 658–671.

[39] Schuller D, Alves H, Dequin S, Casal M. Ecological survey of *Saccharomyces cerevisiae* strains from vineyards in the Vinho Verde Region of Portugal. FEMS Microbiology Ecology. 2005; 51: 167–177.

[40] Beltran G, Torija MJ, Novo M, Ferrera N, Poblet M, Guillamóna JM, Rozès N, Mas A. Analysis of yeast populations during alcoholic fermentation: A six-year follow-up study. Systematic and Applied Microbiology. 2002; 25: 287–293.

[41] Nikolaou E, Andrighetto C, Lombardi A, Litopoulou-Tzanetaki E, Tzanetakis N. Heterogeneity in genetic and phenotypic characteristics of *Saccharomyces cerevisiae* strains isolated from red and white wine fermentations. Food Control. 2007; 18: 1458–1465.

[42] Lopandic K, Tiefenbrunner W, Gangl H, Mandl K, Berger S, Leitner G, Abd-Ellah GA, Querol A, Gardner RC, Sterflinger K, Prillinger H. Molecular profiling of yeasts isolated during spontaneous fermentations of Austrian wines. FEMS Yeast Research. 2008; 8: 1063–1075.

[43] Ganga MA, Martinez C. Effect of wine yeast monoculture practice on the biodiversity of non-*Saccharomyces* yeasts. Journal of Applied Microbiology. 2004; 96: 76–83.

[44] Schuller D, Casal M. The genetic structure of fermentative vineyard-associated *Saccharomyces cerevisiae* populations revealed by microsatellite analysis. Antonie van Leeuwenhoek. 2007; 91: 137–150.

[45] Maturano YP, Rodriguez ALA, Toro ME, Nally MC, Vallejo M, Castelanos FLI, Combina M, Vazquez F. Multi-enzyme production by pure and mixed cultures of *Saccharomyces* and non-*Saccharomyces* yeasts during wine fermentation. International Journal of Food Microbiology. 2012; 155: 43–50.

[46] Reynolds AG, editor. Managing Wine Quality. Viticulture and Wine Quality. Boca Raton, FL: CRC Press; 2010.

[47] Chasseriaud L, Miot-Sertier C, Coulon J, Iturmendi N, Moine V, Albertin W, Bely M. A New method for monitoring the extracellular proteolytic activity of wine yeasts during alcoholic fermentation of grape must. Journal of Microbiological Methods. 2015; 119: 176–179.

[48] Jolly NP, Varela C, Pretorius IS. Not your ordinary yeast: non-*Saccharomyces* yeasts in wine production uncovered. FEMS Yeast Research. 2014; 14: 215–237.

[49] Pretorius. Tailoring wine yeast for the new millennium: novel approaches to the ancient art of winemaking. Yeast. 2000; 16: 675–729.

[50] Perpetuini G, Di Gianvito P, Arfelli G, Schirone M, Corsetti A, Tofalo R, Suzzi G. Biodiversity of autolytic ability in flocculent *Saccharomyces* cerevisiae strains suitable for traditional sparkling wine fermentation. Yeast. 2016; 33: 303–312.

[51] Maqueda M, Zamora E, Álvarez ML, Ramírez M. Characterization, ecological distribution, and population dynamics of *Saccharomyces Sensu Stricto* killer yeasts in the spontaneous grape must fermentations of Southwestern Spain. Applied and Environmental Microbiology. 2012; 78: 735–743.

[52] Suarez-Lepe JA, Morata A. New trends in yeast selection for winemaking. Trends in Food Science and Technology. 2012; 23: 39–50.

[53] Quintela S, Villaran MC, Lopez de Armentia I, Elejalde E. Ochratoxin A removal in wine: a review. Food Control. 2013; 30: 439–445.

[54] Cappello M, Bleve G, Grieco F, Dellaglio F, Zacheo G. Characterization of *Saccharomycescerevisiae* strains isolated from must of grape grown in experimental vineyard. Journal of Applied Microbiology. 2004; 97: 1274–1280.

[55] Nikolaou E, Soufleros E, Bouloumpasi E, Tzanetakis N. Selection of indigenous *Saccharomyces cerevisiae* strains according to their oenological characteristics and vinification results. Food Microbiology. 2006; 23: 205–211.

[56] Lopes C, Rodriguez ME, Sangorrin M, Querol A, Caballero A. Patagonian wines: implantation of an indigenous strain of *Saccharomyces cerevisiae* in fermentations conducted in traditional and modern cellars. Journal of Industrial Microbiology and Biotechnology. 2007; 34: 139–149.

[57] Verginer M, Leitner E, Berg G. Production of volatile metabolites by grape-associated microorganisms. Journal of Agricultural and Food Chemistry. 2010; 58: 8344–8350.

[58] Campisano A, Antonielli L, Pancher M, Yousaf S, Pindo M, Pertot I. Bacterial endophytic communities in the grapevine depend on pest management. PLoS One. 2014; 9: e112763.

[59] Du Toit M, Pretorius IS. Microbial spoilage and preservation of wine: using weapons from nature's own arsenal-a review. African Journal of Enology and Viticulture. 2000; 21: 74–96.

[60] Barnett JA, Delaney MA, Jones E, Magson AB, Winch B. The numbers of yeasts associated with wine grapes of Bordeaux. Archiv für Mikrobiologie. 1972; 83: 52–55.

[61] Renouf V, Claisse O, Lonvaud-Funel A. Understanding the microbial ecosystem on the grape berry surface through numeration and identification of yeast and bacteria. Australian Journal of Grape and Wine Research. 2005; 11: 316–327.

[62] David V, Terrat S, Herzine K, Claisse O, Rousseaux S, Tourdot-Maréchal R, Masneuf-Pomarede I, Ranjard L, Alexandre H. High-throughput sequencing of amplicons for monitoring yeast biodiversity in must and during alcoholic fermentation. Journal of Indian Microbiology and Biotechnology. 2014; 41: 811–821.

[63] Gayevskiy V, Goddard MR. Geographic delineations of yeast communities and populations associated with vines and wines in New Zealand. The ISME Journal. 2012; 6: 1281–1290.

[64] Francesca N, Chiurazzi M, Romano R, Aponte M, Settanni L, Moschetti G. Indigenous yeast communities in the environment of "Rovello bianco" grape variety and their use in commercial white wine fermentation. World Journal of Microbiology and Biotechnology. 2010; 26: 337–351.

[65] Clavijo A, Calderón IL, Paneque P. Diversity of *Saccharomyces* and non-*Saccharomyces* yeasts in three red grape varieties cultured in the Serranía de Ronda (Spain) vine-growing region. International Journal of Food Microbiology. 2010; 143: 241–245.

[66] Sabate J, Cano J, Esteve-Zarzoso B, Guillamón JM. Isolation and identification of yeasts associated with vineyard and winery by RFLP analysis of ribosomal genes and mitochondrial DNA. Microbiological Research. 2002; 157: 267–274.

[67] Nisiotou AA, Nychas GJE. Yeast populations residing on healthy or *Botrytis*-infected grapes from a vineyard in Attica, Greece. Applied and Environmental Microbiology. 2007; 73: 2765–2768.

[68] Barata A, González S, Malfeito-Ferreira M, Querol A, Loureiro V. Sour rot damaged grapes are sources of wine spoilage yeasts. Fems Yeast Research. 2008; 8: 1008–1017.

[69] Barata A, Seborro F, Belloch C, Malfeito-Ferreira M, Loureiro V. *Ascomycetous* yeast species recovered from grapes damaged by honeydew and sour rot. Journal of Applied Microbiology. 2008; 104: 1182–1191.

[70] Raspor P, Milek DM, Polanc J, Smole Mozina S, Cadez N. Yeasts isolated from three varieties of grapes cultivated in different locations of the Dolenjska vine growing region, Slovenia. International Journal of Food Microbiology. 2006; 109: 97–102.

[71] Subden RE, Husnik JI, van Twest R, van der Merwe G, van Vuuren HJJ. Autochthonous microbial population in a Niagara Peninsula icewine must. Food Research International. 2003; 36: 747–751.

[72] Baffi MA, dos Santos Bezerra C, Arévalo-Villena M, Briones-Pérez AI, Gomes E, Da Silva R. Isolation and molecular identification of wine yeasts from a Brazilian vineyard. Annals of Microbiology. 2011; 61: 75–78.

[73] Combina M, Elía A, Mercado L, Catania C, Ganga A, Martinez C. Dynamics of indigenous yeast populations during spontaneous fermentation of wines from Mendoza, Argentina. International Journal of Food Microbiology. 2005; 99: 237–243.

[74] Yanagida F, Ichinose F, Shinohara T, Goto S. Distribution of wild yeasts in the white grape varieties at Central Japan. The Journal of General and Applied Microbiology. 1992; 38: 505–509.

[75] Chavan P, Mane S, Kulkarni G, Shaikh S, Ghormade V, Nerkar DP, Shouche Y, Deshpande MV. Natural yeast flora of different varieties of grapes used for winemaking in India. Food Microbiology. 2009; 26: 801–808.

[76] Li SS, Cheng C, Li Z, Chen, J-Y, Yan B, Han B-Z, Reeves M. Yeast species associated with wine grapes in China. International Journal of Food Microbiology. 2010; 138: 85–90.

[77] Setati ME, Jacobson D, Bauer FF. Sequence-based Analysis of the *Vitis vinifera* L. cv cabernet sauvignon grape must mycobiome in three South African vineyards employing distinct agronomic systems. Frontiers in Microbiology. 2015; 6: 1358.

[78] Csoma H, Zakany N, Capece A, Romano P, Sipiczki M. Biological diversity of *Saccharomyces* yeasts of spontaneously fermenting wines in four wine regions: comparative genotypic and phenotypic analysis. International Journal of Food Microbiology. 2010; 140: 239–248.

[79] Tufariello M, Chiriatti MA, Grieco F, Perrotta C, Capone S, Rampino P, Tristezza M, Mita G, Grieco F. Influence of autochthonous *Saccharomyces cerevisiae* strains on volatile profile of Negroamaro wines. LWT: Food Science and Technology. 2014; 58: 35–48.

[80] Borrull A, Lopez-Martinez G, Miro AE, Salvado Z, Poblet M, Cordero-Otero R, Rozes N. New insights into the physiological state of *Saccharomyces cerevisiae* during ethanol acclimation for producing sparkling wines. Food Microbiology. 2016; 54: 20–29.

[81] Aponte M, Blaiotta G. Selection of an autochthonous *Saccharomycescerevisiae* strain for the vinification of "Moscato di Saracena", a southern Italy (Calabria Region) passito wine. Food Microbiology. 2016; 54, 30–39.

[82] Sipiczki M, Romano P, Lipani G, Miklos I, Antunovics Z. Analysis of yeasts derived from natural fermentation in a Tokaj winery. Antonie van Leeuwenhoek. 2001; 79: 97–105.

[83] Chatonnet P, Dubourdieu D, Boidron JN. The influence of *Brettanomyces/Dekkera* sp. yeasts and lactic acid bacteria on the ethylphenol content of red wines. American Journal of Enology and Viticulture. 1995; 46: 463–468.

[84] Crouigneau AA, Feuillat M, Guilloux-Benatier M. Influence of some factors on autolysis of *Oenococcus oeni*. Vitis. 2000; 39: 167–171.

[85] Fornachon JCM. Influence of different yeasts on the growth of lactic acid bacteria in wine. Journal of the Science of Food and Agriculture. 1968; 19: 374–378.

[86] Caruso M, Fiorel C, Contursi, M, Salzano G, Paparella A, Romano P. Formation of biogenic amines as criteria for the selection of wine yeasts. World Journal of Microbiology and Biotechnology. 2002; 18: 159–163.

[87] Torrea D, Ancin C. Content of biogenic amines in a Chardonnay wine obtained through spontaneous and inoculated fermentations. Journal of Agriculture and Food Chemistry. 2002; 50: 4895–4899.

[88] Jolly NP, Augustyn OPH, Pretorius IS. The effect of non-*Saccharomyces* yeasts on fermentation and wine quality. South African Journal of Enology and Viticulture. 2003; 24: 55–62.

[89] Zott K, Thibon C, Bely M, Lonvaud-Funel A, Dubourdieu D, Masneuf-Pomared I. The grape must non-*Saccharomyces* microbial community: Impact on volatile thiol release. International Journal of Food Microbiology. 2011; 151: 210–215.

[90] Romano, P., Fiore, C., Paraggio, M., Caruso, M., Capece, A. Function of yeast species and strains in wine flavor. International Journal of Food Microbiology. 2003; 86 (1–2): 169–180.

[91] Crepin L, Sanchez I, Nidelet T, Dequin S, Camarasa C. Efficient ammonium uptake and mobilization of vacuolar arginine by *Saccharomycescerevisiae* wine strains during wine fermentation. Microbial Cell Fact. 2014; 13: 1–13.

[92] Mendoza LM, Nadra MCM, Farias ME. Kinetics and metabolic behavior of a composite culture of *Kloeckera apiculata* and *Saccharomycescerevisiae* wine related strains. Biotechnology Letters. 2007; 29: 1057–1063.

[93] Rozanov AS, Kotenko AV, Akberdin IR, Peltek SE. Recombinant strains of *Saccharomyces cerevisiae* for ethanol production from plant biomass. Russian Journal of Genetics: Applied Research. 2015; 5: 375–382.

[94] Paciello L, Zueco J, Landi C. On the fermentative behavior of auxotrophic strains of *Saccharomyces cerevisiae*. Electronic Journal of Biotechnology. 2014; 17: 246–249.

[95] Masneuf-Pomarede I, Bely M, Marullo P, Albertin W. The genetics of non-conventional wine yeasts: current knowledge and future challenges. Frontiers in Microbiology. 2015; 6: 1563.

Chapter 11

Non-*Saccharomyces* Yeasts: Biotechnological Role for Wine Production

Margarita García, Braulio Esteve-Zarzoso and
Teresa Arroyo

Additional information is available at the end of the chapter

http://dx.doi.org/10.5772/64957

Abstract

Non-*Saccharomyces* yeasts play a substantial role in the early stages of wine fermentation. With the increase in alcohol concentration, indigenous or commercial strains of *Saccharomyces cerevisiae* take over and complete the transformation of the grape must sugars into ethanol, CO_2, and other secondary metabolites. The presence of non-*Saccharomyces* during the fermentation has an impact on the wine composition, and consequently, their contribution during the fermentation process cannot be ignored. The new challenges to enhance the appeal and value of wine elaborated by traditional technology are being achieved by selecting and using autochthonous non-*Saccharomyces* and *Saccharomyces* strains that may enhance regional identity of wines. Greater understanding of yeast biochemistry and physiology is enabling the selection and development of yeast strains that have defined specific influences on process efficiency and wine quality. The aim of this chapter was to show the different aspects of non-*Saccharomyces* species that may play a positive incidence in the biotechnological process to conduct to wine elaboration.

Keywords: non-*Saccharomyces* yeast, wine production, aroma, enzymes, mixed culture

1. Introduction

Since 1866 when Louis Pasteur first elucidated the bioconversion of grape juice into wine, this complex biochemical process and the role of the yeast therein have been studied continuously. *Saccharomyces cerevisiae* has often been the wine yeast that has received the most attention in the wine elaboration. This yeast is not only related with the conversion of grape sugar into alcohol and CO_2 in the fermentation process which also plays an important role in the generation of secondary metabolites, just as the production of varietal wine aromas from grape aroma precursors [1]. The spontaneous fermentation of the wine is produced by a mixture of yeast species simultaneously present in the grape must [2]. Although the control of *S. cerevisiae* (inoculated or native) in the fermentation is predicted and desired, the presence of indigenous non-*Saccharomyces* yeasts in the must leads to a competitive situation for the nutrients contained in the must. These non-*Saccharomyces* yeasts are perfectly adapted to the specific environment and frequently are in greater number of *S. cerevisiae* [3].

Yeast and bacterial microbial dynamics have been studied since the 1970s [4]. Using classical methods, microbial count and diversity is determined by employing synthetic culture media containing agar. The biodiversity in the complex ecosystems is impossible to characterize with precision using classical microbiological culture-dependent methods. When using enrichment methods and growth on culture medium, the microbiota naturally present in the sample are open to undergo important changes because of the ability of the certain species take control of the system and overcome other microbial components [5]. Due to this, populations less numerous or sensitive to the stress conditions are hard to find again and identify. Therefore, the use of culture-dependant methods could cause a misidentification of the microbial ecology of complex ecosystems [6].

Since the end of the 1990s, molecular techniques have helped provide a good overview of microbial ecology. These methods, generally named culture-independent methods, are used for the identification of microorganisms directly in the system through the study of their DNA and RNA without the need for isolation and cultivation. There are several advantages of the direct characterization of wine microbial DNA against to culture-dependent methods. Firstly, not all microbial populations are able to grow in enrichment media due to injury, lack of appropriate nutrients, or presence of viable but not culturable states. Secondly, the direct analyses allow saving time in comparison with the enrichment methods. This advantage could enable winemakers to use microbial detection data during the fermentation process, being able to anticipate possible spoilage problems in the wine. Furthermore, the DNA-based studies allow processing a larger numbers of samples than plating methods [7].

Nevertheless, in the past few years, successful culture-independent methods such as denaturing gradient gel electrophoresis (DGGE), real-time PCR, fluorescence in situ hybridization (FISH), or Fourier transform infrared spectroscopy (FT-IR) have been described. The PCR-DGGE method was first developed for the study of the microbial ecology in the environmental samples but soon found application in food microbiology [8]. This method is based on the separation of same length DNA fragments, but of different sequences. PCR-DGGE method has reported detection limits between 10^2 CFU/mL in pure cultures and 10^4 CFU/mL in wine or must samples [9].

In recent years, scientists have used real-time quantitative PCR (QPCR) to detect and quantity microorganisms in different alimentary environments [10]. The advantages of QPCR are the low detection level, often as low as one cell per mL, the speed by which assays are performed, and the ability to quantify yeasts present following alcoholic fermentation.

http://dx.doi.org/10.5772/64957

Fluorescence in situ hybridization (FISH) is a very promising technique for wine ecology studies for its simplicity and rapidity as well as the ability to observe the cell morphology by a microscope and the high sensitivity obtained using a flow cytometer [11].

Fourier transform infrared (FT-IR) spectroscopy is used to identify isolates according to the different components of the cell [12]. The relative success of this method is directly dependent on the complexity within a reference spectral library, identification results on genus and species level. Due to high automation and cost efficiency, this high-throughput method gives much deeper insights into functional yeast diversity during wine production.

Fortunately, DNA-based approaches have largely helped to clarify modern taxonomy. DNA sequence analysis is now widely used in the identification and classification of yeasts and other fungi therefore helping to reassign species within genus level because of the new species that are now discovered [13]. In food and beverage industry, these name changes influence our ability to notify the identity of spoilage microorganisms and to be in accordance with the regulations governing the presence of certain microorganisms in this industry. Current taxonomies recognize 149 yeast genera comprising nearly 1500 species [14]. Of these, more than 40 species have been isolated from grape must [15].

2. Use of non-*Saccharomyces* yeasts in the wine production

Non-*Saccharomyces* yeasts were originally seen as responsible for microbial spoilage in wine production due to their isolation from spoiled wines [16]. Traditionally, the use of non-*Saccharomyces* in the wine elaboration has not been usual due to preceding investigations showed that several species produce high levels of undesirable compounds that affect the wine quality such as acetoin, ethyl acetate, acetic acid, and acetaldehyde [17]. Unfortunately, this exclusion of non-*Saccharomyces* yeasts from the fermentation process may result in a loss of complexity and wines lacking distinctive characteristics.

The initial belief that all non-*Saccharomyces* yeasts died soon after the commencement of an alcoholic fermentation due to the rising ethanol concentration and added SO_2 has not been sustained by later research [18]. Currently, the cellar technology and hygiene in modern cellars have improved greatly; as a consequence of this, the use of SO_2 has been significantly reduced. For these reasons, the survival of a higher number and diversity of non-*Saccharomyces* yeasts has increased. The great number of non-*Saccharomyces* yeasts reported in recent literature has also been influenced for the use of modern laboratory techniques that have made the detection of non-*Saccharomyces* yeasts easier [19].

During fermentation, and more evident in spontaneous fermentations, which lack the initial high-density inoculum of *S. cerevisiae*, there is a sequential succession of yeasts. Initially, species of *Hanseniaspora* (*Kloeckera*), *Rhodotorula*, *Issatchenkia*, *Pichia*, *Debaryomyces*, *Zygosaccharomyces*, *Torulaspora*, *Schizosaccharomyces*, *Candida*, *Metschnikowia*, and *Cryptococcus* are found at low levels in fresh must [20]. Of these, the most common yeast present in the highest numbers is *Hanseniaspora uvarum*, followed by different *Candida* spp. This is normally more evident in red must than white, probably as a result of the higher pH in red wine. However, *Hanseniaspora* may sometimes be absent or present at low levels [15].

Despite the sustained presence of certain non-*Saccharomyces* yeasts, the majority are not possible to recovery on plates during the early stages of a vigorous fermentation. This might be due to their slow growth and inhibition by the combined effects of SO_2, pH, increase in ethanol and oxygen deficiency [21]. This is consistent with their oxidative or weak fermentative metabolism. Nutrient limitation and size or dominance of *S. cerevisiae* inoculum can also have a suppressive effect, sometimes separate from temperature or ethanol concentration [22]. It has been reported that *Torulaspora delbrueckii* and *Lachancea thermotolerans* are less tolerant to low oxygen levels, and this, rather than ethanol toxicity, affects their growth and leads to their death during fermentation [23]. It was also shown that a cell–cell contact mechanism in the presence of high concentrations of viable *S. cerevisiae* yeasts played a role in the inhibition of these two non-*Saccharomyces* species [24]. But these mechanisms are not corroborated by Pérez-Nevado et al. [25] and Wang et al. [26]. Both authors were able to show that some metabolites produced by *S. cerevisiae* may be the responsible of the inhibitory effect on the growth of non-*Saccharomyces* wine yeasts. Recently, Branco et al. [27] showed that one derived peptide of the GAPDH could be the responsible of this inhibitory effect.

Species	Metabolites and/or physical properties	References
Candida stellata or *Starmerella bacillaris* (formerly *Candida zemplinina*)	Acetic acid	[9, 35, 42, 43]
	Dodecanoic acid	
	Ethyl octanoate	
	Glycerol	
	Higher alcohols	
	2-Methyl propanoic acid	
	Succinic acid	
	Terpenoids	
Schwanniomyces vanrijiae (formerly *Debaryomyces vanriji*)	Esters	[44]
	Medium-chain fatty acids	
	Terpenoids	
Hanseniaspora uvarum	Acetate and ethyl esters	[9, 47]
	Sulfur compounds	
	Higher alcohols	
	Medium-chain fatty acids	
Hanseniaspora vineae	2-Phenylethyl acetate	[48]
Hanseniaspora guilliermondii	Acetate and ethyl esters	[47]
	Acetone	
	Heavy sulfur compounds	
Hanseniaspora osmophila	Ethyl acetate	[49, 50]

Species	Metabolites and/or physical properties	References
	Ethyl lactate	
Wickerhamomyces anomalus (formerly *Pichia anomala*)	Acetate and ethyl esters	[37, 50, 51]
	Higher alcohols	
Issatchenkia orientalis	Color	[41]
	Methanol	
Lachancea thermotolerans (formerly *Kluyveromyces thermotolerans*)	Glycerol	[17, 38]
	Lactic acid	
	2-Phenylethanol	
Metschnikowia pulcherrima	Esters	[35, 46]
Meyerozyma guilliermondii (formerly *Pichia guilliermondii*)	Color	[53]
Pichia kluyveri	Thiols	[52]
Pichia fermentans	Acetaldehyde	[29, 49]
	2,3-Butanediol	
	Ethyl esters	
	Higher alcohols	
	Polysaccharides	
Schizosaccharomyces pombe	Maloalcoholic deacidification	[39, 40]
	Pyruvic acid	
	Color	
	Propanol	
Torulaspora delbrueckii	Succinic acid	[35–37, 66]
	Linalool	
	Polysaccharides	
Zygosaccharomyces bailii	Polysaccharides	[49]
Zygosaccharomyces florentina (formerly *Zygosaccharomyces florentinus*)	Polysaccharides	[50]

Table 1. Metabolites produced and/or physical properties enhanced in wines fermented with non-conventional yeasts as single or co-fermentations compared with pure fermentation with *Saccharomyces cerevisiae*.

The non-*Saccharomyces* yeasts have an important influence on wine flavor depending of species and strain specific involved in the fermentation process. These yeasts present particular metabolic features that affect to the organoleptic characteristics in the wine including the capacity to secrete enzymes and metabolites related with the primary and secondary aroma of wines, low volatile acidity, release of mannoproteins, or increase in wine color stability [28]. Should be note their potential positive contributions to wine quality, while keeping fermentation kinetics and consistency under control, many researchers have postulated co-inoculation or sequential inoculation of *S. cerevisiae* or closely related species with one or more non-*Saccharomyces* strains (**Table 1**).

3. Influence of non-*Saccharomyces* yeasts in mixed fermentations

In recent years, re-evaluation of the role of non-*Saccharomyces* yeasts in winemaking has resulted in several studies that have looked at the use of controlled fermentations using *Saccharomyces* along with non-*Saccharomyces* yeast species from the winemaking [29, 30]. The mixed fermentations are used as a biotechnological tool in order to enhance especial and specific characteristics of a wine and thus improve their complexity. Indeed, the application of mixed and controlled fermentations in the wine elaboration has changed the standardized way to make wine and can improve the quality of the final product. These fermentations consist in controlled inoculations of *S. cerevisiae* starter culture and non-*Saccharomyces* strains. The non-*Saccharomyces* application could enhance the analytical composition of wine by taking advantages of diverse metabolic pathways of these yeasts.

The use of controlled mixed fermentations of non-*Saccharomyces* yeasts together with *S. cerevisiae* can be suggested as an useful tool for wine production which allows reproduce microbiological and technical aspects that really occur in the spontaneous fermentation, as well as an increase in the wine aroma complexity owing to a more complex synthesis of aromatic compounds [31]. This practice has also been reported as being able to increase some desirable metabolites, such as some acetate esters [32] and glycerol [33]. Moreover, some non-*Saccharomyces* yeasts have been reported as being able to release more polysaccharides than *S. cerevisiae* strains [34].

The influence of multistarter fermentation practices on final wine composition and on growth and death rates of the *S. cerevisiae* and non-*Saccharomyces* strains have been investigated. *Torulaspora delbrueckii*, one of the few non-*Saccharomyces* yeast species currently commercialized, is reported to have a positive effect on the taste and aroma of alcoholic beverages [35] and exhibits low production of acetaldehyde, acetoin, acetate, and ethyl acetate [36].

Recent studies using strains of *T. delbrueckii* and *L. thermotolerans* as starter cultures together with *S. cerevisiae* (specifically in mixed and sequential fermentations) have generated important changes in the wine composition. *Torulaspora delbrueckii* produced a reduction in acetic acid content and *L. thermotolerans* produced a reduction in acetaldehyde concentration and increase in titratable acidity [37, 38].

Other studies have been carried out with the aim of de-acidifying the grape must or wine through malic acid degradation using mixed fermentations of *Schizosaccharomyces pombe* and *S. cerevisiae*; some species of *S. pombe* had been recognized to improve some of sensory parameters of the wine, especially those related to wine color stability due to the correlation between the amount of pyruvic acid released into the medium and the formation of vitisin A (a pyranoanthocyanin, natural polyphenol, found in grapes) [39, 40]. *Issatchenkia orientalis* is one of the indigenous yeasts present in the wine. The strain KMBL 5774 of *I. orientalis* isolated from Korean wine pomace can degrade malic acid thus could be important in decreasing malic acid content in the wine and be useful the wine industry by this attribute [41].

The use of *Starmerella bacillaris* (formerly *Candida stellata*) yeast in mixed fermentation with *S. cerevisiae* starter cultures has been widely investigated these last few years, and several studies have shown an increase in glycerol content in mixed wines. Glycerol is related with the mouthfeel and complexity of wine flavor. The analytical and organoleptic profile of the wine was improved without any negative analytical profile in the fermentation of grape musts with mixed fermentations of *S. bacillaris* and *S. cerevisiae* starter strains. Also, it has been observed that the inoculation of grape must with pure cultures of *S. bacillaris* can result in the production of high concentrations of acetaldehyde and acetoin [42, 43]. In fact, in comparison with *S. cerevisiae* inoculated fermentations, the use of co-fermentations (mixed or sequential inoculations), continuous fermentation, and immobilized cells can contribute the following: (1) complementary consumption of glucose and fructose; (2) enhanced glycerol and succinic acid concentrations; and (3) no increases in acetaldehyde and acetoin contents, due to the presence of the *S. bacillaris* based on the existence of acetaldehyde exchange between the two species without any increment in its levels. The practice of multistarter fermentation can also be used to improve the complexity of organoleptic properties of a wine contributing to the enjoyment of wine. Garcia et al. [44] were able to verify an increase in geraniol production in mixed cultures of *Debaryomyces vanriji* and *S. cerevisiae*. This enhancement was due to the high levels of β-glucosidase activity exhibited by this non-*Saccharomyces* strain.

Another non-*Saccharomyces* yeast with several interesting features from wine industry is *Metschnikowia pulcherrima*. This yeast is generally predominant during the initial stages of alcoholic fermentation and also shown some significant effect in wine composition. In particular, *M. pulcherrima* is a high producer of β-glucosidase [45], and its presence in mixed cultures can provide important improvements in the wine such as decrease in the volatile acidity and increase in the production of medium-chain fatty acids, higher alcohol, esters, terpenoids, and glycerol. Some authors have also reported that *M. pulcherrima* can produce a reduction in the titratable acidity of the final wines. Depending on the initial acidity level of the grape must, this effect could be taken positively or negatively. It has also been reported that *M. pulcherrima* has a higher capacity to release polysaccharides from yeast cell walls during fermentation process compared to *S. cerevisiae*. More recently, sequential fermentations with *M. pulcherrima* and *S. cerevisiae* have shown that a reduction in ethanol concentration is occurring in this type of culture [46].

Yeasts belonging to the *Hanseniaspora* (*Kloeckera*) genus are the non-*Saccharomyces* yeasts found in the highest numbers in grape must. Due to their ability to produce unpleasant compounds, such as acetic acid and ethyl acetate, these apiculate yeasts have long been considered as spoilage yeasts particularly during the early stages of wine fermentation. Mixed fermentation trials in presence of *H. uvarum* and *S. cerevisiae* starter cultures have presented increases in isoamyl acetate [47], while use of *Hanseniaspora osmophila* provides improvements in 2-phenylethyl acetate production [48–50]. According to a report by Kurita [51], mixed inoculations using *Wickerhamomyces anomalus* (formerly *Pichia anomala*) resulted in positive enhancement of isoamyl acetate. Higher concentrations in the varietal thiols have been shown in mixed fermentations with *S. cerevisiae* and another strain belonging to the *Pichia kluyveri* species [52].

Benito et al. [53] have studied the hydroxycinnamate decarboxylase (HCDC) activity of *Meyerozyma guilliermondii* (formerly *Pichia guilliermondii*) in mixed and sequential fermentations with *S. cerevisiae*. *Meyerozyma guilliermondii* with HCDC activity can be used to decarboxylate hydroxycinnamic acids and form vinylphenols that condense with grape anthocyanins to produce vinylphenolic pyranoanthocyanin adducts—molecules that show great color stability.

4. Highlights produced by non-*Saccharomyces* yeasts

4.1. Enzymes with oenological interest

Over the last several decades, the utilization of enzymes has become more important in winemaking. Enzymatic treatments of grapes, musts, and wines are nowadays useful for multiple positive aims, reduction of times maceration, clarification and filtration, increase in free and press juice yields, improvements in color and aroma extraction as well as wine stability [54]. The enzymes are proteins usually produced by bacteria or by filamentous fungi [55]. These proteins are very valuable tools for the winemakers; they now strengthen the use of endogenous enzymes over commercial exogenous enzymes. The production of extracellular hydrolytic enzymes by indigenous yeast could be notable and a better understanding to their benefit of wine production is required. Moreover, wine yeast has a decisive role in the production of commercial enzymes to be used in the wine elaboration process [56]. The principal wine yeast, *Saccharomyces cerevisiae*, is not notable as a significant producer of extracellular enzymes, although some strains have been mentioned to degrade polygalacturonate [57].

On the contrary than *Saccharomyces* species, the non-*Saccharomyces* yeasts produce and secrete several enzymes as well as esterases, glycosidases, lipases, β-glucosidases, proteases, cellulases, etc., to the periplasmic space and the medium, where they have the capacity to bind with grape precursors compounds to produce aroma active compounds and thus play an important role in varietal aroma and flavor profiles [58].

Terpenoids, fatty acid esters, higher alcohols, glycerol, acetaldehyde, acetic acid, and succinic acid are some metabolic products generated from non-*Saccharomyces* growth. There is a distinction in types of flavors according to their origin, the primary flavor of wine comes naturally from grapes, each grape variety offers a unique set of aromas and flavors, and the fermentation process creates a group of bouquets that are commonly referred to secondary flavor [59]. Several flavor and aroma compounds are present in grapes as glycosidic precursors without sensory properties. The β-glucosidase enzyme might hydrolyze these compounds to form free volatiles increasing the flavor and aroma of wine and contributing to the higher fruit-like characteristic of final product; this enzyme is not encoded by the *S. cerevisiae* genome [60]. Instead, certain non-*Saccharomyces* genera as *Debaryomyces, Hansenula, Candida, Pichia,* and *Hanseniaspora* (*Kloeckera*) have different degrees of β-glucosidase activity which can have on the sensory character of wines [61].

Proteolytic and pectinolytic (polygalacturonase) are other extracellular enzymatic activities produced by non-*Saccharomyces* yeasts which may also be beneficial to winemaking. For example, proteolytic activity of some non-*Saccharomyces* yeast reduces the protein concentration of the grape juice by approximately one-third with accompanying increase in protein stability of the final product. The protein haze reduction is one of most significant changes for alcoholic beverages manufacturers. Protein precipitation in bottled wines especially in whites and red with low amounts of polyphenols causes protein haze where a coagulation of proteins occurs in alcoholic beverage with unfavorable storage conditions. These denatured proteins can either flocculate into a hazy suspension or form sediments in bottle [62]. Species found to produce the greatest number of extracellular enzymes are *C. stellata, H. uvarum*, and *M. pulcherrima* [56].

Non-*Saccharomyces* yeasts have also been reported to affect the concentration of polysaccharides in wine [49]. An enzymatic degradation happens in the dead yeast cells; the cells components such as proteins, nucleic acids, and lipids are broken down into smaller compounds as amino acids, peptides, fatty acids and nucleotides and also occur in releasing soluble polysaccharides (mannoproteins) from the cell wall. Most of these products have flavor impact or flavor-enhancing potential [63], but their specific contributions to wine character require more focused research and will depend on the extent to which the wine is exposed to the lees [64]. Moreover, there is evidence that peptides released during yeast autolysis could have antioxidant and other bioactive properties. Polysaccharides improve the sensory properties in wines, as they can positively influence mouthfeel (texture) by increasing its viscosity and mouth-filling [65]. Some non-*Saccharomyces* yeasts die in the early stages of fermentation process and they can also be a nutrient source that *S. cerevisiae* used to ferment optimally. Charoenchai et al. [58] reported the effect of nitrogen sources on the production of extracellular proteases by non-*Saccharomyces* wine yeasts. From 26 yeast strains, protease activity was observed in strains of *M. pulcherrima, Kloeckera apiculata*, and *W. anomalus*. Also, *T. delbrueckii* has reported as releasing higher amount of polysaccharides, Gonzalez-Royo et al. [66] have found that the high content of polysaccharides obtained by sequential culture with *T. delbrueckii* has a positive effect on the foam properties in sparkling wines.

The role of pectinases in winemaking has been evaluated by Canal-Llaubères [67]. Some of the applications in mash treatment are to improve juice extraction, clarification process, filterability, and also color extraction. The use of pectolytic enzymes for maceration may also accelerate the extraction of phenolic compounds, reducing the maceration time needed for high quality of wine [68]. The addition of fungal pectinase preparations is a normal practice in wine industry even though pectin esterase and polygalacturonase enzymatic activities increase during grape ripening and are produced by non-*Saccharomyces* yeasts present in must.

The accumulation of esters in wine is known to be a result of the balance between the yeast's ester-synthesizing enzymes and hydrolysis reactions involving esterases (responsible for cleavage and in some cases, formation of ester bonds). The production of extracellular esterases in *Saccharomyces* wine yeasts is well known [69], but the situation for non-*Saccharomyces* needs further investigation. Yeast esterases studied include those of the genus *Brettanomyces* [70] and *Rhodotorula mucilaginosa* [71]. Also, one strain of *Debaryomyces hansenii* has been reported as producing strain of an esterase enzyme [72].

The lipids proceeding from the grape or from autolytic activity of yeasts can be degraded by lipases. After this enzymatic reaction, free fatty acids would be released into the juice or wine, which can lead to changes in wine quality. While properties of lipoxygenase and peroxide-cleaving enzymes from grapes have been well established [73], the knowledge about lipase enzyme production by non-*Saccharomyces* yeasts is not well documented yet. Ratledge and Tan [74] reported data about the production of extracellular lipases by yeasts, only a single species of *Yarrowia lipolytica* (formerly *Candida lipolytica*) and *Saccharomycopsis lipolytica*.

Hanseniaspora and *Torulaspora* genera are reported as good producers of enzymes such as β-glucosidases, pectinases, proteases, and those involved in xylan degradation [58, 75]. However, the secretion of each enzyme is not characteristic of a particular genus or species, but depends on the yeast strain analyzed [76].

4.2. Use of lower ethanol efficiency yeasts

In recent decades, the increasing alcohol level in wine is one of the most important challenges facing in the enological industry. The problem is related in part to global warming, which results in modifications of fruit maturation patterns, as well as a lack of balance between sugar accumulation and the phenolic ripeness of berries [77]. Fermentations with higher initial sugar content combined with high final ethanol concentration may have impact on microbiological, technological, sensorial, and financial aspects of winemaking. Higher sugar level delivers shifts in alcohol, altering flavors and mouthfeel. Musts with higher sugar concentrations cause a stress response in yeast leading to an increased formation of fermentation co-products, such as acetic acid. Also, this increasing sugar content leads to delay harvest period so as to insure appropriate aromatic and phenolic maturity. On the commercial side, excess ethanol can get worse sensory quality of wine, discourage consumers, because of the health effects associated with the excessive alcohol consumption or become a drawback in the global market, due to regulations and taxes associated with the alcohol content of beverages. All these reasons stimulate the creation of strategies directed to reduce alcohol level in wine.

Researchers, engineers, and oenologists are working together to develop approaches to limit ethanol content of wines, targeting almost all the steps in the production cycle [78], including among other examples, grapevine clonal selection, vineyard management, winemaking practices adapted to unripe grapes [79], use of lower ethanol efficiency yeast strains [80] or metabolic inhibitors [81] and partial dealcoholization by physical means [82]. While some of these technologies are still in need of fundamental research, others are in several stages of regulatory support and implementation by the industry.

The development of low-alcohol yeasts is a current challenge in wine industry. During the last years, researchers have been investigating *S. cerevisiae* metabolism to reduce the yield ethanol/sugar consumed. Two approaches were used as follows: metabolic engineering strategies diverting sugar metabolism towards products other than ethanol [Genetically Modified Organisms (GMO) strategy] [83] and more recently an adaptive evolution-based strategy [84]. An alternative to these approaches is to select low-ethanol producers in *S. cerevisiae* species by screening wild yeast population or to use breeding strategies.

http://dx.doi.org/10.5772/64957

Given the vast potential for diverse wine relevant phenotypes among non-*Saccharomyces* yeast, it has been proposed that strains able to utilize oxygen grape sugars could be used to decrease ethanol concentration in wine [85]. Unlike *S. cerevisiae*, which favors fermentative metabolism over aerobic respiration when sugar concentration exceeds 10 g/L (due to the Crabtree effect), many non-*Saccharomyces* yeast are able to use oxygen for growth regardless of sugar concentration and thus divert carbon into other metabolites and therefore away from ethanol formation.

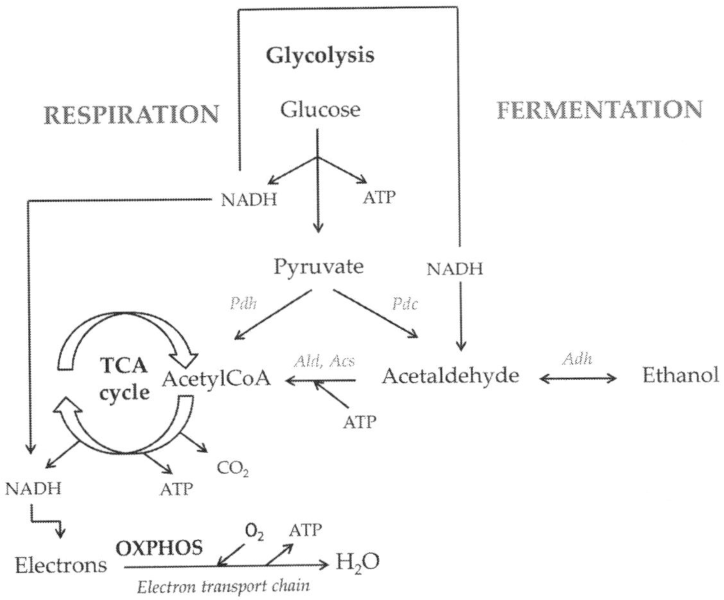

Figure 1. Yeast energy metabolism.

Respiration and fermentation are two pathways for ATP production from glucose used by yeasts. Both pathways start with glycolysis, the major process for sugar degradation where the breakage of one glucose molecule results in the production of two molecules of pyruvate and ATP. In fermentation, pyruvate is finally transformed into ethanol by pyruvate decarboxylase (*Pdc*) and alcohol dehydrogenase (*Adh*) enzymes. This process does not produce additional ATP but the NADH that is released in glycolysis is recycled by *Adh* into NAD^+, and thus, alcoholic fermentation can occur in the absence of oxygen. In respiration, pyruvate is transformed into acetyl-coenzyme A by pyruvate dehydrogenase (*Pdh*) which is then oxidized to CO_2 through the TCA cycle and oxidative phosphorylation (OXPHOS), where it yields additional ATP but requires oxygen. At abundant levels of sugars and oxygen, Crabtree-positive yeasts use fermentation and respiration simultaneously. Once glucose has been depleted in the environment, one way to generate ATP is recycling the ethanol accumulated.

This process causes a loss in terms of ATP because of the conversion of ethanol to acetyl-CoA carried out by aldehyde dehydrogenase (*Ald*) and acetyl-CoA synthetase (*Acs*) requires one additional ATP per ethanol recycled (**Figure 1**).

Yeasts can be classified depending on the way they regulate their respiro-fermentative metabolism. Crabtree-positive yeasts could ferment under aerobic conditions only if sugar concentration is above certain thresholds. The prime example of Crabtree-positive species is *S. cerevisiae*, yeast with clear preference towards fermentative metabolism. This Crabtree-positive character has allowed to *S. cerevisiae* adapt in sugar rich environments [86]. In contrast, the extent of fermentative metabolism for Crabtree-negative species would be very limited whenever enough oxygen is available [87]. *Hanseniaspora uvarum* and *Candida utilis* are examples of Crabtree-negative yeasts [88]. In spite of his preference for respiratory metabolism, some Crabtree-negative yeasts, such as *Kluyveromyces lactis* and *Kluyveromyces marxianus*, can grow in the absence of oxygen [89]. Finally, some yeast species are not able to ferment sugars and are obligate aerobes.

It is thought that redox balance in the metabolism of sugars generates the production of metabolic by-products as acetic acid, ethanol, and glycerol. In recent years, it is trying to take the control of metabolic systems in order to redirect carbon flux towards desirable compounds release, for example, glycerol overproduction. An added benefit of this approach is that enhanced glycerol concentrations can have a favorable influence on wine by enhancing its sweetness, smoothness, and overall body [90].

Several yeast strains, including *M. pulcherrima*, *K. lactis*, and *Candida sake* isolates, were found to be good candidates to develop fermentation procedures aiming at reducing alcohol content in wine by respiration. Results of previous work also indicated that, besides the study of yeast ability in sugar respiration metabolism under aerated winemaking conditions, it is necessary to find a compromise between ethanol yield, acetic acid production, and growth performance in grape must. Differences of up to one order of magnitude in acetic acid yield were found among the different yeast strains studied [91].

4.3. Bioprotection by non-*Saccharomyces* yeasts

Vinification process is composed by different and delicate steps as growing, harvesting, fermentation, and aging and storage in the winery. Unsuitable precautions or poor practice during any of these steps can lead to growth of wine spoilage organisms and consequent production losses. The major microorganisms involved in wine spoilage are acetic acid bacteria from genera *Acetobacter*, *Gluconobacter*, and lactic acid bacteria from *Leuconostoc*, *Lactobacilli*, and *Pediococcus* genera [92], whereas the yeasts involved in wine spoilage mainly are from genera *Dekkera*/*Brettanomyces*, *Pichia*, *Zygosaccharomyces*, and *Candida*, usually isolated from wines with aroma defects [93]. Bacterial wine spoilage imparts mousy taint, bitterness, geranium notes, volatile acidity, oily and slimy-texture, and overt buttery characters to the wine [92], whereas the common spoilage effects due to yeasts are off odors, off-tastes, film formation, cloudiness or haziness, sediments, and gas production in bottled wines [93].

Traditionally, sulfur dioxide (as potassium metabisulphite), sorbic acid, fumaric acid, and dimethyl dicarbonate (DMDC) are used for preservation of different wines in various countries. Due to these drawbacks and growing consumer bias against chemical preservatives, research efforts are directed towards use of different physical methods and exploitation of natural antimicrobial compounds obtained from plants, animals, and microorganisms for wine preservation. Many studies have demonstrated the potential of natural products such as hydroxycinnamates and organic acids [94], chitosan [95], nisin [96], lysozyme [97], antimicrobial peptides [98], killer toxins [99], natamycin [100], β-glucanases [101], bovine lactoferrin-derived peptides [102], carvacrol and thymol [103], and vitamin K5 [104] for the control of wine spoilage yeasts and bacteria.

One of the biological mechanisms for the regulation of population dynamics in several microbial ecosystems is the production of toxins capable of kill or inhibit other microorganisms, taxonomically related or not to the producing strains. The toxins synthesized by yeasts, known as killer factor, are proteins or glycoproteins whose action is mediated by specific receptors in the cell wall of the sensitive microorganism. The killer character, first reported on the decade of the 1960s in a *S. cerevisiae* strain, is well distributed among other yeast genera as *Candida, Hansenula, Pichia, Debaryomyces, Ustilago, Cryptococcus, Metschnikowia, Williopsis, Kluyveromyces*, and *Zygosaccharomyces* [105].

Saccharomyces cerevisiae's killer toxins and their relevance in winemaking have been thoroughly investigated in the literature. However, these killer toxins exhibit narrow spectra of activity limited to the other strains of *S. cerevisiae* [106] except for the Klus killer toxin and the killer toxin from *S. cerevisiae* strain Y500-4 L that are active against a few non-*Saccharomyces* species and are therefore unsuitable as agents to prevent the development of spoilage yeasts. Therefore, the differences in stationary phase cell concentrations between yeast species in wine fermentations may be due to the fact that the non-*Saccharomyces* yeasts are more sensitive to certain growth-inhibitory compounds than *S. cerevisiae*.

The killer toxins secreted by the yeast species *Pichia membranifaciens, Kluyveromyces wickerhamii, Metschnikowia pulcherrima*, and *Wickerhamomyces anomalus* (formerly *Pichia anomala*) and the filamentous fungus *Ustilago maydis* have been specifically investigated for their killer activity against *Brettanomyces bruxellensis* [107]. These killer toxins successfully inhibited the growth of *B. bruxellensis* in wine and grape juice. Furthermore, the killing activity of certain non-*Saccharomyces* killer toxins has been demonstrated against the apiculate yeast *Hanseniaspora uvarum* [108] and also against the grapevine pathogen *Botrytis cinerea* [109].

5. Concluding remarks

Strain selection is of key importance, as not all strains within a species will necessarily show the same desirable characteristics [110]. The accepted list of desirable characteristics as pertaining to the wine yeast *S. cerevisiae* will not necessarily apply to non-*Saccharomyces* yeasts. These wine yeasts will necessarily have a different list of desired characteristics. Thorough briefings and assistance of wine producers will have to accompany any new non-*Saccharomyces* technology for wine production. However, the aims submitted by Pretorius et al. [111] and other authors advocate the use of selected non-*Saccharomyces* yeasts as able to consume grape juice sugars, enhance production of desirable volatile esters, enhance liberation of grape terpenoids, and produce glycerol to improve wine flavor and other sensory properties. A modern approach to multispecies inoculations backed by science and rigorous research is essential to help winemakers achieve their primary objective of attaining conversion of grape sugar to alcohol and carbon dioxide, at a controlled rate and without the development of off-flavors.

Author details

Margarita García[1*], Braulio Esteve-Zarzoso[2] and Teresa Arroyo[1]

*Address all correspondence to: margarita_garcia_garcia@madrid.org

1 Agroalimentary Department, Madrid Institute for Research and Rural Development in Food and Agriculture (IMIDRA), Alcalá de Henares, Spain

2 Department of Biochemistry and Biotechnology, Rovira i Virgili University, Tarragona, Spain

References

[1] Swiegers JH, Pretorius IS. Yeast modulation of wine flavour. Advances in Applied Microbiology. 2005;57:131–175. doi:10.1016/S00652164(05)57005-9

[2] Fleet GH. Which yeast species really conducts the fermentation? In: Williams PJ, Davidson DM, Lee TH, editors. 7th Australian Wine Industry Technical Conference; 13–17 August 1989; Adelaide, Australia; 1990. p. 153–156

[3] Cray JA, Bell ANW, Bhaganna P, Mswaka AY, Timson DJ, Hallsworth JE. The biology of habitat dominance; can microbes behave as weeds? Microbial Biotechnology. 2013;6:453–492. doi:10.1111/1751-7915.12027

[4] Barnett JA, Delaney MA, Jones E, Magson AB, Winch B. The numbers of yeasts associated with wine grapes of Bordeaux. Archives of Microbiology. 1972;83:52–55

[5] Pace NR. A molecular view of microbial diversity and the biosphere. Science. 1997;276:734–740. doi:10.1126/science.276.5313.734

[6] Hugenholtz P, Goebel BM, Pace NR. Impact of culture- independent studies on the emerging phylogenetic view of bacterial diversity. Journal of Bacteriology. 1998;180: 4765–4774

[7] Cocolin L, Campolongo S, Alessandria V, Dolci P, Rantsiou K. Culture independent analyses and wine fermentation: an overview of achievements 10 years after first application. Annals of Microbiology. 2011;61:17–23. doi:10.1007/s13213-010-0076-6

[8] Ercolini D. PCR-DGGE fingerprinting: novel strategies for detection of microbes in food. Journal of Microbiological Methods. 2004;56:297–314. doi:10.1016 /j.mimet. 2003.11.006

[9] Andorrà I, Landi S, Mas A, Guillamón JM, Esteve-Zarzoso B. Effect of oenological practices on microbial populations using culture-independent techniques. Food Microbiology. 2010;25:849–856. doi:10.1016/j.fm.2008.05.005

[10] Blackstone GM, Nordstrom JL, Vickery MC, Bowen MD, Meyer RF, DePaola A. Detection of pathogenic *Vibrio parahaemolyticus* in oyster enrichments by real-time PCR. Journal Microbiological Methods. 2003;53:149–55. doi:10.1111/j.1472-765X.2010.02895.x

[11] Wang C, Esteve-Zarzoso B, Cocolin L, Mas A, Rantsiou K. Viable and culturable populations of *Saccharomyces cerevisiae, Hanseniaspora uvarum* and *Starmerella bacillaris* (synonym *Candida zemplinina*) during barbera must fermentation. Food Research International. 2015;78:195–200. doi:10.1016/j.foodres.2015.10.014

[12] Gerhards D, Lehnigk C, Buchl N, Wenning M, Scherer S. Biodiversity and development of yeasts from Riesling vines to wines monitored by FT-IR. In: Darriet P, Geny L, Lonvaud A, Lucas P, de Revel G, Teissedre P-L, editors. 9th International Symposium of Oenology-Bordeaux. Dunod: Paris; 2011. p. 263–268

[13] Hittinger CT, Rokas A, Bai FY, Boekhout T, Gonçalves P, Jeffries TW, Kominek J, Lachance MA, Libkind D, Rosa CA, Sampaio JP, Kurtzman CP. Genomics and the making of yeast biodiversity. Current Opinion in Genetics and Development. 2015;35:100–109. doi:10.1016/j.gde.2015.10.008

[14] Kurtzman CP, Fell WF, Boekhout T, Robert V. Methods for Isolation, Phenotypic Characterization and Maintenance of Yeasts. In: Kurtzman CP, Fell JW, Boekhout T, editors. The Yeasts: A Taxonomic Study. 5th ed. Amsterdam: Elsevier Science Publishers; 2011. p. 87–110

[15] Jolly NP, Augustyn OPH, Pretorius IS. The role and use of non-*Saccharomyces* yeasts in wine production. South African Journal of Enology and Viticulture. 2006;27:15–39

[16] Van Kerken AE. Contribution to the ecology of yeasts occurring in wine [thesis]. Bloemfontein, South Africa: University of the Orange Free State; 1963

[17] Comitini F, Gobbi M, Domizio P, Romani C, Lencioni L, Mannazzu I, Ciani M. Selected non-*Saccharomyces* wine yeasts in controlled multistarter fermentations with *Saccharomyces cerevisiae*. Food Microbiology. 2011;5:873–882. doi:10.1016/j.fm.2010.12.001

[18] Querol A, Jimenez M, Huerta T. Microbiological and enological parameters during fermentation of musts from poor and normal grape harvests in the region of Alicante

(Spain). Journal of Food Science. 1990;55:1603–1606. doi:10.1111/j. 1365-2621.1990.tb03580.x

[19] Jolly NP, Varela C, Pretorius IS. Not your ordinary yeast: non-*Saccharomyces* yeasts in wine production uncovered. FEMS Yeast Research. 2014;14:215–237. doi: 10.1111/1567-1364.12111

[20] Bisson LF, Kunkee R. Microbial interactions during wine production. In: Zeikus JG, Johnson EA editors. Mixed Cultures in Biotechnology. New York: McGraw-Hill; 1991. p. 37–68

[21] Heard GM, Fleet GH. The effects of temperature and pH on the growth of yeast species during the fermentation of grape juice. Journal Applied Bacteriology. 1988; 65: 23–28. doi:10.1111/j.1365-2672.1988.tb04312.x

[22] Granchi L, Ganucci D, Messini A, Rosellini D, Vicenzini M. Dynamics of yeast populations during the early stages of natural fermentations for the production of Brunello di Montalcino wines. Food Technology Biotechnology. 1998;36:313–318.

[23] Lachance MA, Kurtzman CP. *Lachancea* Kurtzman (2003). In: Kurtzman CP, Fell JW, Boekhout T, editors. The Yeasts: A Taxonomic Study. 5th ed. Amsterdam: Elsevier Science Publishers; 2011. p. 511–519

[24] Nissen P, Nielsin D, Arneborg N. Viable *Saccharomyces cerevisiae* cells at high concentrations cause early growth arrest of non-*Saccharomyces* yeasts in mixed cultures by a cell-cell contact-mediated mechanism. Yeast. 2003;20:331–341

[25] Pérez-Nevado F, Albergaria H, Hogg T, Girio F. Cellular death of two non-*Saccharomyces* wine-related yeasts during mixed fermentations with *Saccharomyces cerevisiae.* International Journal of Food Microbiology. 2006;108:336–345. doi:10.1016/ j.ijfoodmicro. 2005.12.012

[26] Wang C, Mas A, Esteve-Zarzoso B. Interaction between *Hanseniaspora uvarum* and *Saccharomyces cerevisiae* during alcoholic fermentation. International Journal of Food Microbiology. 2015;206:67–74. doi:10.1016/j.ijfoodmicro.2015.04.022

[27] Branco P, Francisco D, Chambon C, Hébraud M, Arneborg N, Almeida MG, Caldeira J, Albergaria H. Identification of novel GAPDH-derived antimicrobial peptides secreted by *Saccharomyces cerevisiae* and involved in wine microbial interactions. Applied Microbiology and Biotechnology. 2014;98:843–853. doi:10.1007/ s00253-013-5411-y

[28] Ciani M, Comitini F, Mannazu I, Domizio P. Controlled mixed culture fermentation: a new perspective on the use of non-*Saccharomyces* yeasts in winemaking. FEMS Yeast Research. 2010;10:123–133. doi:10.1111/j.1567-1364.2009.00579.x

[29] Clemente-Jiménez JM, Mingorance-Cazorla L, Martínez-Rodríguez S, Las Heras-Vázquez FJ, Rodríquez-Vico F. Influence of sequential yeast mixtures on wine fermen-

tation. International Journal of Food Microbiology. 2005;98:301–308. doi:10.1016/j.ijfoodmicro.2004.06.007

[30] Lema C, Garcia-Jares C, Orriols I and Angulo L. Contribution of *Saccharomyces* and non-*Saccharomyces* populations to the production of some compounds of Albarino wine aroma. American Journal of Enology and Viticulture. 1996;47:206–216

[31] Sun SY, Gong HS, Jiang XM, Zhao YP. Selected non-*Saccharomyces* wine yeasts in controlled multistarter fermentations with *Saccharomyces cerevisiae* on alcoholic fermentation behaviour and wine aroma of cherry wines. Food Microbiology. 2014;44:15–23. doi:10.1016/ j.fm.2014.05.007

[32] Rojas V, Gil JV, Piñaga F, Manzanares P. Acetate ester formation in wine by mixed cultures in laboratory fermentations. International Journal of Food Microbiology. 2003;86:181–188. doi:10.1016/S0168-1605(03)00255-1

[33] Contreras A, Hidalgo C, Henschke PA, Chambers PJ, Curtin C, Varela C. Evaluation of non-*Saccharomyces* yeasts for the reduction of alcohol content in wine. Applied and Environmental Microbiology. 2014;80:1670–1678. doi:10.1128/AEM.03780-13

[34] Giovani G, Rosi I, Bertuccioli M. Quantification and characterization of cell wall polysaccharides released by non-*Saccharomyces* yeast strains during alcoholic fermentation. International Journal of Food Microbiology. 2012;160:113–118. doi:10.1016/ j.ijfoodmicro .2012.10.007

[35] Sadoudi M, Tourdot-Marechal R, Rousseaux S, Steyer D, Gallardo-Chacon JJ, Ballester J, Vichi S, Guerin-Schneider R, Caixach J, Alexandre H. Yeast-yeast interactions revealed by aromatic profile analysis of Sauvignon Blanc wine fermented by single or co-culture of non-*Saccharomyces* and *Saccharomyces* yeasts. Food Microbiology. 2012;32:243–253. doi:10.1016/j.fm.2012.06.006

[36] Bely M, Stoeckle P, Masneuf-Pomarède I, Dubourdieu D. Impact of mixed *Torulaspora delbrueckii-Saccharomyces cerevisiae* culture on high-sugar fermentation. International Journal of Food Microbiology. 2008;122:312–320. doi:10.1016/ j.ijfoodmicro. 2007.12.023

[37] Izquierdo Cañas PM, Palacios Garcia AT, Garcia Romero E. Enhancement of flavour properties in wines using sequential inoculations of non-*Saccharomyces* (*Hansenula* and *Torulaspora*) and *Saccharomyces* yeast starter. Vitis. 2011;50:177–182

[38] Gobbi M, Comitini F, Domizio P, Romani C, Lencioni L, Mannazzu I, Ciani M. *Lachancea thermotolerans* and *Saccharomyces cerevisiae* in simultaneous and sequential co-fermentation: a strategy to enhance acidity and improve the overall quality of wine. Food Microbiology. 2013;33:271–281. doi:10.1016/j.fm.2012.10.004

[39] Morata A, Benito S, Loira I, Palomero F, Gonzalez MC, Suarez-Lepe JA. Formation of pyranoanthocyanins by *Schizosaccharomyces pombe* during the fermentation of red must. International Journal of Food Microbiology. 2012;159:47–53. doi:10.1016/j.ijfoodmicro. 2012.08.007

[40] Benito S, Palomero F, Morata A, Calderón F, Palmero D, Suárez-Lepe JA. Selection of appropriate *Schizosaccharomyces* strains for winemaking. Food Microbiology. 2014;42:218–224. doi:10.1016/j.fm.2014.03.014

[41] Kim DH, Hong YA, Park HD. Co-fermentation of grape must by *Issatchenkia orientalis* and *Saccharomyces cerevisiae* reduces the malic acid content in wine. Biotechnology Letters. 2008;30:1633–1638. doi:10.1007/s10529-008-9726-1

[42] Ciani M, Ferraro L. Combined use of immobilized *Candida stellata* cells and *Saccharomyces cerevisiae* to improve the quality of wines. Journal of Applied Microbiology. 1998;85:247–254. doi:10.1046/j.1365-2672.1998.00485.x

[43] Jolly NP, Augustyn OPH, Pretorius IS. The effect of non-*Saccharomyces* yeasts on fermentation and wine quality. South African Journal of Enology and Viticulture. 2003;24:55–62

[44] Garcia A, Carcel C, Dalau L, Samson A, Aguera E, Agosin E, Gunata Z. Influence of mixed culture with *Debaryomyces vanriji* and *Saccharomyces cerevisiae* on the volatiles in a Muscat wine. Journal of Food Science. 2002;67:1138–1143. doi:10.1111/j.1365-2621.2002. tb09466.x

[45] Rodríguez ME, Lopes CA, Valles S, Giraudo MR, Caballero A. Selection and preliminary characterization of β-glycosidases producer Patagonian wild yeasts. Enzyme and Microbial Technology. 2007;41:812–820. doi:10.1016/j.enzmictec.2007.07. 004

[46] Rodríguez ME, Lopes CA, Barbagelata R, Barda NB, Caballero AC. Influence of *Candida pulcherrima* Patagonian strain on alcoholic fermentation behaviour and wine aroma. International Journal of Food Microbiology. 2010;138:19–25. doi:10.1016/j.ijfoodmicro.2009.12.025

[47] Moreira N, de Pinho PG, Santos C, Vasconcelos I. Volatile sulphur compounds composition of monovarietal white wines. Food Chemistry. 2010;123:1198–1203. doi:10.1016/j.foodchem.2010.05.086

[48] Medina K, Boido E, Fariña L, Gioia O, Gomez ME, Barquet M, Gaggero C, Dellacassa E, Carrau F. Increased flavour diversity of Chardonnay wines by spontaneous fermentation and co-fermentation with *Hanseniaspora vineae*. Food Chemistry. 2013;141:2513–2521. doi:10.1016/j.foodchem.2013.04.056

[49] Domizio P, Romani C, Lencioni L, Comitini F, Gobbi M, Mannazzu I & Ciani M. Outlining a future for non-*Saccharomyces* yeasts: selection of putative spoilage wine strains to be used in association with *Saccharomyces cerevisiae* for grape juice fermentation. International Journal of Food Microbiology. 2011;147:170–180. doi:10.1016/j.ijfoodmicro. 2011.03.020

[50] Domizio P, Romani C, Comitini F, Gobbi M, Lencioni L, Mannazzu I, Ciani M. Potential spoilage non-*Saccharomyces* yeasts in mixed cultures with *Saccharomyces cerevisiae*. Annals of Microbiology. 2011;61:137–144. doi:10.1007/s13213-010-0125-1

[51] Kurita O. Increase of acetate ester-hydrolysing esterase activity in mixed cultures of *Saccharomyces cerevisiae* and *Pichia anomala*. Journal of Applied Microbiology. 2008;104:1051–1058. doi:10.1111/j.1365-2672.2007.03625.x

[52] Anfang N, Brajkovich M, Goddard MR. Co-fermentation with *Pichia kluyveri* increases varietal thiol concentrations in Sauvignon blanc. Australian Journal of Grape and Wine Research. 2009;15:1–8. doi:10.1111/j.1755-0238.2008.00031.x

[53] Benito S, Morata A, Palomero F, Gonzalez MC, Suarez-Lepe JA. Formation of vinylphenolic pyranoanthocyanins by *Saccharomyces cerevisiae* and *Pichia guillermondii* in red wines produced following different fermentation strategies. Food Chemistry. 2011;124:15–23. doi:10.1016/j.foodchem.2010.05.096

[54] Rogerson FSS, Vale E, Grande HJ, Silva MCM. Alternative processing of port-wine using pectolytic enzymes. Ciência e Tecnologia de Alimentos. 2000;2:222–227

[55] Esteve-Zarzoso B, Manzanares P, Ramón D, Querol A. The role of non-*Saccharomyces* yeasts in industrial winemaking. International Microbiology. 1998;1:143–148

[56] Strauss MLA, Jolly NP, Lambrechts MG, Van Rensburg P. Screening for the production of extracellular hydrolytic enzymes by non-*Saccharomyces* wine yeasts. Journal of Applied Microbiology. 2001;91:182–190. doi:10.1046/j.1365-2672.2001.01379.x

[57] McKay AM. Degradation of polygalacturonic acid by *Saccharomyces cerevisiae*. Letters in Applied Microbiology. 1990;11:41–44. doi:10.1111/j.1472-765X.1990.tb00132.x

[58] Charoenchai C, Fleet GH, Henschke PA, Todd BEN. Screening of non-*Saccharomyces* wine yeast for the presence of extracellular hydrolytic enzymes. Australian Journal of Grape Wine Research. 1997;3:2–8. doi:10.1111/j.1755-0238.1997.tb00109.x

[59] Lambrechts MG, Pretorius IS. Yeast and its importance to wine aroma. South African Journal of Enology and Viticulture. 2000;21:97–129

[60] Ubeda-Iranzo JF, Briones-Perez AI, Izquierdo-Canas PM. Study of the oenological characteristics and enzymatic activities of wine yeasts. Food Microbiology. 1998;15:399–406. doi:10.1006/fmic.1998.0183

[61] Cordero-Bueso G, Esteve-Zarzoso B, Cabellos JM, Gil-Díaz M, Arroyo T. Biotechnological potential of non-*Saccharomyces* yeasts isolated during spontaneous fermentations of Malvar (*Vitis vinifera cv. L.*). European Food Research and Technology. 2012.236:193–207. doi:10.1007/s00217-012-1874-9

[62] Pocock KF, Waters EJ. Protein haze in bottled white wines: How well do stability tests and bentonite fining trials predict haze formation during storage and transport? Australian Journal of Grape Wine Research. 2006;12:212–220. doi:10.1007/s00217-012-1874-9

[63] Charpentier C, Nguyen Van Long T, Bonaly R, Feuillat M. Alteration of cell wall structure in *Saccharomyces cerevisiae* and *Saccharomyces bayanus* during autolysis. Applied Microbiology and Biotechnology. 1986;24:405–413

[64] Perez-Serradilla JA, Luque de Castro, MD. Role of lees in wine production: a review. Food Chemistry. 2008;111:447–456. doi:10.1016/j.foodchem.2008.04.019

[65] Vidal S, Francis L, Williams P, Kwiatkowski M, Gawel R, Cheynier W, Waters E. The mouth-feel properties of polysaccharides and anthocyanins in a wine like medium. Food Chemistry. 2004;85:519–525. doi:10.1016/S0308-8146(03)00084-0

[66] Gonzalez-Royo E, Pascual O, Kontoudakis N, Esteruelas M, Esteve-Zarzoso B, Mas A, Canals JM, Zamora F. Oenological consequences of sequential inoculation with non-*Saccharomyces* yeasts (*Torulaspora delbrueckii* or *Metschnikowia pulcherrima*) and *Saccharomyces cerevisiae* in base wine for sparkling wine production. European Food and Research Technology. 2015;240:999–1012. doi:10.1007/s00217-014-2404-8

[67] Canal-Llaubères RM. Enzymes in winemaking. In: Fleet GH editors. Wine Microbiology and Biotechnology. Philadelphia: Hardwood Academic Publishers. 1993. p. 477–506

[68] Romero-Cascales I, Ros-García JM, López-Roca JM, Gómez-Plaza E. The effect of a commercial pectolytic enzyme on grape skin cell wall degradation and colour evolution during the maceration process. Food Chemistry. 2012;130:626–631. doi:10.1016/j.foodchem. 2011.07.091

[69] Dubourdieu D, Darriet P, Ollivier C, Boidron JN, Ribéreau-Gayon P. Role of the yeast Saccharomyces cerevisiae in enzymatic hydrolysis of terpenic heterosides of grape juice. Compts Rendus de l'Academie des Sciences Paris. 1988;306:489–493

[70] Spaepen M, Verachtert H. Esterase activity in the genus *Brettanomyces*. Journal of the Institute of Brewing. 1982;88:11–17

[71] Lee H, To RJB, Latta RK, Biely P, Schneider H. Some properties of extracellular acetylxylan esterase produced by the yeast *Rhodotorula mucilaginosa*. Applied and Environmental Microbiology. 1987;53:2831–2834

[72] Besançon X, Ratomahenina R, Galzy P. Isolation and partial characterization of an esterase (EC 3.1.1.1) from a *Debaryomyces hansenii* strain. Netherland Milk and Dairy Journal. 1995;49:97–110

[73] Valentin G, Crouzet J. Peroxide cleavage enzyme in grapes. In: Ribèreau-Gayon P, Lonvaud A, editors. Actualités Oenologiques 89. Paris: Bordas, 1989. p. 133–138

[74] Ratledge C, Tan KH. Oils and fats: production. Degradation and utilization by yeasts. In: Verachtert H, De Mot R editors. Yeast Biotechnology and Biocatalysis. New York: Marcel Dekker. 1990. p. 223–254

[75] Pérez G, Fariña L, Barquet M, Boido E, Gaggero C, Dellacassa E, Carrau F. A quick screening method to identify β-glucosidase activity in native wine yeast strains:

application of esculin glycerol agar (EGA) medium. World Journal of Microbiology and Biotechnology. 2011;27:47–55. doi:10.1007/s11274-010-0425-4

[76] Ganga MA, Martínez C. Effect of wine yeast monoculture practice on the biodiversity of non-*Saccharomyces* yeasts. Journal of Applied Microbiology. 2004;96:76–83

[77] Mira de Orduña, R. Climate change associated effects on grape and wine quality and production. Food Research International. 2010;43:1844–1855. doi:10.1016/j.foodres. 2010. 05.001

[78] Schmidtke LM, Blackman JW, Agboola SO. Production technologies for reduced alcoholic wines. Journal of Food Science. 2012;71:25–41. doi:10.1111/j.1750-3841.2011. 02448.x

[79] Kontoudakis N, Esteruelas M, Fort F, Canals JM, Zamora F. Use of unripe grapes harvested during cluster thinning as a method for reducing alcohol content and pH of wine. Australian Journal of Grape and Wine Research. 2011;17:230–238. doi:10.1111/j. 1755-0238.2011.00142.x

[80] Kutyna DR, Varela C, Henschke PA, Chambers PA, Stanley GA. Microbiological approaches to lowering ethanol concentration in wine. Trends Food Science & Technology, 2010;21:293–302. doi:10.1016/j.tifs.2010.03.004

[81] Abalos D, Vejarano R, Morata A, Gonzalez C, Suarez-Lepe JA. The use of furfural as a metabolic inhibitor for reducing the alcohol content of model wines. European Food Research and Technology. 2011;232:663–669. doi:10.1007/s00217-011-1433-9

[82] Teissedre PL. Alcohol reduction in wine. Villenave d'Ornon: Vigne et Vin Publications Internationales. 2013. vol. 1. 108 p.

[83] Ehsani M, Fernandez MR, Biosca JA, Julien A, Dequin S. Engineering of 2,3-butanediol dehydrogenase to reduce acetoin formation by glycerol- overproducing, low-alcohol *Saccharomyces cerevisiae*. Applied Environmental Microbiology. 2009;75:3196–3205. doi: 10.1128/AEM.02157-08

[84] Cadière A, Ortiz-Julien A, Camarasa C, Dequin S. Evolutionary engineered *Saccharomyces cerevisiae* wine yeast strains with increased in vivo flux through the pentose phosphate pathway. Metabolic Engineering. 2011;13:263–271. doi:10.1016/j.ymben. 2011.01.008

[85] Gonzalez R, Quiros M, Morales P. Yeast respiration of sugars by non-*Saccharomyces* yeast species: a promising and barely explored approach to lowering alcohol content of wines. Trends in Food Science and Technology. 2013;29:55–61. doi:10.1016/ j.tifs. 2012.06.015

[86] Piškur J, Rozpedowska E, Polakova S, Merico A, Compagno C. How did *Saccharomyces* evolve to become a good brewer? Trends in Genetics. 2006;22:183–186. doi:10.1016/j.tig. 2006.02.002

[87] de Deken RH. The Crabtree effect: a regulatory system in yeast. Journal of General Microbiology. 1966;44:149–156.

[88] Venturin C, Boze H, Moulin G, Galzy P. Influence of oxygen limitation on glucose metabolism in *Hanseniaspora uvarum* K5 grown in chemostat. Biotechnology Letters. 1995.17:537–542.

[89] Contreras A, Hidalgo C, Schmidt S, Henschke PA, Curtin C, Varela C. The application of non-*Saccharomyces* in fermentations with limited aeration as a strategy for the production of wine with reduced alcohol content. International Journal of Food Microbiology. 2015;205:7–15. doi:10.1016/j.ijfoodmicro.2015.03.027

[90] Eustace R, Thornton R. Selective hybridization of wine yeasts for higher yields of glycerol. Canadian Journal of Microbiology. 1987;33:112–117

[91] Quiros M, Rojas V, Gonzalez R, Morales P. Selection of non-*Saccharomyces* yeast strains for reducing alcohol levels in wine by sugar respiration. International Journal of Food Microbiology. 2014;181:85–91. doi:10.1016/j.ijfoodmicro.2014.04.024

[92] Bartowsky EJ, Henschke PA. Acetic acid bacteria spoilage of bottled red wine—a review. International Journal of Food Microbiology. 2008;125:60–70. doi:10.1016/j.ijfood micro.2007.10.016

[93] Loureiro V, Malfeito-Ferreira M. Spoilage yeasts in the wine industry. International Journal of Food Microbiology. 2003;86:23–50. doi:10.1016/S0168-1605(03) 00246-0

[94] Stead D. The effect of hydroxycinnamic acids on the growth of wine- spoilage lactic acid bacteria. Journal of Applied Microbiology. 1993;75:135–141. doi:10.1111/j. 1365-2672. 1993.tb02758.x

[95] Gómez-Rivas L, Escudero-Abarca B, Aguilar-Uscanga MG, Hayward-Jones P, Mendoza P, Ramírez M. Selective antimicrobial action of chitosan against spoilage yeasts in mixed culture fermentations. Journal of Industrial Microbiology and Biotechnology. 2004;31:16–22. doi:10.1007/s10295-004-0112-2

[96] Rojo-Bezares B. Antimicrobial activity of nisin against *Oenococcus oeni* and other wine bacteria. International Journal of Food Microbiology. 2007;116:32–36. doi:10.1016/ j.ijfoodmicro.2006.12.020

[97] Gerbaux V, Villa A, Monamy C, Bertrand A. Use of lysozyme to inhibit malolactic fermentation and to stabilize wine after malolactic fermentation. American Journal of Enology and Viticulture. 1997;48:49–54

[98] Enrique M, Ibanez A, Marcos JF, Yuste M, Martinez M, Valles S, et al. b-Glucanases as a tool for the control of wine spoilage yeasts. Journal of Food Science. 2010;71:41–45. doi:10.1111/j.1750-3841.2009.01448.x

[99] Mehlomakulu NN, Setati ME, Divol B. Characterization of novel killer toxins secreted by wine-related non-*Saccharomyces* yeasts and their action on *Brettanomyces* spp.

International Journal of Food Microbiology. 2014;188:83–91. doi:10.1016/j.ijfoodmicro. 2014.07.015

[100] Thomas LV, Ingram RE, Bevis HE, Brightwell P, Wilson N, Delves-Broughton J. (2005). Natamycin control of yeast spoilage of wine. Food Protection Trends. 2005;25:510–517

[101] Enrique M, Marcos JF, Yuste M, Martínez M, Valles S, Manzanares P. Antimicrobial action of synthetic peptides towards wine spoilage yeasts. International Journal of Food Microbiology. 2007;118:318–325. doi:10.1016/j.ijfoodmicro.2007.07.049

[102] Enrique M, Marcos JF, Yuste M, Martinez M, Valles S, Manzanares P. Inhibition of the wine spoilage yeast *Dekkera bruxellensis* by bovine lactoferrin-derived peptides. International Journal of Food Microbiology. 2008;127:229–234. doi:10.1016/ j.ijfoodmicro. 2008.07.011

[103] Chavan PS, Tupe SG. Antifungal activity and mechanism of action of carvacrol and thymol against vineyard and wine spoilage yeasts. Food Control. 2014;46:115–120. doi: 10.1016/j.foodcont.2014.05.007

[104] Miranda JM, Jorge F, Dominguez L, Cepeda A, Franco CM. In vitro growth inhibition of food-borne pathogens and food spoilage microorganism by vitamin K5. Food Bioprocess Technology. 2011;4:1060–1065. doi:10.1007/s11947-010-0413-6

[105] Magliani W, Conti S, Gerloni M, Bertolotti D, Polonelli L. Yeast killer systems. Clinical Microbiology Reviews. 1997;10:369–400

[106] Gutiérrez AR, Epifanio S, Garijo P, López R, Santamaría P. Killer yeasts: incidence in the ecology of spontaneous fermentation. American Journal of Enology and Viticulture. 2001;52:352–356

[107] Santos A, Navascués E, Bravo E, Marquina D. *Ustilago maydis* killer toxin as a new tool for the biocontrol of the wine spoilage yeast *Brettanomyces bruxellensis*. International Journal of Food Microbiology. 2011;145:147–154. doi:10.1016/j.ijfoodmicro. 2010.12.005

[108] Comitini F, Ciani M. The zymocidial activity of *Tetrapisispora phaffii* in the control of *Hanseniaspora uvarum* during the early stages of winemaking. Letters Applied of Microbiology. 2010;50:50–56. doi:10.1111/j.1472-765X.2009.02754.x

[109] Santos A, Marquina D. Killer toxin of *Pichia membranifaciens* and its possible use as a biocontrol agent against grey mould disease of grapevine. Microbiology. 2004;150:2527–2534. doi:10.1099/mic.0.27071-0

[110] Fleet GH. Wine yeasts for the future. FEMS Yeast Research. 2008;8:979–995. doi:10.1111/ j.1567-1364.2008.00427.x

[111] Pretorius IS, Curtin CD, Chambers PJ. The winemaker's bug: from ancient wisdom to opening new vistas with frontier yeast science. Bioengineered Bugs. 2012;3:147–156. doi:10.4161/bbug.19687

Chapter 12

Aroma Compounds in Wine

Fengmei Zhu, Bin Du and Jun Li

Additional information is available at the end of the chapter

http://dx.doi.org/10.5772/65102

Abstract

Volatile aroma compounds are very important to grape wine quality. In order to understand the flavor of wine, a multitude of scientific investigations was carried out and a number of appropriate analytical tools for flavor study were developed in the past few decades. This chapter deals with major achievements reported in wine aroma and flavor. Firstly, we illustrate the existing knowledge on aroma compounds contributing to wine flavor, as well as the types of wine aroma compounds. Furthermore, the main factors affecting flavor quality in wine are discussed. Finally, the genomics and biotechnology of wine flavor are also summarized. This chapter broadens the discussion of wine aroma compounds to include more modern concepts of biotechnology and also provides relevant background and offers directions for future study.

Keywords: grape wine, aromatic compounds, terpenes, free volatile aroma compounds, bound volatile aroma compounds

1. Introduction

Wine aroma can be perceived by nose or in the mouth via postnasal way [1], and is a direct function of the chemical composition of the wine. Perceived flavor is the result of complex interactions between all the volatile and nonvolatile compounds present in wine [2]. The aroma of the wine consists of 1000 aroma compounds [3]. The diversity of aromatic compounds in wine is immense and ranges in concentration from several mg l^{-1} to a few ng l^{-1} [4]. Wine flavor can be divided into classes: varietal aroma, typical of grape variety; prefermentative aroma, originated during grape processing; fermentative aroma, produced by yeast and bacteria during alcoholic and malolactic fermentations (MLFs); postfermentative aroma, which is due to transformations that occurred during conservation and aging of wine [5]. This chapter will provide an overview of wine aroma and flavor starting with occurrence. The main focus of the review will describe the types of aroma compounds, as well as the main factors of wine aroma compounds in wine. More detail will be given on genomics and biotechnology. The sensory perception will also be discussed.

2. Occurrence of aroma compounds contributing to wine flavor

Knowledge of the volatile composition of a wine is of great interest, since these compounds are highly related to beverage flavor [6]. Although hundreds of chemical compounds have been identified in grapes and wines, only a few compounds actually contribute to sensory perception of wine flavor [7]. The perception of wine flavor and aroma is the result of a multitude of interactions between a large number of chemical compounds and sensory receptors [8]. Higher alcohols, acids and esters are quantitatively dominant in wine aroma and are important in the sensory properties and quality of wine [9]. Small amounts of higher alcohols contribute positively to wine quality, while excessive amounts may detract from quality [10]. Esters contribute to wine odor and relatively concentrations of fatty acids give an appreciable strong odor [11].

The main compounds responsible for the most intense aromas in Sauvignon Blanc wines have been assumed to be methoxypyrazines and varietal thiols in the Marlborough region [12].

Darici et al. [13] evaluated the aroma of a Turkey Çalkarası rosé wine by sensory evaluation analysis, gas chromatography-olfactometry (GC-O) and four quantitative methods. The volatile compounds of the wine were dominated by fresh fruit, floral and red fruit notes. An extract obtained using a dynamic headspace technique was subjected to GC-O. The aroma showed a complex profile with 28 compounds determined above their odor threshold.

3. Types of wine aroma compounds

Aroma styles are particularly important for wine exports, and preferences in aroma profiles can differ between markets in various countries [14].

3.1. Wine aroma formed by yeast during fermentation

Wine quality is closely related to microbial ecology of fermentation. Yeasts contribute to wine aroma by producing volatile metabolites with different flavor profiles. The aromatic profile of wine has been studied in relation to the amount of assimilable nitrogen available from the yeast present in the must [15]. Gil et al. [11] isolated and analyzed the aroma compounds of wines inoculated with pure and mixed cultures of apiculate and *Saccharomyces* yeasts by gas chromatography with flame ionization and mass spectrometry (MS). The findings showed that samples fermented with mixed cultures produced a higher concentration of selected compounds and higher total amounts of alcohols and acids, by comparison, wines produced with pure cultures of *Saccharomyces* spp. Apiculate yeasts are essential in the chemical composition and quality of wine.

The microflora, and especially the yeast, is related with fermentation, conduces to wine aroma by mechanisms: firstly by utilizing grape juice constituents and biotransforming them into aroma- or flavor-impacting components; secondly by bringing enzymes that transform neutral grape compounds into flavor-active compounds and lastly by the de novo synthesis of many flavor-active primary and secondary metabolites [8]. Joseph et al. [16] undertook a survey of 95 Brettanomyces strains and identified whether strains consistently give positive aroma characteristics using a solid phase microextraction with gas chromatography (SPME GC-MS) analysis coupled with olfactory analysis. None of the strains yielded universally positive aromas for the evaluators. The results showed that 22 compounds were identified as having an impact on aroma, including the well-known ethylphenols and vinylphenols, as well as several fatty acids, alcohols, esters, terpenes and an aldehyde.

3.2. Wine aroma formed during alcoholic fermentation

Spontaneous wine fermentations are often unpredictable, resulting in undesirable traits that occasionally lead to spoilage [17]. Zhang et al. [18] studies the effects of three commercial maceration enzymes on aroma compounds of Cabernet Sauvignon wine during alcohol fermentation. The results showed that maceration enzymes could have a significant effect on the formation of aroma compounds. Moreover, the presence of non-*Saccharomyces* species at the onset of alcoholic fermentation may have a greater potential to contribute to the liberation of some aglycons (mainly terpenes) from the flavorless precursor glycoside during fermentation [19].

3.3. Wine aroma formed during amino acid metabolism

The most important flavor and aroma compounds formed from amino acids are higher alcohols and their associated esters and volatile acids. These odd-related products are produced from valine, leucine and isoleucine. It has been shown that the varietal aroma character of certain cultivars could be partially explained by the amino acid composition of the grape must [20]. Although yeast strains differ greatly in their ability to use nitrogen and amino acids, various studies have shown that nitrogen supplementation in the form of assimilable nitrogen and amino acids influences the volatile aroma profile of the wine [8].

There are different ways that amino acids can be metabolized into aroma compounds. The first way is Ehrlich reaction. In this process, amino acids are catabolized into higher alcohols. The Ehrlich reaction also impacts directly or indirectly on the synthesis of other aroma compounds [21]. Secondly, the sulfur-containing amino acids can have a positive impact on the aroma of wine. For example, 3-mercaptohexanol can impart fruity flavors to a wine. Another way is called Maillard reaction. Cysteine can form various odor-impacting compounds through this reaction, in which a chemical reaction between amino and carbonyl groups takes place to form new compounds [22].

3.4. Wine aroma formed during malolactic fermentation

The secondary malolactic fermentation (MLF) is principally a deacidification step, which is used to manicure the acidity of certain wine types and confer added microbial stability to the product [19]. This process is normally carried out by lactic acid bacteria isolated from wine, including *Oenococcus oeni*, *Lactobacillus* spp., *Leuconostoc* spp. and *Pediococcus* spp. [23]. It has been shown that lactic acid bacteria can influence the aroma by producing volatile metabolites and modifying aroma compounds derived from grapes and yeasts during MLF. Moreover, MLF can enhance the fruity aroma and buttery note but reduce vegetative, green/grassy aroma of wine. In recent studies, lactic acid bacteria can also influence wine aroma by producing additional oak-derived compounds [24]. In addition, MLF is important in wines from warmer region because it changes the composition of the wine and improves its organoleptic quality.

4. Main factors affecting flavor quality in wine

Wine aroma is generated through an immensely complex interaction of various classes of aroma compounds and various environmental and biological factors.

4.1. Effect of weather on wine aroma compounds

There are many factors that can influence flavor profile or wine style.

Mendez-Costabel et al. [25] evaluated the impact of winter rainfall on the main compounds responsible for green aromas in grapes and wines during the 2009 and 2010 seasons in California. Fruit and wine components were dramatically effected by the absence of rainfall in both years. Wine descriptive analysis showed that the lack of rainfall produced wines perceived as less green and of more intense fruit attributes in the first season. Due to the reduction in vine growth, however, the same treatment produced wines less intense in fruit aromas and of bad tannin quality in the following season. These results showed that if the rainfall level is below normal, the positive effect on fruit and wine composition achieved.

4.2. Effect of soil type on wine aroma compounds

Soil type, though closely related to soil water status, has an independent effect on grape aroma quality [26]. Falcao et al. [27] assessed soil characteristics of the different four sites in Santa Catarina State, Brazil. The results indicated that vineyard location had a strong influence on the volatile wine fraction. The varietal volatile compounds were a key factor in differentiating wines according to the sites. In addition, Ribereau-Gayon et al. [28] reported that soil has a decisive influence on methoxypyrazine concentrations due to its effect on veg-

http://dx.doi.org/10.5772/65102

etative growth in Merlot, Cabernet Franc and Cabernet Sauvignon wines of the Bordeaux region. The results showed that grapes grown on well-drained gravelly soils have lower concentrations than those grown on limestone or clay-silt soils.

4.3. Effect of vineyard management practices on wine aroma compounds

4.3.1. Effect of water on wine aroma compounds

Bonada et al. [29] assessed the sensory and compositional characteristics of grapes and wines from a field trial where water deficit factor was directly manipulated. The results indicated that the effect of water deficit leading to colorful and flavorsome wines rich in phenolic substances may not be held under high temperature.

4.3.2. Effect of copper on wine aroma compounds

Martins et al. [30] investigated the effect of the application of Bordeaux mixture in the vineyard on the copper concentration in the must, and the consequences for the volatile composition of wine. The results showed that the concentration of copper decreased from 31.4 to 12.6 mg/L during fermentation, as measured by atomic absorption spectrometry. Promoted copper concentration caused a significant decrease in the concentration of higher alcohols, including isoamyl alcohol, and of esters of organic acids, including ethyl lactate, as analyzed by liquid–liquid microextraction gas chromatography (GC)-flame ionization detection and solid phase extraction GC-ion trap-mass spectrometry (MS) analysis. By contrast, the ethyl acetate and linalool concentration rose dramatically.

4.3.3. Effect of nitrogen fertilization on wine aroma compounds

For example, Mendez-Costabel et al. [31] investigated the effect of two irrigation levels and a higher than standard nitrogen fertilization on the concentration of both 3-isobutyl-2-methoxypyrazine (IBMP) and six C6 compounds during fruit development. The results showed that deficit irrigation increased fruit color, quercetin glycosides and phenol-free glucose glycosides (i.e., aroma precursors), decreased vine yield and increased concentration of IBMP during fruit maturation.

4.4. Effect of aging and maturation on wine aroma compounds

The release of aroma precursors can occur during wine aging, under mild acidic conditions [32]. Higher alcohols are important as precursors for ester formation during aging. The flavor-active metabolites that have an impact on wine perception are derived from the grapes and from microorganisms during fermentation, as well as from chemical processes during production and maturation [2]. The final aroma and flavor profile is furthermore strongly dependent on all aspects of postfermentation treatments such as filtration and maturation strategies, including aging in wooden containers [8].

5. Free and bound volatile aroma compound

Aroma compounds in wine are typically found both as "free" and "bound" to a sugar moiety. In wine, a large proportion of aroma compounds are found in the bound form [33]. Potentially volatile terpenes (PVTs) are more responsive to viticultural and oenological practices than free volatile terpenes (FVTs). Many lactic acid bacteria possess catalytic enzymes capable of liberating grape-derived aroma compounds from their natural nonaromatic glycosylated state [34]. The classes of enzymes (β-glucosidase, proteases, esterases, citrate lyases and phenolic acid decarboxylases) can possibly hydrolyze flavor precursors and so influence wine aroma [35–37]. The hydrolysis steps are shown in **Figure 1**.

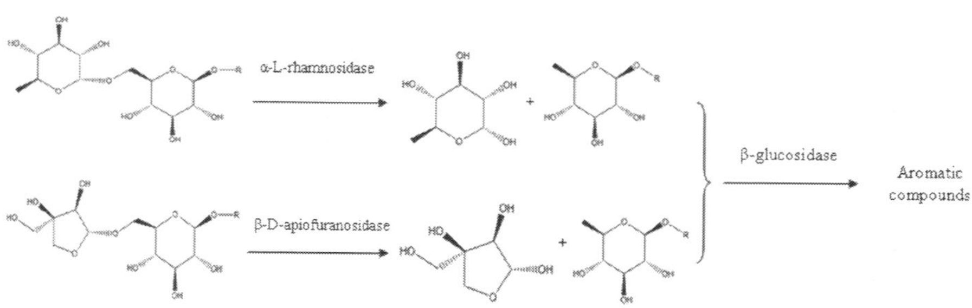

Figure 1. The hydrolysis scheme of glycosidic aroma precursors.

6. Genomics and biotechnology of wine flavor

In wine-making industry, some biotechnological techniques have been of fundamental importance [38]. Colagre et al. [38] illustrated the importance of the characteristics of wine yeast and of a genetic improvement program for the wine industry, as well as the complexity of the genetic modifications of commercial wine yeast strains. Moreover, commercial preparations of glycosidases, usually sourced from *Aspergillus* sp., can be used to liberate more flavor aglycons into the wine, but may not function well under key wine conditions of low pH, ethanol content or residual sugar content [39]. An alternative approach to preparing a crude or pure extract of glycosidases for addition to wine is to express the appropriate enzyme gene in *S. cerevisiae*. For example, the rbaA gene (α-L-rhamnosidase) from *Aspergillus aculeatus* has been successfully expressed in conjunction with the Candida molischiana β-D-glucosidase in an industrial wine yeast strain to increase the pool of linalool, nerol and α-terpineol in Muscat wine [40].

7. Aroma compounds and sensory perception

Aroma compounds play an important role in the quality of wine because those compounds produce an effect on sensory senses [5].

The aroma of wines is the result of the contribution of some hundreds of volatile compounds and it is an important factor to consider in their sensorial quality [6]. Once the wine has been made, the appreciation of wine requires various senses: firstly to observe the color and appearance, secondly to judge the wine bouquet, thirdly to taste the wine itself and fourthly to enjoy the mouthfeel and aftertaste [7].

Quantitative descriptive analysis is one of the most comprehensive and informative tools used in sensory analysis. Tomasino et al. [41] identified aroma compounds of major sensory significance in New Zealand Pinot Noir wines using canonical correlation analysis and addition/omission tests. There are some similar researches in Merlot and Cabernet Sauvignon wine. The effect of benzaldehyde, ethyl octanoate and 2-phenyl ethanol on the aroma of Pinot Noir wine was investigated. The results showed that the contribution of these compounds is similar in a range of red wines. Both ethyl decanoate and ethyl octanoate played an important role acting as aroma enhancer compounds. Moreover, Welke et al. [42] reported the quantitative determination of volatile compounds of Chardonnay wines using HS-SPME-GC × GC/TOFMS along with the determination of odor activity value (OAV) and relative odor contribution (ROC) of volatiles.

On the other hand, sensory analysis involves the detection and description of qualitative and quantitative sensory compounds of a product by a trained panel of judges [43]. Wine tasting and perception is therefore largely a subjective experience, and simple factors such as the absence or presence of saliva greatly influence the release of aroma compounds from both red and white wines [44].

8. Conclusions and future prospectives

This review summarizes the aroma compounds contributing to wine flavor and the types of these aroma compounds. The specific compounds in wine are not clear. It is important to study the chemical and biochemical changes during fermentation and storage. Nevertheless, further research into the sensory impact of wine aroma compounds both alone and in mixtures will be focused.

Acknowledgements

This research was supported by Natural Science Foundation of Hebei Province, P.R. China (C2014407059 and 2015407059); Natural Science Foundation of China (31470542 and 31570374).

Author details

Fengmei Zhu[1], Bin Du[2] and Jun Li[1*]

*Address all correspondence to: trueyeoman@163.com

1 College of Food Science and Technology, Hebei Normal University of Science and Technology, Qinhuangdao, P.R. China

2 Analysis and Testing Center, Hebei Normal University of Science and Technology, Qinhuangdao, P.R. China

References

[1] Francis IL, Newton JL. Determining wine aroma from compositional data. Australian Journal of Grape and Wine Research. 2005;11(2):114–126.

[2] Fairbairn SC, Smit AY, Jacobson D, Prior BA, Bauer FF. Environmental stress and aroma production during wine fermentation. South African Journal for Enology & Viticulture. 2014;35(2):168–177.

[3] Tao YS, Li H. Active volatiles of cabernet sauvignon wine from Changli county. Natural Science. 2009;1:176–182.

[4] Zhang M, Pan Q, Yan G, Duan C. Using headspace solid phase micro-extraction for analysis of aromatic compounds during alcoholic fermentation of red wine. Food Chemistry. 2011;25(2):743–749.

[5] Vilanova M, Genisheva Z, Masa A, Oliveira JM. Correlation between volatile composition and sensory properties in Spanish Albariño wines. Microchemical Journal. 2010;95(2):240–246.

[6] Palomo ES, García-Carpintero EG, Viñas MAG. Aroma fingerprint characterisation of La Mancha red wines. South African Journal for Enology & Viticulture. 2015;36:117–125.

[7] Polásková P, Herszage J, Ebeler SE. Wine flavor: chemistry in a glass. Chemical Society Reviews. 2008;37(11):2478–2489.

[8] Styger G, Prior B, Bauer FF. Wine flavor and aroma. Journal of Industrial Microbiology & Biotechnology. 2011;38(9):1145–1159.

[9] Stashenko H, Macku C, Shibamato T. Monitoring volatile chemicals formed from must during yeast fermentation. Journal of Agricultural and Food Chemistry. 1992;40(11): 2257–2259.

http://dx.doi.org/10.5772/65102

[10] Rapp A, Mandery H. Wine aroma. Experientia 1986;42:873–884.

[11] Gil JV, Mateo JJ, Jiménez M, Pastor A, Huerta T. Aroma compounds in wine as influenced by apiculate yeasts. Journal of Food Science. 1996;61(6):1247–1250.

[12] Jouanneau S, Weaver RJ, Nicolau L, Herbst-Johnstone M, Benkwitz F, Kilmartin PA. Subregional survey of aroma compounds in Marlborough Sauvignon Blanc wines. Australian Journal of Grape and Wine Research. 2012;18(3):329–343.

[13] Darici M, Cabaroglu T, Ferreira V, Lope R. Chemical and sensory characterisation of the aroma of Çalkarası rosé wine. Australian Journal of Grape and Wine Research. 2014;20(3):340–346.

[14] King ES, Osidacz P, Curtin C, Bastlan SEP, Francis IL. Assessing desirable levels of sensory properties in Sauvignon Blanc wines-consumer preferences and contribution of key aroma compounds. Australian Journal of Grape and Wine Research. 2011;17(2): 169–180.

[15] Montevecchi G, Masino F, Simone GV, Cerretti E, Antonelli A. Aromatic profile of white sweet semi-sparkling wine from Malvasia di Candia Aromatica grapes. South African Journal for Enology & Viticulture. 2015;36(2):267–276.

[16] Joseph CML, Albino EA, Ebeler SE, Bisson LF. *Brettanomyces bruxellensis* aroma-active compounds determined by SPME GC-MS olfactory analysis. American Journal of Enology & Viticulture. 2015;66(3):379–387.

[17] Álvarez-Pérez JM, Álvarez-Rodríguez ML, Campo E, Sáenz de Miera LE, Ferreira V, Hernández-Orte P, Garzón-Jimeno E, Coque JJR. Selection of *Saccharomyces cerevisiae* strains applied to the production of prieto picudo ros wines with a different aromatic profile. South African Journal for Enology & Viticulture. 2014;35:242–256.

[18] Zhang MX, Qu WJ, Zhang H, Han FL, Duan CQ. Effect of maceration enzymes on the formation of aroma compounds during Cabernet Sauvignon alcohol fermentation. AgroFood Industry Hi-tech. 2007;18(3):5–7.

[19] Bartowsky DJ, Henschke PA, Hoj PB, Pretorius IS. Chasing wine aroma – Does *Oenococcus oeni* have the potential to release aroma compounds from authentic grape precursors? Wine Industry Journal. 2004;19(2):24–31.

[20] Hernández-Orte P, Cacho JF, Ferreira V. Relationship between varietal amino acid profile of grapes and wine aromatic compounds. Experiments with model solutions and chemometric study. Journal of Agricultural & Food Chemistry. 2002;50(10):2891–2899.

[21] Hazelwood LA, Daran JM, van Maris AJ, Pronk JT, Dickinson JR. The Ehrlich pathway for fusel alcohol production: a century of research on *Saccharomyces cerevisiae* metabolism. Applied & Environmental Microbiology. 2008;74(8):2259–2266.

[22] Marchand S, Reve, GD, Bertrand A. Approaches to wine aroma: release of aroma compounds from reactions between cysteine and carbonyl compounds in wine. Journal of Agricultural & Food Chemistry. 2000;48(10):4890–4895.

[23] Liu SQ. A review: malolactic fermentation in wine–beyond deacidification. Journal of Applied Microbiology. 2002;92(4):589–601.

[24] de Revel G, Martin N, Pripis-Nicolau L, Lonvaud-Funel A, Bertrand A. Contribution to the knowledge of malolactic fermentation influence on wine aroma. Journal of Agricultural & Food Chemistry. 1999;47(10):4003–4008.

[25] Mendez-Costabel MP, Wilkinson KL, Bastian SEP, Jordans C, McCarthy M, Ford CM, Dokoozlian N. Effect of winter rainfall on yield components and fruit green aromas of *Vitis vinifera* L. cv. Merlot in California. Australian Journal of Grape and Wine Research. 2014;20(1):100–110.

[26] González-Barreiro C, Rial-Otero R, Cancho-Grande B, Simal-Gándara J. Wine aroma compounds in grapes: a critical review. Critical Reviews in Food Science & Nutrition. 2013;55(2):202–218.

[27] Falcao LD, Revel GD, Perello MC, Riquier L, Rosier JP. Volatile profile characterization of young cabernet-sauvignon wines from a new grape growing region in Brazil. International Journal of Vine and Wine Sciences. 2008;42(3):133–146.

[28] Ribereau-Gayon P, Glories Y, Maujean A, Dubourdieu D. Handbook of Enology (Volume 2). The Chemistry of Wine and Stabilization and Treatment. John Wiley & Sons, Ltd, New York. 2006.

[29] Bonada M, Jeffery DW, Petrie PR, Moran MA, Sadras VO. Impact of elevated temperature and water deficit on the chemical and sensory profiles of Barossa Shiraz grapes and wines. Australian Journal of Grape and Wine Research. 2015;21(2):240–253.

[30] Martins V, Teixeira A, Gerós H. Changes in the volatile composition of wine from grapes treated with Bordeaux mixture: a laboratory-scale study. Australian Journal of Grape and Wine Research. 2015;21(3):425–429.

[31] Mendez-Costabel MP, Wilkinson K, Bastian LSEP, Jordans C, McCarthy M, Ford CM, Dokoozlian NK. Effect of increased irrigation and additional nitrogen fertilisation on the concentration of green aroma compounds in *Vitis vinifera* L. Merlot fruit and wine. Australian Journal of Grape and Wine Research. 2014;20(1):80–90.

[32] Bisotto A, Julien A, Rigou P, Schneider R, Salmon JM. Evaluation of the inherent capacity of commercial yeast strains to release glycosidic aroma precursors from Muscat grape must. Australian Journal of Grape & Wine Research. 2015;21:194–199.

[33] Hjelmeland AK, Ebeler SE. Glycosidically bound volatile aroma compounds in grapes and wine: a review. American Journal of Enology & Viticulture. 2014;66:1–11.

[34] Grimaldi A, Bartowsky E, Jiranek V. A survey of glycosidase activities of commercial wine strains of *Oenococcus oeni*. International Journal of Food Microbiology. 2005;105(2): 233–244.

[35] Boido E, Lloret A, Medina K, Carrau F, Dellacassa E. Effect of beta-glycosidase activity of *Oenococcus oeni* on the glycosylated flavor precursors of tannat wine during malolactic fermentation. Journal of Agricultural & Food Chemistry. 2002;50(8):2344–2349.

[36] Ugliano M, Genoves A, Moio L. Hydrolysis of wine aroma precursors during malolactic fermentation with four commercial starter cultures of *Oenococcus oeni*. Journal of Agricultural & Food Chemistry. 2003;51(17):5073–5078.

[37] Mtshali PS, Divol B, Van RP, Du TM. Genetic screening of wine-related enzymes in lactobacillus species isolated from South African wines. Journal of Applied Microbiology. 2010;108(4):1389–1397.

[38] Colagre O, Silva A, Fumi MD. Recent applications of biotechnology in wine production. Biotechnology Progress. 1994;10(1):2–18.

[39] Winterhalter P, Skouroumounis GK. Glycoconjugated aroma compounds: occurrence, role and biotechnological transformation. In: Advances in Biochemical Engineering/ Biotechnology. T. Sheper (Ed.). Springer-Verlag. 1997;74–99.

[40] Manzanares P, Orejas M, Gil V, Graaff LHD. Construction of a genetically modified wine yeast strain expressing the *Aspergillus aculeatus* rhaA gene, encoding an α-l-rhamnosidase of enological interest. Applied & Environmental Microbiology. 2003;12:7558–7562.

[41] Tomasino E, Harrison R, Breitmeyer J, Sedcole R, Sherlock R, Frost A. Aroma composition of 2-year-old New Zealand Pinot Noir wine and its relationship to sensory characteristics using canonical correlation analysis and addition/omission tests. Australian Journal of Grape and Wine Research. 2015;21(3):376–388.

[42] Welke JE, Zanus M, Lazzarotto M, Zini CA. Quantitative analysis of headspace volatile compounds using comprehensive two-dimensional gas chromatography and their contribution to the aroma of chardonnay wine. Food Research International. 2014;59(4): 85–99.

[43] Meilgard MC, Civille GV, Carr BT. Sensory evaluation techniques, 1999; CRC Press, Florida.

[44] Genovese A, Piombino P, Gambuti A, Moio L. Simulation of retronasal aroma of white and red wine in a model mouth system. Investigating the influence of saliva on volatile compound concentrations. Food Chemistry. 2009;114(1):100–107.

Chapter 13

Influence of Yeasts in Wine Colour

Morata Antonio, Loira Iris and
Suárez Lepe Jose Antonio

Additional information is available at the end of the chapter

http://dx.doi.org/10.5772/65055

Abstract

Colour is the first impression that the consumer receives from wine and it influences the taste. Colour gives an idea about wine quality, age, oxidation and structure, so it has an important repercussion on the consumer perception of wine. Yeasts promote the formation of stable pigments by the production and release of fermentative metabolites affecting the formation of vitisin A and B type pyranoanthocyanins. The hydrox- and ycinnamate decarboxylase activity showed by some yeast strains produces highly reactive vinylphenols stimulating the formation of vinylphenolic pyranoanthocyanins from grape anthocyanin precursors during fermentation. Some yeasts also influence the formation of polymeric pigments by unclear mechanisms that can include the production of linking molecules such as acetaldehyde. Grape anthocyanins adsorbed in yeast cell walls during fermentation are removed from wine after racking processes affecting final pigment content. Moreover, the intensive use of non-*Saccharomyces* yeasts in current oenology makes it interesting to assess the effect of new species in the improvement of wine colour.

Keywords: *Saccharomyces*, non-*Saccharomyces*, anthocyanins, pyranoanthocyanins, polymeric pigments, wines

1. Introduction

In the past, the formation and evolution of wine colour was conditioned by the anthocyanin composition of the grape variety, the degree of extraction during winemaking and the physicochemical evolution of the pigments during tank, barrel and bottle ageing. In these last processes, the influence of grape proanthocyanins, other flavonoids and oxygen affects the formation of stable polymeric pigments [1].

Grape pigments are monomeric anthocyanins glycosylated in position 3. The colour of anthocyanins is strongly dependent on pH, SO_2 levels and hydration. According to these factors, several anthocyanin derivatives can be found (**Figure 1**). Moreover, the solubility of anthocyanins is affected by the polarity of the medium. During fermentation, alcohol production reduces must polarity and the concentration of anthocyanins decreases.

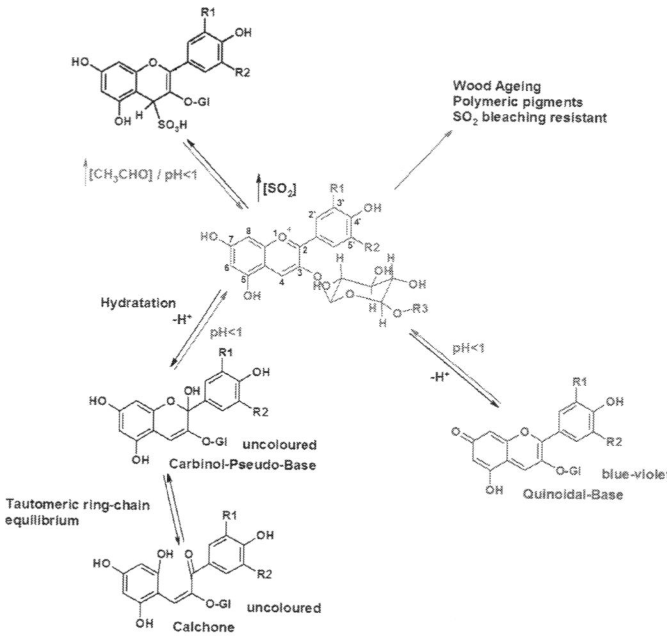

Figure 1. Reactions that can affect anthocyanins during winemaking and the effects on their structure.

Since 1990s the role of yeast in colour stability has been studied deeply. In fact, anthocyanin insolubilisation during fermentation is a consequence of yeast metabolism. However, that is not the only contribution of yeasts to wine colour, the production of derived pigments during fermentation using yeast metabolites as precursors or by means of yeast enzymatic activities are also major concerns that have been analysed in detail in the last decades [2, 3].

Pigment adsorption by yeast cell walls reduces the concentration of anthocyanins affecting wine colour, especially in low colour varieties. This property has been used traditionally to reduce pigment contents in the production of white sparkling wines from red varieties (*blanc de noirs*).

The formation of polymeric pigments has been considered as a way of colour stabilization during ageing. Grape anthocyanins condense with other flavonoids forming polymeric structures (n × [C6-C3-C6]). Sometimes, flavonoids can be linked by some intermediate molecules like acetaldehyde. The polymerization reactions occur during ageing and are affected by precursor levels, oxygen levels and pH. Currently, some evidences point up that yeast can also promote the formation of polymeric pigments during fermentation.

The strong effect of pH in anthocyanins colour intensity can cause slight modifications in wine colour by some yeast species able to either degrade or produce organic acids during fermentation.

Some of these processes with effect on colour can also be enhanced by the use of non-*Saccharomyces* yeasts during fermentation.

2. Formation of pyranoanthocyanins with yeast metabolites

Some yeast metabolites can react with grape anthocyanins during fermentation forming derived pigments with enhanced stability in oenological conditions and slightly different chromatic features. The formation of vitisin A, a pyranoanthocyanin pigment formed by chemical reaction between malvidin-3-*O*-glucoside (malvidin or m3g), a major grape anthocyanin, and pyruvic acid released by yeast, is produced during fermentation. It can be also produced during ageing. In a similar way, vitisin B is formed during fermentation by the reaction between acetaldehyde and m3g (**Figure 2**).

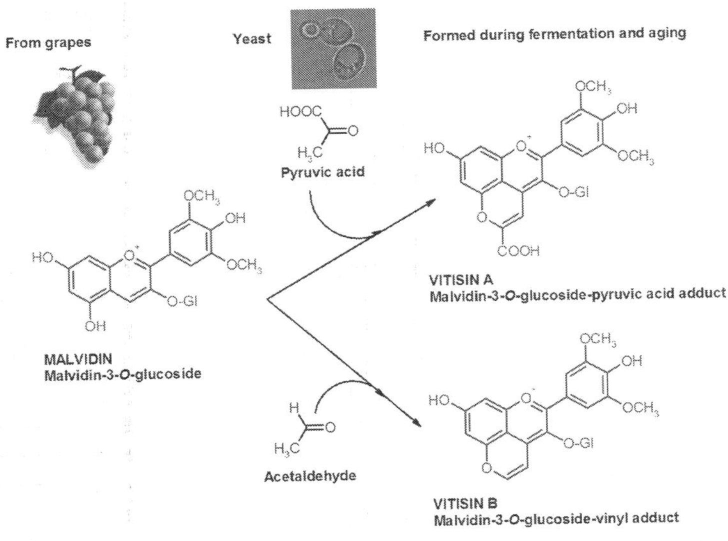

Figure 2. Formation of vitisins A and B by chemical reaction between malvidin and pyruvic acid or acetaldehyde, respectively.

Vitisins A and B are stable pyranoanthocyanin in oenological conditions with stable colour intensity under variable pH [4]. They are also more resistant to oxidative damage probably because of the higher number of resonant forms they have due to the double pyranose ring structure. Moreover, they are not sensitive to sulphur dioxide bleaching because position C4 is fully saturated being unable to react with the bisulphite ion (**Figure 3**).

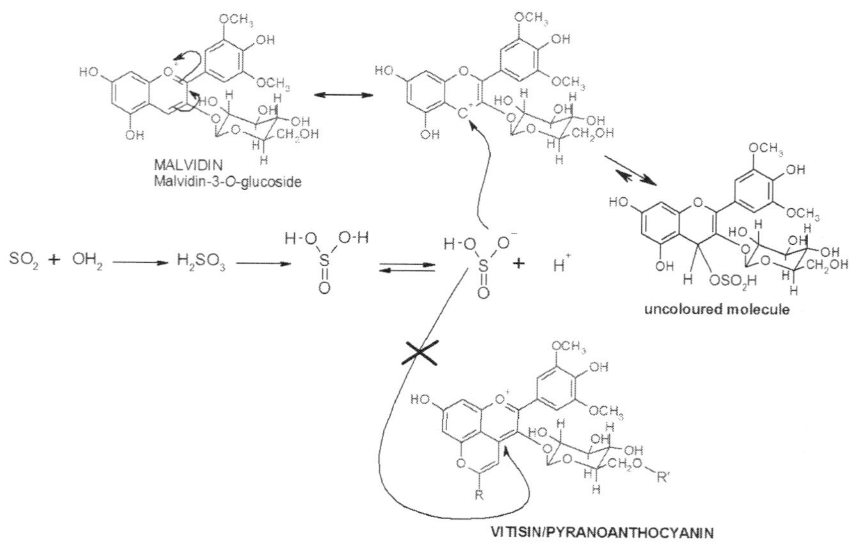

Figure 3. Dis-colouration reaction in anthocyanins by bisulphite attack. This reaction is not possible in pyranoanthocyanins because C4 is saturated.

Chromatic properties of vitisins are slightly different from grape anthocyanins. The maximum of absorption in visible spectra for malvidin is approximately 528 nm, but vitisin A shows a maximum of 515 nm and vitisin B of 495 nm (**Figure 4**). That means that vitisins are red-orange pigments and consumers currently prefer red-bluish colours in wines. However, the normal evolution in wine colour during ageing is from purple to red-orange, and, in this situation, colour of vitisins can be better integrated in wine appearance, indeed, they can even improve it. The stability of vitisins that makes them more persistent during ageing must be also considered.

Vitisins are usually analysed by HPLC-DAD separation and quantitation, using an external standard of malvidin-3-*O*-glucoside. Traditional LC separations are done in C18 reverse phase columns and the elution is produced with a gradient of water and methanol or water and acetonitrile. To keep anthocyanins in the cationic form (pyrilium form), the use of formic acid as the pH regulator is very convenient. Also, it is important to ensure a good ionization in electrospray when ESI-MS analysers are used. In the past, 30 cm columns with 5 μm particles were used, thus obtaining good separation times near or higher than one hour. The newer low size particle columns (1.8 or 2.6 μm) can be used to perform the wine anthocyanins separations in a few minutes. In some chromatographic conditions, vitisins can co-elute with acetylated delphinidin. Co-elution of vitisins A and B is also frequent. The separation of these compounds can be assured by managing both solvent gradient and formic acid concentration. Anthocyanins and vitisins are identified by UV-visible spectra. The quantification is done by interpolation in a calibration curve of an external standard of malvidin-3-*O*-glucoside.

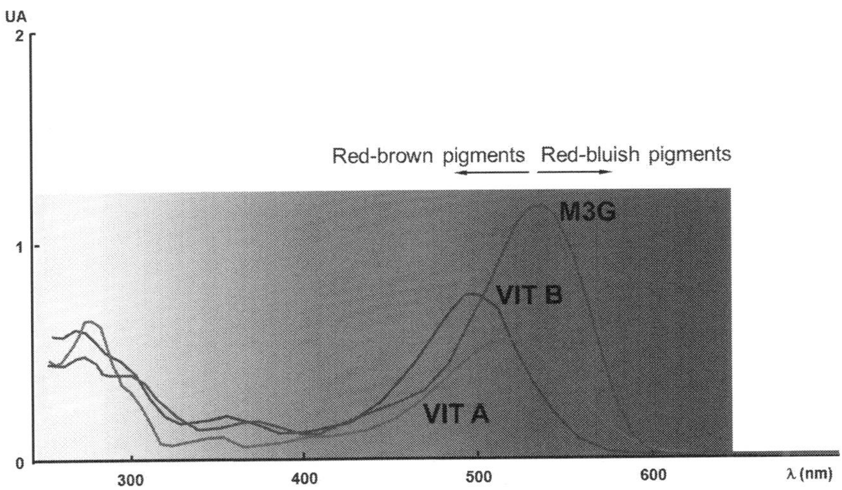

Figure 4. UV-visible spectra of malvidin, and vitisins A and B.

Vitisins A and B can be also identified by mass spectrometry (MS) by the specific fragmentation patterns; in fact, MS facilitates the full identification of these pigments after LC separation. The *m*/*z* for molecular ion (M^+) of vitisin A is 561 and the aglycon fragment is 399 mass units. The fragment is also common for acetyl and coumaroyl vitisins A. M^+ of Vitisin B is 517 and the corresponding aglycon fragment is 355. This last mass moiety also appears in MS spectra of acetyl and coumaroyl vitisins B.

Concentration of vitisins in wines could range from traces to a few mg/l. The amount formed during fermentation can be improved by using selected *Saccharomyces cerevisiae* strains with suitable production of pyruvic acid and acetaldehyde [5]. We observed that the release of pyruvate and acetaldehyde by yeasts during fermentation can be correlated with the formation of vitisin A and B, respectively ($R^2 > 0.8$) [5]. Oenological parameters such as pH, temperature or SO_2 concentration also affect the formation of vitisins during fermentation [6].

The production of pyruvate and acetaldehyde behaves differently in *Saccharomyces*. Usually, the maximum concentration of pyruvic acid is reached on the fourth/fifth day of fermentation and, then, this value decreases to reach a stable concentration towards the end of fermentation. Since pyruvate is an intermediate compound in many metabolic routes, a possible explanation to this behaviour is that at the beginning of fermentation, when enough nutrients are available, the yeast produces and releases it in big amounts. However, at the end of alcoholic fermentation, when nutrients become scarce, the yeast may take the pyruvate previously released into the fermentation medium for use in metabolic processes. The utility of acetaldehyde is less and is basically released progressively from the beginning to the end of fermentation. The patterns of excretion of pyruvate and acetaldehyde influence the formation of the respective vitisins A and B. Maximum concentrations of vitisin A are found after the maximum production of pyruvate in the middle of fermentation, and the higher concentration of vitisin B is produced at the end of fermentation [5, 6].

Selected yeast strains of *S. cerevisiae* with higher production of pyruvate and acetaldehyde increase the formation of vitisins A and B because of the higher concentration of precursors (**Figure 5**). The use of specific strains with high release rates of precursors, along with an optimization of the oenological conditions (pH, SO_2 and temperature) to favour the chemical reactions of condensation between malvidin and either pyruvate or acetaldehyde, is a way to modulate the vitisin formation during fermentation.

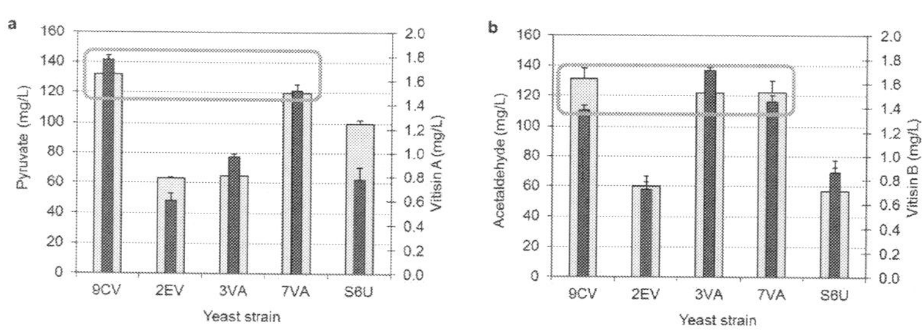

Figure 5. Production of pyruvate (a) and acetaldehyde (b) by several strains of *S. cerevisiae* and the corresponding formation of vitisins A and B respectively. In blue columns precursor (pyruvate or acetaldehyde) in red ones (vitisin A or B).

Recently, we have observed that some non-*Saccharomyces* are able to develop some specific metabolic processes affecting the global excretion of pyruvate. *Schizosaccharomyces pombe* is a fission yeast with asexual reproduction by bipartition instead of budding, and also able to develop singular catabolic pathways like maloalcoholic fermentation. *S. pombe* yeast is able to ferment sugar concentrations representing more than 13% (v/v) of potential ethanol, so it is useful to perform complete wine fermentations and it can also be used in mixed or sequential fermentations. The slower speed of fermentation and the production of excessive amounts of acetic acid, when *S. cerevisiae* is used as the sole fermentative yeast, could present some drawbacks. We have observed a release of pyruvate during the fermentation with *S. pombe* in average higher than with *S. cerevisiae* strains, probably due to its metabolic peculiarities. Thereby, when used to ferment red grapes, the formation of vitisin A and its derivatives is greater than with *S. cerevisiae* [7]. In some cases, the levels can overcome 10 mg/L. Fermentation with *S. pombe* can be used to enhance the formation of vitisin A-type pigments (**Figure 6**) using either pure or sequential cultures with *S. cerevisiae.*

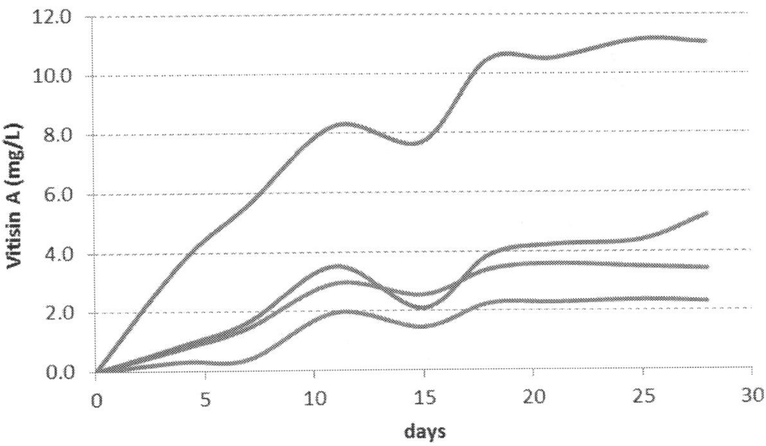

Figure 6. Formation of vitisin A during the fermentation with *S. pombe*. Red lines minimum and maximum values formed during fermentation by *S. pombe* (four strains). Blue line minimum and maximum values formed by *S. cerevisiae* (two strains).

Vitisins are formed by chemical condensation of malvidin with pyruvate and acetaldehyde so the addition of these precursors in wines enhances their formation. The addition of pyruvate is especially effective [8]. However, pyruvate and acetaldehyde are not allowed as oenological additives. Moreover, the use of acetaldehyde has many other effects on reactions between phenolic compounds promoting the condensation between tannins and the precipitation of phenols and pigments.

3. Hydroxycinnamate decarboxylase activity influence on the formation of vinylphenolic pyranoanthocyanins

Vinylphenolic pyranoanthocyanins (VPAs) are also stable pigments with similar properties to vitisins. They were discovered initially in pinotage (*Vitis vinifera* L.) wines and the derived pigment from malvidin and caffeic acid was called pinotin A (malvidin-3-*O*-glucoside-4-vinylcatechol) [9]. A mechanism of formation by chemical reaction between caffeic acid and malvidin with an uncoloured intermediate that recovers colour by slow oxidation under the typical conditions of wine barrel ageing was proposed (**Figure 7**) [10]. Due to their slow formation, these pigments were suggested as age markers in barrel ageing.

Later, the formation of these pigments derived from malvidin and hydroxycinnamic acids (HCAs) (either caffeic, ferulic or *p*-coumaric) was observed during fermentation when yeasts with hydroxycinnamate decarboxylase (HCDC) activity were used [6, 11]. HCDC+ strains are able to decarboxylate HCAs in must during fermentation producing highly reactive vinylphenols (VPh) that spontaneously condense with malvidin and other grape anthocyanins forming VPAs (**Figure 8**) [11]. These compounds can be separated and analysed by LC-ESI/MS in wines. In typical LC conditions to separate anthocyanins, VPAs appear at the end of the chromatogram because of their high apolar structures with five aromatic rings. UV and MS key parameters of VPAs are detailed in **Table 1**.

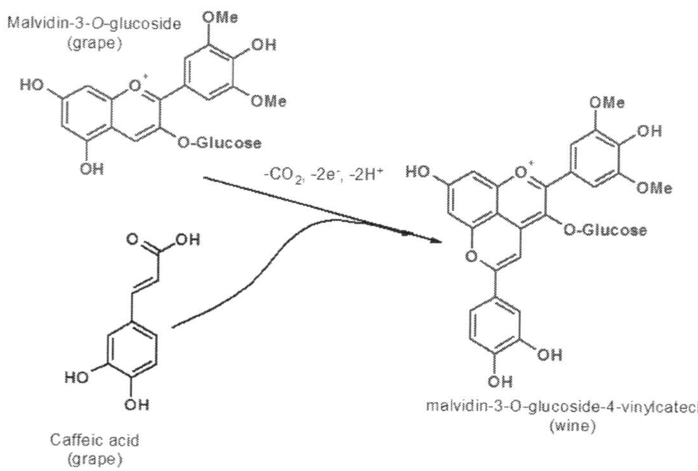

Figure 7. Formation of pinotin A from both malvidin and caffeic acid during wine ageing by a physicochemical process.

The use of HCDC+ strains of *Saccharomyces* and non-*Saccharomyces* during fermentation is a way to promote the formation of stable VPAs during fermentation. Some non-*Saccharomyces* strains (*P. guillermondii, S. pombe*) have been described as HCDC+ and they can favour the formation of VPAs at a higher concentration than *S. cerevisiae* [7, 12]. When low fermentative non-*Saccharomyces* are used to ferment must, they should be used in mixed or sequential cultures with *Saccharomyces* to ensure complete sugar depletion [13].

The selection of HCDC+ and determination of the intensity of enzymatic activity can be tested by using a fermentative medium with hydroxycinnamic acids. The use of *p*-coumaric acid is a good option. The medium can be prepared using grape must from fresh grape or a dilution of concentrated must with 220 g/l of sugars, the pH can be adjusted at 3.5 and a suitable concentration of *p*-coumaric acid is 50 mg/l. The evolution in the degradation of *p*-coumaric acid to 4-vinylphenol can be checked by LC-DAD [14]. The yeast strain can be considered HCDC+ when more than 10% of *p*-coumaric acid is transformed into 4-vinylphenol. The greater the degradation of *p*-coumaric acid and, subsequently, the formation of 4-vinylphenol, the higher the HCDC+ activity of the yeast. It is convenient to use HCDC negative and positive controls during the experiment. The test can be also performed with other grape HCAs such as either caffeic or ferulic.

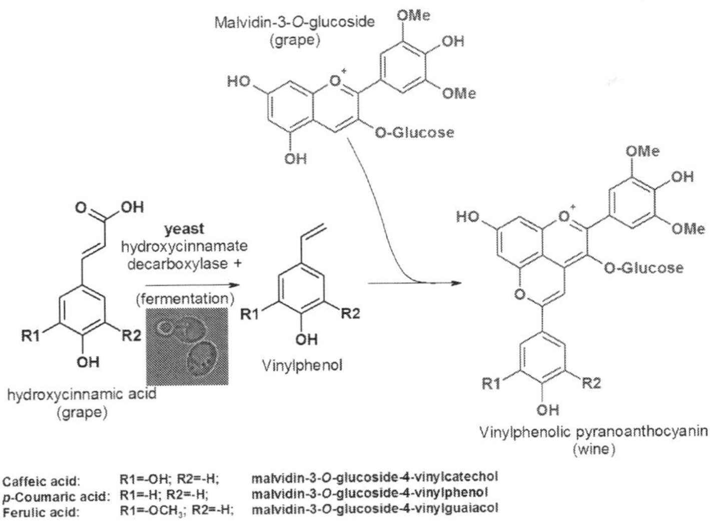

Figure 8. Enzymatic decarboxylation of grape hydroxycinnamic acids by HCDC+ yeast strains and subsequent chemical condensation with malvidin.

Compound	Molecular ion $[M]^+$ (*m/z*)	Aglycone fragment (*m/z*)
Malvidin-3-*O*-glucoside-4-vinylcatechol	625	463
Malvidin-3-*O*-glucoside-4-vinylguaiacol	639	477
Petunidin-3-*O*-glucoside-4-vinylphenol	595	433
Malvidin-3-*O*-glucoside-4-vinylcatechol	625	463
Malvidin-3-*O*-glucoside-4-vinylphenol	609	447
Malvidin-3-*O*-(6-acetyl)-glucoside-4-vinylphenol	651	447
Malvidin-3-*O*-(6-p-coumaroyl)-glucoside-4-vinylphenol	755	447
Malvidin-3-*O*-glucoside-4-vinylcatechol	625	463
Malvidin-3-*O*-glucoside-4-vinylphenol	609	447
Malvidin-3-*O*-glucoside-4-vinylguaiacol	639	477
Malvidin-3-*O*-(6-acetyl)-glucoside-4-vinylguaiacol	681	477
Malvidin-3-*O*-(6-p-coumaroyl)-glucoside-4-vinylguaiacol	785	477

Table 1. Vinylphenolic pyranoanthocyanin pigments identified by HPLC/ESI-MS in musts containing extra hydroxycinnamic acids fermented by selected yeast strains.

HCAs are precursors of ethyl phenols (EPs) in wines, controversial off-smells that highly degrade the wine quality. During barrel ageing, some spoilage yeasts are able to transform grape HCAs into ethyl phenols by means of two enzymatic steps. First, an HCDC activity transform HCAs into VPhs and later a vinylphenol reductase enzyme induce the reduction of VPhs in EPs. Sensory threshold of EPs is very low, about 400 ppb of 4-ethylphenol can be perceptible in wines, although it also depends on wine polyphenolic structure. Higher concentrations can strongly depreciate the wine quality. Formation of VPAs by yeasts during fermentation is a natural way to block VPhs and, consequently, to reduce EP precursors of wines (**Figure 9**).

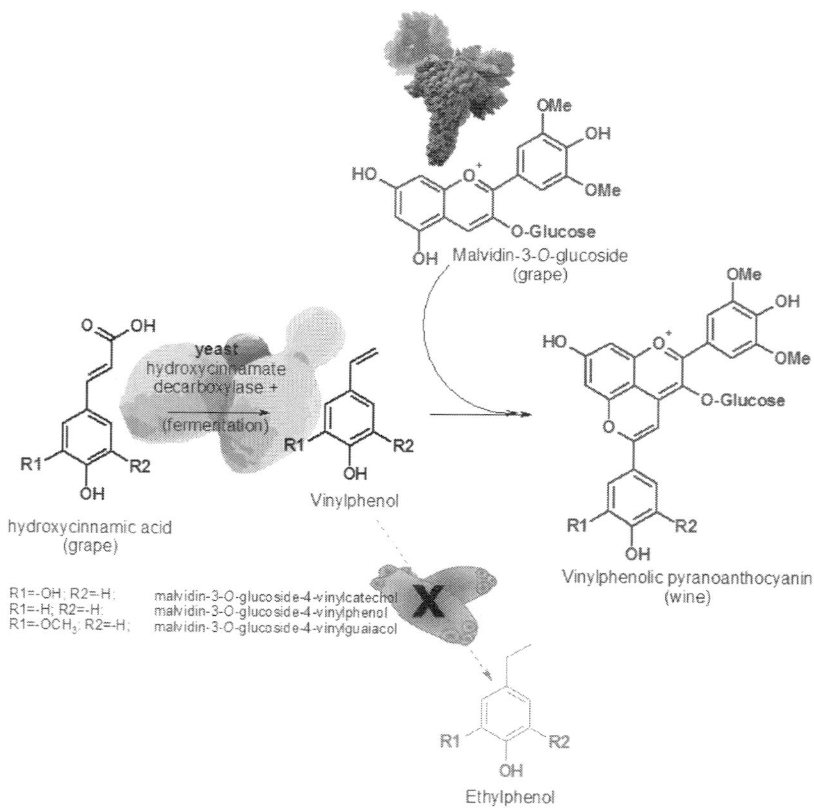

Figure 9. Formation of VPAs by a mix mechanism of enzymatic decarboxylation of HCAs by yeast and chemical condensation with wine anthocyanins.

A reduction in the amount of HCAs, correlated with the amount of VPAs that were formed, has been observed when the effect of ferment with HCDC+ *S. cerevisiae* yeast has been studied. After massive contamination with *Dekkera bruxellensis* (10^6 cfu/ml), the fermentations performed with HCDC+ *Saccharomyces* finished with 470 ppb of 4-ethylphenol and quite closed to the sensory threshold. However, the fermentations by HCDC- *Saccharomyces* reached 1100 ppb meaning near three folds the sensory threshold [15].

Other problem in the formation of ethylphenols in wines is that the levels of tartaric esters of HCAs (TE-HCAs) are frequently higher than free HCAs. These esters can release free HCAs during storage and barrel ageing increasing the amount of ethylphenol precursors. The use of cinnamyl esterases enzymes (CEs) during maceration is a way to release free HCAs. If, at the same time, fermentation is done with HCDC+ yeasts, the HCAs can be used to form VPAs reducing the precursors of EPs. This is a natural enzymatic-biologic-chemical way to decrease the precursors of ethylphenols protecting wines against ulterior contaminations by *Brettanomyces/Dekkera* yeasts (**Figure 10**) [14].

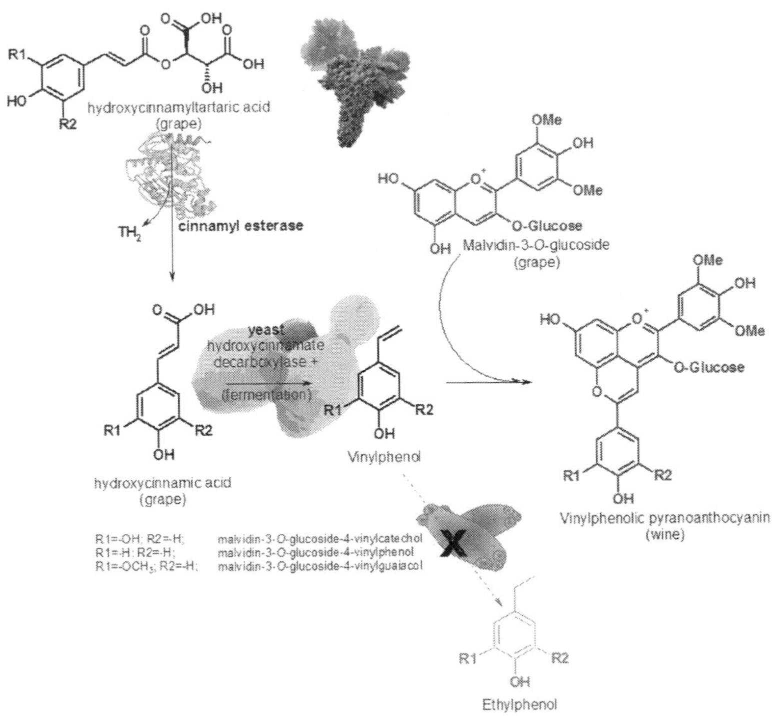

Figure 10. Mobilization of TE-HCAs to form stable VPAs decreasing the amount of 4-ethylphenol precursors during fermentation.

The simultaneous use of CEs and HCDC+ yeasts promotes the formation of stable pigments decreasing at the same time the amount of EPs precursors. We can observe the effect on scale fermentations when HCDC+ (*S. cerevisiae* strains 7VA and TP2A16) and HCDC- (*S. uvarum* strain S6U) yeasts are used in fermentations with or without CEs (**Figure 11**).

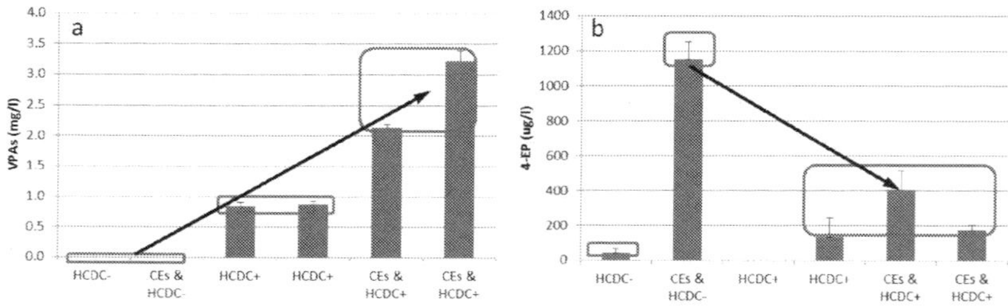

Figure 11. (a) Formation of stable pigments from grape anthocyanins (VPAs) during fermentation with and without CEs. HCDC+ yeast with hydroxycinnamate decarboxylase activity. Two *S. cerevisiae* yeasts were used 7VA and TP2A16. HCDC- yeast without hydroxycinnamate decarboxylase activity. (b) Formation of 4-EP after *Dekkera* contamination (10^6 cfu/ml) in wines from musts treated with and without CEs and fermented by yeasts HCDC- and HCDC+.

4. Formation of polymeric pigments

It is known that the wine colour evolves from red-bluish to red-orange during the ageing and this phenomenon is affected by oxygen levels and temperature (**Figure 12**). During barrel ageing, microoxygenation through the porous surface of wood promotes the browning of the wine pigments, and, at the same time, helps to modulate the aromatic profile and causes tannins smoothing. It is also known that long reductive ageing, as happens in vintage Porto wines, helps to keep red-bluish pigments and to preserve initial colour. During ageing, the colour of wine, initially due to grape anthocyanins, is being substituted by polymeric pigments; these pigments could be responsible of 50% of the colour density after the first year [1].

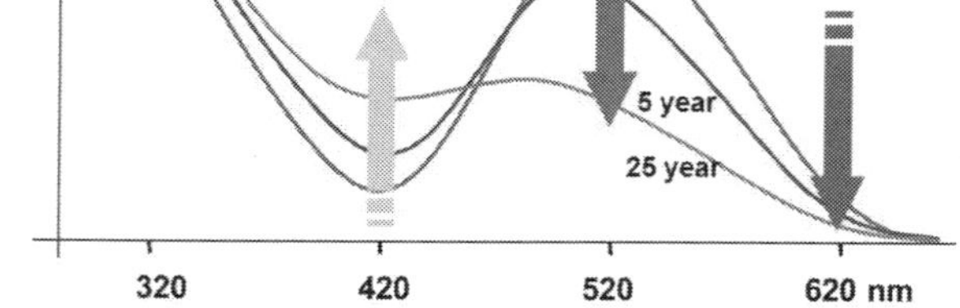

Figure 12. Typical wine colour evolution during barrel ageing.

http://dx.doi.org/10.5772/65055

Polymeric pigments are formed by more than one flavonoid unit, which means compounds with structure $n \times$ [C6-C3-C6]. In this polymer, the colour is due to the anthocyanin moiety but the non-anthocyanin fraction is affecting the colour tonality. Several structures have been proposed for oligomeric pigments found in wines in which anthocyanin can be either the beginning or the end of the chain with structures that can reach up to six flavonoid monomers, the non-pigment moiety is (epi) catechin [16]. Also oligomers formed by 2 or 3 anthocyanin units can be found in grapes [17].

Polymeric pigments show red-orange colours and higher stability against both oxidative damage and SO_2 bleaching, so they are really important in the colour of aged red wines. The analysis of these pigments can be done by LC-MS, capillary electrophoresis and gel electrophoresis. When wine anthocyanins are separated by mass/charge ratio in gel electrophoresis, the monomeric grape anthocyanins run faster being easily separated in the gel front and polymeric pigments are delayed at the end, because all wine anthocyanins have a positive charge in the pyrilium ring, but the mass increase strongly in the oligomers depending on the number of flavonoid units. A red-bluish or red colour in anthocyanin monomers and a red-orange colour in oligomeric pigments can be observed (**Figure 13**).

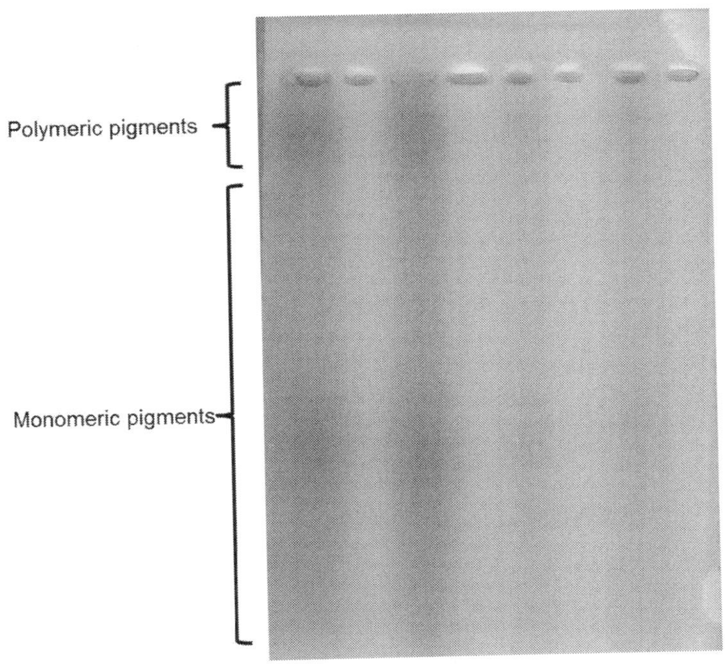

Figure 13. Gel electrophoresis of wine anthocyanins.

Capillary electrophoresis (CE) has also been used to separate polymeric anthocyanins. This technique is quite similar to gel electrophoresis, however, it improves the resolution that is possible to get in traditional gel technique. Although LC separations are usually preferred for separation of monomeric anthocyanins, its performance is worst to identify and separate polymeric pigments. However, CE is good for charged compounds such as anthocyanins that can be easily separated according to the charge/mass ratio.

Formation of polymeric pigments has been traditionally considered as a natural chemical process produced during ageing and promoted in acidic media and under oxidative conditions of barrel ageing. However, recently, the role of yeast in the formation of polymeric pigments during must fermentation has been considered. Moreover, some polymeric pigments can be formed faster because of the connection between anthocyanin and catechins or procyanidins by acetaldehyde bridges. When musts supplemented with catechin and procyanidin B2 were fermented by several selected *Saccharomyces* yeast strains, it was possible to find four dimeric pigments derived from catechin (CA1, CA2, CA3, C4) and also 1 oligomeric compound derived from Procyanidin B2 [18]. Probably, the amount and rate of acetaldehyde produced by the yeasts can affect the formation of these compounds during fermentation. These pigments show a shoulder at 460 nm characteristic of the ethyl linkage in indirect anthocyanin-flavanol condensations [19]. Pigments CA1 and CA2 were identified by their molecular ion *m/z* 809, so both are enantiomers of a dimer formed by malvidin-3-*O*-glucoside and catechin linked by an acetaldehyde bridge [20]. In the same way, CA3 and CA4 were enantiomers of petunidin-3-*O*-glucoside and catechin linked by acetaldehyde [21], and their molecular ion was *m/z* 795. Oligomeric pigment P1 was formed in fermentations that were added with procyanidin B2, its molecular ion had an *m/z* ratio of 1097 corresponding to the malvidin-3-*O*-glucoside-acetaldehyde-procyanidin B2 adduct [22]. The concentration of catechin dimers was higher than procyanidin B2 oligomers, ranging between 0.4 and 1.6 mg/l depending on the yeast strain used to ferment the must [18]. The use of specific strains of *S. cerevisiae* could be an interesting tool to promote the formation of polymeric pigment during fermentation to improve colour stability.

We are also studying the role of non-*Saccharomyces* in the formation of polymeric pigments during fermentation and, so far, it is possible to observe that some species can promote the formation of polymers better than *S. cerevisiae* [23]. Maybe, in the future, the use of mixed or sequential fermentations involving non-*Saccharomyces* yeasts and also the use of polymers precursors could be a way to increase stable pigments in wines.

5. Pigment adsorption in yeast cell walls

During fermentation, yeasts are able to adsorb the molecules in external cell wall surface. The adsorption of anthocyanins [24, 25], phenols [26, 27], aromatic compounds [28] and toxic molecules [29, 30], have been previously reported. In the exponential fermentation, yeast population range 10^8–10^9 cfu/ml, and considering the typical elliptic geometry and size of *S. cerevisiae*, it provides a specific surface of around 10 m^2/l of must [24]. So, yeast cells are able to adsorb big amounts of pigments during must fermentation.

http://dx.doi.org/10.5772/65055

Moreover, cell adsorption is a strain-dependant phenomena being possible to select yeasts with lower anthocyanin adsorption than others [24]. The adsorption of anthocyanins in cell walls is not yet well understood, but probably depends on cell wall surface structure and composition. It has been observed that the amount of each anthocyanin type molecule adsorbed on cell walls is affected by the polarity of the anthocyanin. Polarity of grape anthocyanins is affected by B ring substitution pattern (methoxylation makes anthocyanin more apolar, hydroxylation makes it more polar), and the type of acylation: none, acetylation, coumaroylation or caffeoylation in decreasing order of polarity. It has been observed that apolar anthocyanins can be more strongly adsorbed than polar ones.

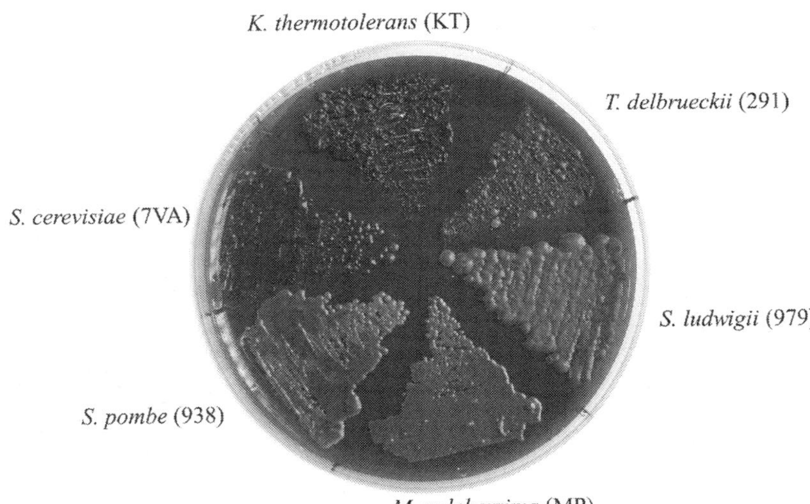

Figure 14. Several yeast species growing in YEPD-agar medium supplemented with anthocyanins.

The selection of yeasts with low anthocyanin adsorption helps to keep more anthocyanins in solution, what means wines with higher amount of anthocyanins. Of course, this will be especially interesting for those grape varieties in which production of anthocyanins is low (Pinot noir, Grenache) or in regions where the synthesis of anthocyanins is inhibited by unsuitable climatic conditions. Although global adsorption in *S. cerevisiae* strains have been evaluated in the range 1.6–5.8% of total anthocyanins (average 3.5%) [25], the range for coumaroylated anthocyanins, that are more strongly adsorbed, is 8–28%, moreover, these pigments show the red-blue colours to improving the wine tonality.

To evaluate the ability to adsorb anthocyanins by yeast, two kinds of procedures have been used. The first one, that is fast and easy to apply, is the plating in agar medium enriched in grape anthocyanins. The medium is a YEPD-agar but supplemented with a high concentration of anthocyanins extracted from grape skins [18, 31]. When yeasts colonies grow and develop in plate surface adsorb anthocyanins from the surrounding medium, and this adsorption is proportional to the affinity of their cell walls to link anthocyanins. So, more pigmented colonies are coming from strains with strong anthocyanin adsorption (**Figure 14**). This technique allows to perform a fast screening for selecting either yeast strains or species with low anthocyanin adsorption.

The second procedure to evaluate anthocyanins adsorption by yeasts, that is more precise but more difficult and tedious to apply, is the recovery of anthocyanins adsorbed from cell walls and the characterization and quantitation of them by LC-DAD or LC-ESI/MS [24, 25]. The procedure requires separating all the lees from fermentation and it is possible to do it when red wines are fermented without skins maceration. If winemaking is done with skin maceration, what is the usual industrial process, which is very difficult to separate the yeast lees? But, it is easy to produce a red must with enough colour and tannins to make fermentation in absence of skins by using accelerated maceration, for example: heating and pressing the grapes, freezing or using ultrasounds. Thus, the lees can be separated from the wine by centrifugation at the end of fermentation.

Later, the lees can be washed with water or water-ethanol (88/12, v/v) solutions to remove anthocyanins that are not strongly adsorbed in cell wall but only in suspension in the surrounding medium among cells. It is not easy to evaluate what is the degree of extraction in this preliminary clean up and maybe some of the removed anthocyanins can be partially/weakly retained on cell walls. The following step, that is particularly delicate, is the separation of anthocyanins from cell walls. Yeast cell wall is a thick layer that externally covers the cell, and is involved in relationship function and giving resistance to osmotic pressure. It is formed by globular mannoproteins sustained by a net of fibrillar polysaccharides mainly formed by β-glucans and chitin [32, 33]. Anthocyanins might be forming links by polar interactions with these cell wall components, but the nature of this process is not yet well described.

Separation of anthocyanins from cell walls must be done by using solvents. The use of formic acid-methanol mixtures has been described [25]. Several washings are necessary to remove most of the adsorbed compounds and some anthocyanins can still remain linked after extraction. The detachment process can be facilitated by applying energy in form of shaking, ultrasounds or temperature. After each washing, supernatant is recovered by centrifugation at 3000 × *g*. Supernatants are collected together, concentrated under vacuum and analysed by LC-DAD to identify and quantify anthocyanins.

6. Biological ageing: ageing on lees

Yeasts are used not only to ferment musts but also in the ageing process of wines. Many traditional wines as Sherry-flor wines, natural sparkling wines and barrel fermented and aged Chardonnays improve their quality after a long period together with the yeast lees produced during fermentation. Along the biological ageing, many cell metabolites and structural components are released into the wine improving sensory quality. Also, the ageing on lees (AOL) technique can be used during barrel maturation of red wines [34]. AOL has a reductive role because yeast lees are oxygen consumers and, moreover, some cell wall constituents such as glutathione (GSH) are antioxidant compounds with a protective effect on aromatic compounds and anthocyanins.

The simultaneous use of barrel ageing and AOL reduce the oxidative degree partially preserving anthocyanins from oxidative degradations [34]. Yeast selection is also a tool to get better strains to improve the wine quality and to protect pigments during AOL [35]. The use of non-*Saccharomyces* yeasts can also be a tool to improve the performance of AOL regarding the release of yeast polysaccharides and colour stabilization [36]. Osmophilic yeasts, able to grow in media with high sugar concentration, frequently show a thick cell wall that releases polysaccharides faster during ageing on lees. Some species like *S. pombe* have a cell wall configured in a double-layer structure with higher thickness than *S. cerevisiae* and some specific polymers.

7. Future trends

Probably, the future of red winemaking will be the separation of maceration and fermentation by means of fast macerations (minutes-hours) using new technologies such as high hydrostatic pressures [37], pulsed electric fields [38], irradiation [39] and ultrasounds, among others, to ensure enough amounts of anthocyanins and tannins in the must. In this situation, the fermentation will be produced in absence of skins and seeds and at low temperature to preserve sensory quality. The use of selected strains of *Saccharomyces* and, especially, new non-*Saccharomyces* will be a complementary biotechnology to improve wine quality and wine colour by promoting the formation of both pyranoanthocyanins and polymeric pigments, and also reducing the adsorption of anthocyanins in cell walls.

Acknowledgements

This work was funded by the Spanish Ministry of Economía y Competitividad, Project AGL2013-40503-R.

Author details

Morata Antonio*, Loira Iris and Suárez Lepe Jose Antonio

*Address all correspondence to: antonio.morata@upm.es

Department of Chemistry and Food Technology, Technical University of Madrid, Madrid, Spain

Chapter 14

New Trends in *Schizosaccharomyces* Use for Winemaking

Ángel Benito, Fernando Calderón and
Santiago Benito

Additional information is available at the end of the chapter

http://dx.doi.org/10.5772/64807

Abstract

Several researchers are studying the winemaking potential of non-*Saccharomyces* yeast strains in order to improve wine quality. For that purpose, yeast species such as *Torulaspora delbrueckii, Lachancea thermotolerans, Metschnikowia pulcherrima, Candida zemplinina, Kloeckera apiculata, Hansenula anomala* and *Pichia guilliermondii* were studied in the past. Yeasts from the genus *Schizosaccharomyces* have been traditionally studied from a winemaking point of view due to its rapid malic acid deacidification, by converting malic acid to ethanol and CO_2. Nevertheless, during the last 5 years, it has been discovered that *Schizosaccharomyces pombe* possesses several remarkable metabolic properties different from its traditional malic acid deacidification that may be useful in modern quality winemaking, including a malic dehydrogenase activity, high autolytic polysaccharides release, ability of gluconic acid reduction, urease activity in order to avoid ethyl carbamate formation, elevated production of pyruvic acid related to colour improvement, and low production of biogenic amines.

Keywords: *Schizosaccharomyces*, malic acid, pyruvic acid, ethyl carbamate, biogenic amines

1. Introduction

In modern traditional winemaking, *Saccharomyces cerevisiae* has been considered as the main species used in the production of quality wines. The incidences of non-selected *Saccharomyces* or non-*Saccharomyces* opportunistic yeasts during fermentations were usually related to off-flavours such as high levels of acetic acid, ethyl phenols and great levels of higher alcohols. On the other hand, at present, scientists and winemakers have started to believe in the helpful effect of some non-*Saccharomyces* in winemaking in matters such as aroma complexity [1–6].

The main problem about using non-*Saccharomyces* in oenology is their inefficiency to finish alcoholic fermentation in a proper way. So, most of the time, it is required the combined use of *Saccharomyces cerevisiae* strains during alcoholic fermentations in order to ensure a proper fermentation end without any residual sugar at industrial levels. The production of remarkable metabolites by non-*Saccharomyces* in higher amounts than *S. cerevisiae* such as glycerol, pyruvic acid and mannoproteins has awakened especial interest during the last few years [3, 7]. The better performance of enzymatic activities by non-*Saccharomyces* such as the type glycosidase or β-lyase is a relatively new issue in modern oenology. The use on non-*Saccharomyces* also looks to be the only microbiology way to get wines with lower alcohol content in warm areas.

Some studies have analysed the use and influence of different non-*Saccharomyces* species in wine quality. Some of this yeast species are *Kloeckera apiculata, Hanseniaspora uvarum, Hanseniaspora vineae, Torulospora delbrueckii, Metschnikowia pulcherrima, Starmerella bacillaris, Zygosaccharomyces bailii, Pichia guilliermondii, Schizosaccharomyces pombe, Lachancea thermotolerans* and *Hansenula anomala* [1]. **Table 1** summarizes the main quality improvements about using these different yeast species in winemaking.

Yeast species	Oenological interest
Kloeckera apiculata	Aroma complexity
Hanseniaspora vineae	Aroma complexity, high 2-phenyl-ethyl acetate production ,biogenic amines reduction
Torulospora delbrueckii	Aroma complexity, acetic acid reduction.
Metschnikowia pulcherrima	Aroma complexity, increase in esters, terpenes and thiols.
Starmerella bacillaris	
Zygosaccharomyces bailii	Polysaccharides increase
Pichia guilliermondii	Formation of high stability colour compounds
Pichia kluyveri	Aroma complexity, increase in varietal thiols and esters.
Lachancea thermotolerans	Acidification, L-lactic acid production
Hansenula anomala	Decreased of C6 alcohols
Schizosaccharomyces pombe	Deacidification, L-Malic acid consumption
Candida stellata	High glycerol production

Table 1. Summary of oenological interest of some non-*Saccharomyces* species.

The chance to modify the flavour and elegance of fermented beverages through different fermentation methodologies is increasing the awareness in researching most imaginable blends of non-*Saccharomyces* and *Saccharomyces* [8]. Regarding this matter, most scientific trials performed fermentations with non-*Saccharomyces* strains by their own, with mixed fermentations (synchronized) and sequential inoculation, comparing them against an alcoholic fermentation performed by *S. cerevisiae* by itself. Most studies testimony sequential inoculation as the finest option in winemaking.

Among non-*Saccharomyces* yeast genera, *Schizosaccharomyces* has been traditionally used to reduce acidity in wines presenting high levels of malic acid. This fact is related to its unique ability to transform L-malic acid into ethanol [9–11]. Nevertheless, novel uses of these *Schizosaccharomyces* species related to different abilities not so well studied until the last few years have been developed to increase wine quality and food safety [12–14]. **Figure 1** summarizes these new uses. One of this novel uses is the performance of specific *Schizosaccharomyces* mutants to decrease the original content of gluconic acid from rotten grape juices [15]. Other modern use is its application in ageing over lees, thanks to their superior polysaccharide release [16]. *S. pombe* metabolism also offers a method of increasing the pyranoanthocyanin content in red wines [12]. *Schizosaccharomyces* is also of great interest in food safety. The urease activity of *Schizosaccharomyces* reduces high ethyl carbamate content in wine by reducing the concentration of urea (main precursor of ethyl carbamate) [13, 14, 17]. *Schizosaccharomyces* can also reduce biogenic amines contents avoiding the classical malolactic fermentation performed by lactic bacteria.

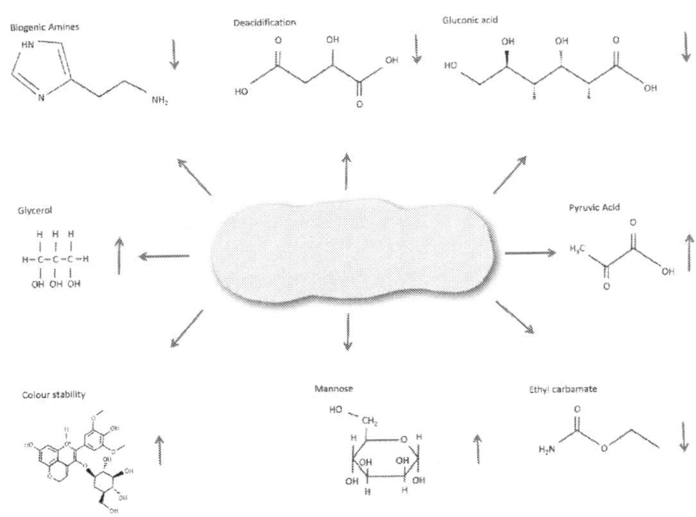

Figure 1. Summary of the new uses of *Schizosaccharomyces*.

The use of the genus *Schizosaccharomyces* in winemaking was approved by the International Organisation of Vine and Wine (Resolution OENO/MICRO/97/75/phase 7). However, *Schizosaccharomyces* was not commonly used due to specific off-flavours associated with the metabolism of non-selected wild strains of this genus [12]. Indeed, *Schizosaccharomyces* has been described to be isolated from wines showing strong organoleptic and chemical deviations such as high levels of acetic acid, sulfidric acid, acetaldehyde, acetoin and ethyl acetate [12]. Due to the enormous variability in the genetic composition of any species such as *S. pombe* [18], recent selection processes have been performed in order to obtain proper strains for winemaking purposes [19, 20]. The last studies regarding *Schizosaccharomyces* genus have demonstrated that it is possible to produce quality wines through the combination of wild *Schizosaccharomyces* strains with selected *Saccharomyces* strains or more recently through the use of selected *Schizosaccharomyces* strains that are able to perform by themselves a complete proper fermentation process, especially under very acidic conditions. This study aims to show the main new potential of this genus in modern winemaking.

2. Physiology, morphology and taxonomy of *Schizosaccharomyces* genus

In the past, Lodder and Kreger van Rij documented four species belonging to *Schizosaccharomyces*: *Schizosaccharomyces pombe* Lindner (1883), *Schizosaccharomyces octosporus Beijerinck* (1894), *Schizosaccharomyces japonicus* var. versatilis Wickerhan and Duprat (1945) and *Schizosaccharomyces malidevorans* Rankine and Fornachon (1964) [12]. Nowadays, it is believed that the genus *Schizosaccharomyces* is a compound of three species: *S. pombe*, *S. octosporus* and *S. japonicus*. They have been mainly classified according to the principle that involves the number of spores per ascus and their ability to ferment Sucrose and Raffinose [12] (**Table 2**). Most of the time, their presence is related to hot climate regions.

	Fermentation		Assimilation	
	Sucrose	Raffinose	Sucrose	Raffinose
S. pombe	+	+	+	+
S. japonicus	+	+	+	+
S. octosporus	+	+	+	+

Table 2. Summary of fermentation and assimilation properties of species from *Schizosaccharomyces* genus.

The species *S. pombe* is usually long rectangular cells of about 2–4 × 5–18 μm (**Figure 2**). They commonly appear as single cells or in pair groups. *S. pombe* is a sporulating species. It can reproduce asexually by binary fission (**Figure 2**) when it forms a septum in the midpoint of the cell. *S. pombe* is not able to assimilate nitrates, it does not have α-glucosidase activity and it cannot breaking down arbutin by enzymatic activity. The species possesses urease positive activity. It reacts with diazonium blue that makes it possible to distinguish it from other basidiomycetous (**Figure 3**). It possesses a high fermentative power, producing 11°–13° of alcohol in anaerobiosis and 14–15.5° with slight aeration.

S. pombe can metabolize malic acid and to convert it into ethanol and CO_2. In the past, a strain of *S. Pombe* was denominated—*Schizosaccharomyces acidovorans* (acidodevoratus)—by Chalenko (1941) [12] due to its special ability to eliminate most of the malic acid from growing media, nowadays this ability is highly strain-dependent [19, 20].

Schizosaccharomyces genus owns a cell structure known as *Schizosaccharomyces*-type that is very particular among ascomycetes. It is richer in polysaccharides and α-galactomannose than any other known yeast species [16, 21].

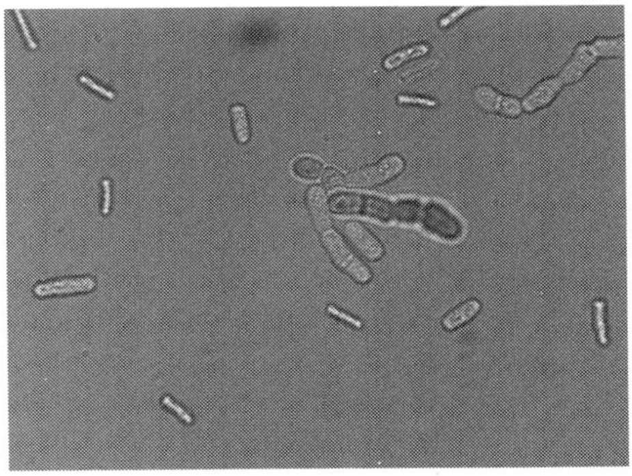

Figure 2. Details of *Schizosaccharomyces pombe*.

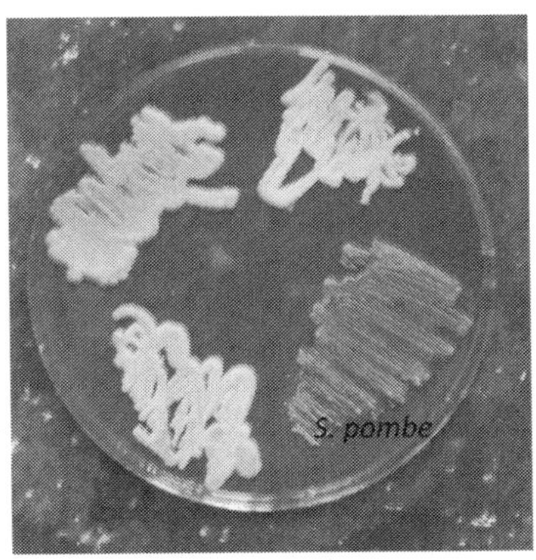

Figure 3. Details of *Schizosaccharomyces pombe* reaction with diazonium blue that makes it possible to distinguish it from other yeast specie.

3. Physiological and biochemical properties of *Schizosaccharomyces* genus

Due to its peculiar fission cell division (**Figure 2**), *Schizosaccharomyces* genus is considered as a model organism to study this phenomenon for molecular and genetic microorganism cycle studies [18]. On the other hand, until the last decade, just little information had been published related to factors that influence its growth, survival and biochemical activities of these microorganisms in industry processes. Most of the few studies available are focus on *S. pombe*. During the last few years, several studies have been performed, especially for wine-making industry, but further studies are needed.

This genus is facultative anaerobic and able to metabolize hexose sugars such as glucose and fructose or disaccharides such as sucrose. They mainly ferment by means of glycolytic pathway producing ethanol and carbon dioxide as main products; several secondary metabolites are also produced. *Schizosaccharomyces* was occasionally described as a higher producer of hydrogen sulphide when it was compared to *S. cerevisiae* [12]. This genus has also been described as a high glycerol producer [22].

Schizosaccharomyces genus is notable known among yeast genera due to its high capacity to metabolize malic acid into ethanol during anaerobic fermentation processes. A NAD-dependent malic enzyme decarboxylates malate to pyruvate. Later, pyruvate is decarboxylated to ethanol that is finally reduced to ethanol (**Figure 4**). A proton-dicarboxylate symport was proved for the transportation of L-malic acid into *S. pombe,* and the presence of glucose is required for malic acid metabolism.

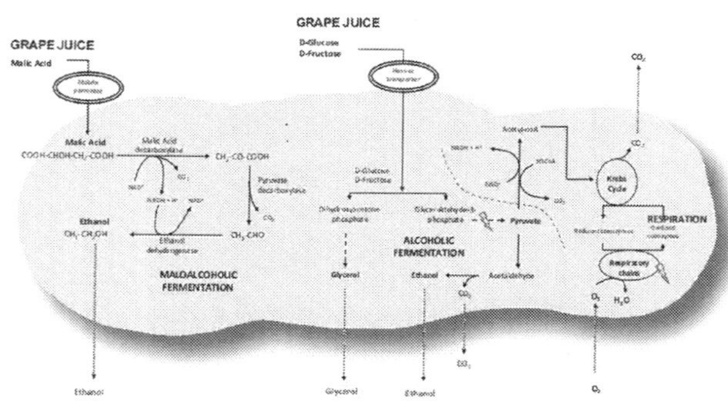

Figure 4. Summary of main metabolic routes performed by *Schizosaccharomyces pombe.*

During the last few years, it was suggested that extracellular amylases, pectolytic enzymes and proteases were not produced by *Schizosaccharomyces* spp., but some new studies start to show that those activities could be strain-dependent. Nevertheless, *S. pombe* has been used to degrade starch with plasmids carrying the glucoamylase gene of *Sacch. diastaticus. Schizosaccharomyces octosporus* produces an extracellular lipase that can hydrolyse lard to produce significant quantities of stearic acid, but the lipid degrading ability of other species of *Schizosaccharomyces* is not known.

Schizosaccharomyces pombe has been reported as being able to develop at higher temperatures than *S. cerevisiae,* up to 35°C, data from other *Schizosaccharomyces* species do not have been reported [12]. Other data indicate that *S. japonicus* is skilful of growing at 37°C, but earlier literature reports the growth of *S. pombe* and *S. octosporus* at this temperature. Nevertheless, the optimum fermentation temperatures are reported to be between 24 and 30°C [23], further studies are needed to determinate the minimum and maximum performing temperatures of this genus. The pH influence is not very well known. It grows properly in pH close to 7, but it is usually isolated from grape juices with pH 2.9–3.1, probably due to the inhibition of other competitor microorganisms that cannot develop as so low pH such as *S. cerevisiae*.

The *Schizosaccharomyces* genus shows a special ability to develop in food media of high sugar content and osmotic pressure [24]. Some authors have reported it as osmotolerant yeasts, capable of growth in the presence of 50% glucose (and possibly 60% glucose) at water activity (aw) values as low as 0.78 [24]. This ability has been described as species dependent for *S. pombe* and *S. octosporus*, minimum a_w values of 0.89–0.90 with glycerol, glucose and fructose are described as stressing levels; in the case of *S. japonicus*, those levels are higher up to 0.92–0.94. Conversely, the genus is less resistant to high salt concentrations [19] and does not develop at a_w levels less than 0.95 of this solute. Most *S. pombe* strains are incapable of growing in the presence of 3% NaCl, pH 5.5. Growth in low a_w environments is accompanied by the production of intracellular glycerol as a compatible solute.

Schizosaccharomyces genus has been widely described to be higher tolerant to several stabilizers than other microorganisms such as *S. cerevisiae* or *Dekkera bruxellensis* [24]. Some of those preservatives are acetic acid, sulphur dioxide, benzoate or sorbate that are normally used during food processing (**Table 3**). This genus looks to be notably higher tolerant to these preservers than *S. cerevisiae*. Opposition to inactivation by heat treating was studied for *S. pombe*. About 99% of the population suspended in phosphate buffer, pH 6.5, containing 48% sucrose (a_w 0.95), was destroyed at 65°C in 3 min (D65 1.99 min). Quicker death proportions were achieved when sucrose was absent from the buffer. Nevertheless, higher thermotolerance is achieved by yeast pre-exposition to minor heat (40°C). This phenomenon makes the production of intracellular trehalose that is used as a thermoprotectant agent.

S. pombe	**Temperature**	**37°C**
	Acetic acid	1% v/v
	SO_2	120 mg/kg
	Benzoate	>600 mg/L
	Sorbate	>600 mg/L
	Actidione	>100 mg/L

Table 3. Summary of several resistance factors of *S. pombe*.

4. *Schizosaccharomyces* strain isolation

Schizosaccharomyces genus strains have been occasionally isolated from fermented drinks and similar products such as wine, must, grapes and beer. However, most of the isolates related to genus *Schizosaccharomyces* have been reported in foods containing high sugar levels, such as dried fruit, sweets, molasses and honey [20, 25]. Indeed, no yeast species belonging to *Schizosaccharomyces* genus are included in the 20 Food-Borne Yeasts most frequently described [24, 25] (**Figure 5**). This lack of yeast strains from this genus in nature has avoided the obtaining of commercial strains with proper industrial abilities and free of collateral effects. On the other hand, there is an especial interest from oenological industry yeast manufacturers to get strains able to perform properly at industry level, as a result of the OIV's approval of "Deacidification by Schizosaccharomyces" as an authorized/recommended practice (Resolution OENO/MICRO/97/75/Stage 7). However, at this moment, there are no any commercial strains selected after performing an appropriate selection process, due to their low presence in grape juices and the absence of an adequate method for isolating strains from this complicate genus.

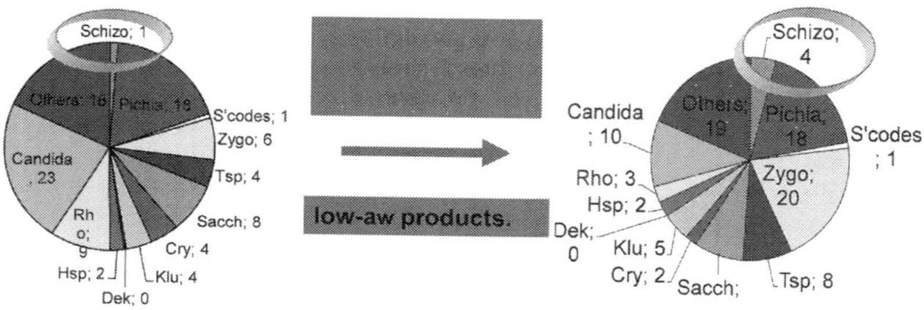

Figure 5. Estimated frequencies (%) of yeast species in fruit and beverages and in high sugar concentrated products. Estimated from [18].

In the past, some authors suggested the use of culture media rich in tryptone glucose yeast extract agar combined with some antibacterial antibiotics, such as streptomycin gentamicin, oxytetracycline and chloramphenicol. It was also suggested the use of sugar and acetic acid in high concentrations as selective agents. Lysine as a selective source of nitrogen was suggested as a way to inhibit the growth of the main yeast species competitor *S. cerevisiae*. Our personal experience showed us that, in spite of using these culture media, there are too many false positives produced by competitor microorganisms; this fact makes impossible to isolate an elevated number of *Schizosaccharomyces* strains. So, in spite of the fact that most important yeasts posse an specific selective-differential method, no specific culture medium has been described until the last few years for isolating yeasts of the genus *Schizosaccharomyces* [24], this media has appeared as a consequence to the great interest an demand that these species have awakened in oenological industry.

A novel and specific isolation method for *Schizosaccharomyces* is nowadays in process of patenting. It has been developed and optimized during the last few years in order to isolate and to select strains of *S. pombe* (**Figure 6**). This new method uses a differential selective medium that contains selective factors such as actidione antibiotic. This antibiotic has been described before in most differential selective media regarding the genera *Brettanomyces/Dekkera*. Among the reported false positives for those media, the genus *Schizosaccharomyces* appeared on some occasions.

http://dx.doi.org/10.5772/64807

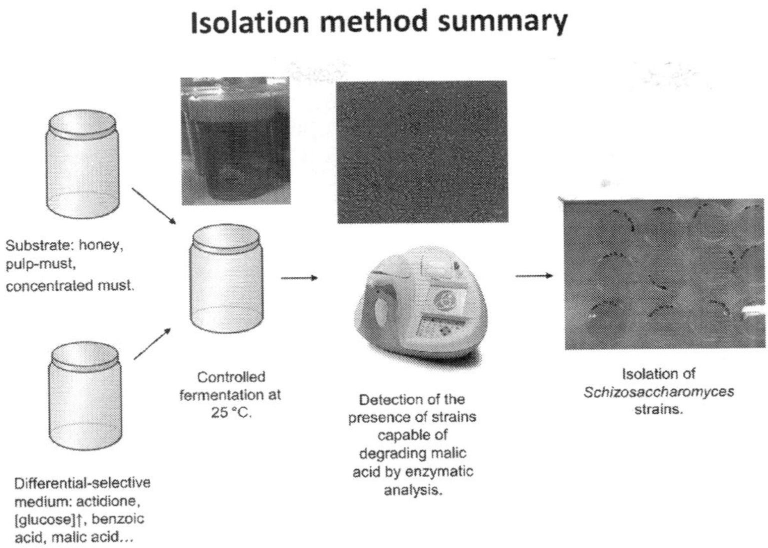

Figure 6. Summary of the specific isolation method developed to isolate *Schizosaccharomyces* strains.

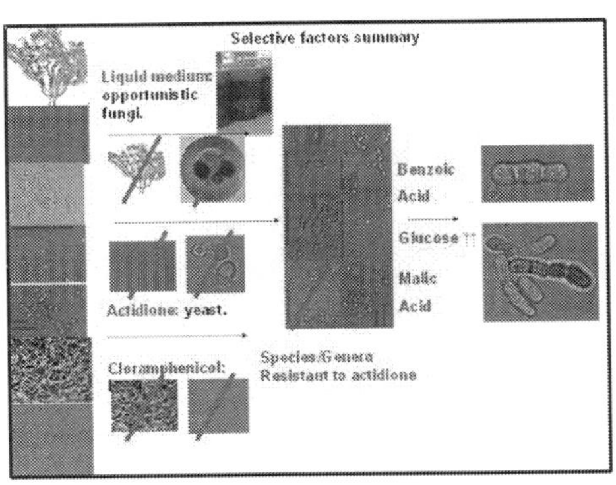

Figure 7. Summary of the main points regarding *Schizosaccharomyces pombe* selective-differential media.

Even though other false positive remain in the described method due to their resistance to actidione (**Figure 7**). Those can also be avoided throw the use of other inhibitor agents such as benzoic acid, acetic acid or high sugar concentration (**Table 3**). To improve the differentiation process, malic acid is commonly used as it makes possible to identify the presence of micro-organism able to degrade it throw pH control or enzymatic analyse. This methodology (**Figure 6**) has allowed generating *Schizosaccharomyces* collections of hundreds of different strains [12].

5. *Schizosaccharomyces* selections for winemaking

The isolation methodology explained above has allowed obtaining representative universes of *Schizosaccharomyces* genus that made it possible to perform basic selection processes to get strains with especial aptitude for winemaking [20]. It has been observed that just a small percentage up to 5% of strains could perform a proper alcoholic fermentation process without collateral effects.

Some of the basic parameters studied in this initial selection processes have been correct sugar consumption, moderate acetic acid production, complete malic acid degradation, glycerol production and the correct sensory profile of the wines produced with these strains.

5.1. Sugar consumption

Most recent studies report that *S.pombe* ferments great quantities of sugars during alcoholic fermentation up to 240 g/L. However, most studies report a slower kinetic metabolism than that described for *S. cerevisiae* [12]. Some authors have reported differences up to 2–4 days to complete an alcoholic fermentation process when duration was compared against a *S. cerevisiae* control [9, 19, 22]. Nevertheless, we must consider that the second fermentation performed by lactic bacteria in red wines is not needed in the case of wines made by *Schizosaccharomyces* yeasts [7]. This process usually takes long time than the alcoholic fermentation by yeasts and the risk of deviations is higher.

5.2. Malic acid consumption

Malic acid consumption has been reported in most studies regarding *S. pombe* to be completed in most cases. Nevertheless, great differences regarding different kinetics depending on the strains have been reported [19, 20]. On some occasion, especially in very acidic musts with malic acids contents over 6 g/L [9] from northern regions, the deacidified wines were preferred by the testing panels due to the excessive acidity described for the controls performed by regular *S. cerevisiae* without malate dehydrogenase activity. Increments of about 0.4 in pH were produced after malic acid consumption by *S. pombe* [7, 9].

5.3. Acetic acid production

Acetic acid is the factor that has showed the biggest variety among the studied strains in most studies [19, 20]. On some occasions, values over 1 g/L have been reported [9]. These values are not compatible with quality wines. Nevertheless, in other studies, moderate levels have been reported. According to the last studies, we can report that it is possible to select *S. pombe* strains that produce wines with as low content of acetic acid as regular wines performed by *S. cerevisiae* if they are properly selected [19, 20]. Another option to reduce this possible collateral effect was to combine the use of *S. pombe* with other yeast species that produce lower levels of acetic acid such as *L. thermotolerans* [7].

5.4. Urease activity/ethyl carbamate reduction

Schizosaccharomyces genus has been described among the few yeast species that can develop urease activity [17, 19, 25]. This enzymatic activity was observed in several fermentation trials [4, 10, 12–14] were *Schizosaccharomyces* fermentations always reported final urea values after fermentation of about 0 mg/L. However, controls regarding *S. cerevisiae* reported notably higher values up to 3 mg/L. The enzymatic activity in winemaking could reduce the initial level of the main precursor (urea) for ethyl carbamate (one of the most toxic compounds reported in wine) formation. Nowadays, ethyl carbamate is a main problem for human health as it is considered a powerful carcinogen with an especial incidence in fermented beverages [19, 27]. It is also a very important problem regarding to wine exportations as several countries have already set specific limits for this toxicological compound that varied from 10 to 30 μm.

5.5. Pyruvic acid

Fermentations performed by *Schizosaccharomyces* always were reported as higher producers of pyruvic acid than *S. cerevisiae*. Nevertheless, important differences have been observed depending on the *Schizosaccharomyces* strains [12, 13]. Maximum values during the first days of alcoholic fermentation up to about 0.5 g/L have been achieved by *Schizosaccharomyces* fermentations while maximum values up to about 0.1 g/L have been reported for specific *S. cerevisiae* strains selected according to this criterion [6]. The oenological meaning about producing high levels of pyruvic acid is related to the strong correlation between the amount of pyruvic acid released during alcoholic fermentation by yeasts and the formation of vitisin A [15, 16]. Vitisin A is known as a very stable coloured pigment that directly influences wine colour quality and stability. This parameter is considered nowadays an important criterion in red wine yeast selection processes. Until now, *Schizosaccharomyces* genus is the highest producer yeast of pyruvic acid in winemaking.

5.6. Glycerol

One of the first experiments involving *Schizosaccharomyces* [7] indicated that *Schizosaccharomyces pombe* possesses a highly developed glyceropyruvic pathway compared to other yeast species. This fact explains also the greater production of pyruvic acid explained before. Modern trials have reported glycerol productions up to 10 g/L [13] and values higher than 1 g/L when they were compared to *S. cerevisiae* controls [4, 6]. Increased glycerol content is described as one of the main contributions of some non-*Saccharomyces* strains in winemaking [5] because it directly influences positively to the mouth-feel. Even though other yeast species such as *Candida stellata* have been described as higher producer of glycerol [5], the use of *Schizosaccharomyces* could be interesting in order to improve this quality parameter.

5.7. Ethanol

Many winegrowing areas observe an increase in the alcohol content of their wines, on some occasions to more than 14% by volume. This phenomenon may become increasingly common due to the effects of climate change. Several practices are proposed to decrease the ethanol levels in fermented beverages either completely or partially, for instance the use of great temperatures to drive off the ethanol, chemical extraction, cryoconcentration, filtration using semipermeable membranes and supercritical fluids extraction [26]. Some authors have reported that some non-*Saccharomyces* types of yeast produce lower ethanol levels than *Saccharomyces*. *Schizosaccharomyces* has been described in some occasions as lower producer of ethanol than *S. cerevisiae* in amounts of about 0.4-0.2% vol [2, 3, 7, 9]. However, other authors have reported no significant differences when compared to *S. cerevisise* strains selected for its high-developed glyceropyruvic pathway [19]. The sugar metabolism can be used to produce different compounds other than ethanol, such as glycerol or pyruvic acid, or to increase the yeast biomass [12]. Other authors observed lower final ethanol levels using other non-*Saccharomyces* species under very specific settings related to high aeration conditions [2]. In those cases, reductions higher than 1 or 2% vol in ethanol can be achieved. This reduction is higher and more efficient than those described for *Schizosaccharomyces*, so *S.pombe* could be interesting when the needed reduction in ethanol is about 0.5% vol in ethanol.

5.8. Biogenic amines

Biogenic amines are other toxicological compounds that can appear in wine. Several authors have described harmful effects in human beings produced by biogenic amines [7, 17]. For these reasons, this topic is considered a serious matter in food safety that must be considered. Several countries have established legal limits. A histamine value of 2 mg/L is considered the most restrictive level in some countries. Several trials performed by *Schizosaccharomyces* show that *S. pombe* does not produce higher levels of biogenic amines than *S. cerevisiae* [7, 17]. Reduction in biogenic amines that come from spoilage grapes such as cadaverine have been reported to decrease for *Schizosaccharomyces* in quantities up to a few mg/L. Similar processes have been described before for other yeast species [7]. However, the main use of *Schizosaccharomyces* about reducing/avoiding biogenic amine formation is based on the fact that most biogenic amines are produced during wine ageing and especially during malolactic fermentation [7, 17], as they are compounds produced primarily by lactic acid bacteria. Thus,wines fermented by *Schizosaccharomyces* do not need malic acid consumption by lactic bacteria any more.This fact notably decreases the risk of biogenic amines formation [17].

5.9. Volatile aroma

Schizosaccharomyces strains were not used in the past because of specific off-flavours commonly associated with the metabolism of non-selected strains. In the past, *Schizosaccharomyces* was commonly isolated from wines suffering from organoleptic and chemical faults through the appearance of sulfidric acid, acetic acid, acetaldehyde, acetoin and ethyl acetate [12]. First, control fermentations performed by mixed non-selected *Schizosaccharomyces* strains combined with *S. cerevisiae* fermentations reported higher concentrations of acetaldehyde, propanol and 2,3-butanediol up to several mg/L [9]. Nevertheless, the last fermentations performed by selected *Schizosaccharomyces* strains show lower levels in higher alcohols than the non-selected *Schizosaccharomyces* and the *S. cerevisiae* controls [7, 18]. In those studies, the tested *S. pombe* strains produced also less esters than the *S. cerevisiae* strains. Similar effects have been reported for other non-*Saccharomyces* [1]. This finding could be of interest in facilitating the making of wines with varietal character for specific grape varieties or to increase wine complexity avoiding the influence of higher alcohols or esters. No differences between selected *S. pombe* and *S. cerevisiae* strains have been observed with respect to compounds considered negative, such as ethyl acetate and diacetyl when selected strains have been employed [19, 20]. Nevertheless, compounds such as acetaldehyde and ethyl acetate show significant differences depending on the different *S. pombe* strains [20].

5.10. Gluconic acid

Rotten grape musts contain high concentrations of gluconic acid formed by fungal attacks and acetic bacteria from rotten grapes. This fact drastically reduces the quality of wines made from those grapes. The sensory properties of wines are considerably altered by the presence of gluconic acid, which decreases the wine's microbiological stability and raises long-term storage problems that can be solved only by reducing its concentration in the wine. Specific *S. pombe* strains have been used to remove gluconic acid in 50% rates from wines up to 2.5 g/L obtaining in the end better final wines. This fact produced a good influence in volatile compounds spoilage by gluconic acid influence [15].

5.11. Polysaccharides

The methodology of ageing over lees is nowadays considered important in the making of red wines because it has been probed that it produces high quality wines with peculiar identity. It allows winemakers to produce new different wines in a market that shows great homogeneity. This methodology, however, demands appreciable investment in resources and is not free of problems. Many research groups are now working on how to minimize these difficulties, and on how to obtain balanced products quicker and simpler.

Schizosaccharomyces pombe is yeast species whose cell wall has a particular structure and composition owed to the presence of polysaccharides and sugar derivatives that are unusual within the family Saccharomycetaceae. The main difference between *Saccharomyces cerevisiae* and *Schizosaccharomyces pombe* is the possession of a-galactomannose rather than mannose, along with the presence of α-(1=>3) glucan. The use of *Schizosaccharomyces pombe* in over-lees ageing has reported faster release kinetics and increases of about 100 mg/L in polysaccharides pullulans than *S. cerevisiae* in 142 days [16].

5.12. Sensory impact

Wines developed by *Schizosaccharomyces* usually show big differences in the sensory perception of acidity when they are compared to *S. cerevisiae* controls (**Figure 8**) [7, 9, 19]. This fact is related with the malic acid consumption and pH increased explained above. In some occasions, wines fermented by *Schizosaccharomyces* have been described as sweeter than those fermented by *S. cerevisiae* in spite of the fact that all wines did not contain any residual sugar [9, 19]. This phenomenon is explained by the new balance generated between acidity, sweetness, bitterness and salty perception when acidity is highly reduced. Severe faults have been reported in trials involving non-selected *S. pombe* strains regarding to high acetic acid, reduction and sulfidric acid characters [19, 22]. Nevertheless, in modern fermentation trials, some selected *S. pombe* have received the best scores when compared against non-selected *Schizosaccharomyces* strains and *S. cerevisiae* strains when fermentations took place in high acidic musts [19, 22]. The preference commonly has been related to excessive high acidity for the tasters due to high levels of malic acid up to 6 g/L and in the case of the non-selected *Schizosaccharomyces* strains due to their several collateral effects.

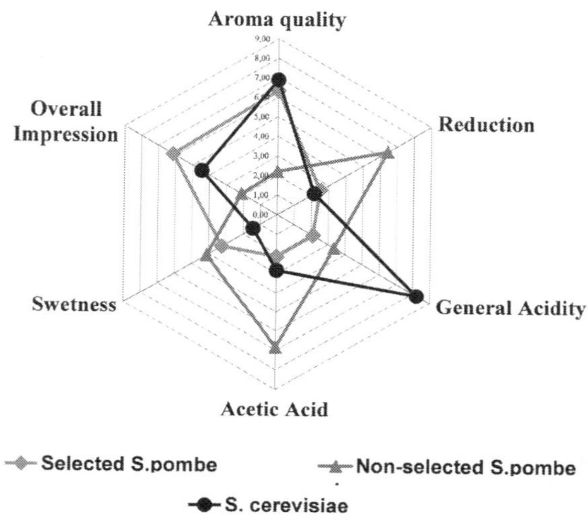

Figure 8. Summary of sensory profiles performed by Selected *S. pombe*, non-selected *S. pombe* and selected *S. cerevisiae* strains from very acidic grape musts.

5.13. Combination with other yeast species

New trends involving mixed fermentations between *S. pombe* and *L. thermotolerans* have been recently performed in warm viticulture regions with few malic acid content and high pH of about 4. The objective of this combination is to avoid biogenic amines formation during malolactic fermentation at high pH. In that case, *L. thermotolerans* is used in order to avoid an excessive deacidification through lactic acid formation. These wines showed lower final levels of biogenic amines up to almost 2 mg/L than the controls that underwent malolactic fermentation [7]. The pH was also reduced in 0.25 instead of increasing.

6. Conclusion

There are many new uses related to *Schizosaccharomyces* genus that can be applied in modern enology different from the classic malic acid deacidification. These new applications are not only related to improve wine quality. They can also improve food safety parameters doing the act of drinking wine a healthier habit. Last few studies have demonstrated that it is possible to make quality wine by using *Schizosaccharomyces* genus when selected strains are used. These strains can be used to produce wines with low levels of malic acid, acetic acid, gluconic acid, ethyl carbamate and biogenic amines, and with an appropriate volatile aroma profile.

Author details

Ángel Benito, Fernando Calderón and Santiago Benito*

*Address all correspondence to: santiago.benito@upm.es

Department of Chemistry and Food Technology, Polytechnic University of Madrid, University Campus S/N, Madrid, Spain

References

[1] Benito S, Hofmann T, Laier M, Lochbühler BC, Schüttler A, Ebert K, Fritsch S, Röcker J, Rauhut D. Effect on quality and composition of Riesling wines fermented by sequential inoculation with non-Saccharomyces and *Saccharomyces cerevisiae*. European Food Research and Technology. 2015; 241: 707–717. DOI: 10.1007/s00217-015-2497-8

[2] Belda I, Navascués E, Marquina D, Santos A, Calderón F, Benito S. Dynamic analysis of physiological properties of *Torulasporadelbrueckii* in wine fermentations and its incidence on wine quality. Applied Microbiology and Biotechnology. 2015; 99: 1911–1922. DOI: 10.1007/s00253-014-6197-2

[3] Benito A, Calderón F, Palomero F, Benito S. Quality and composition of airen wines fermented by sequential inoculation of *Lachancea thermotolerans* and *Saccharomyces cerevisiae*. Food Technology and Biotechnology. 2016; 54(2) 135–144. DOI: 10.17113/ftb.54.02.16.4220

[4] Rojas V, Gil JV, Piñaga F, Manzanares P. Studies on acetate ester production by non-*Saccharomyces* wine yeasts. International Journal of Food Microbiology. 2001; 70(3): 283–289. DOI: 10.1016/S0168-1605(01)00552-9

[5] Ciani M, Comitini F, Mannazzu I, Domizio P. Controlled mixed culture fermentation: a new perspective on the use of non-*Saccharomyces* yeasts in winemaking. FEMS Yeast Research. 2010; 10(2): 123–133. DOI: 10.1111/j.1567-1364.2009.00579.x

[6] Comitini F, Gobbi M, Domizio P, Romani C, Lencioni L, Mannazzu I, Ciani M. Selected non-Saccharomyces wine yeasts in controlled multistarter fermentations with *Saccha-*

romyces cerevisiae. Food Microbiology. 2011; 28(5): 873–882. DOI: 10.1016/j.fm.2010.12.001

[7] Benito A, Calderón F, Palomero F, Benito S. Combine use of selected *Schizosaccharomyces pombe* and *Lachancea thermotolerans* yeast strains as an alternative to the traditional malolactic fermentation in red wine production. Molecules. 2015; 20: 9510–9523. DOI: doi:10.3390/molecules20069510

[8] Jolly NP, Varela C, Pretorius IS. Not your ordinary yeast: non-Saccharomyces yeasts in wine production uncovered. FEMS Yeast Research. 2014; 14: 215–237. DOI: 10.1111/1567-1364.12111

[9] Benito S, Palomero F, Morata A, Calderón F,Palmero D, Suárez-Lepe JA. Physiological features of *Schizosaccharomyces pombe* of interest in making of white wines. European Food Research and Technology. 2013; 236: 29–36. DOI: 10.1007/s00217-012-1836-2

[10] Maconi E, Manachini PL, Aragozzini F, Gennari C, Ricca GS. A study of the maloalcoholic fermentation pathway in *Schizosaccharomyces pombe*. Biochemical Journal. 1984; 217(2): 585–588. DOI: 10.1042/bj2170585

[11] Taillandier P, Strehaiano P. The role of malic acid in the metabolism of *Schizosaccharomyces pombe*: substrate consumption and cell growth. Applied Microbiology and Biotechnology. 1991; 35(4): 541–543. DOI: 10.1007/BF00169765

[12] Benito S, Palomero F, Calderón F, Palmero D, Suárez-Lepe JA. Schizosaccharomyces. In: Batt CA, Tortorelo ML, editors. Encyclopedia of Food Microbiology. 2nd ed. Amsterdam: Elsevier; 2014. p. 365–370.

[13] Lubbers MW, Rodriguez SB, Honey NK, Thornton RJ. Purification and characterization of urease from *Schizosaccharomyces pombe*. Canadian Journal of Microbiology. 1996; 42(2): 132–140. DOI: 10.1139/m96-021

[14] Bacanamwo M, Witte CP, Lubbers M, Polacco J. Activation of the urease of *Schizosaccharomyces pombe* by the UreF accessory protein from soybean. Molecular Genetics and Genomics. 2002; 268(4): 525–534. DOI: 10.1007/s00438-002-0769-z

[15] Peinado RA, Moreno JJ, Maestre O, Mauricio JC. Removing gluconic acid by using different treatments with a *Schizosaccharomyces pombe* mutant: effect on fermentation by products. Food Chemistry. 2007; 104: 457–465. DOI: 10.1016/j.foodchem.2006.11.070

[16] Palomero F, Morata A, Benito S, Calderón F, Suárez-Lepe JA. New genera of yeasts for over-lees aging of red wine. Food Chemistry. 2009; 112: 432–441. DOI: 10.1016/j.foodchem.2008.05.098

[17] Benito S. The use of *Schizosaccharomyces* yeasts in order to reduce the content of biogenic amines and ethylcarbamate in wines. Journal of Food Processing & Technology. 2015; 6: 9. DOI: 10.4172/2157-7110.S1.026

[18] Jeffares DC, Rallis C, Rieux A, Speed A, Převorovský M, Mourier T, et al. The genomic and phenotypic diversity of *Schizosaccharomyces pombe*. Nature Genetics. 2015; 47: 235–241. DOI: 10.1038/ng.3215.

[19] Benito A, Jeffares D, Palomero F, Calderón F, Jurg J, Bai Y, Benito S. Selected *Schizosaccharomyces pombe* strains have characteristics that are beneficial for winemaking.2016; 11(3) e0151102. DOI: 10.1371/journal.pone.0151102 Accepted.

[20] Benito S, Palomero F, Calderón F, Palmero D, Suárez-Lepe JA. Selection of appropriate Schizosaccharomyces strains for winemaking. Food Microbiology. 2014; 42: 218–224. DOI: 10.1016/j.fm.2014.03.014

[21] Belda I, Navascués E, Marquina D, Santos A, Calderón F, Benito S. Outlining the influence of non-conventional yeasts in wine ageing over-lees. Yeast. 2016. Accepted.

[22] Benito S, Palomero F, Gálvez L, Morata A, Calderón F, Palmero D, Suárez-Lepe JA. Quality and composition of red wine fermented with *Schizosaccharomyces pombe* as sole fermentative yeast, and in mixed and sequential fermentations with *Saccharomyces cerevisiae*. Food Technology and Biotechnology. 2014; 52(3): 376–382.

[23] Benito S, Palomero F, Morata A, Calderón F, Suarez-Lepe JA. New applications for *Schizosaccharomyces pombe* in the alcoholic fermentation of red wines. International Journal of Food Science and Technology. 2012; 47: 2101–2108. DOI: 10.1111/j.1365-2621.2012.03076.x

[24] Benito S, Gálvez L, Palomero F, Calderón F, Morata A, Palmero D, Suárez-Lepe JA. *Schizosaccharomyces* selective differential media. African Journal of Microbiology Research. 2013; 7(24): 3026–3036. DOI: 10.5897/AJMR2013.5684

[25] Deák T. Handbook of Food Spoilage Yeasts. 2nd ed. CRC Press, Boca Raton, Florida.

[26] Loira I, Morata A, González C, Suárez-Lepe JA. Selection of glycolytically inefficient yeasts for reducing the alcohol content of wines from hot regions. Food and Bioprocess Technology. 2012; 5(7): 2787–2796. DOI: 10.1007/s11947-011-0604-9

[27] Xia C, Hong M, YunFeng Z, Ning WY. Research progress on toxicity and contamination of ethyl carbamate in fermented foods. Journal of Food Safety and Quality. 2014; 5(9): 2617–2262. DOI: 10.5897/AJMR2013.5684

Chapter 15

The Use of Indigenous Yeast to Develop High-Quality Patagonian Wines

Silvana María Del Mónaco, Yolanda Curilen, Ramona Del Carmen Maturano, Sebastián Mario Ezequiel Bravo, Adriana Beatriz Simes and Adriana Catalina Caballero

Additional information is available at the end of the chapter

http://dx.doi.org/10.5772/64750

Abstract

In young wines, the compounds responsible for wine *flavor* come from two possible origins: grapes and microorganisms involved in winemaking. Yeasts play the most important role in flavor influence because of their role in conducting the alcoholic fermentation (FA), the key process of winemaking. Ecological studies show that yeast diversity is significantly influenced by geographical and technological features of each particular winegrowing region. Wines from Argentine have achieved high-quality certifications, and particularly, in the Comahue region, wine production is mainly oriented to young red wines varieties, some of which found in this region optimal ecological condition to express all their enological potential. Despite this, the need to satisfy the demands of an increasingly competitive and globalized international market and the consumer demand for new wine styles with the best quality/price ratio imposes the regional productive sector new challenges that require technological innovation. The use of starter cultures developed from indigenous yeast isolated from our region, specially selected for its enological properties, appears as a valuable tool for differentiation, diversification, and quality improvement of wines. In this context, conventional and non-conventional yeasts were isolated and selected over the years and used for vinifications in red grape varieties (Pinot noir and Malbec). Assays were carried out at laboratory and pilot scale, in the 2010–2015 vintages. The experiences developed along the years contribute to a better understanding of the processes involved in the production of improved wines by autochthonous strains, an important practice to develop a more competitive regional wine industry.

Keywords: wine flavor, fermentation, *Saccharomyces cerevisiae*, non-*Saccharomyces* yeasts, mixed starters, microbial interactions, indigenous biota

1. Introduction

Wine chemical composition is the foundation for its sensorial features, as color, appearance, body, flavor (aroma and taste), as well as its mouth and palate sensations [1]. Among these, wine flavor is a key attribute for quality and choice from consumers. The flavor of wine is a sensory perception that varies with the individual, the context of the consumer experience, and the chemical composition of the product [2, 3]. The wine chemical composition is determined by many factors such as grape variety, the geographical and viticultural conditions of grape cultivation, the microbial ecology of the grape and fermentation processes, and winemaking practices [4]. In young wines (without aging), like the ones mainly produced in the Patagonia Argentina, the compounds responsible for wine flavor come from two possible origins: grapes and microorganisms involved in winemaking, mostly yeast and to a lesser extent lactic acid bacteria (LAB) [5]. Grapes contribute with varietal aroma compounds, such as floral monoterpenes or volatile thiols, among others [1]. Among the microorganisms, yeasts play the most important role in flavor influence because of their role in conducting the alcoholic fermentation (AF), the key process of winemaking. During AF yeasts transform grape sugars and other components to ethanol, carbon dioxide, and different primary metabolites that confer a particular character to wine, but they also contribute with minority volatile compounds involved in determining the fermentative or secondary aroma [1, 2, 6, 7]. These compounds (esters, higher alcohols, carbonyls, short-chain fatty acids and sulfur compounds) arise from the metabolism of sugars and amino acids from the must, and their quality and content depends on the ecology wine yeast associated with the process [1]. Finally, when needed, LAB through malolactic fermentation not only provides wine deacidification, but can also enhance its flavor profile [1, 2, 8].

In Argentina, wine production has historically occupied a place of importance in the agricultural industries. The country is currently the fifth largest producer of wines, the seventh consumer, and the tenth largest exporter. There are two ways of imposing a wine in the market, highlighting its grape quality, that is, *Vitis vinifera* variety or alluding to the region from where they were harvested and vinified (*terroir*). While wines from Argentine are sold taking into account the variety, there are regions and subregions in the country where wines achieved high-quality certifications (of origin or geographical designations) [9]. One of them is the Comahue region, located in the Argentinean North Patagonia at 39–40° southern latitude; it is the southernmost winegrowing region of Argentina and one of the most southern regions in the world (**Figure 1**). This region has optimal agro-ecological conditions for high-quality viticulture and a long winemaking tradition [10] so although the wine industry is still a secondary economic activity, it represents a very interesting alternative to diversify the local production.

http://dx.doi.org/10.5772/64750

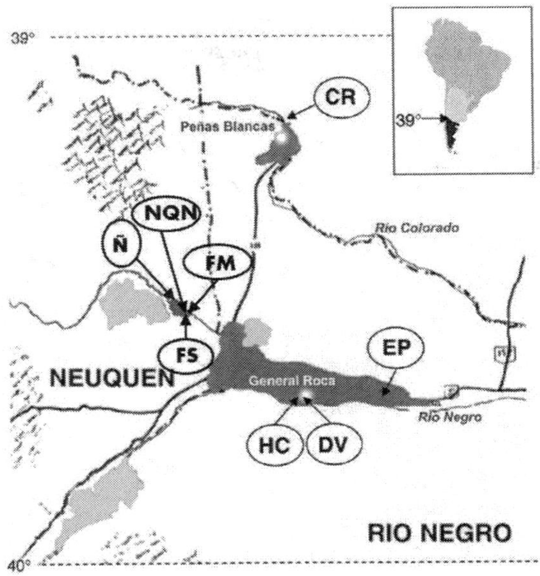

Figure 1. Comahue region (Argentinean North Patagonia). Location of sampled Patagonian cellars and vineyards (EP, DV, HC, and CR in Río Negro Province and FS, FM, NQN, and Ñ in Neuquén Province). Dark gray: cultured vine areas. In the right top corner: South America (dark gray), Argentina (light gray), and Argentinean Patagonia (black). Source: INV.

North Patagonian wine production is mainly oriented to elaboration of young dry wines from red grape varieties (80%) some of each, as Pinot noir and Merlot, have found in this region the optimal ecological conditions to express all their enological potential [11]. Additionally, important volumes of Malbec grapes, the *Vitis vinifera,* L variety emblematic of Argentina, are also vinified [12]. Despite this, the need to satisfy the demands of an increasingly competitive and globalized international market actually oversupplied, and the consumer demand for new wine styles with the best quality/price ratio, impose the national productive sector, and in particular the regional one, new challenges that require technological innovation.

Actually, regional wine production is based on both spontaneous alcoholic fermentations of the grape musts or conducted fermentations using commercial yeast starters. However, commercial starters for alcoholic fermentation found actually in the market are composed by yeast strains isolated from the most important winegrowing areas in the world, except Argentina. Given the significant influence that the biota of yeast has on the aromatic quality of young wines elaborated from aromatically neutral grapes, wine style produced mostly in Patagonia, development of starter cultures consisting of yeast strains isolated from the own region (indigenous yeasts) appears as a valuable tool to improve the quality of Patagonian vitiviniculture, upgrading its capacity of commercial competition in domestic and international market.

1.1. The microbial ecology of Patagonian red winemaking

Winemaking is a complex microbial ecosystem that involves interactions between filamentous fungi, yeasts, and bacteria with different physiological and metabolic characteristics [3, 13, 14]. Ecological studies have shown that this microbial diversity is significantly influenced by geographical and technological features of each particular winegrowing region, which defines the terroir. To understand how this microbial terroir contributes to the natural environment of vineyard and how it imprints differential character of wine required to know all processes associated with winemaking which start at the harvest of grapes and then evolve throughout fermentation process [15–17].

Microbiological studies carried out during several years in the Patagonian region allowed to characterize the biota associated with grapes [18–22], cellars [23, 24], and red vinification environments [19–22, 25, 26], (**Figure 2**). Yeasts associated with spontaneous wine fermentations come from two possible origins: grapes and cellar surfaces. Several factors such as development stage and sanitary state of the berries, the climate (particularly temperature and rainfall), water availability, direct exposure to sunlight, use of agrichemicals, grape vine canopy management system as well as nature, cleaning and sanitization of equipment surfaces, nature, cleaning and sanitization of equipment surfaces, among others, as well as certain enological practices could affect the yeast community composition on grapes and cellar surfaces affecting the kinetics of yeasts growth during fermentation [3, 14, 27]. In this context, in all cases, ripe, whole and healthy grapes from Merlot, Malbec, and Pinot varieties were gathered by random sampling (n = 536) from vineyards associated with cellars which is noted in **Figure 1** for 1993–1998 and 2005–2009 vintages at harvest time, and cellar sampling was carried out on the internal walls of the fermentation vats from the same cellars approximately four weeks before harvest. Yeast samples from grape surfaces were obtained by agitation followed sonication of each grape in pure and sterile water. Additionally, samples (1L) of Malbec, Merlot, and Pinot noir fermentation musts samples were taken in duplicates during spontaneous alcoholic fermentation at the initial (14°Bé), middle (6°Bé), and end (≦2 g/L TRS) stages in the same cellars described above. All yeasts were isolated on GPY agar (composition in g/l: yeast extract 10, glucose 20, peptone 20, and agar 20; pH 4.5 supplemented with 100 ppm of ampicillin) plates, and they were identified according to the methods and keys proposed by Kurtzman and Fell [28] and by PCR-RFLP analysis of the ITS1-5,8S-ITS2 region from the nuclear rDNA gene complex [29]. Results of these studies evidence that yeast biota associated with grape surfaces are mostly aerobic (57%), while the one associated with cellar surface is mostly facultative (64%). These results of yeast diversity obtained from initial musts are consistent with the hypothesis proposed on origin of yeasts musts (**Figure 2**). On the other hand, this study also evidences that *Saccharomyces cerevisiae* occur in extremely low population on healthy, undamaged grapes (1/536 isolates, <0.2%) and it is the major species (45/150 isolates, 30%) together with species of genera *Candida, Kloeckera/Hanseniaspora,* and other sporadic yeasts on winery surface (artificial man-made environmental), hypothesis claimed by Martini school [30–33]. Additionally and in agreement with the reported in bibliography [13, 34–36], even though *S. cerevisiae* is the most important yeast in spontaneous vinifications, other species of yeasts belonging to non-*Saccharomyces* species such as *Kloeckera apiculata/Hanseniaspora* *uvarum, Candida stellata* and *dattila, Pichia kudriavzevii* and *Torulaspora delbrueckii* are also able to remain in Patagonian musts and during fermentation periods in appropriate concentrations to significantly contribute to the sensory quality of the product (**Figure 2**). Hence, the yeast ecology of wine fermentation has been found to be much more complex than assumed dominance of *S. cerevisiae* species, and the metabolic impact of yeasts on wine character is much more diverse than simple fermentation of grape juice sugars [36, 37]. Additionally, the incorporation of molecular methods (mitADN RFLP using Hinf I) together with killer biotype in these studies demonstrated the existence of a wide variety of individuals (strains) within indigenous populations of *S. cerevisiae* [38] and other different species of yeast such as *K. apiculata, Metschnikowia pulcherrima*, and *Pichia guilliermondii* [21], as well as *P. kudriavzevii* [39]. This intraspecific variability was significantly influenced by geographical or/and technological specific factors of productive region and it is very important from the enological point of view. Similar results have been reported from other winegrowing regions of the world [40, 41].

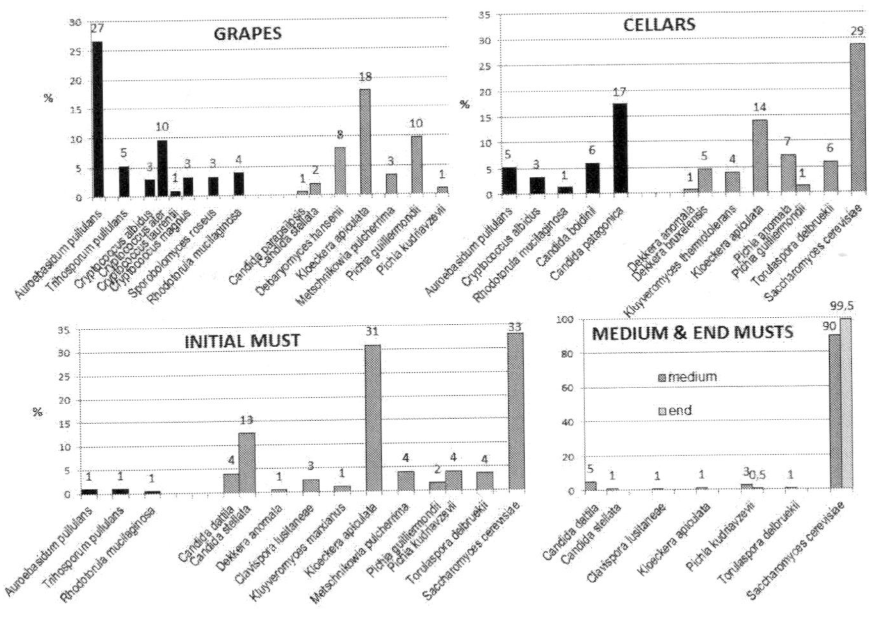

Figure 2. Diversity of yeasts associated with Grape, n = 536; and cellar surfaces, n = 150 (up) and musts from initial (14°Bé, n = 322), middle (6°Bé, n = 320), and end (0°Bé, n = 397) stages from spontaneous fermentations (down). Black bars, aerobic yeasts (respiratory strict metabolism), and gray bars, facultative yeasts (respiratory and fermentative metabolism).

With this greater knowledge, alcoholic fermentation is now seen as a key process where winemakers can creatively engineer wine character and value through better yeast management and can strategically tailor wines to a changing market [17].

Malolactic fermentation (MLF) is other opportunity to modulate the aroma of wine [1, 42]. MLF, the decarboxylation of L(−)malic acid to L(+)lactic acid, is an important secondary fermentation carried out by lactic acid bacteria (LAB) during the vinification of most red wine styles as those elaborated in Patagonia. Malic acid, together with tartaric acid, is the most important constituents of organic nonvolatile acid fraction in grapes and grape musts, accounting for 90% of the titratable acidity and imbalances in this fraction can affect the physicochemical and sensory properties of wine, mainly mouthfeel [43–48]. In addition to deacidification, MLF can increase microbiological stability [13, 49] and enhance wine flavor and/or complexity [1, 50] but off-flavors as well as dangerous health compounds could also be formed [45]. Although *Oenococcus oeni* is the major species during malolactic fermentation other LAB species belonging to *Lactobacillus, Leuconostoc* and *Pediococcus* genera can grow in the wine [8, 22, 34, 45, 51, 52] and their contribution on wine quality should not be underestimated [34, 53]. Recent red wine trials carried out at laboratory scale have shown that strains of *Lactobacillus plantarum* have the potential to conduct an efficient MLF and also produce desirable sensory attributes in red wines [54, 55]. However, at industrial-scale, spontaneous MLF is a very difficult process because different factors associated with winemaking, as yeast-

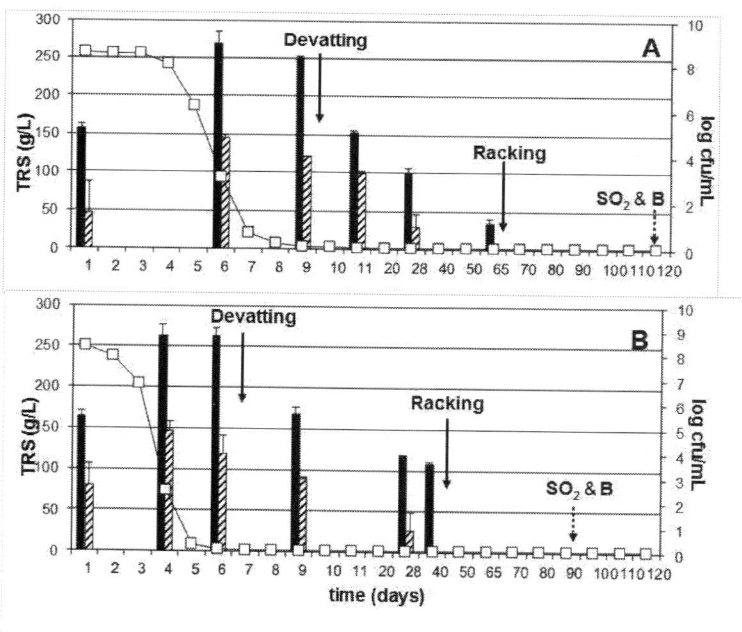

Figure 3. Time course of fermentative processes and growth kinetics of yeasts (black bars) and lactic acid bacteria (stripped bars) in spontaneous (a) and guided (b) Pinot noir vinifications carried out at pilot scale. SO_2 and B: Sulfur dioxide adding and bottling. Extracted from Curilén et al. [22].

http://dx.doi.org/10.5772/64750

LAB antagonistic interactions, inhibit lactic bacteria growth [8, 22, 34, 56, 57] and the use of commercial starters to induce and guide the process is not always effective [58]. Various studies have been done to attempt an understanding of the interaction between yeast and bacteria [59–64]. The degree and type of interactions vary from one pair to another and seem to be closely related to the chosen yeast strain.

Several factors such as grapevine variety, vineyard agricultural practice, temperature, humidity and berry maturity degree, among others, may affect organic nonvolatile acid concentration in grape musts [44, 65, 66]. In particular, L(−)malic acid content, directly related to respiratory quotient of berries, is higher in grape musts from cooler regions than the ones from warmer regions [67]. In the Comahue region, one of the southernmost winegrowing regions of the world, malic acid concentrations account for the 56% of red grape must titratable acidity reaching the 66% in Pinot noir [19] the emblematic regional vine variety [11]. Additionally, to its contribution to wine acidity, malic acid represents a fermentable substrate for other microorganisms which can spoil the wine before and after bottling [68]. Without adjustment of acidity, the wines will be regarded as unbalanced or spoilt [1]. For these reasons, MLF is a routine enological practice in the Patagonian red winemaking and yeast-LAB interactions are a great concern for winemakers and researchers.

Organic acid [*]	Grapes	Wines				P value[*]
		GF		NF		
		Running wine	Bottled	Running wine	Bottled	
L(−)malic	2.20 ± 0.66[a]	2.08 ± 0.02[ab]	2.01 ± 0.01[b]	1.16 ± 0.02[c]	0.97 ± 0.01[d]	<0.001
DL lactic	Nd	0.84 ± 0.00[a]	0.58 ± 0.00[b]	0.97 ± 0.09[c]	1.12 ± 0.02[d]	<0.001
L(+)lactic	Nd	0.25 ± 0.03[a]	0.30 ± 0.01[a]	0.55 ± 0.04[b]	0.75 ± 0.04[c]	<0.001
Citric	0.78 ± 0.11[a]	1.06 ± 0.04[a]	nd[b]	0.83 ± 0.14[a]	nd[b]	<0.001
Succinic	Nd	1.50 ± 0.12[a]	0.96 ± 0.16[b]	1.07 ± 0.13[ab]	0.86 ± 0.05[b]	<0.001

[#]g L^{-1}; [+]mg L^{-1}; nd: not detected; *one-way ANOVA and Tukey's test n = 2.

Table 1. Organic acids composition of Patagonian Pinot noir grapes and wines obtained from guided (GF) and spontaneous (NF) winemaking (extracted from Curilén et al. [22]).

In Patagonian winegrowing region, potential yeast-LAB interaction was studied in Pinot noir wine fermentations carried out at pilot [22] and industrial scales [8]. In all cases, malolactic fermentations were carried out spontaneously, whereas alcoholic fermentations were carried out in both, spontaneous (NF, carried out by indigenous yeast biota) and guided (GF, carried out by *S. cerevisiae* F15, Laffort) forms. All microbiological and physicochemical processes were characterized (**Figure 3** and **Table 1**). Musts and wines samples were appropriately diluted, and they were spread in duplicate onto MRS plus tomato agar (total LAB) supplemented with 100 ppm cycloheximide. Organic acid were quantified by HPLC and enzymatically (L and D malic acids). Results evidence that the numerical dominance of commercial starters at the GF beginning affects their fermentation kinetic and yeast diversity. Although AF as well as MLF showed normal kinetics for the assayed scale and both were complete fermentations, inoculation with commercial *S. cerevisiae* F15 starter significantly affected the time course of the global fermentative process, which was faster in GF than in NF (**Figure 3**). **Figure 3** also shows no significant changes in total yeast biota extension when similar stages between both fermentations are compared, but their qualities at initial stages were significantly different because all non-*Saccharomyces* yeasts were eliminated by starter inoculation (data not shown). Like for yeasts, no changes were observed in the LAB biota extension (**Figure 3**), but its quality was significantly affected by the inoculation. At the beginning of both fermentations, LAB biota were mostly facultatively homofermentative (62%) with *Lb. plantarum* as the major species (8 of 13 isolates), and minor heterofermentative LAB isolates were presumptively identified as belonging to *Leuconostoc* (1) and *Oenococcus oeni* (1) [22]. However, NF LAB evolved to a mostly heterofermentative biota (≥66% in average) from middle stage up to racking, whereas the GF LAB one was always mostly facultatively homofermentative (≥50% in average). These differences in LAB biota between NF and GF were consistent with the data obtained for FML evolution from chemical analysis evidencing antagonist interaction between *S. cerevisiae* F15 and LAB biota. In this context, yeast-LAB interactions should be included in the selection criteria for Patagonian wine yeasts.

2. Development of patagonian yeast starters

The practical consequence of studies on microbial ecology in winemaking was the development of starter cultures of AF. The use of commercial starter cultures for the FA in enology, with selected cultures of *S. cerevisiae* for the inoculation of fruit juice, has been applied since the 60s from the last century, and it has been one of the most important technological advances in the wine industry [69, 70].

The inoculation of grape musts with commercial starters was an enological practice strongly resisted by wine producers from Europe. The solidest argument for this resistance referred to the sensory quality standard, where flattened aromatic profiles were produced in each wine fermented with those starters. In modern wineries, the use of commercial starter culture to steer fermentations is being doubted, since they often lack of some advantageous enological traits, which are present when the spontaneous fermentation is ruled by indigenous populations [6]. For that reason, the exploitation of indigenous strains biodiversity has great importance for the characterization and selection of strains with peculiar phenotypes [17, 36]. While the preservation of spontaneous microflora is essential to obtain the typical flavor and aroma of wines deriving from different grape varieties [5, 13], the development of starter from *S. cerevisiae* strains indigenous of each winegrowing region ended this controversy [26]. The advantages to use starters of indigenous *S. cerevisiae* strains, better adapted to the ecological and technological features of each particular winegrowing area preserving its own natural biodiversity, are now recognized by all wine producers, including the European ones. However, the knowledge generated in recent years in the field of enological microbiology and described above, evidence that during the process, certain non-*Saccharomyces* species can also contribute significantly to the sensory characteristics of wines by producing compounds that impact on the varietal aroma and in its taste. The potential of using them in winemaking as stand-alone, single starter cultures, or together with *Saccharomyces* yeast cultures as mixed starter cultures could be especially interesting in elaboration of wines from *Vitis vinifera* varieties aromatically neutral, such as mostly vinified in Comahue region, whose aroma is developed exclusively during fermentation. For that reason, in the last years, there has been an increasing interest for the selection of strains of non-*Saccharomyces* yeasts for the winemaking industry, mainly due to their ability to enhance the analytical composition of the wines, particularly for their effects on wine composition, flavor, and aroma [39, 71–81].

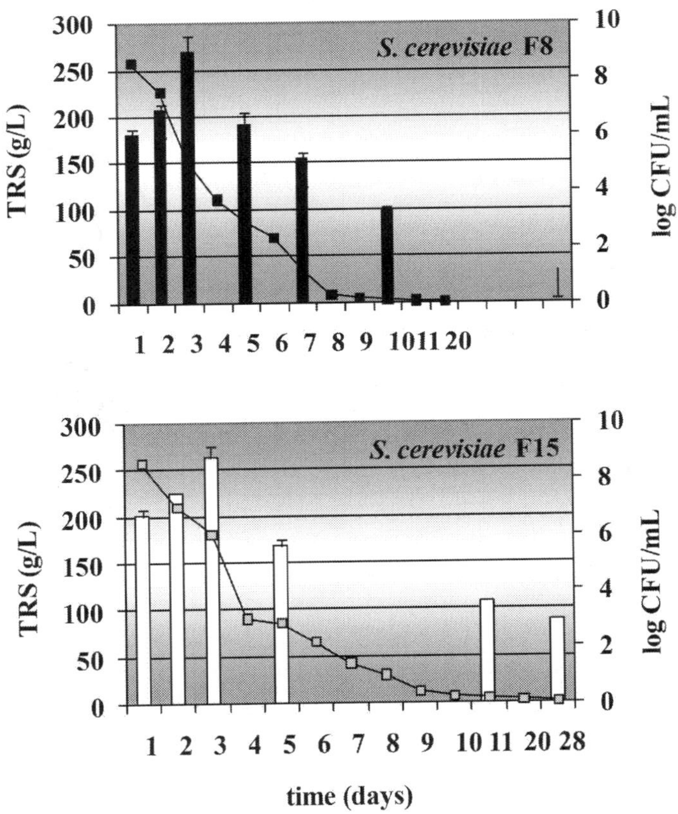

Figure 4. Yeast growth (bars) and total reducing sugars evolution (squares) during pilot scale Pinot noir vinifications guided by indigenous F8 (upper panel, black symbols) and commercial F15 (bottom panel, white symbols) *S. cerevisiae* strains [26].

Microbiological studies carried out during several years in the Patagonian region described above allowed constituting an important collection of organisms relevant for enological application and wine starter elaboration. Our enology applied studies were divided into two basic objectives: isolation of indigenous *S. cerevisiae* strains and isolation of local non-*Saccharomyces* strains with potential use in enology for develop pure and mixed indigenous starters to conduct AF. From those studies two yeast strains were selected as follows: an indigenous *S. cerevisiae* named F8 and a *P. kudriavzevii* strain called ÑNI15 which could metabolize malic acid and control wine acidity. Vinification studies were carried out using Pinot noir, Merlot, and Malbec grape musts as substrates and alcoholic fermentations were guided by with those strains, using a commercial *S. cerevisiae* (F15, Laffort) strain as control for comparison.

2.1. *Saccharomyces cerevisiae* indigenous starter

When indigenous F8 and commercial F15 *S. cerevisiae* strains were evaluated, the initial cellular densities and biomass evolution were similar in Pinot noir fermentations (**Figure 4**) as well as in Merlot and Malbec fermentations (data not shown). Additionally, fermentations were mostly completed to dryness (TRS ≤ 2 g/L), but the sugar consumption rates during dryness stages, such as it is displayed in **Figure 4** for Pinot noir, were ever higher in F8-guided fermentations than in F15 fermentations. As a consequence, fermentative processes guided by the indigenous starter were faster than those guided by the commercial starter.

In order to evaluate the capacity of the indigenous starter to dominate the fermentations, the dynamics of the *S. cerevisiae* populations were determined by means of mtDNA-RFLP analysis. The results obtained from these studies, and partially showed in **Figure 5**, evidence that indigenous F8 and commercial F15 *S. cerevisiae*, were the strains mostly found at the initial and final stages of their respective fermentations proving their very good and similar implantation capabilities.

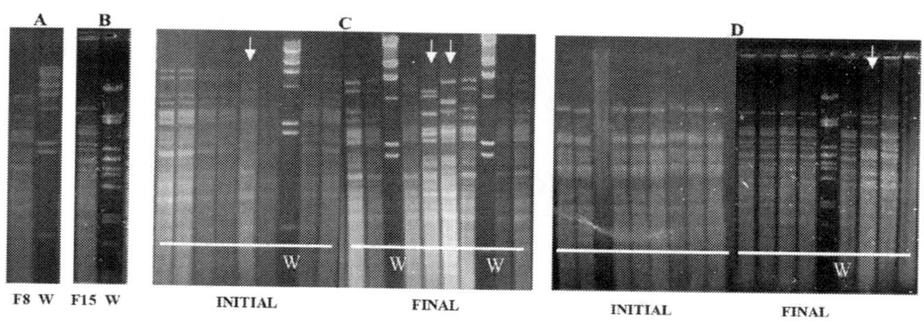

Figure 5. mtDNA-RFLP patterns of indigenous (a) and commercial (b) starters and of *S. cerevisiae* isolates obtained from F8 (c) and F15 (d) guided Pinot noir fermentations (2013 Vintage) at the initial and final stages. Arrows indicate isolates with mtDNA-RFLP patterns different from the inoculated starters. W = molecular weight marker [26].

On the other hand, physicochemical analysis of the wines obtained during different years from Pinot noir, Merlot, and Malbec varieties were highly similar between both inoculated strains, where every product was considered acceptable for local young wines [26]. Nonetheless, sensorial analysis carried out by experts and consumers using qualitative and quantitative tests, respectively, showed significant differences between F8 and F15 wines.

Qualitative analysis was performed by a panel of experts using descriptive tests. As a whole, the global quality scores obtained in this analysis by F8 Pinot noir (68 = good) and Merlot (6.6 = pleasant) wines were higher than those obtained by F15 Pinot noir (52 = correct) and Merlot (5.7 = slightly pleasant) wines. Particular descriptions evidenced that Pinot noir F8 wines had good color intensities (showing a red color typical for the variety) and aromas of red fruits (cherries) with notes of sherry. In mouth, they were described as middling fruity, slightly rusty, sweet alcoholic. Meanwhile, Pinot noir F15 wines showed limpid and bright aspect and an intense reduced aroma that did not disappear with agitation. In mouth, they were perceived as slightly fruity and bitter, astringent, and tannic. On the other hand, both F8 and F15 Merlot wines showed a limpid and bright aspect and an intense brick red color, but the F8 aroma was more intense than the F15 aroma, being both aromas of medium quality. Pepper, red fruits, butter, leather, spice, and vanilla were the aromatic descriptors highlighted in the former and green pepper, cooked red fruits, spices, and pepper were described for the later. In the mouth, both wines showed good acidity and body, and they were persistent. However, F8 wines were described as round and equilibrated while F15 wines showed a tart taste [26].

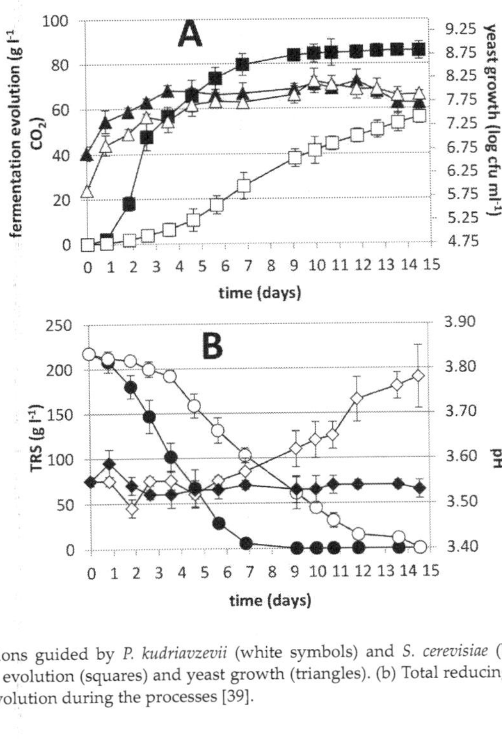

Figure 6. Microvinifications guided by *P. kudriavzevii* (white symbols) and *S. cerevisiae* (black symbols) indigenous strains. (a) Fermentation evolution (squares) and yeast growth (triangles). (b) Total reducing sugars (TRS) (circles) and pH values (diamonds) evolution during the processes [39].

At last, F8 Pinot noir and F8 Merlot wines were the favorite for the consumers ($p < 0.05$) with 72 favorable responses out of 119 questioned and 13 favorable responses out of 17 questioned, respectively, compared with their F15 controls according to the paired-preference test.

2.2. Non-Saccharomyces indigenous starter

One of the purposes of the study was to select autochthonous yeasts with metabolic ability to degrade L-malic acid for its potential use in equilibrated young wine elaboration. A total of 57 Patagonian non-*Saccharomyces* yeast of enological origin were identified by conventional molecular methods and tested in their capability to grow at the expense of L-malic acid. An isolate, noted as *P. kudriavzevii* ÑNI15, was able to degrade L-malic acid in microvinifications, increasing the pH 0.2–0.3 units with a minimal effect on the acid structure of wine. Additionally, this isolate was a weak producer of ethanol, an important producer of glycerol (10.41 ± 0.48 g l^{-1}), a producer of acceptable amounts of acetic acid (0.86 ± 0.13 g l^{-1}), as well as it was able to improve the sensorial attributes of wine increasing its fruity aroma [39].

Figure 6 shows the results obtained from vinifications guided by pure cultures indigenous *P. kudriavzevii* ÑNI15 and *S. cerevisiae* F8, carried out at laboratory and using synthetic musts with similar amino and organic acids composition to Patagonian Pinot noir juice as substrate. An acceptable yield in biomass (**Figure 6a**) and similar end sugar concentrations (**Figure 6b**) were observed in both microvinifications; however, the fermentative efficiency (**Figure 6a**) as well as the sugar consumption rate (**Figure 6b**) were higher for *S. cerevisiae* than for *P. kudriavzevii*. A noteworthy fact, and in agreement with what was reported in the broth assays, is that *P. kudriavzevii* was again capable of raising significantly the medium pH with a minimal effect on acid structure of the wine, whereas in the *S. cerevisiae*, culture pH was maintained constant along the fermentation (**Figure 6b**).

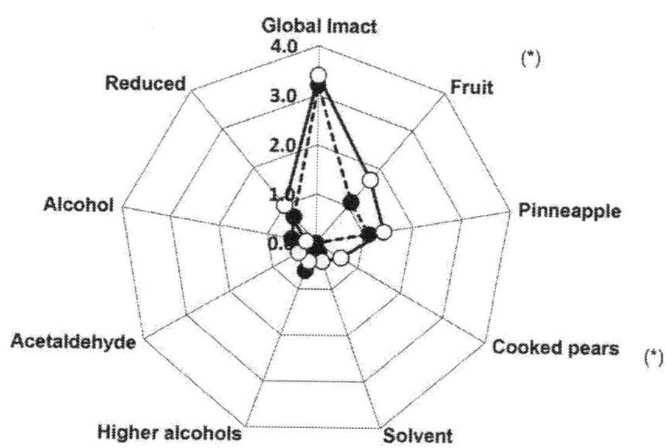

Figure 7. Sensory quality of microvinification wines obtained from *P. kudriavzevii* (white circles) or *S. cerevisiae* (black circles) inoculation. ANOVA and Tukey's test, n = 12. Asterisks indicate statistic differences ($p < 0.05$) [39].

http://dx.doi.org/10.5772/64750

Finally, sensorial analysis evidenced significantly differences in aromatic perception between *P. kudriavzevii* and *S. cerevisiae* wines. These differences were in favor of the former, which showed a higher fruity and cooked pears aroma than the latter (**Figure 7**).

3. Conclusions

The extension of the selection of yeast for enological use among *Saccharomyces* and non-*Saccharomyces* species led to the finding of yeast strains with novel and interesting enological characteristics which could have significant implications in the production quality improved Patagonian young wines. Results presented show that *S. cerevisiae* F8 strain drives red vinifications improving the quality of the local fermented products. On the other hand, the use of *P. kudriavzevii* ÑNI15 as wine starter would eliminate the cultural and cellar operations undertaken to adjust the musts acidity improving wine quality and reducing production costs. The co-inoculation of *S. cerevisiae* F8 and *P. kudriavzevii* ÑNI15 in local musts implies the enological potential of using these strains to formulate a regional starter culture for the production of well-balanced and physicochemical stable Patagonian young red wines.

Acknowledgements

This work was supported by grants from Universidad Nacional del Comahue (Programa de Investigación 04/L003 Desarrollo de Tecnologías y de Productos de Interés para la Industria Agroalimentaria) and MINCyT (PICT SU 2804/12 Levaduras y Bacterias Lácticas para la diferenciación de vinos Patagónicos).

Author details

Silvana María Del Mónaco[1*], Yolanda Curilen[1,2], Ramona Del Carmen Maturano[1], Sebastián Mario Ezequiel Bravo[1,2], Adriana Beatriz Simes[2] and Adriana Catalina Caballero[1,2]

*Address all correspondence to: silmdm@yahoo.com

1 Enology Group, Institute of Investigation and Development in Process Engineering, Biotechnology and Alternative Energies (PROBIEN), National Council of Scientific and Technical Investigations (CONICET), Neuquén City, Neuquén Province, Argentina

2 Faculty of Science and Food Technology, National University of Comahue, Villa Regina City, Río Negro Province, Argentina

References

[1] Swiegers, J.H., Bartowsky E.J. Henschke, P.A., Pretorius I.S. Yeast and bacterial modulation of wine aroma and flavour. Australian Journal of Grape and Wine Research. 2005;11:139–173.

[2] Lambrechts, M.G., Pretorius, I.S. Yeast and its importance. South African Journal of Enology and Viticulture. 2000;21:97–129.

[3] Fleet, G.H. Yeast interactions and wine flavour. International Journal of Food Microbiology. 2003;86:11–22.

[4] Cole, V.C., Noble, A.C. Flavour chemistry and assessment. In: Lea, A.G.H., Piggott, J.R., editors. Fermented Beverage Production. London: Blackie Academic and Professional; 1997. p. 361–385.

[5] Renouf, V., Claisse, O., Lonvaud-Funel, A. Understanding the microbial ecosystem on the grape berry surface through numeration and identification of yeast and bacteria. Australian Journal of Grape Wine and Research. 2005;11:316–327.

[6] Fleet, G.H., Heard, G. Yeasts-growth during fermentation. In: Fleet, G.H., editors. Wine Microbiology and Biotechnology. Chur, Switzerland: Harwood Academic Publishers; 1993. p. 27–54.

[7] Romano, P., Fiore, C., Paraggio, M., Caruso, M., Capece, A. Function of yeast species and strains in wine flavour. International Journal of Food Microbiology. 2003;86:169–180.

[8] Curilén, Y., Carreño, V., Camacho, E., del Mónaco, S., Semorile, L., Caballero, A. Influence of yeast biota on spontaneous malolactic fermentations associated with Patagonian winemaking. In: Proceedings of 37th World Congress of Vine and Wine: Southern Vitiviniculture, a Confluence of Knowledge and Nature; 9–14 November 2014; Mendoza (ARGENTINA). 514:2014. p. 1–6.

[9] Llorente, A., Casazza, M. Southern Winegrowing Region from Argentina. El vino y su industria. 2005;40:7–13.

[10] Le Guillou, B. Río Negro. Vines in the oases from Argentine Patagonia. La Vigne. 2000;116:112–113.

[11] Weizman, D. The Argentine map of the senses. Rumbos. 2009;326:18–24.

[12] Catania C., Avagnina S. Sensory interpretation of Wine . In: Argentina's wine terroir. INTA y Editora Andina Sur.; 2010. p. 111–121.

[13] Pretorius, I.S. Tailoring wine yeast for the new millennium: novel approaches to the ancient art of winemaking. Yeast. 2000;16:675–729.

[14] Barata, A., Malfeito-Ferreira, M., Loureiro, V. The microbial ecology of wine grape berries. International Journal of Food Microbiology. 2012;153:243–259.

[15] Van Leeuwen, C., Seguin, G. The concept of terroir in viticulture. Journal of Wine Research. 2006;17:1–10.

[16] Bokulich, N.A., Ohta, M., Richardson, P.M., Mills, D.A. Monitoring seasonal changes in winery-resident microbiota. PLoS One. 2013; 8:e66437.

[17] Pinto, C., Pinho, D., Cardoso, R., Custódio, V., Fernandes, J., Sousa, S., Pinheiro, M., Egas, C., Gomes, A.C. Wine fermentation microbiome: a landscape from different Portuguese wine appellations. Frontiers in Microbiology. 2015;6(905):1–13.

[18] van Broock, M., Zajonskovsky, I., Assadourian, M., Lavalle L., Caballero de Castro, A. Wine yeasts associated to merlot type grapes from North Patagonian region. An ecological study. In: Proceedings 10th International Biotechnology Symposium and 9th International Symposium on Yeasts. Sydney, Australia. 1996. p. 13–5.

[19] Caballero, A., Sangorrín, M., Lopes, C.A., Rodríguez, M.E., Lavalle, T.L., Zajonskovsky, I.E., Arcucci, G., Mariconda, L., Méndez, P. Diversity of indigenous yeasts of Southern Winegrowing Region. El vino y su industria. 2004;19:56–68.

[20] Caballero, A., Sangorrin, M., Lopes, C., Rodriguez, M.E., Zajonskovsky, I., Barbagelata, R., Lavalle, T.L. Vineyards and cellars from Neuquén, new source of yeast for the development of regional starter cultures. Proyectos Federales de Innovación productiva. Agencia Promoción Científica y Técnica. Informe Final: MECyT Resol 8 Nº1028/04. Universidad Nacional del Comahue; 2008.

[21] Rodríguez, M.E., Lopes, C.A., van Broock, M., Vallés, S., Ramón, D., Caballero, A.C. Screening and typing of Patagonian wine yeasts for glycosidase activities. Journal of Applied Microbiology. 2004;96:84–95.

[22] Curilén, Y.L., Barda, N.B., Barbagelata, R.J., Bravo Ferrada, B., Gallina, M., Semorile L.C., Caballero, A.C. Commercial yeast inoculation effect on kinetic and microbiology of fermentative process and its relation to the sensorial quality of wine. A preliminary study in patagonian pinot noir winemaking. In: Facultad de Ciencias de la Alimentación, UNER, editors. Actas del XII Congreso CYTAL—AATA; October 2009; Entre Ríos, Argentina. Sección 3, Alimentos Fermentados y Bebidas:2009. p. 6.

[23] Lopes, C.A., Sangorrín, M.P., Marongiú, A.R., Caballero, A.C. Yeasts killer as potential control agents for contaminants from Patagonian cellars. Enología. 2006a;4:44–48.

[24] Sangorrín, M.P., Lopes, C.A., Guiraudo, M.R., Caballero, A.C. Diversity and killer behaviour of indigenous yeasts isolated from the fermentation vat surfaces in four Patagonian wineries. International Journal of Food Microbiology. 2007;119:351–357.

[25] Lopes, C.A., Rodríguez, M.E., Sangorrín, M., Querol, A., Caballero, A.C. Patagonian wines: the selection of an indigenous yeast starter. Journal of industrial microbiology and biotechnology. 2007;34:539–546.

[26] del Mónaco, S.M., Bravo, S.M.E., Curilén, Y.L., Carreño, V.A., Caballero, A.C. A regional starter for high quality wines: an Argentinean Patagonia experience. Bulletin de L'OIV. 2014a;87:217–222.

[27] Fleet, G.H., Prakitchaiwattana, C., Beh, A.L., Heard, G.M. The yeast ecology of wine grapes. In: Ciani M, editor. Biodiversity and Biotechnology of Wine Yeasts. Kerala, India: Research Signpost; 2002. p. 1–18.

[28] Kurtzman, P.C., Fell, J.W., editors. The Yeasts—A Taxonomic Study, 4th ed. Amsterdam. Elsevier; 1998.

[29] Esteve-Zarzoso, B., Belloch, C., Uruburu, F., Querol, A. Identification of yeasts by RFLP analysis of the 5.8S rRNA gene and two ribosomal internal transcribed spacers. International Journal of Systematic Bacteriology. 1999;49:329–337.

[30] Martini, A. Biotechnology of natural and winery-associated strains of *Saccharomyces cerevisiae*. International Microbiology. 2003;6:207–209.

[31] Schuller, D., Alves, H., Dequin, S., Casal, M. Ecological survey of *Saccharomyces cerevisiae* strains from vineyards in the Vinho Verde Region of Portugal. FEMS Microbiology and Ecology. 2005;51:167–177.

[32] Santamaría, P., Garijo, P., López, R., Tenorio, C., Gutiérrez, A.R. Analysis of yeast population during spontaneous fermentation. Effect of the age of the cellar and the practice of inoculation. International Journal of Food Microbiology 2005;103:49–56.

[33] Mercado, L., Dalcero, A., Masuelli, R., Combina, M. Diversity of *Saccharomyces* strains on grapes and winery surfaces: analysis of their contribution to fermentative flora of Malbec wine from Mendoza (Argentina) during two consecutive years. Food Microbiology. 2007;24:403–412.

[34] Costantini, A., García-Moruno, E., Moreno-Arribas, M.V. Biochemical Transformations Produced by Malolactic Fermentation. In: Moreno-Arribas, M.V., Polo, M.C., editors. Wine Chemistry and Biochemistry. Berlin, Germany. Springer Science+ Business Media; 2009. p. 27–49.

[35] Jolly, N.P., Varela, C., Pretorius, I.S. Not your ordinary yeast: non-*Saccharomyces* yeasts in wine production uncovered. FEMS Yeast Research. 2014;14:215–237.

[36] Capozzi, V., Garofalo, C., Chiriatti, M.A., Grieco, F., Spano, G. Microbial terroir and food innovation: the case of yeast biodiversity. Microbiological Research. 2015;181:75–83.

[37] Fleet, G.H. Wine yeasts for the future. FEMS Yeast Research. 2008;8: 979–995.

[38] Lopes, C.A., Lavalle, T.L., Querol, A., Caballero, A.C. Combined use of killer biotype and mtDNA-RFLP patterns in a Patagonian Wine *Saccharomyces cerevisiae* diversity study. Antonie van Leeuwenhoek. 2006b;89:147–156.

[39] del Mónaco, S.M., Barda, N.B., Rubio, N.C., Caballero, A.C. Selection and characterization of a Patagonian *Pichia kudriavzevii* for wine deacidification. Journal of Applied Microbiology. 2014b;117:451--464.

[40] Vigentini, I., De Lorenzis, G., Fabrizio, V., Valdetara, F., Faccincani, M., Panont, C.A., Picozzi, C., Imazio, S., Failla, O., Foschino, R. The vintage effect overcomes the terroir effect: a three year survey on the wine yeast biodiversity in Franciacorta and Oltrepò Pavese, two northern Italian vine-growing areas. Microbiology. 2015;161:362–373.

[41] Garofalo, C., Russo, P., Beneduce, L., Massa, S., Spano, G., Capozzi, V. Non-*Saccharomyces* biodiversity in wine and the 'microbial terroir': a survey on Nero di Troia wine from the Apulian region, Italy. Annals of Microbiology. 2016;66:143–150.

[42] Lonvaud-Funel, A. Lactic acid bacteria in the quality improvement and depreciation of wine. Antonie van Leeuwenhoek. 1999;76:317–331.

[43] Beelman, R.B., Gallander, J.F. Wine deacidification. Advances in Food Research. 1979;25:1–53.

[44] Ruffner, H.P. Metabolism of tartaric and malic acids. Vitis. 1982;21:247–259.

[45] Henick-Kling, T. Malolactic fermentation. In: Fleet, G.H., editors. Wine Microbiology and Biotechnology. Berlin: Springer; 1993. p. 286–326.

[46] Radler, F. Yeasts-metabolism of organic acids. In: Fleet, G.H., editors. Wine Microbiology and Biotechnology. Switzerland: Harwood Academic Publishers; 1993. p. 165–182.

[47] Gao, C., Fleet, G.H. Degradation of malic and tartaric acids by high density cell suspensions of wine yeasts. Food Microbiology. 1995;12:65–71.

[48] Gawel, R., Francis, L., Waters, E.J. Statistical correlations between the in-mouth textural characteristics and the chemical composition of Shiraz wines. Journal of Agricultural and Food Chemistry. 2007;55:2683–2687.

[49] Delcourt, F., Taillandier, P., Vidal, F., Strehaiano, P. Influence of pH, malic acid and glucose concentrations on malic acid consumption by *Saccharomyces cerevisiae*. Applied Microbiology and Biotechnology. 1995;43:321–324.

[50] Hernandez-Orte P., Cersosimo M., Loscos N., Cacho J., Garcia-Moruno E., Ferreira V. Aroma development from non-floral grape precursors by wine lactic acid bacteria. Food Research International. 2009;42:773–781.

[51] Bravo-ferrada, B.M., Delfederico, L., Hollmann, A., Valdés la hens, D., Curilén, Y., Caballero, A., Semorile, L. *Oenococcus oeni* from patagonian red wines: isolation, characterization and technological properties. International Journal of Microbiology Research. 2011;3:48–52.

[52] Valdés La Hens, D., Bravo-Ferrada, B.M., Delfederico, L., Caballero, A., Semorile, L. Patagonian red wines: PCR-DGGE analysis with two targeted genes revealed the prevalence of *Lactobacillus plantarum* and *Oenococcus oeni* during spontaneous malolactic fermentations. Australian Journal of Grape and Wines Research. 2015;21:49–56.

[53] Henick-Kling, T., Sandine, W.E., Heatherbell, D.A. Evaluation of malolactic bacteria isolated from Oregon Wines. Applied and Environmental Microbiology. 1989;55:2010–2016.

[54] Lerm, E., Engelbrecht, L., du Toit, M. Selection and characterisation of *Oenococcus oeni* and *Lactobacillus plantarum* South African wine isolates for use as malolactic fermentation starter cultures. South African Journal Enology and Viticulture. 2011;32:280–295.

[55] Bravo-Ferrada, B., Hollmann, A., Delfederico, L., Valdés La Hens, D., Caballero, A., Semorile, L. Patagonian red wines: selection of *Lactobacillus plantarum* isolates as potential starter cultures for malolactic fermentation. World Journal of Microbiology and Biotechnology. 2013;29:1537–1549.

[56] Wibowo, D., Eschenbruch, R., Davis, C.R., Fleet, G.H., Lee, T.H. Occurrence and growth of lactic acid bacteria in wine: review. American Journal of Enology and Viticulture. 1985;36:302–313.

[57] Thornton, R.J., Rodriguez, S.B. Deacidification of red and white wines by a mutant of *Schizosaccharomyces malidevorans* under commercial winemaking conditions. Food Microbiology. 1996;13:475–482.

[58] Coucheney, F., Desroche, N., Bou, M., Tourdot-Marèchal, R., Dulau, L., Guzzo, J.A new approach for selection of Oenococcus oeni strains in order to produce malolactic starters. International Journal of Food Microbiology. 2005;105:463–470.

[59] Henick-Kling, T., Park, Y.H. Considerations for the use of yeast and bacterial starter cultures: SO_2 and timing of inoculation. American Journal of Enology and Viticulture. 1994;45:464–469.

[60] Rossi, I., Fia, G., Canuti, V. Influence of different pH values and inoculation time on the growth and malolactic activity of a strain of *Oenococcus oeni*. Australian Journal of Grape Wine Research. 2003;9:194–199.

[61] Arnink, K., Henick-Kling, T. Influence of *Saccharomyces cerevisiae* and *Oenococcus oeni* strains on successful malolactic conversion in wine. American Journal of Enology and Viticulture. 2005;56:228–237.

[62] Guilloux-Benatier, M., Remize, F., Gal, L., Guzzo, J., Alexandre, H. Effects of yeast proteolytic activity on *Oenococcus oeni* and malolactic fermentation. FEMS Microbiology Letters. 2006;263:183–188.

[63] Jussier, D., Morneau, A.D., Mira de Orduña, R.M. Effect of simultaneous inoculation with yeast and bacteria on fermentation kinetics and key wine

parameters of cool-climate Chardonnay. Applied and Environmental Microbiology. 2006;72:221–227.

[64] Osborne, J.P., Edwards, C.G. Inhibition of malolactic fermentation by *Saccharomyces* during alcoholic fermentation under low- and high nitrogen conditions: a study in synthetic media. Australian Journal of Grape Wine Research. 2006;12:69–78.

[65] Flanzy, C. Mundi Prensa Libros S.A., editor. Oenology: Scientific and technological foundations. Madrid, España: AMV; 2000.

[66] Volschenk, H., Viljoen-Bloom, M., Subden, R.E, Van Vuuren, H.J.J. Malo-ethanolic fermentation in grape must by recombinant strains *of Saccharomyces cerevisiae.* Yeast. 2001;18:963–970.

[67] Ribéreau-Gayon P., Dubourdieu D., Doneche B., Lonvaud A. Wine chemistry. Stabilization and treatments. In: Tratado de enología, 1ª ed. Buenos Aires, Argentina: Editorial Hemisferio Sur y Mundi Prensa; 2003. p. 3–49.

[68] du Toit, M., Pretorius, I.S. Microbial spoilage and preservation of wine: using weapons from nature's own arsenal—a review. South African Journal of Enology and Viticulture. 2000;21:74–96.

[69] Molina, A.M., Guadalupe, V., Varelab, C., Swiegers, J.H., Pretorius, I.S., Agosin E. Differential synthesis of fermentative aroma compounds of two related commercial wine yeast strains. Food Chemistry. 2009;117:189–195.

[70] Sun, S.Y., Gong, H.S., Jiang, X.M., Zhao, Y.P. Selected non-*Saccharomyces* wine yeasts in controlled multistarter fermentations with *Saccharomyces cerevisiae* on alcoholic fermentation behaviour and wine aroma of cherry wines. Food Microbiology. 2014;44:15–23.

[71] Ciani, M., Maccarelli, F. Oenological properties of non-*Saccharomyces* yeasts associated with wine-making. World Journal of Microbiology and Biotechnology. 1998;14:199–203.

[72] Soden, A., Francis, I.L., Oakey, H., Henschke, P.A. Effects of co-fermentation with *Candida stellata* and *Saccharomyces cerevisiae* on the aroma and composition of Chardonnay wine. Australian Journal of Grape and Wine Research. 2000;6:21–30.

[73] Ciani, M., Beco, L., Comitini, F. Fermentation behaviour and metabolic interactions of multistarter wine yeast fermentations. International Journal of Food Microbiology. 2006;108:239–245.

[74] Ciani, M., Comitini, F., Mannazzu, I., Domizio, P. Controlled mixed culture fermentation: a new perspective on the use of non-*Saccharomyces* yeasts in winemaking. FEMS Yeast Research. 2010;10:123–133.

[75] Rodríguez, M.E., Lopes, C.A., Barbagelata, R.J., Barda, N.B., Caballero, A.C. Influence of *Candida pulcherrima* Patagonian strain on alcoholic fermentation

behaviour and wine aroma. International Journal of Food Microbiology. 2010;138:19–25.

[76] Benito, S., Morata, A., Palomero, F., Gonzalez, M.C., Suarez-Lepe, J.A. Formation of vinylphenolic pyranoanthocyanins by *Saccharomyces cerevisiae* and *Pichia guillermondii* in red wines produced following different fermentation strategies. Food Chemistry. 2011;124:15–23.

[77] Comitini, F., Gobbi, M., Domizio, P., Romani, C., Lencioni, L., Mannazzu, I., Ciani, M. Selected non-*Saccharomyces* wine yeasts in controlled multistarter fermentations with *Saccharomyces cerevisiae*. Food Microbiology. 2011;28:873–882.

[78] Jolly, N.P., Augustyn, O.P.H. and Pretorius, I.S. The role and use of non-*Saccharomyces* yeasts in wine production. South African Journal of Enology and Viticulture. 2006;27:15–39.

[79] Morata, A., Benito, S., Loira, I., Palomero, F., Gonzalez, M.C., Suarez-Lepe, J.A. Formation of pyranoanthocyanins by *Schizosaccharomyces pombe* during the fermentation of red must. International Journal of Food Microbiology. 2012;159:47–53.

[80] Di Maio, S., Genna, G., Gandolfo, V., Amore, G., Ciaccio, M., Oliva, D. Presence of *Candida zemplinina in* Sicilian musts and selection of a strain for wine mixed fermentations. South African Journal of Enology and Viticulture. 2012;33:80–87.

[81] Contreras, A., Hidalgo, C., Schmidt, S., Henschke, P.A., Curtin, C., Varela, C. The application of non-*Saccharomyces* yeast in fermentations with limited aeration as a strategy for the production of wine with reduced alcohol content. International Journal of Food Microbiology. 2015;205:7–15.

Chapter 16

Wine Lees: Traditional and Potential Innovative Techniques for their Exploitation in Winemaking

Giovanna Fia

Additional information is available at the end of the chapter

http://dx.doi.org/10.5772/65043

Abstract

Aging of wine on lees enhances the sensorial characteristics of wine. Only a small part of the lees produced in winemaking are used in traditional aging. Most are collected and then distilled or processed to obtain low quality wine. For these reasons, lees are currently an undervalued by-product of winemaking. A new technique was tested on an industrial scale to provide wine from lees of different origin. After racking, the lees were collected in an innovative steel system and processed by cycles of mixing in controlled condition of temperature and micro-oxygenation. The processing technique contributed to improving the chemical characteristics of wine from the lees. Wines obtained from the treated lees were characterized by color intensity, total polyphenols and total polysaccharides significantly higher compared to those from not treated lees, used as control. The addition of small quantities of wine from lees to a base wine led to a significant decrease of astringency and increase in body, overall aroma, olfactory intensity, and sweet sensation. The obtained results indicate that the proposed method could be an effective tool to exploit lees on a winery scale.

Keywords: lees, wine, colour, astringency, body, aroma

1. Introduction

Wine lees are defined as 'the residue that forms at the bottom of vessel containing wine, after fermentation, during the storage or after authorized treatments, as well as the residue obtained following the filtration or centrifugation of this product' (EEC regulation No. 337/79). The word 'lees' is used also to define the heterogeneous matrix that is deposited during pre-fermentative decantation of white must. Over the course of winemaking, the lees are progressively depleted of their constituents due to decantation phases and their composition varies depending on the origin (grape variety, vinification step and type of operation). A scheme of lees origin during both red and white winemaking processes is shown in **Figure 1**.

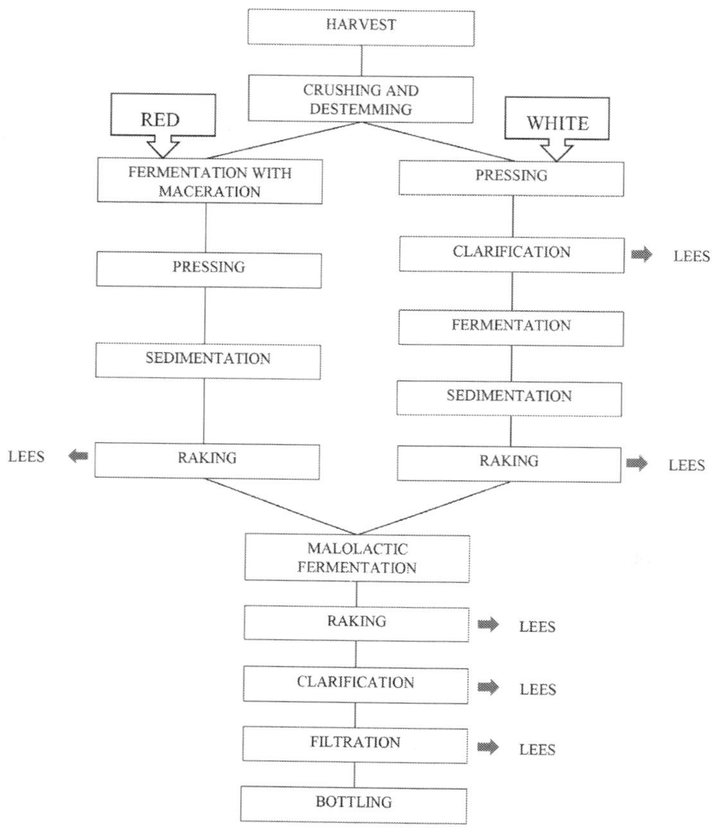

Figure 1. General scheme of lees origin during red and white winemaking processes. The lees come from several steps of racking and other operations of the production chain in which they are separated from the wine and treated as a by-product.

In Tuscany alone, in the past 2 years, the mean production of wine before lees separation was more than 2 million hectolitres and we estimate that about 5% of this volume is lees. The majority of lees produced at industrial level are collected at the wineries after racking and other operations and then disposed of. Most of the lees are sent to distilleries but this practice is currently not convenient for farms. From a financial standpoint, European support for distilleries is reduced at the production level. Alternative destinations—agronomic and energetic—are now authorized to tentatively solve the problem of lees disposal, but they could

http://dx.doi.org/10.5772/65043

generate new problems of sustainability in the long term [1, 2]. Only a small portion of lees from winemaking processes are used in traditional practice of wine aging. It is known from several studies about wine aging on lees that this matrix is complex and rich of wine active compounds and its application in winemaking improves wine characteristics. For these reasons, lees are currently an undervalued by-product of the winemaking industry.

The oenological practice of aging wine on lees is traditionally performed with lees originated after alcoholic fermentation and some racking operations. In general, these lees are a low density matrix, mainly composed of microorganisms (yeasts and bacteria) related to the winemaking process. Tartaric acid and inorganic matter are a minor part of lees composition. Some particular types of wine are left in contact with their lees during aging. For example, sparkling wine elaborated through a classic method is aged on lees after foaming. Currently, this technique is used still in wine, more widely in white than in red. The major effect of aging on lees in white wine is the reduction of oxidative phenomena of colour and aroma [3]. In red wine, colloids of lees prevent the precipitation of complexes that are formed between tannins and anthocyanins, and the practice of aging on lees can lead to a better stability of wine colour [4]. Contact with lees produces a less astringent wine, with slightly less colour intensity. Body and roundness of the wine are enhanced by contact with lees. The aromatic and phenolic fraction is deeply modified during the aging due to the complex interactions between lees and wine compounds. In general, traditional ageing of wine on lees is described as a technique that enhances the sensorial characteristics of wine. The oenological potential of lees originates in their complex composition and properties exploited by a traditional approach [5]. During aging, autolysis of yeast cells modifies the composition of lees in terms of wine active compounds [6, 7]. Indeed, mannoproteins, polysaccharides, lipids, volatile compounds and enzymes are released by yeast into the wine during post-fermentative contact.

Despite the positive effects described above, some problems can arise during aging on lees. Sulphur odours and off-smell can appear in wine during aging on lees, primarily due to oxygen consumption by dead yeasts and development of spoilage microorganisms. Lees can be responsible for the presence of precursors and enzymes that, under favourable conditions, can lead to the synthesis of biogenic amines [8]. These compounds are responsible for commonly reported disagreeable odours and are also a risk for consumers due to their physiological effects [9]. On the other hand, lees are described as a matrix that can play a role in the removal of undesirable compounds of wine such as volatile phenols and residues form treatments [5]. Over the course of traditional aging, lees are periodically re-suspended by stirring (a practice of French origin called 'batonnage') to increase the amount of macromolecules extracted into the wine. To prevent the appearance of reduction notes in wine, aging on less is traditionally performed in oak barrels. The micro-oxygenation technique, which consists in the addition of small and controlled doses of oxygen to wine aging in steel tank, may be an effective way to manage this practice. The use of adequate doses of sulphur dioxide is useful to prevent microbiological alterations.

The traditional technique of aging wine on lees involves a considerable demand of winery resources, a potential disadvantage. Indeed, long times of wine storage, frequent 'batonnages' and close monitoring of the evolution of the product are required during aging on lees to achieve positive results.

With the aim of improving the efficiency of aging wine on lees, several techniques were previously investigated. Enzyme-assisted and physical treatments, such as ultrasound and microwave, were employed during on lees processing to maximize the extraction of wine active compounds from yeast cells [10, 11]. The results of these studies concern the improvement of traditional technique by the addition of lyophilized yeast cells at different doses. The exploitation of the large amount of lees generated by winemaking processes still remains an unsolved problem.

Recently, a new technique was tested on an industrial scale for the management of liquid lees collected after the first and second racking during the winemaking process to obtain high-quality wine [12, 13]. The achievement of this objective allows wine producers to develop an efficient and sustainable strategy for the exploitation of lees at the winery.

2. Materials and methods

2.1. Processing system

An innovative vertical tank (under patent) was used (**Figure 2a**). It is made of stainless steel (25 hL capacity), designed to maintain the temperature by means of cooling bands that cover 75% of the outer surface and insulation (60 mm) on the entire surface. Inside the system, there are several mixing means splined to a shaft. Movement of the means on the shaft puts the system into action. Each mixing mean includes two coplanar agitators. The first agitator moves the oenological matrix downward while the second moves the oenological matrix upward, producing a gentle swirl. Depending on the work phases, the speed of rotation is variable and the direction of the mixing means can be either clockwise or counterclockwise. The system is equipped with several accessories for the optimization of loading, draining, pump-over, micro-oxygenation of the product and discharge of semi-solid residue. During the process, a device automatically controls operations and product temperature. The processing system is suitable for the innovative lees management technique and for traditional oenological practices such as pre- and post-fermentative maceration, clarification and storage. Trials were conducted in 2012, 2013 and 2014. A semi-automatic prototype was used in 2012.

2.2. Lees

At the winery, lees of different origin were collected after fermentation and used for the trials. Test A was performed with lees from Sangiovese (80%) and Cabernet Franc (20%) red grapes, vintage 2012. Lees from Sangiovese grapes (100%) were used for the test C, vintage 2013. In 2013, several varieties of red and white grapes (Cabernet Franc, Syrah, Merlot, Montepulciano, Cabernet Sauvignon and Viognier) were vinified separately, and the collected lees were used to perform test F. In 2014, lees derived from different winemaking processes of white grapes

(Chardonnay, Viognier, Bellone and Incrocio Manzoni) were collected at the winery and then processed with the aid of the processing system.

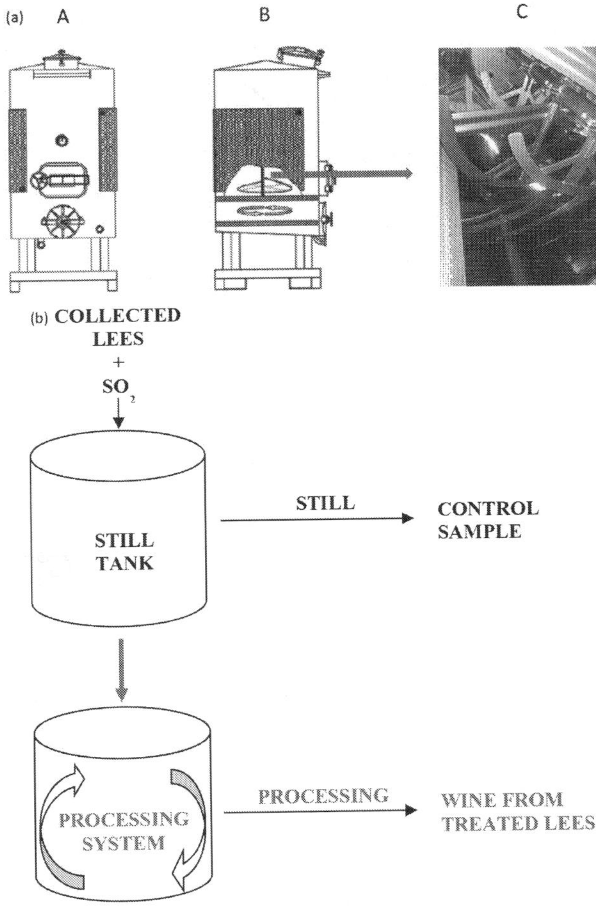

Figure 2. (a) Processing system (front view A; side view B) used to perform treatment and a particular (C) of the inside with stirring device. (b) Scheme of the processing technique conducted using the innovative processing system. A β-glucanase commercial preparation was added at the dose of 10 g/hL directly into the system for tests C and F. Samples C and F were micro-oxygenated at the dose of 3 mg/L/month.

2.3. Processing techniques

The red grapes were harvested at maturity in excellent health and vinified depending on the winery protocol. After completion of alcoholic and malolactic fermentations, the wines were kept in a tank for sedimentation and then racked. A scheme of the processing technique is shown in **Figure 2**.

The lees were collected in a steel tank and added with SO_2 (60 mg/L). The lees, with density varying from 1.1 to 1.4 g/L, were pump-overed for 30 min and transferred to the processing system. All tests were performed at 22°C. For test A (vintage 2012), the lees were processed for 30 days by stirring every 2 days for 30 min. For the analyses, samples were taken at the beginning and at the end of treatment, immediately after mixing. In 2013, the lees were stirred every 8 h for 10 min for 7 days (tests C and F) and micro-oxygenated at the dose of 3 mg/L/month. Then, the lees were kept still at 20°C, with micro-oxygenation (3 mg/L/month) until 30 days. Samples were taken from the system at the beginning (0), after 3 and 7 days, immediately after mixing and 1 month later (30 days). A β-glucanase commercial preparation (10 g/hL) was added to the lees at the beginning for tests C and F. Control samples CA, CC and CF were obtained from the tanks where lees were maintained still at 20°C, without providing O_2 and analysed after 30 days (**Figure 2**). Analysis of wine from the lees was performed on clear supernatant after centrifugation of the samples for 10 min, at 7000 rpm at 4°C.

In 2014, white grapes of different varieties (Chardonnay, Viognier, Bellone and Incrocio Manzoni) were harvested at maturity and vinified separately, according to the winery protocol. After the end of fermentations, the wines were decanted and then racked. The lees, collected in a steel tank, were added with SO_2 (60 mg/L) and transferred to the processing system. The lees were stirred every 8 h for 10 min for 15 days at 22°C and 3 mg/L/month of O_2 were provided. The lees were then kept still inside the system at 20°C for sedimentation. After racking, the obtained wine was aged in oak barrels for 6 months and then used to blend a white wine.

2.4. General analyses

Reducing sugar, total acidity and pH of wine were evaluated according to the official or usual methods recommended by the International Organization of the Vine and Wine (OIV) [14]. Analyses were performed in duplicate.

2.5. Phenolic indexes and wine colour

The total polyphenol index (TPI_{280}) was determined by measuring the absorbance at 280 nm of a 1:50 dilution of wine using a 10-mm quartz cuvette [15]. Total flavonoids (TF), total anthocyanins (TA) and total flavanols (TFn) were estimated on an acquired spectrum between 230 and 700 nm of the wine diluted 50-fold in acidic ethanol solution (ethanol:H_2O:HCl = 70:30:1), according to the method described by Di Stefano et al. [16]. Wine colour at 420, 520 and 620 nm was measured in cuvettes with 1 mm of optic pathway; colour intensity (I) was calculated as the sum of A420, A520 and A620 and hue (H) as the ratio A420/520 and expressed as absorbance unit (Au) [17]. The contribution of co-pigmented anthocyanins to the total wine colour at pH 3.6 (Copig %), the degree of anthocyanin polymerization (Pol %), the monomeric anthocyanins (Mon %) and the estimate cofactors (C) were determined [18]. Gelatin index (G) was evaluated as described by Glories [19]. All measurements of the absorbance values were obtained with a Perkin Elmer Lambda 10 Spectrophotometer (Massachusetts, USA).

2.6. Astringency mucin index

Wine polyphenols were purified as described by Condelli et al. [20]. A 10 g C18 cartridge (Bond Elut C18, Varian) was activated with 25 mL of methanol and 50 mL of distilled water. Ten millilitres of wine was loaded on the cartridge. Carbohydrates and acidic compounds were washed with 10 mL of 0.01% H_2SO_4 solution. Phenolic compounds were eluted with 15 mL of methanol and a rotary evaporator was used to remove solvents. The phenolic extract was dissolved in 10 mL of 1% ethanol.

The astringency mucin index (AMI) was determined by reacting standard protein, mucin from pig stomach (Sigma, Milan, Italy), with phenols dissolved in 1% ethanol according to Monteleone et al. [21]. The AMI was expressed in terms of nephelometric turbidity units (NTU). The phenol-mucin sample (NTUS) consisted of phenolic extracts (2 mL) mixed with 0.5 mL of mucin solution (0.2% w/v, in citrate phosphate buffer, pH 3.5). The mucin solution (0.5 mL) mixed with 1% ethanol (2 mL) and phenolic extracts (2 mL) mixed with citrate phosphate buffer (0.5 mL) were used as control samples (mucin control-NTUM and phenol control-NTUP, respectively). After 1 min of reaction at 37°C, the turbidity value of all mixes was measured by a Hach 2001N Laboratory Turbidimeter (Hach Co., Loveland, CO) in nephelometric turbidity units. Wine sample phenolic extracts were tested in duplicate. The following formula was used to compute AMI: AMI = NTUS – (NTUM + NTUP).

2.7. Total proteins

Proteins were isolated by precipitation with ethanol [22, 23]. Protein content was evaluated on the ethanol extract. To extract phenols, absolute ethanol (30 mL) was added to 10 mL of wine. The sample was kept at 0°C for 1 h, and then it was centrifuged at 12,000 rpm at 4°C for 10 min. The obtained pellet was washed once with ethanol, and the excess of ethanol was gently removed with the aid of a Pasteur pipette and vacuum pump. After drying at 40°C for 1 h, the pellets were re-solubilized in 1.0 mL of distilled water. Proteins were quantified by Protein Assay kit (Bio-Rad, Milan, Italy). The method is based on the absorbance maximum shifts from 465 to 595 nm of an acidic solution of Coomassie Brilliant Blue G-250 when binding to the protein occurs [24]. A standard curve of bovine serum albumin (BSA) was prepared in a range of concentration from 0.2 to 1.4 mg/mL. All assays were performed in duplicate and averaged.

2.8. Total polysaccharides

Total polysaccharide (TP) was evaluated following the method described by Usseglio-Tommaset [25]. Twenty mL of wine was added with 100 mL of absolute ethanol and 1 mL of 37% HCl, diluted 1:1, to avoid tartaric precipitation. Solution was mixed, kept at 4°C for 24 h and then filtered through a previously weighed membrane (pore size of 0.45 μm). Precipitate was washed two times with 20 mL of ethanol. The membrane was dried at 40°C for 1 h, stored in a desiccator for 30 min and then weighed. The amount of total polysaccharides was obtained by subtracting weight of the filter from the weight of filter with the precipitate. All assays were performed in duplicate and averaged. The total polysaccharide content was expressed as milligram per litre of wine (mg/L).

2.9. Sensory analysis

Descriptive sensory analysis was performed following the sensory profile method according to standard ISO U590A1950 (ISO, 1998) by a panel composed of 20 enology and viticulture students from the University of Florence. The descriptors were scored on a scale of 0–10 (0: descriptor was not perceived, 10: high intensity).

2.10. Statistical analysis

Chemical analyses were performed in duplicate, and the data are presented as mean ± standard deviation. Analysis of variance and comparison of treatment means (LSD, 5% level) were performed using Statgraphics Plus 3.1. Principal component analysis (PCA) was performed using Unscrambler X 10.2 software on data of the sensory descriptions to find the dominant sensory attributes of the wines.

3. Results and discussion

3.1. Wine from lees (vintages 2012 and 2013)

Chemical parameters of wine samples A, C and F were used to perform a principal component analysis (**Figure 3**) [13].

Figure 3. The PCA of the chemical parameters evaluated for the samples A, C and F at the beginning (0), after 3 and 7 days of treatment and 30 days later. CA30, CC30 and CF30 are the controls analysed after 30 days. Colour intensity (I) and hue (H), total polyphenol (TPI_{280}), total flavonoids (TF), total anthocyans (TA), total flavanols (TFn), monomeric anthocyanins (Mon %), polymeric anthocyanins (Pol %), copigmented anthocyanins (Copig %), cofactors (C), astringency mucin index (AMI), gelatine index (G), total polysaccharides (TP) and total protein (TPr).

A large part of the variance (72%) is explained by the first two principal components. Samples grouped according to the vintage, variety of grapes and type of treatment of the lees. Control samples (CA30, CC30 and CF30), analysed after 30 days, were separated from those at the start of processing (A0, C0 and F0) and from the treated samples A, C and F collected during processing. CA30 and CF30 resulted on the opposite side of the graph with respect to the samples of wine A30 and F30 originating from treated lees. The treated samples were different from the control for most of the chemical parameters measured, which were significantly higher (**Figure 3**). The evolution of colour intensity and hue, total polyphenol index, total polysaccharides, total protein and astringency mucin index during processing of samples A, C and F are shown in **Tables 1–3**.

Sample	I	H	TPI_{280}	TP (g/L)	TPr (mg/L)	AMI
A0	6.2 ± 0.1b	0.8 ± 0.02b	48.1 ± 0.2c	1.4 ± 0.0a	23.2 ± 1.0c	18.2 ± 0.0c
A30	6.4 ± 0.1c	0.7 ± 0.02a	47.5 ± 0.2b	1.7 ± 0.0b	11.9 ± 2.6b	12.4 ± 0.0b
CA30	3.2 ± 0.2a	0.9 ± 0.02c	39.0 ± 0.2a	1.4 ± 0.0a	4.3 ± 1.7a	8.3 ± 0.4a

CA30 is the control sample after 30 days. Colour intensity (I) and hue (H), total polyphenol (TPI_{280}), astringency mucin index (AMI), total polysaccharides (TP) and total protein (TPr).
Data expressed as mean ± SD. Mean values labelled with different letters indicate significant differences among the samples ($P < 0.05$).

Table 1. Chemical parameters of samples A at the beginning (0) and after 30 days of treatment (A30).

Sample	I	H	TPI_{280}	TP (g/L)	TPr (mg/L)	AMI
C0	3.4 ± 0.2b	0.7 ± 0.02b	37.0 ± 0.2b	1.2 ± 0.0a	15.4 ± 2.1e	7.8 ± 0.5d
C3	3.6 ± 0.1b	0.6 ± 0.02a	38.0 ± 0.2c	1.3 ± 0.0b	9.2 ± 1.2d	5.7 ± 0.6bc
C7	3.6 ± 0.1b	0.6 ± 0.02a	40.0 ± 0.3d	1.1 ± 0.1a	4.1 ± 0.8c	5.0 ± 0.7b
C30	5.3 ± 0.2c	0.6 ± 0.02a	41.0 ± 0.2e	1.7 ± 0.0d	1.6 ± 1.7a	6.2 ± 0.4c
CC30	2.8 ± 0.1a	0.9 ± 0.02c	34.0 ± 0.2a	1.4 ± 0.0c	0.8 ± 1.8a	4.3 ± 0.0a

CC30 is the control sample after 30 days. Colour intensity (I) and hue (H), total polyphenol (TPI_{280}), astringency mucin index (AMI), total polysaccharides (TP) and total protein (TPr).
Data expressed as mean ± SD. Mean values labelled with different letters indicate significant differences among the samples ($P < 0.05$).

Table 2. Chemical parameters of samples C at the beginning (0), after 3 (C3) and 7 (C7) days of treatment and 1 month later (C30).

Sample	I	H	TPI_{280}	TP (g/L)	TPr (mg/L)	AMI
F0	9.4 ± 0.1d	0.6 ± 0.02a	49.3 ± 0.2e	1.1 ± 0.0c	60.2 ± 2.0e	26.9 ± 1.4e
F3	8.8 ± 0.1c	0.6 ± 0.02a	44.9 ± 0.2b	0.9 ± 0.0b	48.6 ± 1.5c	23.1 ± 1.5d
F7	9.3 ± 0.2d	0.7 ± 0.02b	47.2 ± 0.2c	0.9 ± 0.0b	44.8 ± 2.0b	11.4 ± 0.8c
F30	8.1 ± 0.1b	0.7 ± 0.02b	48.4 ± 0.2d	0.8 ± 0.0a	56.8 ± 2.4d	7.8 ± 0.1b
CF30	4.8 ± 0.2a	0.7 ± 0.02b	32.3 ± 0.2a	0.8 ± 0.0a	34.1 ± 0.2a	0.0 ± 0.0a

CF30 is the control sample after 30 days. Colour intensity (I) and hue (H), total polyphenol (TPI_{280}), astringency mucin index (AMI), total polysaccharides (TP) and total protein (TPr).
Data expressed as mean ± SD. Mean values labelled with different letters indicate significant differences among the samples ($P < 0.05$).

Table 3. Chemical parameters of samples F at the beginning (0), after 3 (F3) and 7 (F7) days of treatment and 1 month later (F30).

After 30 days, colour intensity of the processed samples A30 (6.4 Au) and C30 (5.3 A.U) was higher than those of the samples (A0 and C0) analysed at the start (6.2 and 3.4 Au) (**Tables 1** and **2**). On the contrary, sample F30, which maintained a high colour intensity until 7 days (9.3 Au), showed a decrease to 8.1 Au after 30 days. After 1 month, polysaccharide content of samples A and C passed from 1.4 to 1.7 and from 1.2 to 1.7 mg/L, respectively. Conversely, in sample F, polysaccharides decreased from 1.1 to 0.8 (**Tables 1–3**). In all tests, total protein content and astringency mucin index values decreased over the course of processing. The astringency mucin index is a useful tool to predict the strength of perceived astringency, induced by polyphenols of wine [26].

In comparison to the control samples, the intensity of colour of all treated samples after 30 days was higher while hue remained quite stable, indicating that processing contributes to maintaining colour characteristics of the wine (**Tables 1–3**). In all tests, total polyphenol index and protein of control samples analysed after 30 days were at significantly lower levels with respect to the treated samples, indicating a depletion of the wine maintained still on the lees. Total polysaccharides content of the treated samples A and C (**Tables 1** and **2**) was higher than that of the control while it was similar to the control in sample F (**Table 3**). The AMI index of processed samples was always higher compared to that of the control, which was at a very low level. During the treatment, no off-flavours or off-odours arose in wine without further addition of SO_2. Overall results indicated that the processing technique can lead to improving or maintaining the chemical characteristics of wine from the lees.

The release of mannoproteins, parietal polysaccharides of yeast, during aging on lees was previously reported by other authors [5]. Several months were required to obtain a significant increase of these compounds during traditional post-fermentative contact of wine with their lees [5]. With the main objectives of shortening, the aging on lees period and improving the release of polysaccharides from yeast cells, some methods were tested by other authors during traditional processes [10, 11]. The innovative technique described in this work differs from those previously investigated for a general target. Indeed, the new technique aims to exploit most of the liquid lees produced at winery scale by a method of industrial application. Lees with a very high density and complex composition (cells of yeast and bacteria, organic and inorganic matter derived from grapes) were processed. The yield in wine can reach 65% in function of the type of lees processed. The addition of enzymes can improve extraction but the method is efficient also without enzyme addition, as demonstrated by the results obtained for sample A. Extraction of chemical compounds and stability of wine colour can be achieved by means of the type of mixing undertaken by the new processing system and micro-oxygenation of the product.

Our results indicated that the wine from the lees maintained or improved colour characteristics also with enzyme (10 g/hL) addition, contrary to that observed by other authors who used a β-glucanase preparation at the dose of 5 g/hL [10].

Regarding the application of physical treatments to shorten the process of aging on lees, the positive results obtained by ultrasound application on the release of polysaccharides from yeast cells into the wine do not seem related to sensory perception [11]. Moreover, the same authors highlighted some problems of active compounds (anthocyanins and volatile) lost and oxidation during the treatment.

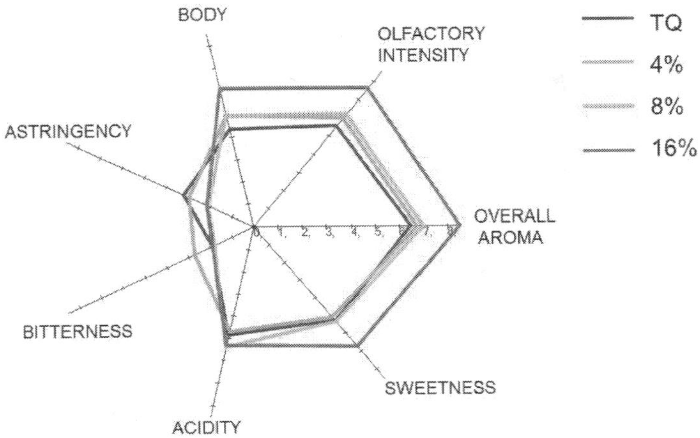

Figure 4. Sensory profiles of a white wine (TQ) and of the same wine added with different percentage of wine from lees.

During processing, chemical analysis of the red wine obtained using the new technique revealed variations in the contribution to wine colour by monomeric, polymeric and co-pigmented anthocyanins. Sensory evaluation showed that in blending with a base wine, the wines obtained with the new technique from white lees could contribute to decrease astringency sensation and enhance body, olfactory intensity, overall aroma and sweet characteristics of the final wine (**Figure 4**). Similar results were also obtained for wine from red lees (data not shown).

The new system proposed in this work has the advantage of its multi-functionality and the proposed technique can contribute to reach the objective of exploiting lees rapidly at the winery. In 2013, with the introduction of the new system, only 7 days of processing led to a significant increase of colour intensity. A decrease of anthocyanins in wines after contact with lees is reported by other authors [5]. Adsorption of anthocyanins on lees is involved in the decrease of these compounds during traditional aging. These phenomena are reversible, and desorption occurs mainly when anthocyanins oxidize or bind tannins. It can be assumed that the increase of colour intensity of wine observed after processing is due to the release of pigments adsorbed on the matrix. The stirring technique and oxygen provided during processing may have contributed to a stabilization of anthocyanins, promoting the reaction with tannins. The reactivity of polyphenols against proteins, measured in terms of AMI, slightly decreased during processing and indicated that the final wine can be perceived as less astringent. However, AMI of the control analysed after 30 days was at a very low level revealing a loss of body of the wine maintained still on the lees [26].

3.2. Wine from lees (vintage 2014)

In 2014, lees originating from vinification of white grapes (Chardonnay, Viognier, Bellone and Incrocio Manzoni) were processed. The obtained white wine was aged for 6 months in oak barrels and then used to blend a base white wine (Viognier, vintage 2013). Wine from the lees showed a total acidity of 5.53 g/L as tartaric acid, pH 3.45 and residual sugar below 1 g/L. Chemical parameters of Viognier were 5.51 g/L total acidity (as tartaric acid), pH 3.35 and residual sugar below 1 g/L. Sensory profiles of Viognier base wine (TQ) and Viognier added with different percentages (4%, 8% and 16%) of the wine from the lees are shown in **Figure 4**.

Astringency of Viognier added with 8% and 16% of wine from lees was significantly lower compared to the Viognier TQ. Moreover, Viognier added with 16% of wine from lees had body, olfactory intensity, overall aroma and sweet sensation significantly higher with respect to the Viognier TQ.

4. Conclusions

The chemical and sensory characteristics of wines obtained from lees following an innovative technique were attained. These characteristics can vary as a function of the type of lees and conditions of processing. Evaluation of the chemical characteristics reveals good quality of the wines from the lees and the sensorial approach confirms and emphasizes this judgment. A high degree of efficiency was achieved by introducing an upgraded industrial system at the winery and optimization of the process (micro-oxygenation and automatic control). In conclusion, the overall results indicate that the proposed method can be an effective tool to exploit the lees on a winery scale.

Acknowledgements

The author like to thank the firm Enoitalia Srl, Cerreto Guidi, Firenze, Italia and Vino Vigna, di Claudio Gori, Empoli, Firenze for the financial support given to carry out this research. The author is also grateful to the farms Ômina Romana, Velletri, Roma and Tenuta Il Poggione, Montalcino, Siena for the availability to perform tests in the cellar.

Author details

Giovanna Fia

Address all correspondence to: giovanna.fia@unifi.it

Department of Management of Agricultural System, Food and Forestry, Florence, Italy

References

[1] Krzywoszynska A. 'Waste? You mean by-products!' From bio-waste management to agro-ecology in Italian winemaking and beyond. The Sociological Review. 2012;60:47–65.

[2] Ruggieri L, Cadena E, Martínez-Blanco J, Gasol CM, Rieradevall J, Gabarrell X, Gea T, Sort X, Sánchez A. Recovery of organic wastes in the Spanish wine industry. Technical, economic and environmental analyses of the composting process. Journal of Cleaner Production. 2009;17:830–838.

[3] Lavigne V, Pons A, Dubourdieu D. Assay of glutathione in must and wines using capillary electrophoresis and laser-induced fluorescence detection. Changes in concentration in dry white wines during alcoholic fermentation and aging. Journal of Chromatography A. 2007;1139:130–135.

[4] Escot S, Feuillat M, Dulau L, Charpentier C. Release of polysaccharides by yeasts and the influence of released polysaccharides on colour stability and wine astringency. Australian Journal of Grape and Wine Research. 2001;7:153–159.

[5] Pérez-Serradilla JA, Luque de Castro MD. Role of lees in wine production: A review. Food Chemistry. 2008;111:447–456.

[6] Rosi I, Gheri A, Domizio P, Fia G. Production of parietal macromolecules by *Saccharomyces cerevisiae* and their influence on malolactic fermentation. In: Proceedings of the Symposium on Colloids and Mouthfeel in Wines; 27–29 May 1999; Montreal. Toulouse Cedex: Lallemand S.A.; 1999;35–39.

[7] Del Barrio-Galán R, Pérez-Magariño S, Ortega-Heras M. Techniques for improving or replacing ageing on lees of oak aged red wines: The effects on polysaccharides and the phenolic composition. Food Chemistry. 2011;127:528–540.

[8] González-Marco A, Ancín-Azpilicueta C. Influence of lees contact on evolution of amines in Chardonnay wine. Journal of Food Science. 2006;71:544–548.

[9] Bauza T, Blaise A, Mestres JP, Teissedre PL, Daumas F, Cabanis JC. Biogenic amines contents and their variation parameters in côtes du Rhone, vallée du Rhone and Provence wines. Science Des Aliments. 1995;15:367–380.

[10] Palomero F, Morata A, Benito S, González M.C, Suárez-Lepe JA. Conventional and enzyme-assisted autolysis during ageing over lees in red wines: Influence on the release of polysaccharides from yeast cell walls and on wine monomeric anthocyanin content. Food Chemistry. 2007;105:838–846.

[11] Liu L, Loira I, Morata A, Suárez-Lepe JA, González MC, Rauhut D. Shortening the ageing on lees process in wines by using ultrasound and microwave treatments both

combined with stirring and abrasion techniques. European Food Research and Technology. 2016;242:559–569.

[12] Fia G, Gori C. Alternative management of lees in the vinification of Sangiovese grape. In: Proceedings of the Third Edition of the International Conference Series on Wine Active Compounds; 26–28 March 2014; Beaune. Beaune: Université de Bourgogne, France; 2014;71–72.

[13] Fia G, Zanoni B, Gori C. A new technique for exploitation of wine lees. Agriculture and Agricultural Science Procedia. 2016;8:748–754.

[14] International Organisation of the Vine and Wine (OIV) [Internet]. 2014. Available from: http://www.oiv.int/oiv/info/frmethodesanalyses.

[15] Ribéreau-Gayon P, Glories Y, Maujean A, Dubordieu D, editors. Handbook of Enology. 2nd edition. Chichester: Wiley; 2006;141 p.

[16] Di Stefano R, Cravero MC, Gentilini N. Metodi per lo studio dei polifenoli dei vini. Enotecnico. 1989;25:83–89 .

[17] Glories Y. La couleur des vins rouges. 2éme partie: mesure, origine et interpretation. Connaisance de la Vigne et du Vin. 1984;18:253–271.

[18] Boulton RB. A method for the assessment of copigmentation in red wines. Presented at the 47th Annual Meeting of American Society for Enology and Viticulture, Reno, NV; June 1996.

[19] Glories Y. Recherchessur la structure et les propiétés des composes phénoliques polymérisés des vins rouges III. Fractionnemant des composes phénoliques avec la polyvinylpyrrolidone (p.v.p.). Connaissance Vigne Vin. 1976;10:51–71.

[20] Condelli N, Dinnella C, Cerone A, Monteleone E, Bertuccioli M. Prediction of perceived astringency induced by phenolic compounds II: Criteria for panel selection and preliminary application on wine samples. Food Quality and Preference. 2006;17:96–107.

[21] Monteleone E, Condelli N, Dinnella C, Bertuccioli M. Prediction of perceived astringency induced by phenolic compounds. Food Quality and Preference. 2004;15:761–770.

[22] Moreno-Arribas MV, Pueyo E, Polo MC. Peptides in musts and wines. Changes during the manufacture of Cavas (sparkling wines). Journal of Agricultural and Food Chemistry. 1996;44:3783–3788.

[23] Moreno-Arribas MV, Pueyo E, Polo MC. Analytical methods for the characterization of proteins and peptides in wines. Analytica Chimica Acta. 2002;458:63–75.

[24] Bradford MM. A rapid and sensitive method for the quantification of microgram quantities of protein utilizing the principle of protein-dye binding. Analytical Biochemistry. 1976;72:248–254.

[25] Usseglio-Tommaset L. Les Colloïdes glucidiques solubles des moûtset des vins. Connaissance Vigne Vin. 1976;10:193–226.

[26] Fia G, Dinnella C, Bertuccioli M, Monteleone E. Prediction of grape polyphenol astringency by means of a fluorimetric micro-plate assay. Food Chemistry. 2009;113:325–330.

Chapter 17

Viticultural and Biotechnological Strategies to Reduce Alcohol Content in Red Wines

Miguel Ángel Olego, José Manuel Álvarez-Pérez, Miguel Javier Quiroga, Rebeca Cobos, Mario Sánchez-García, Jesús Esteban Medina, Sandra González-García, Juan José Rubio Coque and José Enrique Garzón-Jimeno

Additional information is available at the end of the chapter

http://dx.doi.org/10.5772/64982

Abstract

Viticultural and biotechnological strategies are two approaches to deal with higher must sugar levels at harvest time. A wide range of factors could significantly affect sugar accumulation in the grape such as choice of vineyard site, soil composition, irrigation strategy, rootstock, and grape cultivar selection as well as grape yield. In this sense, approaches to canopy management are continually evolving in response to changes in other vineyard management practices; some of these could contribute to reduce soluble sugars on grape berries at harvest time. On the other hand, among possible biotechnological strategies, one of the most relevant is the control of the fermentative process by using selected yeast strains. In this chapter, we will show how some viticultural practices have influenced the accumulation of soluble sugars and other enological parameters in grape berries at harvest time. We will also report how a careful yeast selection and the implementation of different fermentation strategies can also contribute to reduce ethanol content in wines.

Keywords: auxins, microfermentation, total soluble solids, veraison, yeast

With the data set obtained in both experiments with NAA (cvs. Tempranillo, Cabernet Sauvignon, and Syrah), the relationship between YAN, MA, and TSS levels in order to assess the intercorrelations among these must quality parameters was studied (**Figure 3**). From the data in **Figure 3**, it is apparent a possible linear relation between YAN and MA levels. Furthermore, levels of MA are positively correlated with levels of YAN. On the other hand, the findings do not indicate an apparent pattern in case of TSS with any of the other parameters.

Although the mechanisms that control the ripening of the nonclimacteric grape berry are poorly understood [21], the results of this study indicate the ability of NAA to decrease TSS at harvest time. Although the lower levels of TSS in treated berries may be mainly due to a delay in sugar accumulation, these data suggest that auxin treatments may be useful in controlling high must sugar levels at harvest time.

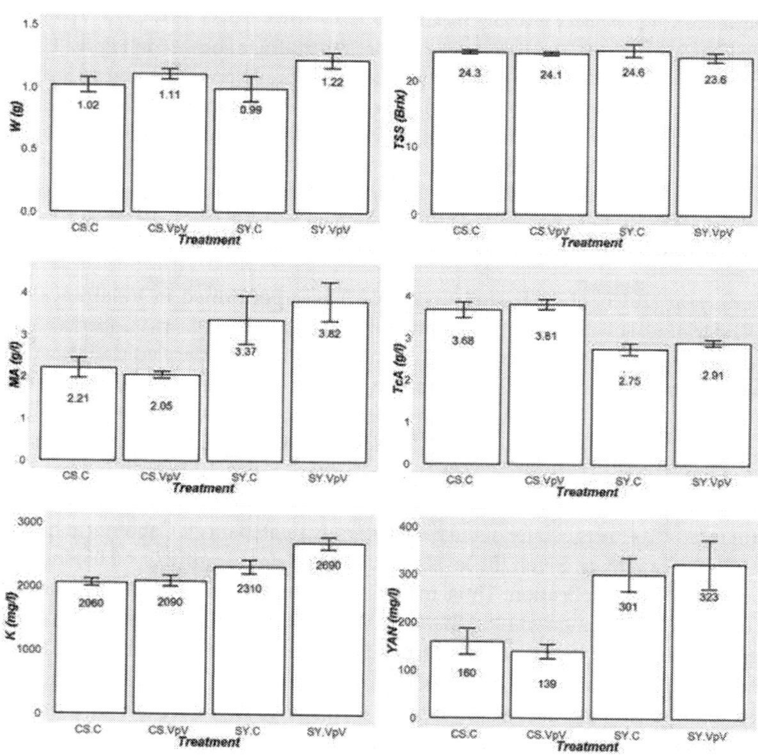

Figure 2. Bar graphs of berry weight (W), total soluble solids (TSS), malic acid (MA), tartaric acid (TcA), potassium in must (K), and yeast assimilable nitrogen (YAN) (2015). Average values are displayed within bar graphs. Standard errors are shown as bars (±1 SE mean). CS.C and SY.C: control in Cabernet sauvignon and Syrah, respectively; CS.VpV and SY.VpV: NAA sprayed 5 days pre- and postveraison in Cabernet sauvignon and Syrah, respectively.

http://dx.doi.org/10.5772/64982

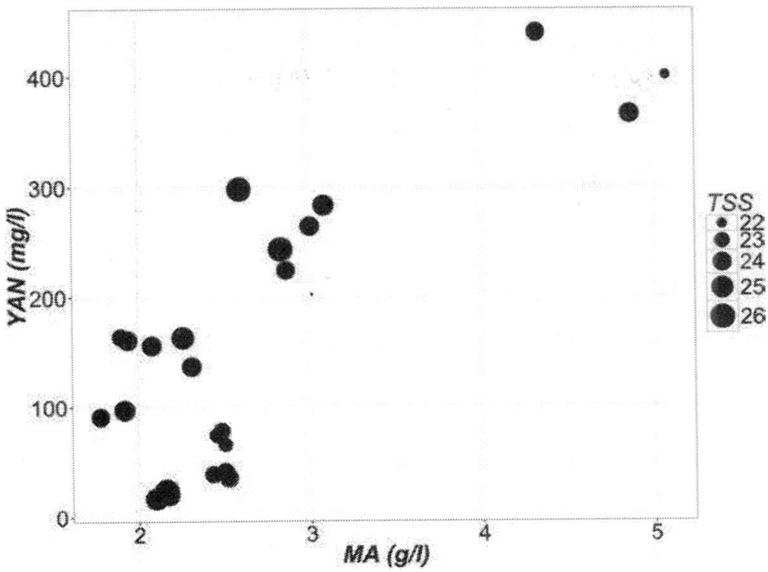

Figure 3. Relationship between malic acid (MA), yeast assimilable nitrogen (YAN), and total soluble solids (TSS) levels in grapes at harvest time. TSS in Brix scale.

3. Effects of Mg^{2+} foliar fertilization on berry sugar content

It must be recognized that grapevine nutrition remains an important part of managing a vineyard since it impacts on berry development and, finally, wine quality is derived to a large degree from berry composition. Some grape growers avoid any fertilizer for fear of overstimulating growth, whereas in other cases vineyard blocks might be fertilized when only specific areas of the block require fertilizer. Therefore, it is important that growers have a sound basis for determining the fertilizer needs of their vines [22]. Elsewhere, since the general relationship between vine nutritional status (in both nutrient macro- and microelements) and grape composition is obscure, further efforts are necessary to acquire greater knowledge in this topic. This is an important knowledge gap because these elements should necessarily influence grape juice quality and, therefore, the vinification process.

It is recognized that plants need K^+ for the formation of sugars and starches, for protein synthesis, and for cell division. Additionally, K^+ also neutralizes organic acids, regulates the activities of other mineral nutrients in plants, activates certain enzymes, and helps adjust water relationships (Hewitt, cited by [20]), but free potassium ions are released when the grape cell membranes are broken during grape processing, and form crystals with tartrate, which drop grape juice and wine acidity [11]. On the basis of the antagonistic interaction between levels of K^+ and Mg^{2+} reported by several authors at the root-soil interface [23, 24], another study was performed during 2014 vintage in the PDO area of Ribera del Duero, Spain. The impact of Mg^{2+} supply on berry chemistry attributes from this trial is shown below.

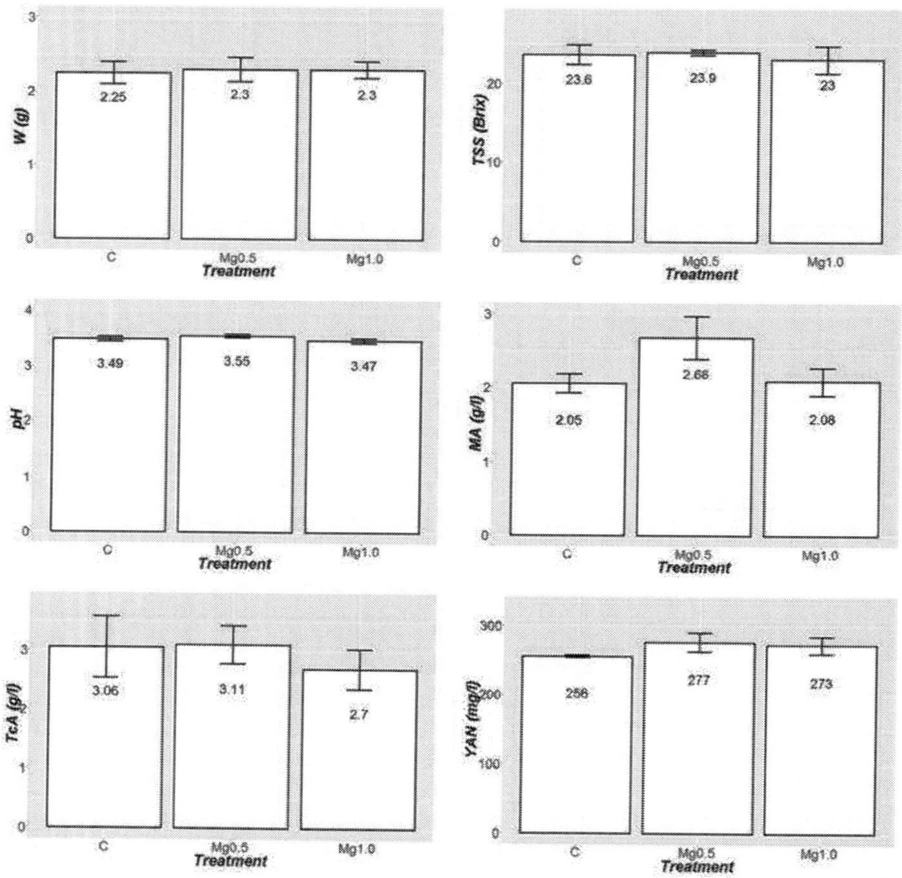

Figure 4. Bar graphs of berry weight (W), total soluble solids (TSS), real acidity (pH), malic acid (MA), tartaric acid (TcA), and yeast assimilable nitrogen (YAN) (2014). Average values are displayed within bar graphs. Standard errors are shown as bars (±1 SE mean). C: control; Mg0.5: foliar Mg at 0.5 kg/ha; Mg1.0: foliar Mg at 1.0 kg/ha.

The cultivar chosen, *Vitis vinifera* L. Tempranillo, is important in Spanish PDOs such as Rioja, Navarra, and Toro, and in many other countries [25]. The trial consisted on a randomized triplicate design with control and a foliar Mg^{2+} spray which was applied to cv. Tempranillo at two doses (0.5 kg/ha (Mg0.5) and 1.0 kg/ha (Mg1.0)) at veraison stage. Each replicate consisted of 10 treated vines (~150 bunches) in each replicate. The treatments were evaluated by data from cluster samples (10 clusters per replicate) at harvest time. One hundred berries from the clusters were removed and weighed (W). The berries were then crushed and several parameters (pH, TSS, MA, TcA, and YAN) were determined in the must. Means and standard errors of all evaluated parameters arranged by treatment are shown in **Figure 4**.

Several *F*-tests were carried out to study the effect of foliar Mg on W, TSS, pH, MA, TcA, and YAN. According to the ANOVA, foliar Mg treatments had no significant effects on any of the must quality parameters evaluated. Although none of TSS differences were statistically significant, as can be seen from **Figure 3** only Mg1.0 treatment showed a lower TSS level than control subplots, whereas Mg0.5 showed a greater TSS level than other treatments. Interestingly, although there are no significant differences with the control subplots in terms of grape weight (W) an increasing trend could be observed in Mg treated subplots (Control (2.25 g ± 0.29 95% CI) < Mg0.5 (2.30 g ± 0.31 95% CI) < Mg1.0 (2.31 g ± 0.22 95% CI)). In a similar way, leaf and shoot removal also affected YAN levels in such way that both Mg treatments may be associated with an increase in this key quality parameter. However, with a small sample size ($n = 9$ and 1 year), caution must be applied in both cases (W and YAN), because it is important to bear in mind the possible bias in the response to a foliar Mg treatment. Thus, if we consider our results collectively, they do not allow us to draw clear conclusions about the impact of foliar Mg treatments on harvest parameters, and therefore on TSS.

4. Effects of leaf removal and lateral shoot removal on berry sugar content

One of the most important and commonly applied summer canopy management operations in viticulture is the removal of leaves [26] and shoots in the fruit zone. Both practices are performed on grapevines to increase air circulation, light exposure, penetration of fungicide sprays, as well as decrease disease incidence. In general, exposing fruit to the sun will increase fruit temperature along with the enzymatic activities therein. Consequently, when compared to shaded fruit, exposed fruit will normally contain higher soluble solids [27]. Nevertheless, it should be noted that these actions on the vine canopy microclimate, which basically depends on the amount and distribution of leaf area in space and its interaction with above-ground climate [28], will have different effects on harvest quality according to the time and the shoot position when they were carried out. Most canopy microclimate components are of different values than those around the canopy, due to attenuation by the canopy. The degree of shading within grapevine canopies can be altered by three principal means: by varying the shoot number, the vine vigor, and/or the training system employed [28]. At the same time, a number of viticultural practices in wine grape improvement programs have been a topic of discussion in the scientific community in order to improve grape quality at harvest: optimum balance in vine pruning, shoot thinning, leaf and lateral shoot removal, early cluster thinning, late cluster thinning, shoot positioning, and tipping or irrigation scheduling.

On the basis of the above, a research was performed during 2014 vintage in the PDO area of Ribera del Duero, Spain. The cultivar chosen was *Vitis vinifera* L. Tempranillo. The trial consisted on a randomized triplicate design with control and two leaf removal treatments both at veraison stage. One of the treatments was a leaf and lateral shoots removal between the clusters positioned at bottom and top in the productive shoots (LRbt), whereas the other one was a leaf and lateral shoots removal below the clusters positioned at bottom in the productive shoots (LRbl). Each replicate consisted of 10 treated vines and 100 berries from the clusters were weighed (W) and then crushed to evaluate several must quality parameters at harvest time (pH, TSS, MA, TcA, and YAN). Means and standard errors of must parameters arranged by treatment are shown in **Table 1**.

Must parameter	Control		LRbt		LRbl	
	Mean	SE	Mean	SE	Mean	SE
W (g)	2.25	0.15	2.50	0.25	2.57	0.06
TSS (Brix)	23.6	1.22	23.1	0.66	23.6	0.81
pH	3.49	0.04	3.46	0.05	3.48	0.06
MA (g/l)	2.05	0.13	1.93	0.18	2.29	0.08
TcA (g/l)	3.06	0.52	2.92	0.19	2.98	0.36
YAN (mg/l)	256	1.68	239	8.85	270	22.4

LRbt: leaf and lateral shoots removal between the clusters positioned at bottom and top in the productive shoots; LRbl: leaf and lateral shoots removal below the clusters positioned at bottom in the productive shoots.

Table 1. Means and standard errors (SE) of berry weight (W), total soluble solids (TSS), real acidity (pH), malic acid (MA), tartaric acid (TcA), and yeast assimilable nitrogen (YAN) (2014).

Several *F*-tests were carried out to study the effect of management practices on W, TSS, pH, MA, TcA, and YAN. According to the ANOVA these viticultural practices had no significant effects on any of the must quality parameters evaluated. However, fruit composition has been shown to be affected by microclimate manipulation. While it is true that leaf and shoot removal in both treatments (LRbt and LRbl) had no significant effect on W, both treatments increased consistently this parameter. In an opposite direction both treatments decreased consistently the real acidity (pH). On the other hand, only LRbt treatment showed a lower TSS level than Control (Control (23.6 °Brix ± 2.39 95% CI) > LRbt (23.1 °Brix ± 1.29 95% CI)). In contrast to earlier findings [28, 29] levels of tartaric acid (TcA) are increased by shade (C > LRbl > LRbt). The latter showed that this decrease in the concentrations of tartaric acid in the shaded berries was due to an increase in berry size. Additionally, in contrast to Petrie et al. [30], levels of pH decreased by shade (C > LRbl > LRbt). It is difficult to explain these controversies with previous studies, but it might be related with the time of carrying out both viticultural practices. In this regard, because when veraison begins to occur green growth slows to a stop (while the vine directs energy toward the grape clusters) [31], the choice of another phenological stage for making these viticultural practices might have a different impact on harvest parameters. The lack of scientific evidence and controversies have been reported by several authors. For instance, whereas Bledsoe et al. [32] found that yield and yield components were not signifi-

http://dx.doi.org/10.5772/64982

cantly affected by the timing of leaf removal in Sauvignon Blanc grapes, Hunter and Visser [33] found that 33% defoliation prior to berries reaching pea size reduced berry size and yield, but had no effect when applied at veraison. Thus, understanding the impact of the timing of leaf and shoot removal on vines is crucial for vineyard managers and winemakers.

5. Biotechnological approaches to reduce the alcohol content in wine

Currently, several different technological strategies are available in order to reduce the ethanol content in final wines. Yeasts are the main microorganisms involved in the ethanol production from grapes and wine production, and accordingly, some of these strategies are based in a different management of wine yeasts, including the isolation of strains with a lower ability to produce ethanol.

Natural screening might be the first attempt to obtain lower ethanol-producing strains. However, this approach is unlikely to succeed because different aspects of the biochemistry, physiology, and genetics of *Saccharomyces cerevisiae*. In fact, this microorganism has evolved by natural selection to boost the ethanol production, even when oxygen is available [1, 34]. On the other hand, an attractive option would be to develop engineered wine yeast with reduced ethanol yield. So far, one of the most promising strategies is to redirect the metabolic flux toward an increased production of glycerol instead ethanol [2, 35]. However, this strategy has shown some unwanted effects like an overproduction of undesirable compounds from an organoleptic point of view [1]. Indeed, glycerol and ethanol produced during alcoholic fermentation are important regulators of the cellular redox balance, and consequently any attempt to redirect carbon flux by gene manipulation would modify the concentration of a range of other metabolites in order to correct the redox imbalance [36]. Acetaldehyde, acetate, succinate, acetoin, diacetyl, and 2,3-butanediol, among others, are some of the compounds to be avoided, since their presence at levels exceeding their sensorial threshold may be detrimental to the final wine quality [1, 36]. Furthermore, the public attitude toward the use of genetically modified organisms (GMOs) in food products and the legal restrictions in their use suggest that novel yeast strains will have to be generated using non-GM approaches.

Microorganisms, and particularly yeast, have a huge ability to adapt rapidly to different environmental conditions. This property has been used in recent years to modify the natural properties of yeasts by conducting adaptive laboratory evolution (ALE) experiments [2]. This approach mimics the natural evolution, by environmental or metabolic constraints, with the main purpose of obtaining improve yeast strains for several biotechnological applications and, of course, in winemaking processes [37–40]. A recent ALE study, by using KCl as osmotic and salt stress agent during 450 generations, achieved a wine with 0.6% (v/v) less ethanol in pilot scale fermentation when it was compared to the previous ancient strain. Besides that, the use of intrastrain hybrids by breeding techniques (a non-GMO technique) has proven the reduction of the alcoholic strength to 1.3% (v/v) [2].

An alternative approach to modify the final alcohol content of wines is related to the performance of modified fermentation procedures. Although *S. cerevisiae* is the main yeast species responsible for conducting the alcoholic fermentation, the contribution of a nonnegligible number of other yeast species associated to the initial stages of the fermentation and their contribution to final sensorial properties are well established [41–43]. These strains are naturally present in sound grapes and might be easily isolated from the grape must at initial fermentation stages.

There are significant differences in sugar metabolism between some of these species and *S. cerevisiae*. The non-*Saccharomyces* strains actually allow an increased breakdown of sugars via respiratory pathways than through fermentation. An enhancement of this respiratory catabolism has been suggested by several authors in order to reduce the amount of sugar conversion into ethanol [44, 45]. For this reason, mixed cultures between a non-*Saccharomyces* strain and a *Saccharomyces cerevisiae* strain might be a good alternative to the previously mentioned approaches. Further, an additional advantage of this strategy would be to use autochthonous yeast strains in order to maintain the typicality of wines elaborated in this way [46].

	Muestra/microvinification					
	Young vineyard		**Middle-age vineyard**		**Old vineyard**	
Fermentation phase	**Yeast analyzed**	**Identification**	**Yeast analyzed**	**Identification**	**Yeast analyzed**	**Identification**
Initial	10	**2** *Hanseniaspora uvarum* **1** *Lachancea thermotolerans* **4** *Metschnikowia aff. fructicola* **3** *Metschnikowia pulcherrima*	10	**2** *Lachancea thermotolerans* **4** *Metschnikowia aff. fructicola* **1** *Metschnikowia chrysoperlae* **3** *Metschnikowia pulcherrima*	10	**1** *Debaryomyces hansenii* **1** *Hanseniaspora uvarum* **3** *Lachancea thermotolerans* **2** *Metschnikowia aff. fructicola* **2** *Metschnikowia pulcherrima* **1** *Kluyveromyces dobzhanskii*
Medium	10	**5** *Hanseniaspora uvarum* **1** *Lachancea thermotolerans* **4** *Saccharomyces cerevisiae*	10	**2** *Hanseniaspora uvarum* **5** *Lachancea thermotolerans* **3** *Saccharomyces cerevisiae*	10	**5** *Hanseniaspora uvarum* **1** *Lachancea thermotolerans* **1** *Saccharomyces cerevisiae* **3** *Saccharomyces bayanus*
Final	10	**10** *Saccharomyces cerevisiae*	10	**8** *Saccharomyces cerevisiae* **2** *Saccharomyces paradoxus*	10	**10** *Saccharomyces cerevisiae*

Table 2. Yeast species isolated at different stages of the fermentation process and identified from spontaneous fermentations of natural "Tinta de Toro" grape juice obtained from vineyards of different ages.

Although high levels of ethanol content in final wines is a worldwide issue, as mentioned earlier, in Spain this problem is still more pronounce in the Denomination of Origin (DO) Toro (Toro, Zamora, Spain), whose wines easily reach and exceed 15–16° alcohol content. For this reason a biotechnological-based approach was developed with the final aim to reduce their ethanol levels. During 2013 vintage, the population of indigenous strains associated to a winery belonged to DO Toro was characterized. Spontaneous fermentations were carried out on natural grape juice ("*Tinta de Toro*" grape variety), obtained from grapes of three different vineyards. Microfermentations were conducted at fixed temperature (21°C) and were daily monitored by measuring their weight loss until completion (constant weight). Yeast isolation was made from three different phases of the fermentation process: initial, medium, and final stage (final wine). Yeast strains were randomly selected for genetic typing. Yeast identification was carried out by RFLP analysis of the 5.8S-ITS-rRNA region amplified by using ITS1 and ITS4 primers [47], and confirmed by sequencing the D1-D2 regions of 26S rDNA using the NL-1 and NL-4 primers [48]. The results are shown in **Table 2**.

S. cerevisiae strain typing was performed by RFLP-mtDNA analysis with *Alu*I restriction enzyme [49]. The RFLP profiles were compared using the InfoQuest FP software package (Bio-Rad, Hercules, CA, USA). Nine different *S. cerevisiae* strains were identified (**Figure 5**), resulting no matches with the commercial strains routinely used in the winery.

The predominant *S. cerevisiae* strain (**Sc-1**, 44% of *S. cerevisiae* total population) and two non-*Saccharomyces* strains, *Lachancea thermotolerans* and *Kluyveromyces dobzhanskii*, were selected to conduct experimental fermentation with mixed cultures. The final aim was to the decrease the ethanol content by increasing the respiratory catabolism of sugars by non-*Saccharomyces* strains in the initial stages of the alcoholic fermentation. Two different methodologies were assayed for both non-*Saccharomyces* strain: coinoculation with the *S. cerevisiae* strain; and sequential-inoculation, by adding first the non-*Saccharomyces* strain and then (15 days later) the *S. cerevisiae* strain. Microvinifications were conducted in triplicate by inoculation with the selected strains. Microfermentations were carried out at 25°C and monitored by measuring the loss of weight, as described above. Once fermentations were completed, yeast cells were removed by centrifugation and analysis. Samples for quantitative analysis were stored at −20°C until analyses were performed.

The analysis of final wines were performed by HPLC using an Agilent 1200 series (Agilent Technologies, Santa Clara, CA, USA) chromatograph equipped with a HyperREZ XP Carbohydrate H^+ column (8 μm particle size, 300 × 7.7 mm) and a HyperREZ XP carbohydrate H^+ Guard pre-column (Thermo Scientific, Waltham, MA, USA), maintained at 50°C. Samples were filtered using 0.45 μm cellulose acetate filters (Costar, Washington, DC, USA) prior to analysis. A refraction index detector (RID) (positive polarity) at a flow rate of 0.8 ml/min with 4 mmol/l H_2SO_4 as mobile phase (injection volume 25 μl) was used to detect glycerol and ethanol. One-way analysis of variance was carried out to determine the influence of the "yeast used" factor on ethanol and glycerol content. The results are shown in **Table 3**. Coinoculation methodology decreased the alcohol level in final wines from 0.55 to 0.62% (v/v) when a *L. thermotolerans* or *K. dobzhanskii* strains had been used, respectively. A two-step (sequential) inoculation strategy achieved a higher reduction in ethanol values: up to 0.79% (v/v) reduction was obtained by using *K. dobzhanskii* strain, whereas a 0.82% (v/v) decrease was obtained by using the *L. thermotolerans* strain. Although some differences in the glycerol content were detected in the final wines, sensory analyses by a panel of expert tasters did not find significant differences in the wines thus elaborated.

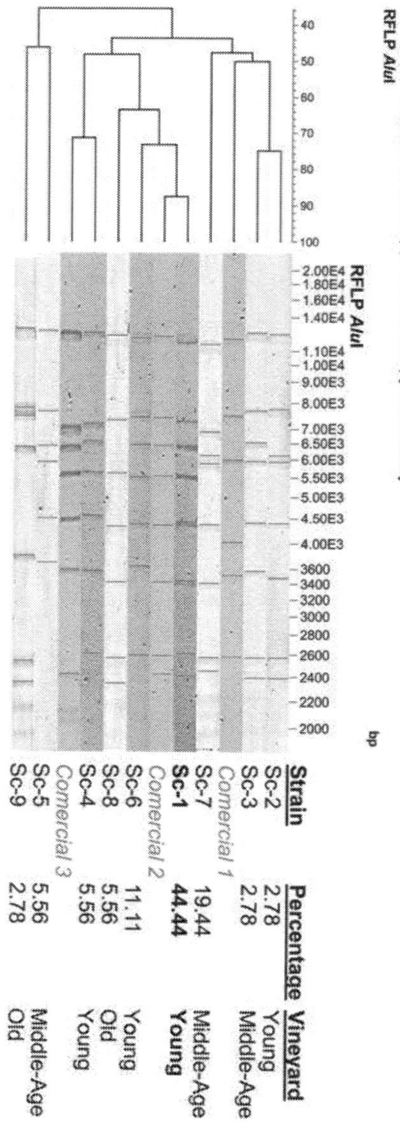

Figure 5. Dendrogram (top) based on RFLP-mtDNA analyses with the restriction endonuclease AluI (center) of all the *S. cerevisiae* strains (bottom) isolated from the three vineyards belonging to the winery (DO Toro, Zamora, Spain). The strain selected for the further oenological approach is highlighted in bold. Commercial strains have not been detected in neither case.

Compound	Control (Sc-1)	Sc-1 + *K. dobzhanskii*		Sc-1 + *L. thermotolerans*	
		Coinoculation	*Two-step inoculation*	*Coinoculation*	*Two-step inoculation*
Ethanol (%, v/v)	12.70 (0.01)[a]	12.08 (0.06)[b,c]	11.90 (0.03)[c]	12.15 (0.09)[b]	11.87 (0.15)[c]
Glycerol (g/l)	4.04 (0.01)[a]	3.93 (0.15)[a]	5.57 (0.29)[b,c]	4.41 (0.16)[a,b]	6.23 (1.26)[c]

Values in parentheses correspond to the standard deviation (SD) in each case. Means with different letters are significantly different according to ANOVA results ($p < 0.05$).

Table 3. Ethanol and glycerol content of the final wines analyzed by HPLC.

Recently, a novel study addressed the same issue with a similar experimental design. In fact, Morales et al. [45] used a *Metschnikowia pulcherrima*/*S. cerevisiae* mixed cultures. The authors used an oxygen flux during the first stages of the fermentation. An alcohol reduction of 2.2% (v/v) was achieved with correct levels of volatile acidity. These data suggest that a higher reduction in the ethanol level could be obtained by increasing aeration at the beginning of the alcoholic fermentation, in order to increase the sugar consumption rate by non-*Saccharomyces* strains (respiration).

Therefore, the implementation at the industrial level of strategies to lower the ethanol content of wine, owing to breakdown of sugars by non-*Saccharomyces* yeasts, appears to be an interesting challenge. However, a further optimization is required in several aspects as yeast strains selection, inoculation protocols, aeration conditions, and other requirements.

6. Conclusion

Taken together the findings showed in this chapter, it has become evident that there are several potential efficient practices to overcome high must sugar levels at harvest time. Most favorable results were obtained by using plant growth regulators (auxins) and yeast selection. The generalizability of these results could be subject to certain limitations. Thus, in the case of auxins the cultivar behaves as an important factor to be taken into consideration, whereas in the case of yeast selection, more research is required to determine the efficacy of implementation at the industrial level.

Acknowledgements

We are grateful to the R Core team for their hard work and nonprofit effort with the software.

Author details

Miguel Ángel Olego[1*#], José Manuel Álvarez-Pérez[2#], Miguel Javier Quiroga[2], Rebeca Cobos[1], Mario Sánchez-García[2], Jesús Esteban Medina[2], Sandra González-García[2], Juan José Rubio Coque[2] and José Enrique Garzón-Jimeno[2]

*Address all correspondence to: molem@unileon.es

1 RGA Bio-Investigación, SL, Laboratorio 29, Instituto de Recursos Naturales, Universidad de León, León, Spain

2 Instituto de Investigación de la Viña y el Vino (IIVV), Universidad de León, León, Spain

#These authors contributed equally to this work

References

[1] Kutyna DR, Varela C, Henschke PA, Chambers PJ, Stanley GA. Microbiological approaches to lowering ethanol concentration in wine. Trends in Food Science and Technology. 2010;1:293–302.

[2] Tilloy V, Ortiz-Julien A, Dequin S. Reduction of ethanol yield and improvement of glycerol formation by adaptive evolution of the wine yeast *Saccharomyces cerevisiae* under hyperosmotic conditions. Applied and Environmental Microbiology. 2014;80:2623–2632.

[3] Matthews MA. Terroir and Other Myths of Winegrowing. Oakland, USA: University of California Press; 2015. 328 p.

[4] Jackson RS. Wine Science: Principles and Applications. 4th ed. London, UK: Elsevier; 2014. 978 p.

[5] Piazzolla F, Pati S, Amodio ML, Colelli G. Effect of harvest time on table grape quality during on-vine storage. Journal of the Science of Food and Agriculture. 2016;96:131–139.

[6] Creasy GL, Creasy LL. Grapes. Wallingford, UK: The Centre for Agriculture and Bioscience International; 2009. 312 p.

[7] Agasse A, Vignault C, Kappel C, Conde C, Gerós H, Delrot, S. Sugar transport and sugar sensing in grape. In: Roubelakis-Angelakis KA, editor. Grapevine Molecular Physiology and Biotechnology. 2nd ed. Dordrecht, The Netherlands: Springer; 2009. pp. 105–139.

[8] Hufnagel JC, Hofmann T. Quantitative reconstruction of the nonvolatile sensometabolome of a red wine. Journal of Agriculture and Food Chemistry. 2008b;56:9190–9199.

[9] Zamboni A, Di Carli M, Guzzo F, Stocchero M, Zenoni S, Ferrarini A, Tononi P, Toffali K, Desiderio A, Lilley KS, Pe ME, Benvenuto E, Delledonne M, Pezzotti M. Identification

of putative stagespecific grapevine berry biomarkers and omics data integration into networks. Plant Physiology. 2010;154:1439–1459.

[10] Varela C, Dry PR, Kutyna DR, Francis IL, Henschke PA, Curtin CD, Chambers PJ. Strategies for reducing alcohol concentration in wine. Australian Journal of Grape and Wine Research. 2015;1:670–679.

[11] Keller M. The Science of Grapevines: Anatomy and Physiology. 2nd ed. London, UK: Academic Press; 2015. 522 p.

[12] Bindon KA, Varela C, Kennedy J, Holt H, Herderich M. Relationships between harvest time and wine composition in *Vitis vinifera* L. Cabernet Sauvignon 1. Grape and wine chemistry. Food Chemistry. 2013;138:1696–1705.

[13] Kontoudakis N, Esteruelas M, Fort F, Canals JM, Zamora F. Use of unripe grapes harvested during cluster thinning as a method for reducing alcohol content and pH of wine. Australian Journal of Grape and Wine Research. 2011;17:230–238.

[14] Fraga H, Malheiro AC, Moutinho-Pereira J, Santos JA. An overview of climate change impacts on European viticulture. Food and Energy Security. 2012;1(2):94–110.

[15] Ozturk B, Anli E. Different techniques for reducing alcohol levels in wine: a review [Internet]. Available from: http://www.bio-conferences.org/articles/bioconf/pdf/2014/02/bioconf_oiv2014_02012.pdf [Accessed: 2016-02-18].

[16] Pötter GH, Daudt CE, Brackamnn A, Leite TT, Penna NG. Partial de foliation on vines and its effects on Cabernet Sauvignon grapes and wines from the southwest of Rio Grande do Sul, Brazil. Ciencia Rural. 2010;40(9):2011–2016.

[17] Davies, C. Understanding and Managing the Timing of Berry Ripening and the Flavor-Ripe/Sugar-Ripe Nexus. Clayton, Australia: CSIRO; 2014. 108 p.

[18] Böttcher C, Harvey K, Forde CG, Boss PK, Davies C. Auxin treatments of pre-veraison grape (*Vitis vinifera* L.) berries delays ripening and increases the synchronicity of sugar accumulation. Australian Journal of Grape and Wine Research. 2011b;17:1–8.

[19] Davies C, Boss PK, Robinson SP. Treatment of grape berries, a nonclimacteric fruit with a synthetic auxin, retards ripening and alters the expression of developmentally regulated genes. Plant Physiology. 1997;115:1155–1161.

[20] Dundon CG, Smart RE, McCarthy MG. The effect of potassium fertilizer on must and wine potassium levels of Shiraz grapevines. American Journal of Enology and Viticulture. 1984;35(4):200–205.

[21] Böttcher C, Keyzers RA, Boss PK, Davies C. Sequestration of auxin by the indole-3-acetic acid-amido synthetase GH3-1 in grape berry (*Vitis vinifera* L.) and the proposed role of auxin conjugation during ripening. Journal of Experimental Botany. 2010;61:3615–3625.

[22] Poling B, Spayd S. The North Carolina Winegrape Grower's Guide. Raleigh, USA: Publications Office, North Carolina University; 2015. 196 p.

[23] Ohno T, Grunes DL. Potassium magnesium interactions affecting nutrient uptake by wheat forage. Soil Science Society of America Journal. 1985;49:685–690.

[24] Huang JW, Grunes DL, Welch RM. Magnesium, nitrogen form, and root temperature effects on grass tetany potential of wheat storage. Soil Science Society of America Journal. 1990;82:581–587.

[25] Walker L. It's time for Tempranillo. Wines & Vines. 2006;May:77–80.

[26] Poni S, Casalin L, Bernizzon F, Civardi S, Intrieri C. Effects of early defoliation on shoot photosynthesis, yield components, and grape composition. American Journal of Enology and Viticulture. 2006;57:397–407.

[27] Reynolds AG. Viticultural and vineyard management practices and their effects on grape and wine quality. In: Reynolds AG, editor. Managing Wine Quality. Volume 1: Viticulture and Wine Quality. Cambridge, UK: Woodhead Publishing Limited; 2010. pp. 365–444.

[28] Smart, RE. Principles of grapevine canopy microclimate manipulation with implications for yield and quality. A review. American Journal of Enology and Viticulture. 1985;36(3):230–239.

[29] Crippen, D, Morrison J. The effects of Sun exposure on the compositional development of Cabernet Sauvignon berries. American Journal of Enology and Viticulture. 1986;37(4):235–242.

[30] Petrie PR, Trought MCT, Howell G, Buchan G. The effect of leaf removal and canopy height on whole-vine gas exchange and fruit development of *Vitis vinifera* L. Sauvignon Blanc. Functional Plant Biology. 2003;30:711–717.

[31] Stanley M. Creating World Class Red Wine. West Chester, USA: Ascension Press; 2014. 150 p.

[32] Bledsoe A, Kliewer W, Marois J. Effects of timing and severity of leaf removal on yield and fruit composition of Sauvignon Blanc. American Journal of Enology and Viticulture. 1988;39:49–54.

[33] Hunter J, Visser J. The effect of defoliation on growth characteristics of *Vitis vinifera* Cabernet Sauvignon II. Reproductive growth. South African Journal of Enology and Viticulture. 1990;11:26–32.

[34] Piškur J, Rozpedowska E, Polakova S, Merico A, Compagno C. How did Saccharomyces evolve to become a good brewer?. Trends in Genetics. 2006;22:183–186.

[35] Varela C, Kutyna DR, Solomon MR, Black CA, Borneman A, Henschke PA, et al. Evaluation of gene modification strategies for the development of low-alcohol-wine yeasts. Applied and Environmental Microbiology. 2012;78:6068–6077.

[36] Schmidtke LM, Blackman JW, Agboola SO. Production technologies for reduced alcoholic wines. Journal of Food Science. 2012;77:25–41.

[37] McBryde C, Gardner JM, Lopes MdeB, Jiranek V. Generation of novel wine yeast strains by adaptive evolution. American Journal of Enology and Viticulture. 2006;57:423–430.

[38] Stanley D, Fraser S, Chambers PJ, Rogers P, Stanley GA. Generation and characterisation of stable ethanol-tolerant mutants of *Saccharomyces cerevisiae*. Journal of Industrial Microbiology and Biotechnology. 2009;37:139–149.

[39] Cadière A, Ortiz-Julien A, Camarasa C, Dequin S. Evolutionary engineered Saccharomyces cerevisiae wine yeast strains with increased in vivo flux through the pentose phosphate pathway. Metabolic Engineering. 2011;13:263–271.

[40] Kutyna DR, Varela C, Stanley GA, Borneman AR, Henschke PA, Chambers PJ. Adaptive evolution of *Saccharomyces cerevisiae* to generate strains with enhanced glycerol production. Applied Microbiology and Biotechnology. 2012;93:1175–1184.

[41] Fleet GH. Yeast interactions and wine flavour. International Journal of Food Microbiology. 2003;86:11–22.

[42] Ciani M, Comitini F, Mannazzu I, Domizio P. Controlled mixed culture fermentation: a new perspective on the use of non-Saccharomyces yeasts in winemaking. FEMS Yeast Research. 2010;10:123–133.

[43] Medina K, Boido E, Fariña L, Gioia O, Gomez ME, Barquet M, et al. Increased flavour diversity of Chardonnay wines by spontaneous fermentation and co-fermentation with Hanseniaspora vineae. Food Chemistry. 2013;141:2513–2521.

[44] Erten H, Campbell I. The production of low-alcohol wines by aerobic yeasts. Journal of the Institute of Brewing. 2001;107:207–215.

[45] Morales P, Rojas V, Quiros M, Gonzalez R. The impact of oxygen on the final alcohol content of wine fermented by a mixed starter culture. Applied Microbiology and Biotechnology. 2015;99:3993–4003.

[46] Álvarez-Pérez JM, Álvarez-Rodríguez ML, Campo E, Sáenz de Miera LE, Ferreira V, Hernández-Orte P, et al. Selection of *Saccharomyces cerevisiae* strains applied to the production of Prieto Picudo rosé wines with a different aromatic profile. South African Journal of Enology and Viticulture. 2014;35:242–256.

[47] White TJ, Bruns T, Lee S, Taylor J. Amplification and direct sequencing of fungal ribosomal RNA genes for phylogenetics. In: Innis MA, Gelfand DH, Sninsky JJ, White TJ, editors. PCR Protocols. A Guide to Methods and Applications. San Diego, USA: Academic Press; 1990. pp. 315–322.

[48] O'Donnell K. Fusarium and its near relatives. In: Reynolds DR, Taylor JW, editors. The Fungal Holomorph: Mitotic, Meiotic and Pleomorphic Speciation in Fungal Systematics. Wallingfork, UK: CAB International; 1993. pp. 225–233.

[49] Querol A, Barrio E, Ramón D. A comparative study of different methods of yeast strain characterization. Systematic and Applied Microbiology. 1992;15:439–446.

Chapter 18

Innovations in the Use of Bentonite in Oenology: Interactions with Grape and Wine Proteins, Colloids, Polyphenols and Aroma Compounds

Milena Lambri, Donato Colangelo,
Roberta Dordoni, Fabrizio Torchio and
Dante Marco De Faveri

Additional information is available at the end of the chapter

http://dx.doi.org/10.5772/64753

Abstract

Bentonite is used in oenology to improve the limpidity and the stability of wine and to predict the formation of deposits in the bottle. The exchangeable cations in its lamellar structures strongly influence some properties, such as the specific surface, the exchange capacity and the adsorption behavior. The unintended use of bentonite for juice settling and/or for wine fining produces jeopardized effects on colloidal and protein stability, the aroma compounds and sensory profiles. The interactions with haze-forming proteins, other colloids, as well as aroma compounds and phenols would have been to discover as the modulation of wine colloids by an adjuvant severely affects the wine resilience and the sensory profile. This chapter reviews several studies that focus on the impact of commercial bentonite samples used for both juice clarification and wine fining on the colloids, proteins, phenols and aroma compounds of white and red wines. Some parameters of practical value, such as the wine heat stability, the concentrations of total and haze-forming proteins and the content of the most relevant aromas, have been assessed to track the effects of bentonite and to achieve findings that are applicable to the field of oenology.

Keywords: bentonite, colloids, proteins, polyphenols, aroma, wine

1. Bentonite

1.1. Chemical structure

Bentonite is a phyllosilicate of the class of dioctahedral smectites [1] with a general composition of (M+y · nH_2O) ($Al^{3+}{}_{2\text{-}}yMg^{2+}y$)$Si^{4+}{}_4O_{10}(OH)_2$.

The structures of phyllosilicates are all based on tetrahedral T and octahedral O sheets arranged in either a 1:1 (kaolinite, dickite and nacrite) or 2:1 (smectites, vermiculite, mica and chlorite) ratio to form an anisotropic TO or TOT layer, respectively (**Figure 1**).

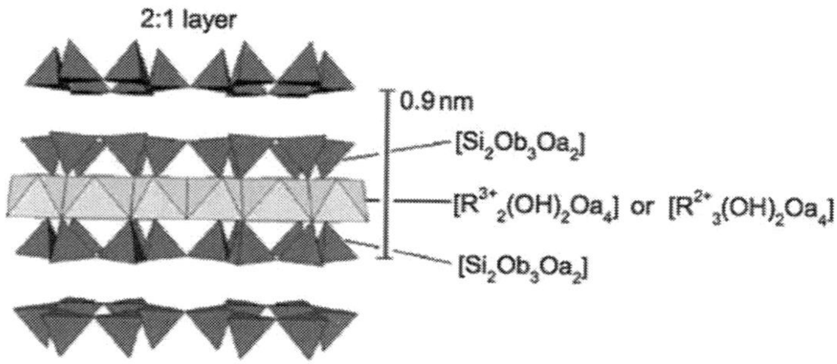

Figure 1. Structure of a phyllosilicate of the class of dioctahedral smectites [1] as bentonite is.

The layers are stacked on top of each other to form what is called a 'particle'. In turn, an assembly of particles is known as an 'aggregate'. Thus, a particle is formed by stacked layers that are separated from each other by an interlayer space. For a single layer, each tetrahedron consists of a cation (Si^{4+}, Al^{3+} and Fe^{3+} are the most common) coordinated to four oxygen atoms and linked to an adjacent tetrahedra by sharing three basal oxygen atoms to form a two-dimensional 'hexagonal' pattern. However, each octahedron consists of a cation (Al^{3+}, Fe^{3+}, Mg^{2+}, or Fe^{2+}) coordinated by six oxygen atoms and linked to a neighbouring octahedra by sharing edges. In the 2:1 structures, such as smectites, one octahedral sheet is sandwiched between two tetrahedral sheets. The surface properties of clay minerals depend on many factors, such as their chemical composition, nature of the surface atoms, extent and type of charge, and type of exchangeable cations. There are two types of surfaces with different properties, the planar surface and the edge surface, which is affected by pH. In a 2:1 clay, such as bentonite, the oxygen atoms of the Si tetrahedron represent the atoms of the basal surface. The charge on the oxygen atoms depends on the difference in electronegativity between oxygen and the other atoms involved in the bond. Thus, the ionicity of the bonds follows the order H–O < Si–O < Al–O < Mg–O < Li–O.

Thus, in smectites, the planar surfaces are negatively charged, with the magnitude of the layer charge ranging from 0.2 to 0.6 per half-unit cell. The type of metals that are bonded to the oxygen atoms governs the hydrophobicity of the surface. In the absence of isomorphous

http://dx.doi.org/10.5772/64753

substitutions, Si–O bonds prevail in the tetrahedra and Al–O bonds prevail in the octahedral sheets, resulting in a hydrophobic surface. When isomorphous substitutions occur, that is substitution of trivalent cations for Si^{4+} in tetrahedral sites and of divalent cations for Al^{3+} in octahedral sites, hydrophilicity is introduced and exchangeable cations are present [1].

Using the electronegativity equalization method (EEM) [2], beidellite and montmorillonite were shown to be the most electronegative and reactive 2:1 clays. The negative charges that result from isomorphous substitutions are balanced by the presence of exchangeable cations in the interlayer space. Na^+, Ca^{2+}, Mg^{2+} and K^+ are the most common ions. The mechanisms involved in the cation hydration for smectites are mainly electrostatic interactions that minimize water–water repulsion [3].

The presence of clustered or 'nanoconfined' water molecules around the exchangeable cations and the water bound to the clay through hydrogen bonds, charge dipole attractions and van der Waals interactions expands the interlayer space. This expansion or swelling depends on the cations present, the ionic strength of the medium and other factors. During hydration, water molecules orient their negative dipole towards the cations, thus weakening the electrostatic interactions with the charged layer and leading to increased separation between the two successive layers. As a result, the cations on the layer surface can be exchanged with the cations in the external solution and, based on the persistence or absence of significant interactions between two successive layers, this separation is called delamination or exfoliation, respectively [4]. In the first case, the smectite increases its volume to maintain its structure, whereas in the second case, isolated layers from the stacks of layers detach from the structure, becoming independently mobile.

Based on their swelling properties, sodium bentonite is different from calcium bentonite because the former is a high-swelling type of clay, while the latter is a low-swelling type. In fact, when sodium bentonite takes up water, exfoliation is more frequent than in the case of calcium bentonite. This difference is due to the type of ion that is intercalated on the surface of the layers. The amount of Na^+ needed for the compensation of the negative charge on the layers is double that needed for Ca^{2+} because the charge of Na^+ is half that of Ca^{2+}. In addition, based on the hydrated ionic radii of Na^+ and Ca^{2+}, the number of water molecules intercalated in the interlayer space during wetting in the presence of Na^+ is higher than that in the presence of Ca^{2+}.

The other type of surface, the edge surface, is characterized by the presence of hydroxyl groups called terminal OH groups. These hydroxyl groups carry a charge that is dependent on the type of metal ion to which they are bonded in the structure and on the pH of the solution. The charging arises from the adsorption or dissociation of protons. At low pH values, the excess protons in the aqueous medium create positive edge charges, whereas negative charges are produced by the dissociation of silica and alumina groups at high pH values [5, 6].

1.2. Extraction sites

In 2009, the worldwide production of bentonite amounted to 9.66 Mt. Approximately 90% of the world production is concentrated in 13 countries, with 55% of the total production in

Greece, the USA and the Commonwealth of Independent States. The different properties of bentonite reflect its different origins (47 countries contribute to the total production). Bentonite deposits were formed by diagenetic or hydrothermal alteration of volcanic glass and by authogenesis of smectite-rich materials in alkaline continental basins [7]. Dioctahedral smectites first become Fe^{3+}-rich and then subsequently become Al^{3+}-rich (montmorillonite) minerals formed in the absence of Mg^{2+}, and the most important deposits are located in Wyoming, Montana, Arizona (USA) [8], India and Europe (Czech Republic, Denmark, Germany and Greece). Regarding the contents of Na^{+} and Ca^{2+}, which are important for the swelling and rheological properties, there is high variability among the clays from different deposits in the same country. Generally, American bentonites are Na^{+} rich, while the European bentonites are Ca rich, with few exceptions.

1.3. Industrial activation treatments

To modify the properties of bentonite, industrial activation processes are performed on bentonites that do not present the optimal characteristics for the use for which they are designed. There are two types of modification processes, mineral activation and organic modification.

In the mineral activation process, wet mud is treated at 80°C in the presence of Na_2CO_3. During the process, Ca^{2+} ions are precipitated as $CaCO_3$. Consequently, the Na/Ca ratio of the clay is modified by substituting the interlayer and surface Ca^{2+} ions for Na^{+} ions. The treatment improves the swelling ability of the slurries and enhances the adsorption of wine proteins [9]. Significant changes in the rheological characteristics (viscosity), pH, cationic exchange capacity (CEC), and structure of bentonite could also be produced after alkaline magnesium activation [10]. Another treatment that is used to improve the adsorbent performance of bentonite is the acid activation treatment with HCl and H_2SO_4, in which the main task is to increase the specific surface area (SSA) and the porosity by degrading the structure of the clay due to the leaching of Al^{3+}, Fe^{2+} and Mg^{2+}. In this case, the exchangeable cations are replaced by H^{+}. Organic modification is widely performed to improve the ability of the clay to remove heavy metals and organic compounds from water [11]; it reduces the total number of interlayer ions (Na, Ca, Mg) by replacing them with organic cations. Two categories of modified bentonite are produced, organoclays [12] and oxide pillared clays [13]. The main modifications induced by these processes are to the surface properties of the clays, which transition from hydrophilic to hydrophobic surfaces. This improves the ability of the modified bentonite to remove organic contaminants, such as phenols, from water [14].

2. Bentonite applications

2.1. Uses in non-food fields

Based on its properties and its relatively low cost of production, bentonite use has attracted interest from researchers in several fields, particularly those in the industrial and environmental fields [15]. Due to its viscosity, thixotropy, plastering ability and plasticity, bentonite is used to grout cracks in rocks, for soil injections, as a thickener in paints and as an additive in ceramics [16]. To reduce gas release from landfill decomposition and to impede water percolation across the landfill, clay liners have been used as substitutes for the compacted soil components normally used [17]. In the context of disposal programmes, bentonite is projected to be used in engineered barriers for high-level waste and spent nuclear fuels [18]. A significant amount of the global bentonite production is destined to iron ore pelletizing, foundry moulding and oil-well drilling processes [19]. Bentonite is also used in animal feed, as a litter adsorbent and as an oil and grease adsorbent. Its importance in animal feed is derived from its ability to adsorb mycotoxins, such as aflatoxins [20]. Bentonite plays an important role in the development of health products [21], such as cosmetics and pharmaceuticals [22]. Moreover, it has been shown to be useful in different medical treatments [23]. Bentonite is also applied to the production of sunscreen lotions, as it acts as physical barriers against UV radiation [24] as well as a thickener in paste masks, eyeliners, nail lacquers, shampoos and toothpastes [25]. Other, less important applications include the production of additives for cement and mortar [26] and the cleaning of wastewaters. Recently, bentonites have been used for enzyme immobilization [27] and as bactericidal materials [28].

2.2. Uses in syrup and juice production

The application of bentonite's characteristics has been investigated in different juice and syrup production processes. Fruit juices are naturally cloudy due to the presence of polysaccharides (pectin, cellulose, hemicellulose, lignin and starch), proteins, tannins and metals [29]. Other molecules, such as flavonoids, polyphenols and organic acids, contribute to the dark colour of sugarcane and other juices. Because these components are present in varying amounts, clarification is one of the key steps in juice processing. The commonly used clarification methods include the use of enzymes, ultrafiltration, gelatine, bentonite, PVPP [30], lime, active carbon [31] and ion exchange resins [32]. Bentonite is cheap, does not pose environmental threats and represents one of the most suitable alternatives for the other methods. In particular, activated Na–Ca bentonite is the type of clay used in beverage technology [32]. It is used as the main clarifying agent or in preclarification steps [33]. In the clarifying step for concentrated must production, the basic treatment is based on the addition of bentonite and gelatine for various times of contact (usually 24 or 48 h). In fruit juice production, bentonite can be added to reduce the brown pigments resulting from enzymatic reactions: positive results concerning the browning index and the haze potential have also been attained for banana juice [34]. Finally, bentonite has been evaluated for use as a moisture-regulating adjuvant in food technology [35].

3. Bentonite in oenology

The addition of bentonite to musts and/or wines is aimed at reducing the protein contents, thus reducing the haze potential of the wine (fining). Fining is defined as the process of addition of substances that induce the precipitation of particles in suspension by promoting their sedimentation.

In oenology, bentonite could also be applied to the must to improve the precipitation of the suspended solids [36]. Clarification results in the variable reductions in the protein contents in the juice, but this aspect is not always directly related to the haze potential of the final wines because other wine components, such as phenolic compounds and non-proteinaceous colloids, and the storage temperature significantly affect the colloidal stability of the final wine [37].

Bentonite is mainly used on white wines, whereas in red wines, the aim of fining is softening, that is the removal of some of the tannins and polyphenols to improve the astringency of the product, and it is performed by applying fining agents other than bentonite (gelatine, albumin, isinglass, skim milk, casein and potassium caseinate) [38].

The main advantages of bentonite are its low cost, effectiveness and availability. Moreover, it is easily removed via sedimentation. Alternative treatments and adjuvants have been sought in oenology because of the drawbacks of bentonite, such as its direct and indirect costs. Indeed, the separation of the lees requires specific plants and is accompanied by the retention of variable quantities of wine (3–10%). The disposal of bentonite represents a limitation exacerbated by the fact that bentonite is not reusable [39], due to the lack of processes that promote the desorption of the adsorbed proteins.

Alternative treatments and adjuvants have been sought in oenology because of the drawbacks of bentonite, such as its direct and indirect costs. Indeed, the separation of the lees requires specific plants and is accompanied by the retention of variable quantities of wine (3–10%). Another concern regards the environmental impact of bentonite wastes, a problem that is exacerbated by the fact that bentonite is not reusable [39]: the enrichment of metals and important alterations of pH have been observed above certain levels of addition of bentonite in poor soils [40]. Even though the enrichment of organic matter by spent bentonite could be promising for poor soils and for applications in new vineyards, the disposal of big volumes bentonite should be, thus, carefully studied.

3.1. International regulations on oenological use of bentonite

According to the EU regulations, there is no upper limit for the addition of bentonite during the winemaking process. On the other hand, The International Oenological Codex [41] establishes the properties of the bentonites that are useful in oenology. The following three classes of bentonite are useful: calcium bentonite, sodium bentonite and activated bentonite, all of which are capable of swelling to different extents and must be dried at 80–90°C after grinding and before their commercialization.

According to the Resolution Oeno 11/2003 [42], the quality control of each bentonite should be based on 5 different test trials prior its application as an adjuvant. Mould contamination is not accepted, and as a result, no odour should be perceived. For the release of cations, the extraction trial should be performed in a tartaric acid solution (5 g/L, pH = 3). This recommendation seems to be insufficient to characterize the safety of a bentonite because the release of cations in wine has been shown to be more intense than extraction with tartaric acid [43].

http://dx.doi.org/10.5772/64753

The metals that are covered by the Resolution Oeno 11/2003 are Pb, Hg, As, Fe, Al, Ca, Mg and Na, with appropriate concentration limits for each metal. However, other metals can be part of the elemental composition of a bentonite, which include significant amounts of Li, Be, Sc, V, Co, Ni, Ga, Ge, As, Sr, Y, Zr, Nb, Mo, Cd, Sn, Sb, Ba, W, Tl and Bi [44]. The enrichment of cations from bentonite in wine partially results from the cation exchange mechanisms on which the adsorption of proteins is based [45] but also from the physicochemical conditions of the medium (pH, ethanol content and presence of other cations). Because the concentration of each cation in wine is important in the context of food safety and for the technological stability [46] and the release of metals from bentonite has been shown to be increased over the last few years [44], the control of the release phenomena has gained importance [45]. Other parameters that are controlled are the presence of large particles, the de-acidification test, and the protein adsorption test [42].

3.2. Physicochemical characteristics of oenological bentonite

To optimize the performances of the fining of wines, the choice of the best bentonite is based on the characterization of its structural properties. The parameters that are considered are as follows:

- the surface charge density (SCD), expressed as meq 100 g^{-1} of bentonite [47];
- the specific surface area (SSA) (m^2 g^{-1}), whose determination by methylene blue titration is reported by Resolution Oeno 11/2003 [42];
- the charge density per surface unit (CDSU) (meq m^{-2}), derived by calculating the ratio between the SCD and the SSA;
- the natural pH, expressed as the pH of the supernatant of a 20% water suspension (w/v) after 1 h of contact;
- the swell index (SI) (mL g^{-1}), expressed as the difference in the volumes of dried bentonite and hydrated bentonite after 24 h of contact with demineralized water (2 g in 100 mL of water).

The SCD is an expression of the substitution of ions into the bentonite structure, which was described in Section 1.1. Because the adsorption of proteins on bentonite is mainly driven by cation exchange [48], the higher the negative charge of the clay particles, the higher the removal of proteins, when all other parameters are equal. However, these data alone are not sufficient to describe the complete performance of a bentonite in wines. Although a correlation between the net charge of the bentonite and the removal of proteins in white wines was observed [49], other factors affect the ability of charged bentonite to remove positively charged colloids [50].

The SSA is an indirect measure of the medium-sized bentonite particles and depends on the grinding process. The ratio between the SCD and the SSA results in the most significant parameter, the CDSU, and does not take into account the dose of the bentonite added to the wine and is only related to the surface of the clay. As shown for the SCD, the CDSU could sort the bentonites in the order of their charges.

Calcium bentonite is usually more acidic than an activated bentonite, which ranges from pH 6.5 to 8.5 and from pH 8.5 to 10.0, respectively, whereas natural sodium bentonite shows a quite wide range of pH values (4.5–10.0) [42]. The natural pH of a bentonite is strongly associated with its rheological properties, as the distribution of charges on the surface and along the edges of the of the platelets results in the organization of the platelets into a structure that confers viscosity to the medium. Currently, there is no global consensus on the type of structure formed, as some researchers [51] support the edge-to-face attraction mode, whereas other authors suggest the formation of a gel produced by long-range electrostatic double-layer repulsion both at the edges and at the faces [52]. The ionic strength and the pH should be considered when ascertaining the predominant type of interaction [53], particularly when the pH is lower than 4.5 [54], as in wine. Below pH 7, the edges of the platelets should be positively charged, allowing edge-to-face attractions, but very acidic pH values also increase the ionic strength of the medium [55].

Therefore, below pH 4.0, the acidic conditions could be so extreme that Al^{3+} and Mg^{2+} are dissolved and the clay structure is subsequently decomposed. In the same range of pH values, the high ionic strength of the medium could result in the compression of the double layers, thus reducing the edge-to-face attraction intensities [56]. The remaining card-house structure is thought to be an important property for the removal of proteins [45]. The capacity of the clay to maintain a residual three-dimensional structure is affected by the pH of the wine and by the natural pH of the bentonite.

4. Bentonite and wine colloids

4.1. Must and wine colloids

The colloidal phenomena are articulated in two steps, which include the formation of small, invisible colloidal particles and their subsequent aggregation and formation of large particles that are able to diffuse light and to precipitate when the their size exceeds a certain limit.

In oenology [38], two groups of colloids may be distinguished.

- Associated colloids are charged particles composed of small molecules that are bound by weak bonds (Van der Waals, hydrogen bonds, hydrophobic interactions, etc.). This type of colloid is naturally present in wine as condensed phenols and colloidal colouring matter.
- Macromolecular colloids, such as polysaccharides and proteins, have high molecular masses and can form covalent bonds.

In addition, polysaccharides could also function as protective colloids by coating the associated colloids and preventing their precipitation. The coating action that is typical of protecting colloids is exerted over a range of concentrations, below which a destabilizing effect takes over. The protective colloids are important in stabilizing different colloidal phenomena in wine, such as the precipitation of ferric phosphate and tartrate.

Colloids are also important for the quality properties required in specific products, such as the foamability of sparkling wines resulting from the presence of hydrophilic and hydrophobic domains in proteins that act as surfactants for the bubble film [57]. Further evidence was provided [58] in Prosecco wine, as high molecular weight glycocompounds showed a high capacity to foam.

Although there is some controversy about the degree of glycosylation of wine proteins [59], a variable fraction of the colloids from red wines, ranging from 4.3 to 5.2%, is composed of arabinogalactans [60], which have a protein content of <10%. These molecules are considered polysaccharides and are not enlisted in the class of wine proteins.

4.2. Must and wine proteins

Among the colloids that are present in musts and wines, proteins are the most important because they have a central role in haze formation.

Previous analyses have identified the most heat-unstable proteins in white wines, such as grape class IV chitinases, β-glucanases and a fraction of thaumatin-like (TL) proteins [61, 62]. These proteins are able to persist throughout the winemaking process because they are resistant to proteolysis and are stable at an acidic pH [63]. Conversely, yeast- and grape-derived high-molecular mass glycoproteins have been found to exert a stabilizing effect in white wines and are considered 'haze-protective factors' [39]. In recent years, several reports have attempted to characterize the factors affecting protein instability and haze formation in white wines. Because wines with different haze potentials typically contain very similar protein fractions [49], one or more non-proteinaceous wine components (sulphate anions, ionic strength, phenolic compounds, organic acids and pH values) are thought to have an impact on haze formation [39].

4.3. Interactions of bentonite with grape and wine proteins

The factors that can affect the capacity of a bentonite to bind proteins are its physical properties, the temperature, the pH and the ethanol content of the medium.

High temperature results in higher adsorption [48, 50], perhaps due to the conformational changes that occur in the proteins at low temperatures.

pH has a greater importance in the mechanisms of protein adsorption because it influences the charges of the proteins and, thus, the efficacy of the cation exchange onto bentonite [45, 49, 50].

The ethanol content of the wine influences the adsorption capacity of bentonite because ethanol molecules could displace water from the platelets and induce considerable swelling. In turn, increased swelling should increase the number of exchangeable cations, resulting in an increased adsorption capacity. Ethanol concentrations >10% in wine model solutions result in a considerable increase in the adsorption capacity [48]. However, the type of protein considered could influence this result because the separation of the clay platelets induced by ethanol molecules is finite. Under these conditions, the adsorption could only be attributed to those proteins that are small enough to enter the clay structure [64].

Various factors must be considered when the physical properties of bentonite are examined. CEC, SI and SSA are indeed directly correlated to the removal of proteins [50]. From a practical point of view, the choice of the correct contact time is important for effective fining operations. The general approach is to use a long period of contact time between the wine and bentonite, which is on the order of days [38]. In literature, the optimal contact time is reported to be shorter [48, 50]: a plateau trend is observed after a few minutes of contact between bentonite and wine, indicating that no further adsorption of proteins occurs.

The total protein content in wine is a useful, but it is not the definitive information that determines the risk of haze formation in a wine. Indeed, the types of proteins and their relative concentrations are more important than the total protein content [49].

An optimal performance by bentonite comprises the removal of those fractions that are heat unstable and responsible for the colloidal casse. Chitinases, β-glucanases and some TL protein isoforms are greatly affected by bentonite, but the more stable invertases require a very large amount of bentonite for removal [61, 65]. Extensive investigations [45, 49] on the effect of different bentonite labels on the removal of specific proteins in Chardonnay, Sauvignon and Erbaluce wines showed that bentonite could selectively remove specific protein bands. As a result, the resistance of the wine to haze formation after bentonite fining resulted in a distributed low haze up to 50°C and high cloudiness in the range of 60–80°C. When the distribution of the proteins was investigated with SDS-PAGE, a substantial but not total reduction in the concentration of the TL proteins was observed for all the fined wines. Small peptides were almost completely removed by all the labels tested, but some bentonites were able to remove some of the invertase bands.

The amount of protein removed by bentonite changes was affected by both the bentonite characteristics and the wine pH [45] over a typical oenological range (3.00–3.60). Low molecular mass proteins are efficiently removed by different Na-bentonites, regardless of the pH, but fewer high and medium molecular mass proteins are removed to different extents, based on the wine pH. Reductions in the amount of the vacuolar invertase (GIN1) and VVTL1 fractions of the TL-proteins are induced by bentonites with natural pH values of <10 and are affected, to a lesser extent, by the negative effect of acidic pH on the card-house-like structure. Moreover, at higher pH values, the poor removal of glycoproteins (YGP1 and Hmp1) contributed to the increased thermal stability of the wine. As a matter of fact, the modifications in the protein profile of a white wine in the pH range of 3.00–3.60 resulted in increased quantities of glycoproteins from grape and yeasts [66]. Minor pH-driven conformational changes are sufficient to weaken the interactions between the glycoproteins and the other wine macromolecules, especially tannins, resulting in increased release of hydrophobic glycoproteins. As a result, the temperature at which wine begins to show turbidity is higher.

http://dx.doi.org/10.5772/64753

4.4. Interactions of bentonite with polysaccharides and tannins

It has been shown that bentonite is capable of removing large phenolic compounds, such as anthocyanin, and phenolic compounds that are bound to proteins [67]. Consequently, bentonite may be responsible for modifying wine colour and astringency. The results of the fining of red wines [68] have shown that the high SCD of the bentonite could reduce the tannin levels to some extent, which is accompanied by a relatively high reduction in the anthocyanin levels. It is likely that the absorbance of positively charged anthocyanin induces a significant change in the properties of the bentonite surface such that tannins could be adsorbed via hydrogen bonds and π–π stacking. On the other hand, no effect of bentonites on the removal of tannins in red wines was observed [69], regardless of the molecular weight of the molecules, although the specific physical properties of the bentonite used were not reported in the study.

5. Bentonite and wine aroma compounds

5.1. Varietal odour-active compounds

Grape aroma has an important contribution to wine flavour and characteristics, particularly in the aromatic cultivars, and it evolves during berry development according to the cultivar. Terpenes, which represent the main family of compounds and are present both as free and as glycosylated terpenoids, are mainly present in the grape skin and are present in high amounts in Muscat grape varieties [70]. Variable concentrations of norisoprenoids, that is C-13 norisoprenoids, such as β-damascenone, can improve the fruity note of wines, even at low concentrations [71], and are responsible for the aroma characteristics of Cabernet Sauvignon [72] and other non-floral grapes [73]. In rare cases, methoxypyrazines and sulphur compounds with thiol functional groups have been identified among the aroma compounds of wines. Methoxypyrazines are often associated with 'green' or 'herbaceous' aromas in Cabernet Sauvignon, Sauvignon Blanc, Cabernet Franc and Merlot Noir, among other varieties [74].

5.2. Fermentative odour-active compounds

The aroma complexity dramatically increases during alcoholic fermentation as a result of the synthesis of important volatile compounds and the release of some varietal aroma precursors [75]. The nature and amount of the synthesized volatile compounds depend on multiple factors, such as the nitrogen content of the must, the temperature of fermentation and the yeast strain. They consist of compounds with a wide range of polarity, solubility and volatility. The volatile compounds synthesized by wine yeasts include higher alcohols, medium- and long-chain volatile acids, acetate esters and ethyl esters and aldehydes, among others [76]. Aroma compounds interact with different macromolecules such as proteins or polysaccharides [77], so fining agents may fix substances that act as support for aromatic components [78].

5.3. Impact of bentonite on varietal odour-active compounds

Protein stability and the presence of intense fine aromas are two important requirements for aromatic white wines. Bentonite treatment is typically performed on wine, but it can be applied to increase the rate of juice settling and to facilitate the precipitation of suspended solids [36].

The use of fining agents to clarify juice can positively or negatively affect the composition of the derived must. The must clarification prior to the onset of alcoholic fermentation improves the sensory characteristics of white wines [79]. The removal of grape solids from must enhances the production of ethyl esters and acetates and limits the release of fusel alcohols during alcoholic fermentation, which results in a global increase in wine aroma quality [36]. Nevertheless, a certain amount of colloids must be present because they confer structure and volume, contribute to the fixation of aromatic compounds and produce bubbles in sparkling wines as a result of their tensioactive properties [64]. Conversely, excessive turbidity in musts induces the presence of an herbaceous aroma in the resulting wines, which promotes an evolution towards reduction. Finally, excessive limpidity can cause a retardation or cessation of the fermentation process [79].

A report [37] studied the fates of proteins and terpenols during the processing of Chambave Muscat grapes. The experiments included bentonite addition to must only (100 g hl), to wine only (100 g hl) and double bentonite addition on must and wine in two vintages (2006 and 2007). The results of the experiments demonstrated that a reduced removal of free terpenols was observed in the samples from the double treatment (must + wine) compared with the wines that were only fined with bentonite after alcoholic fermentation. In general [80], bentonite alone has a small effect on the loss of terpenes, but it removed ethyl esters and fatty acids.

Our unpublished data attempted to optimize the clarification of Muscat Blanc must with Ca or Na bentonite or related mixes in a dose range of 10–100 g hL^{-1} through a central composite design (CCD). Response surface methodology (RSM) estimated the combined effect of the bentonite type and dose on the removal of aglycones and glycosylated aroma compounds. The RSM surface plots predicted the removal of the aglycones from aroma compounds by low doses of Na-bentonite and Ca-bentonite, regardless of the dose. Furthermore, the model estimated the removal of glycosylated aroma compounds, particularly by Ca-bentonite at 50 g hL^{-1}.

5.4. Impact of bentonite on fermentative odour-active compounds

The effects of the clarification/stabilization treatments on the sensory qualities and aroma [36, 78] of wine have been studied, but the origin of these phenomena has rarely been explained.

The effect of the dosing time of bentonite on the aroma profile at both industrial and pilot scales has been evaluated [81]. The results seemed to indicate that the addition of bentonite not only affects the wine aroma by the adsorption of compounds but also by the production of these compounds during fermentation. Bentonite treatments at different stages of fermentation generally affected the production of volatile fermentative compounds. In addition to the possible loss of volatile compounds due to adsorption on bentonite, the effects on the production of fermentative compounds could be related to the variations in the nitrogen composition and other nutrients of musts and wines that are removed by the bentonite [82]. In a model solution, the presence of total and purified proteins (TL proteins and chitinase) and bentonite tended to increase the loss of esters with the longest carbon chains that is ethyl octanoate and ethyl decanoate [80], showing that hydrophobicity can be one of the driving forces involved in the interaction of aroma compounds with both bentonite and proteins.

As a matter of fact, it was demonstrated [83] that the effect of bentonite treatments on the aroma substances in white wine depended on the chemical nature and initial concentration of the volatile compounds and on the abundance and nature of proteins in the wine. In general, when low bentonite concentrations (20 g/hl) are used, the concentrations of most aromatic substances are not significantly affected. Most aroma compounds are removed as an indirect effect of deproteinization; some hydrophilic odour-active compounds undergo weak hydrogen binding with protein surfaces, whereas the more hydrophobic aromatic molecules can bind to interior protein sites with a stronger affinity for hydrophobic substances. Only a few odour-active compounds are directly adsorbed by the bentonite through an adsorption process, which is robustly fitted by the Freundlich equation, with a heterogeneous energy distribution of an infinite number of surface-active sites [84]. Bentonites with a lower SSA value and a greater CDSU value seemed to primarily interact with most of the odour-active compounds through physical mechanisms. In contrast, the clay with a large SSA value and a low CDSU value promoted stronger adsorptions that were probably driven by chemical interactions, particularly for the ethyl esters. For the fermentative odour-active compounds, the differences in the adsorption intensity and capacity mainly depended on the characteristics of the bentonite than on the properties of the substances. When yeast-derived material represents an important fraction of the wine macromolecules, colloids that favour aroma inclusion are held in suspension. In this situation, there may be fewer opportunities for the odour-active substances to be directly adsorbed onto the bentonite sheets [83].

6. Conclusions

Oenologists do not really know which parameters they must focus on to choose the bentonite that will obtain both the desired degree of limpidity and colloidal stability while avoiding undesirable side effects on the phenolic and aroma compounds.

This chapter detailed the structure, composition and non-food uses of bentonite and summarized the most recent scientific research that detected:

- the proteins targeted by the bentonite;
- the effect of the protein content and pH towards bentonite fining;
- the bentonite characteristics that affect juice clarifying and wine fining;
- the side effects of bentonite on polyphenols and colour;

- the interactions of bentonite with the free- and glycosylated-varietal aroma compounds;
- the interactions of bentonite with fermentative aroma by indirect and direct removal.

As wine haze formation is a product of different matrix parameters, including the wine pH and the concentration of different wine components, a detailed knowledge of the relationships between the adjuvant that is most commonly used to attain colloidal stability and the oenological matter is fundamental for process optimization and to increase wine resilience.

Author details

Milena Lambri*, Donato Colangelo, Roberta Dordoni, Fabrizio Torchio and Dante Marco De Faveri

*Address all correspondence to: milena.lambri@unicatt.it

Institute of Enology and Agro-food Engineering, Università Cattolica del Sacro Cuore, Via Emilia Parmense, Piacenza, Italy

References

[1] Bergaya B, Theng KG, Lagaly G, editors. Handbook of Clay Science, Developments in Clay Science. Vol. 1. Amsterdam: Elsevier; 2006. p. 1–1224.

[2] Mortier WJ. Electronegativity equalization and its applications. Structure and Bonding. 1987;66:125–143.

[3] Feller D, Glendening ED, Woon DE, Feyereisen MW. An extended basis-set ab-initio study of alkali-metal cation-water clusters. Journal of Chemical Physics. 1995;103:3526–3542.

[4] Bergaya F, Jaber M, Lambert JF. Clays and clay minerals as layered nanofillers for (bio) polymers. In Averous L, Pollet E, editors. Environmental Silicates Nano-Biocomposites, Green Energy and Technology. London: Springer-Verlag; 2012. p. 41–75.

[5] Tournassat C, Ferrage E, Poinsignon C, Charlet L. The titration of clay minerals. II. Structure-based model and implications for clay reactivity. Journal of Colloid and Interface Science. 2004;273:234–246.

[6] Tournassat C, Greneche JM, Tisserant D, Charlet L. The titration of clay minerals. I. Discontinuous backtitration technique combined with CEC measurements. Journal of Colloid and Interface Science. 2004;273:224–233.

[7] Galán R, Pozo, M. Palygorskite and sepiolite deposits in continental environments. Description, genetic patterns and sedimentary settings. In Galán E, Singer, A. editors.

Developments in Palygorskite- Sepiolite research. A new outlook on these nanomaterials. Developments in Clay Science. 2011;3:125–173.

[8] Grim RE, Güven N. Bentonites, Geology, Mineralogy, Properties and Uses, Development in Sedimentology. Vol. 24. Amsterdam: Elsevier; 1978.

[9] Gougeon RD, Soulard M, Miehé-Brendlé J, Chézeau JM, Le Dred R, Jeandet P, Marchal R. Analysis of two bentonites of enological interest before and after commercial activation by solid Na_2CO_3. Journal of Agricultural and Food Chemistry. 2003;51:4096–4100.

[10] Radojevic Z, Mitrović A. Study of montmorillonite and cationic activators system rheological characteristic change mechanism. Journal of the European Ceramic Society. 2007;27:1691–1695.

[11] Boyd SA. Method of removing hydrocarbon contaminants from air and water with organophilic, quaternary ammonium ion-exchanged smectite clay. US Patent. 1993;5:268,109.

[12] Pal OR, Vanjara AK. Removal of malathion and butachlor from aqueous solution by clays and organically. Separation and Purification Technology. 2001;24:167–172.

[13] Bergaoui L, Mrad I, Lambert JF, Ghorbel A. A comparative study of the acidity toward the aqueous phase and adsorptive properties of Al13-pillared montmorillonite and Al13-pillared saponite. Journal of Physical Chemistry. 1999;103:2897–2902.

[14] Shen YH. Removal of phenolic from water by adsorption and flocculation using organobentonite. Water Research. 2002;36:1107–1114.

[15] Dragović S, Mihailović N, Gajić B. Heavy metals in soils: distribution, relationship with soil characteristics and radionuclides and multivariate assessment of contamination sources. Chemosphere. 2008;72(3):491–495.

[16] Nones J, Riella HG, Trentin AG, Nones J. Effects of bentonite on different cell types: a brief review. Applied Clay Science. 2015;105–110:225–230.

[17] Daniel DE, Scranton HB. Report of 1995 Workshop on Geosynthetic Clay Liners. Cincinnati, OH: EPA 600/R-96/149, U.S. EPA; 1996. p. 228.

[18] Reijonen HM, Russell AW. Bentonite analogue research related to geological disposal of radioactive waste: current status and future outlook. Swiss Journal of Geosciences. 2015;108:101–110.

[19] Karagüzel C, Çetinel T, Boylu F, Çinku K, Çelik MS. Activation of (Na,Ca)-bentonites with soda and MgO and their utilization as drilling mud. Applied Clay Science. 2010;48:398–404.

[20] Eckhardt JC, Santurio JM, Zanette RA, Rosa AP, Scher A, Dal Pozzo M, et al. Efficacy of a Brazilian calcium montmorillonite against toxic effects of

dietary aflatoxins on broilers reared to market weight. British Poultry Science. 2014;55:215–220.

[21] Viseras C, Cerezo P, Sanchez R, Salcedo I, Aguzzi C. Current challenges in clay minerals for drug delivery. Applied Clay Science. 2010;48:291–295.

[22] Allo Wanda A, Haydn HM. Mineralogy, chemistry and potential applications of a white bentonite in San Juan province, Argentina. Applied Clay Science. 2004;25(3):237–243.

[23] Iliescu RI, Andronescu E, Voicu G, Ficai A, Covaliu CI. Hybrid materials based on montmorillonite and citostatic drugs: preparation and characterization. Applied Clay Science. 2011:52:62–68.

[24] López-Galindo A, Viseras C. Pharmaceutical and cosmetic applications of clays. In Wypych F, Satyanarayana KG, editors. Clay Surfaces: Fundamentals and Applications. Amsterdam: Elsevier; 2004. p. 267–289.

[25] Murray HH. Bentonite applications. In Murray HH, editor. Developments in Clay Science. Amsterdam: Elsevier; 2006. p. 111–130.

[26] Krøyer H, Lindgreen H, Jakobsen HJ, Skibsted J. Hydration of Portland cement in the presence of clay minerals studied by 29Si and 27Al MAS NMR spectroscopy. Advances in Cement Research. 2003;15:103–112.

[27] Wicklein B, Darder M, Aranda P, Ruiz-Hitzky E. Bio-organoclays based on phospholipids as immobilization hosts for biological species. Langmuir. 2010;26:5217–5225.

[28] Miyoshi H, Ohno H, Sakai K, Okamura N, Kourai H. Characterization and photochemical and antibacterial properties of highly stable silver nanoparticles prepared on montmorillonite clay in n-hexanol. Journal of Colloid and Interface Science. 2010;345:433–441.

[29] Vaillant F, Millan A, Dornier M, Decloux M, Reynes M. Strategy for economical optimisation of the clarification of pulpy fruit juices using crossflow microfiltration. Journal of Food Engineering. 2001;48:83–90.

[30] Li ZX, Wang GM, Liang Q. Preparation and properties of the novel adsorbent agent for juices clarification. Advanced Materials Research. 2013;791–793:248–251.

[31] Laksameethanasan P, Somla N, Janprem S, Phochuen N. Clarification of sugarcane juice for syrup production. Procedia Engineering. 2012;32:141–147.

[32] Jahed E, Khodaparast MHH, Khaneghah AM. Bentonite, temperature and pH effects on purification indexes of raw sugar beet juice to production of inverted liquid sugar. Applied Clay Science. 2014;102:155–163.

[33] Novotny CJ. Ion exchange. In Chen JCP, editor. Cane Sugar Handbook. New York: Wiley; 1985. p. 591–598.

Chapter 19

The Trends and Prospects of Winemaking in Poland

Alina Kunicka-Styczyńska, Agata Czyżowska, Katarzyna Rajkowska, Agnieszka Wilkowska and Piotr Dziugan

Additional information is available at the end of the chapter

http://dx.doi.org/10.5772/64976

Abstract

Viticulture and winery origins in Poland date to the tenth century, but their tradition has been reborn in the last ten years, resulting in a development of small vineyards producing excellent wines not only for the local market. Due to the cold climate, usually short summers with moderate and low temperatures, the grapes are characterized by lower sugar content and higher acidity compared to those grown in the south of Europe. According to the European Union regulations, Poland was classified as the coldest wine-growing region (A) and officially acknowledged as a wine-producing country. The grapevine cultivars adopted to the harsh climatic conditions give the Polish grape wines some unique sensory features. The most popular varieties of grapes for the production of red wine are Regent, Rondo, Pinot Noir, Maréchal Foch, Cabernet Cortis, Tryumf Alzacji, Cascade and Dornfelder. For white wine production, Solaris, Riesling, Seyval Blanc, Pinot Gris, Johanniter, Jutrzenka, Hibernal, Aurora, Bianka, Traminer, Jutrzenka and Siberia are mostly used in Poland. This chapter presents Polish grape winery with its specificity and prospects for the future. The traditional products of Polish fermentation industry, fruit wines and meads, are also mentioned.

Keywords: Polish wines, wine yeast, L-malic acid decomposition, Polish wine regions

1. Introduction

However, Poland was not worldwide recognized as a wine-producing country, but the art of winemaking has been practiced there since the tenth century. Winery was introduced with Christianity and the first vineyards were cultivated by and wineries were established by Benedictine and Cistercian monks; however, wine was produced for religious purposes mainly.

The fruitful time for the Polish winery was the fourteenth century, during which many wineries were operating mainly in Silesia, Zielona Góra, Poznań, Toruń, Płock, Sandomierz, Lublin and Kraków.

Intensive development of wine making was in the age of enlightenment, when the viticulture and wine production were carried out in the Podole. Besides *Vitis vinifera*, hybrid varieties resistant to adverse climatic conditions were grown. After World War II, according to the authorities, two wine-growing regions were designated: the West (Zielona Góra region and Lower Silesia) and the Central (along the Pilica river). Vineyards planted in the communistic economy, however, have begun to bear losses, and in the 1960s, it was focused on the production of fruit wines. The tradition of viticulture and winery has been reborn in the last ten years, resulting in development of small vineyards producing excellent wines for the local market. Poland is located in the zone of the continental climate, where there are also wine regions such as Burgundy and the Loire Valley, Rioja, Piedmont and most of the vineyards of Austria, the Czech Republic, Slovakia or Romania. The climate in Poland is characterized by significant seasonal and daily fluctuations in temperature with the potential problems of frost and hailstorms during winters and springs. Due to the cold climate, usually short summers with moderate and low temperatures, the grapes are characterized by lower sugars content and higher acidity comparing to these grown in the south of Europe. A wine zone, most suitable for the cultivation of vines, is located on the latitude between 32°00′ and 52°00′ northern hemisphere and between 28°00′ and 42°00′ southern hemisphere [1].

Poland extends from the parallel 49°00′ N (south) to 54°50′ N (north), so about half of the country is situated in this area. At the latitudes between 49°00′ N and 52°00′ N, there are many regions in Europe known for their excellent wines, including some appellations in the region of Champagne (e.g. Reims) or German appellations the Rhine, mozelskie and Franconian. Krakow is located similarly [2]. Climate favourable for viticulture is characterized by an average annual temperature not less than 8°C, the average temperature of the hottest month not less than 17°C and the total active annual temperature 25°C. Some areas of Poland are also characterized by these conditions. The current climate changes are conducive to the development of Polish winemaking. The average annual temperature showed an upward trend (about 0.3°C per decade), transitional periods have been shortened, warm periods have been prolonged, the course of winters became milder, allowing the cultivation of early and very early varieties.

According to the European Union classification of climate for viticulture (Council Regulation (EC) No 479/2008), Poland was classified in the coldest wine-growing region (A) and officially acknowledged as a wine-producing country, altogether with, among others, Germany (except for Baden), the Czech Republic (except for Moravia), Belgium and the Great Britain.

Due to the thermal conditions, Polish territory was divided as follow: Region I—the west and southwest of the country, namely provinces of Lubusz, Lower Silesia, Opole, Silesian and southern parts of the provinces Wielkopolska and Lodz; Region II—threatened with greater extent of cold winters, covers the south and southeast of the country, i.e. the province Małopolska, Podkarpackie, świętokrzyskie and southern parts of the provinces of Lublin and Warsaw; and Region III—the other areas, where viticulture is impossible or very difficult [3].

Nowadays, the following winemaking regions can be distinguished in Poland; the largest and the best known vineyards are located in: (i) area of Zielona Góra (Lubuskie Province); (ii) Lower Silesia; (iii) vicinity of Kraków; (iv) area of Jasło and Krosno in the Podkarpackie Province; and (v) along the Vistula gorge (from Sandomierz to Puławy).

2. Wine yeasts

Fermentation in winemaking can either be natural, conducted by native yeasts present on the fruit skins and in the winery, or by selected yeasts strains. The microflora of fruits vary according to a number of factors, among others, the fruit variety, temperature, climatic conditions, soil, viticultural practices and fungicides applied to vineyards [4]. In view of this variability, to obtain high-quality wines of defined aroma and flavour, the inoculation with pure cultures of selected commercial yeasts is recommended. In addition, starter culture fermentations offer the advantages of a more predictable and rapid process, giving wines with high consistency in quality. Criteria for the selection and development of yeasts for wine fermentation encompass: (i) vigorous and complete fermentation of fruit juice sugars, without excessive yeast growth; (ii) fermentation at low temperatures 10–12°C; (iii) production of ethanol in high concentrations and ethanol tolerance of yeasts; (iv) growth and fermentation in musts containing sulphur dioxide; (v) uniform dispersion and mixing throughout the fermenting juice; (vi) low foaming ability; (vii) effective sedimentation at the end of the fermentation; (viii) balanced array of flavour metabolites; and (ix) genetic stability [4–6].

Moreover, for areas with unfavourable climatic conditions including Poland, the high acidity of fruit wines caused inter alia by the excess of malic and tartaric acids in the raw material is specific. In terms of fruits used in the Polish winemaking industry, both organic acids account for up to 90% of all the acids present in grapes, and malic acid content reaches 94–98% of total acids present in apples and cherries [4, 7]. The high level of organic acids unfavourably influences the organoleptic properties and biochemical, physical and microbial stability of wine. One of the strategies for solving this problem is biological deacidification by wine yeasts with an extended ability to degrade L-malic acid. Demalication process is conducted simultaneously with fermentation. Malate decomposition varies greatly and may reach 48%, depending on the strain. However, industrial wine yeasts *Saccharomyces cerevisiae* and *Saccharomyces bayanus*, with a demalication activity reaching 68%, have also been selected and well-characterized [8, 9]. *Saccharomyces* sensu stricto strains are usually considered as less effective in malic acid metabolizing than *Schizosaccharomyces pombe* and *Zygosaccharomyces bailii*, which are additionally characterized by high resistance to acidity and sulphur dioxide [10, 11]. The rather low malate decomposition in *Saccharomyces* spp. is explained by the absence of an active L-malate transport system, the low substrate affinity of the malic enzyme and the mitochondrial location of two malic isoenzymes [12]. The alternative in reduction of the acidity of musts and wines may be the application of mixed cultures of yeasts, composed of species degrading L-malate and L-tartrate, such as *Schizosaccharomyces malidevorans, Kloeckera* spp., *Candida* spp. and *Hansenula* spp. [13]. However, inoculation of musts with yeasts other than *S. cerevisiae* may be associated with the possible interactions between populations and adverse

avium) in the region Świętokrzyskie; from sour cherries in the province Wielkopolska; from plum in the province Zachodniopomorskie [38].

in Central Poland: Auxerrois, Pinot Gris, Pinot Noir, Riesling and Chasselas Blanc, classified as both wine and table cultivars [26].

[29] Snowarski M. Strefy mrozoodporności nowe mapy. Szkółkarstwo [Internet]. 2014. Available from: www.ogrodinfo.pl/szkolkarstwo/strefy-mrozoodpornosci-nowe-mapy/3 [Accessed: 20.02.2016]

[30] Angelov L, Stalev B, Ochmian I, Mijowska K, Chełpiński P. Comparison of processing fruit quality of several grape varieties cultivated in climatic conditions of Poland and Bulgaria. Folia Pomeranae Universitatis Technologiae Stetinensis. Agricultura, Alimentaria, Piscaria et Zootechnica. 2015;318:5–14.

[31] Ochmian I, Chełpiński P, Rozwarski R, Strzelecki R, Pantecki P, Angelov L, Stalev B. The fruits quality and influence of maceration methods on their fruit juice colour in two cultivars of wine grapes. Folia Pomeranae Universitatis Technologiae Stetinensis. Agricultura, Alimentaria, Piscaria et Zootechnica. 2012;295:35–42.

[32] Chełpiński P, Gembara J, Ochmian I, Grajkowski J. Viticulture in the vicinity of Szczecin. In: III Winery Seminar; 23.01.2009; Sulechów, Poland. Sulechów: Institute of Management and Agricultural Engineering State Higher Vocational School in Sulechów; 2009. p. 64–65.

[33] Tarko T, Duda-Chodak A, Sroka P, Satora P, Jurasz P. Polish wines: Characteristics of cool-climate wines. Journal of Food Composition and Analysis. 2010;23:463–468. doi: 10.1016/j.jfca.2010.01.009

[34] Tarko T, Duda-Chodak A, Sroka P, Satora P, Jurasz E. Physicochemical and antioxidant properties of selected Polish grape and fruit wines. Acta Scientarum Polonorum. Technologia Alimentaria. 2008;7(3):35–45.

[35] Socha R, Gałkowska D, Robak J, Fortuna T, Buksa K. Characterization of Polish wines produced from the multispecies hybrid and *Vitis vinifera* L. grapes. International Journal of Food Properties. 2015;18:699–713. doi:10.1080/10942912.2013.845784

[36] Dobrowolska-Iwanek J, Gąstoł M, Wanat A, Krośniak M, Jancik M, Zagrodzki P. Wine of cool-climate areas in South Poland. South African Journal of Enology and Viticulture. 2014;35:1–9.

[37] Popek S. An attempt to assess the impact of selected methods of clarification on the wine quality. Proceedings of Gdynia Maritime University. 2015;88:82–88.

[38] Byszewska I, Kurpińska G. Polish tastes. Culinary tour of regions. Poznań: Zysk i S-ka; 2012. 280 p.

[39] Bonin S. Effects of magnesium ions on both VHG batch and continuous fruit wine fermentations. Journal of the Institute of Brewing. 2014;120:477–485. doi: 10.1002/jib.170

[40] Satora P, Tarko T, Duda-Chodak A, Sroka P, Tuszyński T, Czepielik M. Influence of prefermentative treatments and fermentation on the antioxidant and volatile profiles of apple wines. Journal of Agricultural and Food Chemistry. 2009;57:11209–11217. doi: 10.1021/jf9025053

[41] Kunicka-Styczyńska A. Glucose, L-malic acid and pH effect on fermentation products in biological deacidification. Czech Journal of Food Sciences. 2009;27:S319-S322.

[42] Kunicka-Styczyńska A, Pogorzelski E. L-malic acid effect on organic acid profiles and fermentation by-products in apple wines. Czech Journal of Food Sciences. 2009;27:S228-S231.

[43] Kunicka-Styczyńska A. In the search for novel wine yeast with deacidification activity. Fermentation Technology. 2012;1(e106) doi:10.4172/2167-7972.1000e106

[44] Pogorzelski E, Czyżowska A, Fajkowski J. Bitterness causes of wines made of concentrated apple juice. Scientific-Technical Magazine for Fermentation- and Fruit & Vegetable Industry. 1995;39:15–17.

[45] Pogorzelski E, Czyżowska A, Jagoda T. Study on selection of gelatin to the clarification of juices and wines. Scientific-Technical Magazine for Fermentation- and Fruit & Vegetable Industry. 1995;39:13–15.

[46] Pogorzelski E, Laskowska A, Czyżowska A. Degradation products of nucleic acids in wines fermented with dried autolysate of sedimented wine yeast. Polish Journal of Food and Nutrition Sciences. 2006;15(56):177–181.

[47] Pogorzelski E, Wilkowska A, Kubus M. Inhibiting effect of tannin in chokeberry must on the winemaking process. Polish Journal of Food and Nutrition Sciences. 2006;15(56): 49–53.

[48] Pogorzelski E, Czyżowska A. Some physical and chemical parameters of polyphenols as parameters to the evaluation of the quality of red wines. Scientific-Technical Magazine for Fermentation- and Fruit & Vegetable Industry. 1997;5:14–16.

[49] Czyżowska A, Pogorzelski E. Changes to polyphenols in the process of production of must and wines from blackcurrants and cherries. Part I. Total polyphenols and phenolic acids. European Food Research and Technology. 2002;214:148–154. doi:10.1007/s00217-001-0422-9

[50] Czyżowska A, Pogorzelski E. Changes of polyphenols in process of production of must and wines from blaccurrants and cherries. Part II. Anthocyanins and flavonols. European Food Research and Technology. 2004;218:355–359. doi:10.1007/s00217-003-0857-2

[51] Cieślak J. Homemade fruit wines, mead, vodkas, liqueurs, cocktails. Warszawa: Oficyna Wydawnicza Watra; 2006. ISBN 978-83-879341-1-8

[52] The culture of wine Travel-Wine-Culinarian [Internet]. Available from: www.winokultura.pl [Accessed: 17.07.2016]

[53] Dz.U. 2011 nr 120 poz. 690. The Act of 12 May 2011 on the product and the bottling of wine products, the marketing of these products and the organization of the wine market [Internet]. 2011-05-12. Available from: www.isap.sejm.gov.pl [Accessed: 7.04.2016]

Section 3

Analysis and Origin Authentication

Chapter 20

Determination of Trace Elements in Wine by Atomic Spectroscopy and Electroanalytical Methods

Niina J. Ronkainen

Additional information is available at the end of the chapter

http://dx.doi.org/10.5772/64932

Abstract

The chapter will outline the basic theory, advantages and disadvantages, experimental considerations and set up of various atomic spectroscopy, and electroanalytical quantification methods and their specific application to trace element determination in wines. The reader will gain an introduction to most popular elemental analysis methods used in beverage analysis. Copper, iron, manganese, and zinc will be used as examples of essential trace elements throughout the chapter that at high levels may affect the properties of wine as well as the sensory experience of the consumer. Furthermore, special considerations that should be given to wine as a sample matrix for quantitative analysis of inorganic elements and the use of standard addition methods will be described.

Keywords: trace element determination, atomic spectroscopy, electroanalytical methods, wine, standard addition methods, stripping voltammetry

1. Introduction

Wine is a widely consumed alcoholic beverage produced by yeast fermentation of natural sugars contained in grape juice. Wine has been produced and consumed for thousands of years and is thought to have originated from current Georgia where wine stained pottery from circa 6000 BC has been found [1]. Drinking wine daily is common in many cultures ranging from countries with old, well-established wine cultures in Europe and America to rapidly growing wine markets such as China. In some countries, alcoholic beverages, including wine, account for >12% of the daily intake of beverages [2]. Although the per capita consumption of wine by Americans was only 9.42 L per person in 2010, the United States is the single largest wine market in the world due to the large number of American consumers [3]. However, Old World wine countries such as France, Italy, and Portugal had the highest per capita wine consumption in 2010 (45.7, 42.15, and 41.81 L, respectively) out of the selected wine producing countries (**Figure 1**) [3]. Algeria and Israel had the lowest wine consumption per person of the selected countries at 0.97 L in 2010 [3].

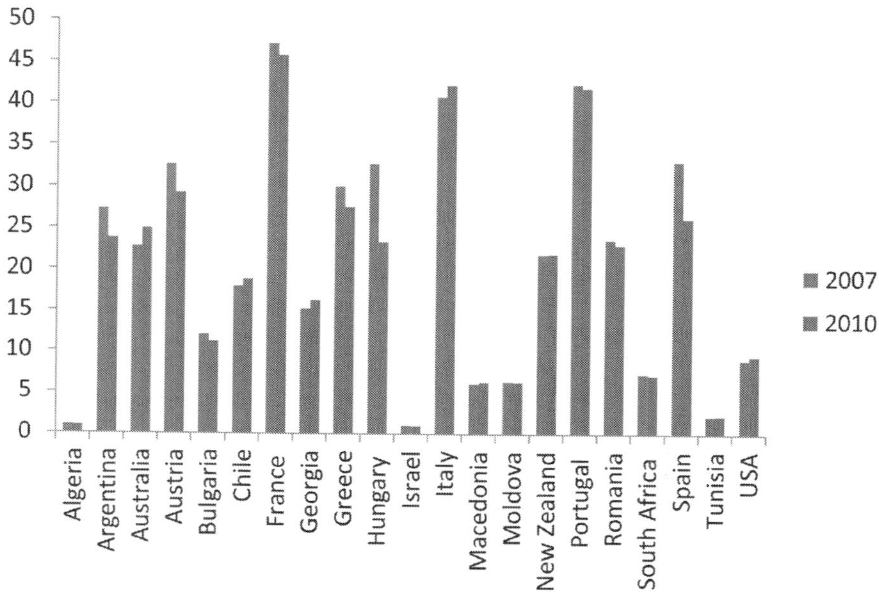

Figure 1. Per capita wine consumption by country (L per person) in 2007 and 2010 from selected wine-producing countries (Wine Institute, Trade Data and Analysis, 2012).

Although fewer studies have been published on trace metal content in alcoholic beverages compared to elemental analysis of foods such as fish, produce, peanuts, and tea, wines are one of the most studied alcoholic beverages. Most studies have focused on quantification of metals such as As, Ni, Cu, Cr, Cd, Pb, Zn, and others in wines from specific countries such as Spain [4, 5], Italy [6–10], Argentina [11–13], Australia [14], Turkey [15–17], Romania [18], and Croatia [19–21] and/or focused on studying regional variation in mineral content from wine-producing regions and grape varieties within a single country.

Wine has significant economic impact and commercial value in addition to its social importance. Wine production is one of the most important agricultural activities in many regions of the world. Wine has a complex sample matrix that contains various components such as macro- and microelements as well as lanthanides that contribute to its nutritional value. Wine composition also greatly affects its quality. The concentration of metals in wine is of great

http://dx.doi.org/10.5772/64932

significance as it affects their conservation and color, in addition to impacting the organoleptic properties of wine, thereby affecting human consumption and the sensory experience [2, 22, 23]. Trace metals affect the organoleptic properties such as aroma, color, flavor, freshness, and taste. Cu, Fe, and Zn, for example, contribute to haze formation and taste effects in wine [24]. However, these minor metals are also favorable for yeast as they are an essential part required for prosthetic group of metalloenzymes that serve as biological catalysts. Other elements such as Ca, K, Mg, and Na are involved in regulating the cellular metabolism of yeasts by helping to maintain adequate ionic balance and pH [25]. Since metal species participate in oxidation-reduction reactions, physical properties of the wine such as color, turbidity, and astringency are largely dependent on metal composition of the wine [23]. Browning, the oxidative spoilage of wine which ultimately results in the loss of aromatic freshness and the appearance of precipitates of condensed phenolic material in the bottled wine, is activated and accelerated by the presence of Fe, Mn, and Cu [2, 22]. The presence and the levels of specific metals can also be used in quality control and the authentication of wines [5, 26]. In addition, metal profile from trace element analysis can act as a fingerprint that may be used to determine the origin, variety, and/or the type of wine and other beverages [27–31]. Therefore, trace element characterization of wine is powerful in fraud detection in commercially sold wines. The trace element composition of wine is influenced by factors such as grape varieties, soil at the vineyard, and viticultural practices [32, 33]. Knowledge of metal concentrations in wine is also of great importance in quality assurance (QA) of branded wines [34].

There are endogenous and exogenous sources of metals in commercial wines. Endogenous metals of natural origin mostly come from the soil on which vines are grown and reach the wine through the harvested grapes [34]. These primary metals make up for the largest part of total metal content in wine [35–37]. Concentrations of these metals are also characteristic of the type of soil at the vineyard, the climate conditions during the growth of the grapes, grape variety, and the maturity of the grapes [38]. The metals of secondary origin are introduced during growth of grapes or at different stages in winemaking starting with harvesting and finishing with storing bottled wine at a cellar. Environmental pollution, addition of fertilizers used in cultivation, as well as application of fungicides and pesticides during the growing season of the grapes increases the amounts of metals such as Cd, Cu, Mn, Pb, and Zn in the finished wine [11, 39–43]. Variation in levels of K, Ca, and Cu in wine may also results from fertilizers applied at the vineyard [44]. The environmental pollution introducing metals into the soil, grapes, and ultimately the wine could be due to nearby industry, waste management practices, traffic, and even mining operations. For example, wines produced at vineyards located near industrial areas or close to major highways often contain higher levels of Cd and Pb due to vehicle exhaust fumes and other emissions to air, water, and/or soil [11, 39, 41, 42]. During winemaking process, the long contact of acidic wine with materials such as aluminum, brass, glass, stainless steel, and wood used in wine-making machinery and pipes, casks, and barrels used for handling and storing wine may also introduce metal contamination to the wine. The steps during the winemaking process often introduce elements such as Al, Cd, Cr, Cu, Fe, and Zn into the finished wine [11, 35]. Metal wine casks at a Sonoma Valley winery are shown in **Figure 2**.

Figure 2. Shows wine storage containers at a Sonoma Valley winery at California (USA).

Many studies have shown that moderate consumption of red wine may improve health and longevity when it is done in combination with a balanced diet [45]. Much attention has been given in the research and medical communities to the properties and activity of polyphenols. Polyphenols have been shown to provide some possible health benefits by acting as antioxidants that helps prevent cell damage. Many of these health benefits of wine are associated with polyphenol antioxidants such as resveratrol (3,5,4′-trihydroxystilbene) that is common in grapes and many other plants [46]. Resveratrol obtained from moderate wine consumption was found to decrease the risk of cardiovascular events and decrease tumor growth in animal models [47, 48]. The powerful antioxidant properties of resveratrol are thought to reduce the oxidation of low-density lipoproteins (LDL) and to inhibit platelet aggregation that is involved in the formation of atherosclerotic lesions [49–51]. However, due to the complex structure, properties, and activity, many of the possible positive effects of polyphenols on animal and human health are not yet fully understood. The structure of resveratrol (3,5,4′-trihydroxystilbene) is shown in **Figure 3** as an example of a potent, well-studied organic antioxidant.

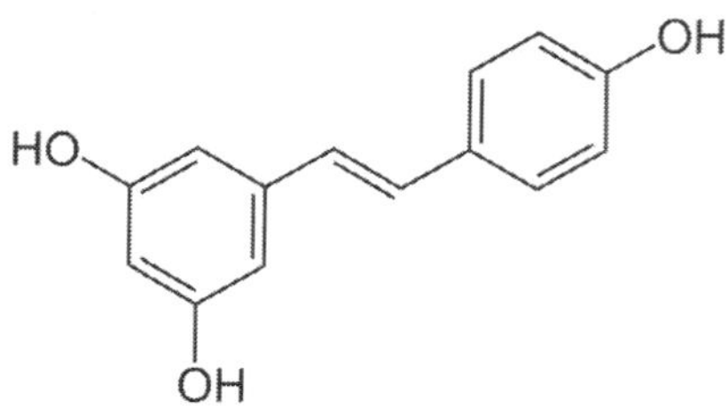

Figure 3. Structure of resveratrol (3,5,4′-trihydroxystilbene), a potent and commonly found polyphenol in red wines with known antioxidant properties.

While wine when used in moderation may also be a good nutritional source for dietary intake of some essential minerals such as iron (Fe) and zinc (Zn) [34, 52], and other metals such as copper (Cu) are considered both an essential and a potentially toxic element. Copper in excess may cause serious health consequences or result in the long-term bioaccumulation and toxicity in the body [53]. Excessive iron intake above the tolerable upper intake level (UL) causes gastrointestinal distress while excess copper intake is likely to result in liver damage. It has been established that even moderate daily consumption of wine contributes significantly to the human nutritional requirements for essential trace elements such as Cr, Co, Fe, Cu, Mn, Mo, Ni, Se, and Zn [54]. The estimated safe intake levels of Cu, Fe, Mn, and Zn established by the US and the European Union (EU) are summarized in **Table 1** [55–57].

Mineral		USA RDA (mg/day)	USA UL (mg/day)	EU PRI (mg/day)	EU SCF (mg/day)
Cu	Adults	0.9	10	1.1	
Fe	Female	18.0	45		15–20
	Pregnant female	27.0	45		30
	Male and postmenopausal female	8.0	45		8–10
Mn	Female	1.8	11[a]		1–10[b]
	Male	2.3	11[a]		1–10[b]
Zn	Female	8.0	40	7	
	Male	11.0	40	9.5	

[a] The estimated safe and adequate dietary intake for Mn was reported as 2–5 mg/day for adults [55].
[b] Scientific Committee on Food of the EU has reported 1–10 mg/day as an acceptable range of Mn intake [56]. The EU does not have a RDA for Mn.

Table 1. Recommended dietary allowance (RDA) and tolerable upper intake level (UL) of selected trace minerals by the Food and Nutrition Board (Washington DC) [57]. Population reference intake (PRI) recommendations by the European Union. The Scientific Committee on Food (SCF) guidelines by the European Union [56].

The World Health Organization and various regulatory agencies have established recommended maximum limits for the concentration of metals in beverages and water utilized in human consumption. In addition, individual countries have rules restricting maximum metal content in wines, which have to be met in order to export wine to these consumer markets. For example, Cu and Zn content must be <5 mg/L by law in wines sold in Australia and Germany [58]. The Office International de la Vigne et du Vin (OIV) sets the limits for Cu at 1 mg/L and Zn at 5 mg/L [58, 59]. Italy has established the following legal levels in wine: 1 mg/mL Br, 1 mg/L Cu, 0.2 mg/L Pb, and 5 mg/L Zn [60]. Wine quality is also greatly affected by their metal content. For example, concentrations of Fe >5 mg/L generally induces haze formation and oxidative spoilage of the wine [2]. Concentration of Cu(II) above 1 μg/mL and Fe(III) ions above 7 μg/mL can give unpleasant, astringent tastes in addition to producing cloudiness in wines that have high pH or high concentration of tannic substances [35, 61]. Typically, the limits on metal concentration imposed for alcoholic beverages are higher than those established for drinking water as the expected intake of alcoholic beverages such as wine is lower [62]. Most of the trace metals are thought to originate from the atmospheric deposition of airborne particulate matter on grapes, composition of the soil, residues of metal containing agrochemical products, transfer of metals from the soil via the roots to the grapes and the wine, vinification methods, and contamination from contact with the metal apparatus during winemaking, production, and packaging processes [2, 54]. For example, some of the copper in finished wine may be due to copper sulfate used for spraying the wines to prevent mildew growth. Most of the iron in wine is determined by the composition of the soil where the grapes for the wine are produced [39]. Elements such as As, Cd, Pb, and Br are considered to be potentially toxic [63].

The ability for discriminating wines regionally within or between countries based on their trace element profile suggests that the elements mainly originate from their movement from the rock to soil and from the soil to grapes used to produce the commercially sold wine [64, 65]. Ability to determine the region of production for a wine greatly benefits the distributors, producers, as well as the consumers. The denomination origin controlled (DOC) system is used in some wine-producing countries to track the origin, to guarantee the wine quality, and to help eliminate fraud [66, 67]. Chemical characterization of finished wine is one of the key requirements to obtaining DOC certification. Wine authenticity may be examined by quantifying suitable geographic tracers in finished wine. For example, the origin may be tracked by examining the specific content of organic constituents that vary regionally such as anthocyanins, flavanols, flavonols, and organic acids [68, 69], by studying the elemental composition profile of inorganic species and by analysis of variation in selected stable isotopes [70]. Chemometric approaches have been applied to large analytical data sets that are generated during wine composition analysis and authenticity determination [12, 21, 66, 67, 70, 71].

The multielemental composition profile in soils appears to be strongly affected by the solubility of inorganic species in the soil. This connection between soil composition and that of finished wine has been confirmed in studies where soil and wine samples were systematically characterized. Strong consensus has been demonstrated between soil and wine samples' element composition while clear differences were shown between the studied regions [70]. In addition to Argentinian wines, statistically significant correlation between vineyard soils and finished Czech wines has been reported [35]. Ultimately, this multielement profile creates a sort of geochemical "fingerprint" that is unique to the vineyard or at least the grape growing region. These regional trace element patterns have been well established in chemical literature using instruments such as inductively coupled plasma-mass spectrometry, ICP-MS [72–84]. Characterization of the geographic origin of wine is even more accurate when natural isotopic abundance ratio of a key element, such as strontium ($^{87}Sr/^{86}Sr$) that is independent of the grape variety, is combined with trace element profile analysis [70].

As described above, concentrations of metals in finished wine are characteristic of the type of soil at the vineyard but are also influenced by the climate conditions during the growth of the grapes, the variety of grapes, and their maturity [38]. An example of a vineyard located in the

http://dx.doi.org/10.5772/64932

northern United States will now be described. **Figure 4** shows the scenery at Chateau Grand Traverse in northern Michigan. The soil at the vineyard consists of loamy sand/glacial till and has a pH of 7.2–8.0. Chateau Grand Traverse vineyards lie directly on the 45° North Parallel at the elevation of 640–850 feet (or 195–259 m). The climate on the Old Mission Peninsula is also moderated by the surrounding deep waters of the Traverse Bay, helping to prevent frost during the growing season. Growing season at the Old Mission Peninsula is usually 145–160 days and the heat units have a 5-year average = 2260° days (+50°F or +10°C). Five-year average rain fall is 24.69″ (or 62.7 cm) with 80.00″ (or 203.2 cm) average snowfall.

Figure 4. Shows the scenery at Chateau Grand Traverse, one of the oldest wineries in northern Michigan. The vineyard is located on a narrow peninsula, north of Traverse City on the shores of Lake Michigan.

The typical composition and properties of wine as a sample matrix for various quantitative analyses will be described in the next section. The sample composition and properties impact important decisions such as what type of sample preparation is needed (if any), should the use of standard addition methods be considered, what is the best instrumental analysis method for characterization of the sample and the quantification of target analytes, etc.

2. Wine as a sample matrix and winemaking treatments

The sample matrix is, by definition, everything in the sample of interest other than the analyte being quantified. Wine is a relatively complex aqueous sample matrix for quantitative chemical analysis due to containing about 12–15% ethanol (by percent volume) and mixture of hundreds of different organic compounds (such as polyhydroxyalcohols, polyphenols, organic acids, polysaccharides, peptides, etc.) in addition to various inorganic species (such as selenium, iron, zinc, nickel, copper, etc.). Major metals in wine, Ca, K, Na, and Mg are typically at levels in the range of 10–10^3 µg/mL [36]. The concentration of K is usually the highest [85]. Al, Cu, Fe, Mn, Rb, Sr, and Zn are called minor metals that are usually found in the range of 0.1–10 µg/mL [34]. Trace metals include Ba, Cd, Co, Cr, Li, Ni, Pb, V, and others in the range of 0.1–100 ng/mL or less [34].

It is also important to note whether or not allowed additives, adjuvants, or fining agents were used in the winemaking process as these do affect the trace-element levels in wine. Ca concentration in wines is often affected by the addition of $CaCO_3$ or $CaSO_4$ during winemaking to deacidify must and wine [11, 43]. Clarifying products such as bentonites are allowed in winemaking. It has been reported that some winemaking treatments such as betonies and yeast hulls can affect the final trace element composition of wine [6]. Bentonite, for example, is a natural highly absorbent clay containing sheet silicates that is widely used in winemaking. Bentonite contains exchangeable cations such as Ca, Mg, and Na. Bentonite, which mostly consists of montmorillonite-type phyllosilicate, acts as a settling aid to clarify wine and to remove amino acids, minerals, polyphenols, and protein, thus minimizing the risk of haze formation in wine. Proteins are depleted in wines due to their adsorption on the surface of the silica layers. Yeast hulls, living and nonliving biomass of *Saccharomyces cerevisiae*, have also been reported to significantly lower heavy metal content of wines by biosorption [86–88].

Nicolini et al. quantified changes of several minerals in red and white wines resulting from the addition of bentonites and yeast hulls using inductively coupled plasma-optical emission spectrometry (ICP-OES) and inductively coupled plasma-mass spectrometry [6]. They studied the changes in trace element composition resulting from Italian wines treated with 10 different types of bentonite at 1 g/L (the highest doses routinely used in winemaking) and two different yeast hulls at doses of 180 and 360 mg/L. The authors found that bentonite fining significantly lowered concentrations of K, Cu, Rb, and Zn [6]. Cu depletion of −43% was observed with bentonite fining. Meanwhile, certain elements such as Ce, Gd, La, Nd, Pr, and Y increased by about one order of magnitude due to bentonite fining that was used. Also, concentrations of Be, Tl, and U increased by about 4–6 times after bentonite fining [6]. Therefore, depending on the specific element, its concentration may increase or decrease in the wine upon treatment with bentonites.

Nonliving yeast cell walls have also been reported to reduce concentrations of certain cations due to biosorption-involving protein-polysaccharide complexes [89]. Treatment of wine by yeast hulls induced significant decreases of Ce, Cu, Fe, La, Sb, U, V, and Y content in red and white wines as reported in [6]. Higher depletion of the elements was seen at the higher doses of 360 mg/L yeast hulls exposed to wines for 3 h.

3. Matrix effect and the use of standard addition methods

Quantification of certain inorganic trace and ultratrace elements such as arsenic, cadmium, and lead in complex sample matrix such as wine is often challenging due to their very low concentrations. Including a preconcentration step prior to quantitative analysis is often necessary for these types of dilute analytes as the analyte concentrations may be below the detection limit of the instrumental analysis technique. The preconcentration step may also help to isolate the analyte of interest from the complex sample matrix, thereby improving selectivity of the detection step and stability of the analyte. Evaporation of some solvent including ethanol and water is usually sufficient to preconcentrate wine samples to a high enough level for quantification of trace elements such as iron, copper, zinc, and manganese by spectroscopy. However, electrolytic deposition is a very efficient method for isolation and preconcentration of many inorganic trace elements prior to quantification using an electrochemical detection method such as stripping voltammetry (SV). Selected examples of spectroscopic and electroanalytical detection methods will be described later in this chapter.

Matrix effect is a detectable change (i.e., an increase or a decrease) in the analytical signal response caused by sample components other than the analyte. Matrix interferences that are sometimes observed in atomic spectroscopy (AS) are often due to differences between surface tension and viscosity of test solutions that are often used to simulate the samples and real samples undergoing quantification. Standard addition is especially useful when the exact sample composition is not known, the composition varies between samples, or the sample is complex and the sample matrix affects the observed analytical signal. Therefore, standard addition methods are well suited for analyzing complex aqueous samples such as wine in which the likelihood of matrix effects is substantial. In standard addition, all samples undergoing analysis are in the same matrix. The standard being added in standard addition procedures is the same substance as the analyte of interest. For example, a certified reference standard purchased from a chemical company for aqueous Zn at ppm levels will be added into aliquots of wines undergoing quantitative analysis for zinc based on standard addition approach. The additions result in an increased signal response such as an increase in absorbance, which is directly proportional to the amount of zinc standard added. The standard addition methods will now be described in more detail.

By following the standard addition approach, sample constituents including the possible matrix effect that may suppress or enhance the signal response should be identical in each case because the standards are prepared in aliquots of the same sample. Standard addition methods are also advantageous when the amount of sample is limited. For example, in clinical laboratory setting, limited sample size could be encountered when blood or urine samples are obtained from premature babies or dehydrated elderly patients. In this scenario, standard additions can be carried out by successive introductions of increments of the standard to a single measured volume of the unknown sample. Signal measurements are then made on the original sample and on the sample plus the standard after each addition. In most cases, the sample matrix is nearly identical after each standard addition, the only difference being the concentration of the analyte that has increased leading to detectable increases in the signal response. For direct metal determination in wines without any pretreatments, the usage of ethanol-containing standards or using the standard addition method is recommended in order to minimize chemical and physical interferences [90–93].

As described, standard addition method can take several forms. For example, the wine sample may be spiked by adding one or more increments of a standard solution containing known concentration of the trace element analyte, such as Zn^{2+}, to sample aliquots containing identical volumes of wine. The standard being added should ideally be relatively concentrated so that the addition of small volumes is sufficient and the sample matrix is not significantly altered. Each solution is then diluted to a constant total volume using deionized water before measurement of the signal. **Figure 5** helps one to visualize stepwise a typical standard addition experiment with constant total volume. This procedure is sometimes called a graphic procedure for standard addition. This graphic procedure is necessary when the analysis method such as atomic spectroscopy consumes some of the prepared solution during the quantification step. Flame atomic absorption, for example, may consume 2–5 mL of the prepared solution when multiple absorbances are obtained in order to determine a more representative mean absorbance value. The sample introduction and atomization steps in flame-atomic absorption spectroscopy (flame-AAS) are very wasteful as the aerosol containing the analyte reaching the flame contains only about 5% of the initial sample. The excess liquid drawn from the sample solution through a capillary tube flows out to drain and ultimately a waste container.

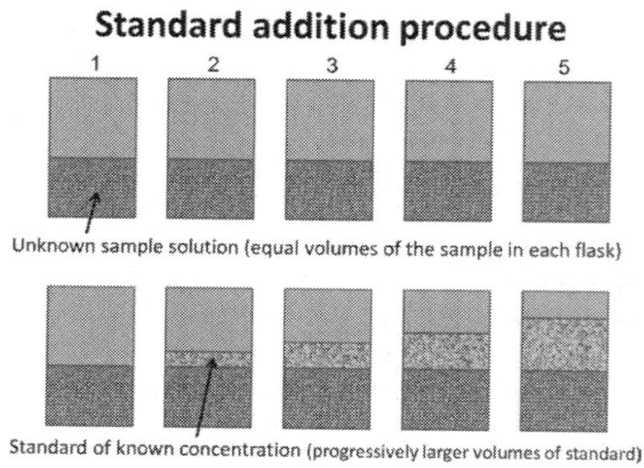

Figure 5. Standard addition procedure shows the addition of equal volume aliquots of the unknown sample (wine) to be quantified to five identical containers, the addition of progressively larger volumes of standard of known concentration and composition (middle layer), and finally, filling the flasks to the same total volume using deionized water prior to the quantitative analysis. Ultimately, all samples undergoing quantitative analysis are in the same sample matrix.

The signal response is determined and recorded for each resulting solution to help quantify the analyte or analytes of interest. It is critical in quantitative analysis, regardless of the detection method chosen, that the signal response to the added analyte is linear. Ideally, each aliquot of added standard in a graphic standard addition procedure should increase the signal by a factor of 1.5–3. The quantification by this procedure usually involves the following steps:

1. Calculate the concentration of added standard for each flask as measured after dilution.

2. Plot analytical signal versus concentration of added standard on a *xy* graph.

3. Add a linear regression line and the equation for the line.

4. The equation for the line ($y = mx + b$) will be used to find the concentration of the analyte in flask 1 by setting y-value equal to zero, and solving for x ($x = -b/m$). The magnitude (absolute value) of the x-axis intercept is the original concentration of the unknown (in units matching the x-axis).

An example of a standard addition plot obtained using graphical standard addition treatment is shown in **Figure 6** for the quantification of Zn in a white wine using flame-atomic absorption spectroscopy. The increases is mean absorbance were proportional to the aliquots of Zn standard added, allowing the reliable determination of Zn in the original wine using linear regression. Each standard addition increased the absorbance signal by a factor of 1.8.

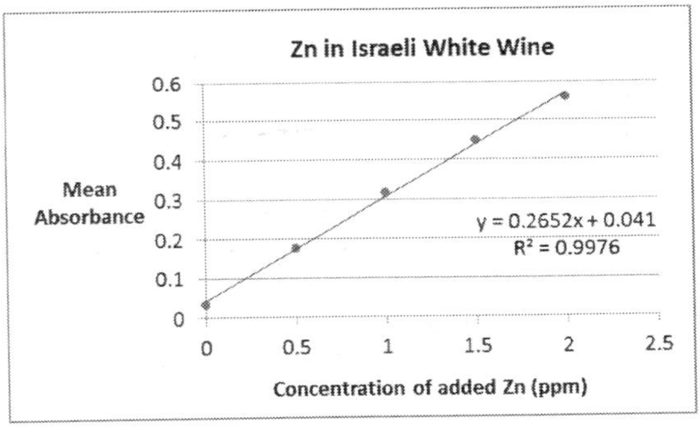

Figure 6. Graphical treatment of standard addition data in quantification of zinc. Concentration of Zn in a white wine from Golan Heights in Israel was determined using 0, 2.5, 5, 7.5, and 10 mL additions 10 ppm zinc standard (from PerkinElmer) to 50 mL volumetric flasks already containing the white wine samples. The first flask with mean absorbance of 0.017 only contained the wine sample and deionized water. Based on the standard addition experiment, this Vintage 2012 Galilee wine from Golan Heights (Israel) contained 1.16 mg/L or ppm of zinc.

4. Common analytical methods for trace element analysis

Various analytical techniques have been used for metal analysis in water, foods, and beverages. The electroanalytical methods include stripping voltammetry and stripping polarography [4, 10, 94–97]. Other analysis methods include X-ray fluorescence [98, 99] and near-IR spectroscopy [14, 29, 30]. Ion chromatography and capillary zone electrophoresis can also be applied to the analysis of metal cations in wines [5, 100, 101].

Atomic absorption and emission spectroscopy methods such as inductively coupled plasma-optical emission spectroscopy, inductively coupled plasma-mass spectroscopy, graphite furnace-atomic absorption spectroscopy (GF-AAS), and flame-atomic absorption spectroscopy are the most commonly used techniques in the determination of metal concentrations due to their high sensitivity and rapid results [7, 14, 30, 58, 102–107]. **Figure 7** shows an F-AAS by PerkinElmer, which is very affordable and reliable for quantification of many trace elements using standard addition methods.

Figure 7. Flame-atomic absorption spectrometer (F-AAS), type AAnalyst 200 by PerkinElmer is shown.

Despite wine being a relatively complex aqueous sample matrix for chemical analysis due to containing ethanol and hundreds of different larger organic compounds (such as polyhydroxyalcohols, organic acids, polyphenols, polysaccharides, peptides, etc.) in addition to various inorganic species, very few of these compounds actually interfere with atomic spectroscopy due to the high temperatures involved in the sample atomization steps [58]. Also, AAS, the official method of analysis for determination of elements such as Fe and Zn by the European Union Regulation [108], and recommended by the OIV and the American Society of Enologists, is affordable, selective, highly sensitive, often capable of direct measurements, and relatively easy to operate. Fe, Cu, Zn, and Mn, for example, are relatively simple to quantify using F-AAS. These four elements have significant impact on wine quality by contributing to haze formation and leading to undesirable changes in wine taste and aroma [23]. These elements may also impact human health via contributing to total dietary intake (in addition to other sources such as multivitamin supplements, food, and other beverages) and ultimately possible bioaccumulation and toxicity in the event of excessive intake [23]. The most common fuel/oxidant combination for quantification of trace elements such as Cu, Fe, Mn, and Zn by F-AAS in aqueous sample matrix is acetylene/air.

Quantitative analysis by atomic spectroscopy consists of an atomization step where the solid sample components are broken into atoms at high temperature (2000–8000 K) flame, furnace, or plasma after liquid portion of the sample has already evaporated. The inorganic analyte species is quantified based on absorption or emission of monochromatic ultraviolet or visible electromagnetic radiation of characteristic wavelength by the gas-phase analyte. The analytes are typically at parts per million (μg/g) to parts per trillion (pg/g) levels for AS.

In ICP-MS, the analyte trace elements are ionized by collisions with excited Ar^+ or energetic electrons in plasma. ICP-MS has detection limits that are 5–7 orders of magnitude lower than atomic spectroscopy based on F-AAS. In addition, the linear range for ICP-MS is 10^8 versus 10^2 for F-AAS. The sample throughput for ICP-MS is greater as most elements may be analyzed in one run and the sample volumes required for analysis are also less. ICP-MS outperforms even other multielement techniques in element analysis due to its extremely low detection limits. However, the purchase cost and maintenance cost of an ICP-MS system is substantially higher and the instrument requires a skilled operator. ICP-MS has been utilized by many especially as an instrumental tool for characterizing wines according to their geographical origin [72–80, 109].

Stripping voltammetry is a very sensitive, low cost, and a popular electroanalytical chemistry method due to incorporating an electrolytic preconcentration step prior to the analysis of trace concentrations of electroactive species in solution. Multiple elements such as Cd and Cu may also sometimes be quantified from the same sample using SV during the same experiment. Extremely low detection limits for metal ions at sub-ppb (or 10^{-10}–10^{-11} M) concentrations have been reported for SV. There are three parts in a stripping experiment: deposition, quiet time, and stripping. During the deposition step, the analyte of interest is accumulated onto the working electrode that may have been chemically modified previously. Deposition of the target species may be increased by forced convection such as stirring the solution, rotation of the working electrode, or creating flow conditions during the deposition step. The required deposition time (of typically 1–10 min) depends on the concentration of the analyte species in the sample. In anodic stripping voltammetry (ASV), the working electrode is held at a potential at least 0.4 V cathodic to the standard potential of the least easily reduced ion being quantified. The stirring is stopped and the system is allowed to reach equilibrium during a quiet time of about 10–15 s. Finally, the accumulated material is reduced or oxidized back into the solution during the stripping step. Change in electrochemical response (i.e., signal) during the stripping step is proportional to the concentration of the analyte that was in or on the working electrode. Anodic stripping voltammetry involves the reduction of a metal ion analyte from the sample during the preconcentration step. SV method for trace metal analysis in dry wines will be described briefly below as an example of sample preparation, solution conditions, and time required for electroanaytical quantification.

Brainina et al. [96] developed and reported a method for quantification of cadmium, copper, lead, and zinc in dry wines by stripping voltammetry with a thick-film modified graphite-containing electrode (TFMGE). The reported method has an advantage of not requiring acid digestion of the wine samples or any other sample preparation approaches for the destruction of organic substances. Supporting electrolyte of 0.5 M HCl was for Cu, Pb, and Cd, and 0.1 M acetate buffer (pH 5.5) and 0.35 M NaCl for Zn. The deposition potentials were −0.8 V for Cu, −1.2 V for Pb and Cd, and −1.4 V for Zn. The wine samples were diluted with the supporting electrolyte prior to deposition step. The deposition times were 30–60 s for Cu, 120 s for Pb and Cd, and 10–30 s for Zn. Results obtained by the TFMGE analysis were compared with quantification of the selected elements in the same samples done using ICP-MS. Of note, the fast and simple analysis method with TFMGE actually provided better reproducibility than methods involving wine decomposition prior to analysis. More recently, Burmakina et al. [110] developed a procedure for determining manganese(II) in wines by SV on a graphite electrode. The detection limit of manganese(II) was 50 μg/L with a linear range 0.1–3 mg/L [110].

5. Conclusions

Wine is a common beverage that has immense economic impact on certain regions of the world. The theory, advantages and disadvantages, experimental considerations and set up of various sensitive atomic spectroscopy, and electroanalytical quantification methods were described in this chapter. Examples of quantification methods suitable for essential trace element as well as the determination of toxic elements generally found at considerably lower concentrations in wine were provided throughout the chapter. Many inorganic elements have been found to affect the properties of wine, how well it stores, as well as the sensory perception of the wine consumer. The unique trace element profile may also be used as a fingerprint to identify the authenticity of the wine if a fraud is suspected or provide information about the growing region of the grapes and the location of the vineyard. The complexity of wine as a sample matrix for quantification of inorganic elements and the use of standard addition methods to improve the reliability of the selected detection methods were described. Examples of common methods, the common origins of the inorganic elements in the finished wine, and reported levels of trace elements in wines were heavily referenced throughout the chapter.

Acknowledgements

The author thanks Benedictine University for financial support of her work.

Author details

Niina J. Ronkainen

Address all correspondence to: nronkainen@ben.edu

Department of Chemistry and Biochemistry, Benedictine University, Lisle, Illinois, USA

References

[1] Keys, D. (28 December 2003). "Now that's what you call a real vintage: professor unearths 8,000-year-old wine." The Independent (London).

[2] Pyrzyńska, K. (2004). Analytical methods for the determination of trace metals in wine. Critical Reviews in Analytical Chemistry, 34(2), 69–83.

[3] Wine Institute. (2012). Per Capita Wine Consumption by the Country. Retrieved July 14, 2016 from: http://www.wineinstitute.org/resources/worldstatistics/article44.

[4] Arcos, M.T., Ancin, M.C., Echeverria, J.C., Gonzalez, A., & Garrido, J.J. (1993). Study of lability of heavy metals in wines with different degrees of aging through differential pulse anodic stripping voltammetry. Journal of Agricultural and Food Chemistry, 41(12), 2333–2339.

[5] Núñez, M., Peña, R.M., Herrero, C., & Garcia-Martin, S. (2000). Analysis of some metals in wine by means of capillary electrophoresis. Application to the differentiation of Ribeira Sacra Spanish red wines. Analysis, 28(5), 432–437.

[6] Nicolini, G., Larcher, R., Pangrazzi, P., & Bontempo, L. (2015). Changes in the contents of micro-and trace-elements in wine due to winemaking treatments. VITIS-Journal of Grapevine Research, 43(1), 41–45.

[7] Dugo, G., La Pera, L., Pellicanó, T.M., Di Bella, G., & D'Imperio, M. (2005). Determination of some inorganic anions and heavy metals in DOC Golden and Amber Marsala wines: statistical study of the influence of ageing period, colour and sugar content. Food Chemistry, 91(2), 355–363.

[8] García-Esparza, M.A., Capri, E., Pirzadeh, P., & Trevisan, M. (2006). Copper content of grape and wine from Italian farms. Food Additives and Contaminants, 23(3), 274–280.

[9] Galgano, F., Favati, F., Caruso, M., Scarpa, T., & Palma, A. (2008). Analysis of trace elements in southern Italian wines and their classification according to provenance. LWT-Food Science and Technology, 41(10), 1808–1815.

[10] Illuminati, S., Annibaldi, A., Truzzi, C., & Scarponi, G. (2014). Recent temporal variations of trace metal content in an Italian white wine. Food Chemistry, 159, 493–497.

[11] Lara, R., Cerutti, S., Salonia, J.A., Olsina, R.A., & Martinez, L.D. (2005). Trace element determination of Argentine wines using ETAAS and USN-ICP-OES. Food and Chemical Toxicology, 43(2), 293–297.

[12] Fabani, M.P., Arrua, R.C., Vasquez, F., Diaz, M.P., Baroni, M.V., & Wunderlin, D.A. (2010). Evaluation of elemental profile coupled to chemometrics to assess the geographical origin of Argentinean wines. Food Chemistry 119, 372–379.

[13] Azcarate, S.M., Martinez, L.D., Savio, M., Camiña, J.M., & Gil, R.A. (2015). Classification of monovarietal Argentinean white wines by their elemental profile. Food Control, 57, 268–274.

[14] Sauvage, L., Frank, D., Stearne, J., & Millikan, M.B. (2002). Trace metal studies of selected white wines: an alternative approach. Analytica Chimica Acta, 458(1), 223–230.

[15] Elci, L., Arslan, Z., & Tyson, J.F. (2009). Determination of lead in wine and rum samples by flow injection-hydride generation-atomic absorption spectrometry. Journal of Hazardous Materials, 162, 880–885.

[16] Aydin, I., Yuksel, U., Guzel, R., Ziyadanogullari, B., & Aydin, F. (2010) Determination of trace elements in Turkish wines by ICP-OES and HG-ICP-OES. Atomic Spectroscopy, 31, 67–71.

[17] Alkis, İ.M., Öz, S., Atakol, A., Yilmaz, N., Anli, R.E., & Atakol, O. (2014). Investigation of heavy metal concentrations in some Turkish wines. Journal of Food Composition and Analysis, 33(1), 105–110.

[18] Geana, I., Iordache, A., Ionete, R., Marinescu, A., Ranca, A., & Culea, M. (2013). Geographical origin identification of Romanian wines by ICP-MS elemental analysis. Food Chemistry, 138(2), 1125–1134.

[19] Klarić, D.A., Klarić, I., Velić, D., & Dragojević, I.V. (2011). Evaluation of mineral and heavy metal contents in Croatian blackberry wines. Czech Journal of Food Sciences, 29(3), 260–267.

[20] Fiket, Ž., Mikac, N., & Kniewald, G. (2011). Arsenic and other trace elements in wines of eastern Croatia. Food Chemistry, 126(3), 941–947.

[21] Leder, R., Kubanovic, V., Petric, I.V., Vahcic, N., & Banovic, M. (2015). Chemometric prediction of the geographical origin of Croatian wines through their elemental profiles. Journal of Food and Nutrition Research, 54(3) 229–238.

[22] Gennaro, M.C., Mentasti, E., Sarzanini, C., & Pesticcio, A. (1986). Undesirable and harmful metals in wines—determination and removal. Food Chemistry, 19, 93–104.

[23] Tariba, B. (2011). Metals in wine-impact on wine quality and health outcomes. Biological Trace Element Research, 144(1–3), 143–156.

[24] Riganakos, K.A., & Veltsistas, P.G. (2003). Comparative spectrophotometric determination of the total iron content in various white and red Greek wines. Food Chemistry, 82(4), 637–643.

[25] Mozaz, S.R., Sotro, A.G., Segovia, J.G., & Azpilicueta, C.A. (1999). Influence of decantation of viura must on the cation content. Evolution during wine fermentation and stabilization. Food Research International, 32(10), 683–689.

[26] Guerrero, M.I., Herce-Paglia, C., Gonzalez, A.G., Heredia, F.J., Troncoso, A.M., & Camean, A.M. (1996). Mineral elements in wine vinegars made by traditional methods. Sciences des Aliments, 16, 143–149.

[27] Li, P., & Hardy, J.K. (1999). Characterization and classification of Ohio wines using multivariate data analysis. Journal of Wine Research, 10(3), 197–206.

[28] Kokkinofta, R., Petrakis, P.V., Mavromoustakos, T., & Theocharis, C.R. (2003). Authenticity of the traditional Cypriot spirit zivania on the basis of metal content using a combination of coupled plasma spectroscopy and statistical analysis. Journal of Agricultural and Food Chemistry, 51, 6233–6239.

[29] Liu, L., Cozzolino, D., Cynkar, W.U., Gishen, M., & Colby, C.B. (2006). Geographic classification of Spanish and Australian Tempranillo red wines by visible and near-

infrared spectroscopy combined with multivariate analysis. Journal of Agricultural and Food Chemistry, 54(18), 6754–6759.

[30] Yu, H., Zhou, Y., Fu, X., Xie, L., & Ying, Y. (2007). Discrimination between Chinese rice wines of different geographical origins by NIRS and AAS. European Food Research and Technology, 225(3–4), 313–320.

[31] Rodrigues, S.M., Otero, M., Alves, A.A., Coimbra, J., Coimbra, M.A., Pereira, E., & Duarte, A.C. (2011) Elemental analysis for categorization of wines and authentication of their certified brand of origin, Journal of Food Composition and Analysis, 24(4–5), 548–562.

[32] Baluja-Santos, C., & Gonzalez-Portal, A. (1992). Application of hydride generation to atomic-absorption spectrometric analysis of wines and beverages: a review. Talanta, 39(4), 329–339.

[33] Augagneur, S., Médina, B., Szpunar, J., & Lobiński, R. (1996). Determination of rare earth elements in wine by inductively coupled plasma mass spectrometry using a microconcentric nebulizer. Journal of Analytical Atomic Spectrometry, 11(9), 713–721.

[34] Pohl, P. (2007). What do metals tell us about wine? Trends in Analytical Chemistry, 26(9), 941–949.

[35] Kment, P., Mihaljevič, M., Ettler, V., Šebek, O., Strnad, L., & Rohlová, L. (2005). Differentiation of Czech wines using multielement composition–a comparison with vineyard soil. Food Chemistry, 91(1), 157–165.

[36] Kunkee, R.E., & Eschnauer, H.R. (2003). Wine. Ullmann's Encyclopedia of Industrial Chemistry. Weinheim, Germany: Wiley-VCH.

[37] Eschnauer, H. (1982). Trace elements in must and wine: primary and secondary contents. American Journal of Enology and Viticulture, 33(4), 226–230.

[38] Ancín, C., Ayestarán, B., Corroza, M., Garrido, J., & González, A. (1996). Influence of prefermentation clarification on the higher alcohol contents of wines. Food Chemistry, 55(3), 241–249.

[39] Galani-Nikolakaki, S., Kallithrakas-Kontos, N., & Katsanos, A.A. (2002). Trace element analysis of Cretan wines and wine products. Science of the Total Environment, 285(1), 155–163.

[40] Lemos, V.A., de la Guardia, M., & Ferreira, S.L. (2002). An on-line system for preconcentration and determination of lead in wine samples by FAAS. Talanta, 58(3), 475–480.

[41] Cvetkovic, J., Arpadjan, S., Karadjova, I., & Stafilov, T. (2006). Determination of cadmium in wine by electrothermal atomic absorption spectrometry. Acta Pharmaceutica, 56(1), 69–77.

[42] Moreno, N.J., & Azpilicueta, C.A. (2007). Binding of oak volatile compounds by wine lees during simulation of wine ageing. LWT-Food Science and Technology, 40(4), 619–624.

[43] Monagas, M., Martín-Álvarez, P.J., Gómez-Cordovés, C., & Bartolomé, B. (2007). Effect of the modifier (Graciano vs. Cabernet sauvignon) on blends of Tempranillo wine during ageing in the bottle. II. Colour and overall appreciation. LWT-Food Science and Technology, 40(1), 107–115.

[44] Díaz, C., Conde, J.E., Estévez, D., Pérez Olivero, S.J., & Pérez Trujillo, J.P. (2003). Application of multivariate analysis and artificial neural networks for the differentiation of red wines from the Canary Islands according to the island of origin. Journal of Agricultural and Food Chemistry, 51(15), 4303–4307.

[45] Pace-Asciak, C.R., Rounova, O., Hahn, S.E., Diamandis, E.P., & Goldberg, D.M. (1996). Wines and grape juices as modulators of platelet aggregation in healthy human subjects. Clinica Chimica Acta, 246(1), 163–182.

[46] Tosun, I., & Inkaya, A.N. (2009). Resveratrol as a health and disease benefit agent. Food Reviews International, 26(1), 85–101.

[47] Alkhalaf, M. (2007). Resveratrol-induced apoptosis is associated with activation of p53 and inhibition of protein translation in T47D human breast cancer cells. Pharmacology, 80(2–3), 134–143.

[48] Sahin, E., Baycu, C., Koparal, A.T., Donmez, D.B., & Bektur, E. (2016). Resveratrol reduces IL-6 and VEGF secretion from co-cultured A549 lung cancer cells and adipose derived mesenchymal stem cells. Tumor Biology 37(6), 7573–7592.

[49] Doll, R., Peto, R., Hall, E., Wheatley, K., & Gray, R. (1994). Mortality in relation to consumption of alcohol: 13 years' observations on male British doctors. British Medical Journal, 309(6959), 911.

[50] Suh, I.L., Shaten, B.J., Cutler, J.A., & Kuller, L.H. (1992). Alcohol use and mortality from coronary heart disease: the role of high-density lipoprotein cholesterol. Annals of Internal Medicine, 116(11), 881–887.

[51] Gaziano, J.M., Buring, J.E., Breslow, J.L., Goldhaber, S.Z., Rosner, B., VanDenburgh, M., Willett, W., & Hennekens, C.H. (1993). Moderate alcohol intake, increased levels of high-density lipoprotein and its subfractions, and decreased risk of myocardial infarction. New England Journal of Medicine, 329(25), 1829–1834.

[52] Companys, E., Naval-Sánchez, M., Martínez-Micaelo, N., Puy, J., & Galceran, J. (2008). Measurement of free zinc concentration in wine with AGNES. Journal of Agricultural and Food Chemistry, 56(18), 8296–8302.

[53] Scheinberg, I.H. (1991). Copper. In: Merian, E., editor. Metals and Their Compounds in the Environment. Weinheim: VCH. Weinheim, Germany: Wiley-VCH, 1299–1308.

[54] Eschnauer, H., & Neeb, R. (1988). Micro-element analysis in wine and grapes in: modern methods of plant analysis. In: Linskens, H.I., Jackson, J.F., editors. Berlin: Springer-Verlag, pp. 67–91.

[55] Freeland-Graves, J. (1994). Derivation of manganese estimated safe and adequate daily dietary intakes. In: Mertz, W., et al., editors. Risk Assessment of Essential Elements. ILSI Press, Washington, DC, USA. pp. 237–252.

[56] SCF (Scientific Committee on Food). (1993). Reports of the Scientific Committee on Food (31st series). Luxembourg: Commission of the European Community, pp. 177–189.

[57] Trumbo, P., Yates, A.A., Schlicker, S., & Poos, M. (2001). Dietary reference intakes: vitamin A, vitamin K, arsenic, boron, chromium, copper, iodine, iron, manganese, molybdenum, nickel, silicon, vanadium, and zinc. Journal of the American Dietetic Association, 101(3), 294–301.

[58] Aceto, M., Abollino, O., Bruzzoniti, M.C., Mentasti, E., Sarzanini, C., & Malandrino, M. (2002). Determination of metals in wine with atomic spectroscopy (flame-AAS, GF-AAS and ICP-AES): a review. Food Additives and Contaminants, 19(2), 126–133.

[59] OIV (2011). Maximum Acceptable Limits of Various Substances Contained in Wine OIV-MA-C1-01: R2011, 1–3. Retrieved July 14, 2016 from: http://www.oiv.int/public/medias/2601/oiv-ma-c1-01.pdf

[60] Italian Republic, December 29th, 1986. Characteristics and limits of some substances contained in wines. Italian Official Gazette, no 13, 17 January, 1987.

[61] Reynard, J.S., Zufferey, V., Nicol, G.C., & Murisier, F. (2011). Soil parameters impact the vine-fruit-wine continuum by altering vine nitrogen status. Journal International des Sciences de la Vigne et du Vin, 45(4), 211–221.

[62] Green, A.M., Clark, A.C., & Scollary, G.R. (1997). Determination of free and total copper and lead in wine by stripping potentiometry. Fresenius' Journal of Analytical Chemistry, 358, 711–717.

[63] Monaci, F., Bargagli, R., & Focardi, S. (2003). Element concentrations in Chianti Classico appellation wines. Journal of Trace Elements in Medicine and Biology, 17, 45.

[64] Hopfer, H., Nelson, J., Collins, T.S., Heymann, H., & Ebeler, S.E. (2015). The combined impact of vineyard origin and processing winery on the elemental profile of red wines. Food Chemistry, 172, 486–496.

[65] Pepi, S., Coletta, A., Crupi, P., Leis, M., Russo, S., Sansone, L., Tassinari, N.,Chicca, M., & Vaccaro, C. (2016). Geochemical characterization of elements in Vitis vinifera cv. Negroamaro grape berries grown under different soil managements. Environmental Monitoring and Assessment, 188(4), 1–15.

[66] Arvanitoyannis, I.S., Katsota, M.N., Psarra, E.P., Soufleros, E.H., & Kallithraka, S. (1999). Application of quality control methods for assessing wine authenticity: use of multivariate analysis (chemometrics). Trends in Food Science and Technology, 10, 321–336.

[67] Saurina, J. (2010). Characterization of wines using compositional profiles and chemometrics. Trends in Food Science and Technology, 29, 234–245.

[68] Makris, D.P., Kallithraka, S., & Kefalas, P. (2006). Flavonols in grapes, grape products and wines: Burden, profile and influential parameters. Journal of Food Composition and Analysis, 19(5), 396–404.

[69] Jaitz, L., Siegl, K., Eder, R., Rak, G., Abranko, L., Koellensperger, G., & Hann, S. (2010). LC–MS/MS analysis of phenols for classification of red wine according to geographic origin, grape variety and vintage. Food Chemistry, 122(1), 366–372.

[70] Di Paola-Naranjo, R.D., Baroni, M.V., Podio, N.S., Rubinstein, H.R., Fabani, M.P., Badini, R.G., Inga, M., Ostera, H.A., Cagnoni, M., Gallegos, E., Gautier, E., Peral-Garcia, P., Hoogewerff, J., & Wunderlin, D.A. (2011). Fingerprints for main varieties of Argentinean wines: terroir differentiation by inorganic, organic, and stable isotopic analyses coupled to chemometrics. Journal of Agricultural and Food Chemistry, 59(14), 7854–7865.

[71] Versari, A., Laurie, V.F., Ricci, A., Laghi, L., & Parpinello, G.P. (2014). Progress in authentication, typification and traceability of grapes and wines by chemometric approaches. Food Research International, 60, 2–18.

[72] Stroh, A., Brückner, P., & Völlkopf, U. (1994). Multielement analysis of wine samples using ICP-MS. Atomic Spectroscopy, 15(2), 100–106.

[73] Baxter, M.J., Crews, H.M., Dennis, M.J., Goodall, I., & Anderson, D. (1997). The determination of the authenticity of wine from its trace element composition. Food Chemistry, 60(3), 443–450.

[74] Greenough, J.D., Longerich, H.P., & Jackson, S.E. (1997). Element fingerprinting of Okanagan Valley wines using ICP-MS: relationships between wine composition, vineyard and wine colour. Australian Journal of Grape and Wine Research, 3(2), 75–83.

[75] Jakubowski, N., Brandt, R., Stuewer, D., Eschnauer, H.R., & Görtges, S. (1999). Analysis of wines by ICP-MS: is the pattern of the rare earth elements a reliable fingerprint for the provenance?. Fresenius' Journal of Analytical Chemistry, 364(5), 424–428.

[76] Marengo, E., & Aceto, M. (2003). Statistical investigation of the differences in the distribution of metals in Nebbiolo-based wines. Food Chemistry, 81(4), 621–630.

[77] Barbaste, M., Medina, B., & Perez-Trujillo, J.P. (2003). Analysis of arsenic, lead and cadmium in wines from the Canary Islands, Spain, by ICP/MS. Food Additives and Contaminants, 20(2), 141–148.

[78] Taylor, V.F., Longerich, H.P., & Greenough, J.D. (2003). Multielement analysis of Canadian wines by inductively coupled plasma mass spectrometry (ICP-MS) and multivariate statistics. Journal of Agricultural and Food Chemistry, 51(4), 856–860.

[79] Coetzee, P.P., Steffens, F.E., Eiselen, R.J., Augustyn, O.P., Balcaen, L., & Vanhaecke, F. (2005). Multi-element analysis of South African wines by ICP-MS and their classification according to geographical origin. Journal of Agricultural and Food Chemistry, 53(13), 5060–5066.

[80] Capron, X., Smeyers-Verbeke, J., & Massart, D.L. (2007). Multivariate determination of the geographical origin of wines from four different countries. Food Chemistry, 101(4), 1585–1597.

[81] Azcarate, S.M., de Araújo Gomes, A., Alcaraz, M.R., de Araújo, M.C.U., Camiña, J.M., & Goicoechea, H.C. (2015). Modeling excitation–emission fluorescence matrices with pattern recognition algorithms for classification of Argentine white wines according grape variety. Food Chemistry, 184, 214–219.

[82] Cheng, J., Zhai, Y., & Taylor, D.K. (2015). Several mineral elements discriminate the origin of wines from three districts in China. International Journal of Food Properties, 18(7), 1460–1470.

[83] Castineira, M.M., Brandt, R., Von Bohlen, A., & Jakubowski, N. (2001). Development of a procedure for the multi-element determination of trace elements in wine by ICP–MS. Fresenius' Journal of Analytical Chemistry, 370(5), 553–558.

[84] Oroian, M. (2015). Romanian white wine authentication based on mineral content. Journal of Agroalimentary Processes and Technologies, 21, 9–13.

[85] Frias, S., Trujillo, J.P., Peña, E., & Conde, J.E. (2001). Classification and differentiation of bottled sweet wines of Canary Islands (Spain) by their metallic content. European Food Research and Technology, 213(2), 145–149.

[86] Volesky, B., May, H., & Holan, Z.R. (1993). Cadmium biosorption by Saccharomyces cerevisiae. Biotechnology and Bioengineering, 41(8), 826–829.

[87] Volesky, B., & Holan, Z.R. (1995). Biosorption of heavy metals. Biotechnology Progress, 11(3), 235–250.

[88] Volesky, B., & May-Phillips, H.A. (1995). Biosorption of heavy metals by Saccharomyces cerevisiae. Applied Microbiology and Biotechnology, 42(5), 797–806.

[89] Blackwell, K.J., Singleton, I., & Tobin, J.M. (1995). Metal cation uptake by yeast: a review. Applied Microbiology and Biotechnology, 43(4), 579–584.

[90] Benítez, P., Castro, R., Jose Antonio Sanchez, P., & Barroso, C.G. (2002). Influence of metallic content of fino sherry wine on its susceptibility to browning. Food Research International, 35(8), 785–791.

[91] Ražić, S., Todorović, M., Holclajtner-Antunović, I., & Stoiljković, M. (1999). Determination of metal traces in wine by argon stabilized dc arc. Fresenius' Journal of Analytical Chemistry, 365(4), 367–370.

[92] Sun, L.X., Danzer, K., & Thiel, G. (1997). Classification of wine samples by means of artificial neural networks and discrimination analytical methods. Fresenius' Journal of Analytical Chemistry, 359(2), 143–149.

[93] Thiel, G., & Danzer, K. (1997). Direct analysis of mineral components in wine by inductively coupled plasma optical emission spectrometry (ICP-OES). Fresenius' Journal of Analytical Chemistry, 357(5), 553–557.

[94] Chen, G.N., Scollary, G.R., & Vicente-Beckett, V.A. (1994). Potentiometric stripping determination of lead, cadmium, and zinc in wine. American Journal of Enology and Viticulture, 45(3), 305–311.

[95] Antonietta Baldo, M. (1997). Determination of lead and copper in wine by anodic stripping voltammetry with mercury microelectrodes: assessment of the influence of sample pretreatment procedures. Analyst, 122(1), 1–5.

[96] Brainina, K.Z., Stozhko, N.Y., Belysheva, G.M., Inzhevatova, O.V., Kolyadina, L.I., Cremisini, C., & Galletti, M. (2004). Determination of heavy metals in wines by anodic stripping voltammetry with thick-film modified electrode. Analytica Chimica Acta, 514(2), 227–234.

[97] Illuminati, S., Annibaldi, A., Truzzi, C., Finale, C., & Scarponi, G. (2013). Square-wave anodic-stripping voltammetric determination of Cd, Pb and Cu in wine: Set-up and optimization of sample pre-treatment and instrumental parameters. Electrochimica Acta, 104, 148–161.

[98] Noble, A.C., Orr, B.H., Cook, W.B., & Campbell, J.L. (1976). Trace elements analysis of wine by proton-induced x-ray fluorescence spectrometry. Journal of Agricultural and Food Chemistry, 24(3), 532–535.

[99] dos Santos, C.E.I., da Silva, L.R.M., Boufleur, L.A., Debastiani, R., Stefenon, C.A., Amaral, L., Yoneama, M.L., & Dias, J.F. (2010). Elemental characterisation of cabernet sauvignon wines using Particle-Induced X-ray Emission (PIXE). Food Chemistry, 121(1), 244–250.

[100] Rovio, S., Sirén, K., & Sirén, H. (2011). Application of capillary electrophoresis to determine metal cations, anions, organic acids, and carbohydrates in some Pinot Noir red wines. Food Chemistry, 124(3), 1194–1200.

[101] Zerbinati, O., Balduzzi, F., & Dell'Oro, V. (2000). Determination of lithium in wines by ion chromatography. Journal of Chromatography A, 881(1), 645–650.

[102] Eschnauer, H., Jakob, L., Meierer, H., & Neeb, R. (1989). Use and limitations of ICP-OES in wine analysis. Microchimica Acta, 99(3–6), 291–298.

http://dx.doi.org/10.5772/64932

[103] Cabrera-Vique, C., Teissedre, P.L., Cabanis, M.T., & Cabanis, J.C. (1997). Determination and levels of chromium in French wine and grapes by graphite furnace atomic absorption spectrometry. Journal of Agricultural and Food Chemistry, 45(5), 1808–1811.

[104] Cabrera-Vique, C., Teissedre, P.L., Cabanis, M.T., & Cabanis, J.C. (2000). Manganese determination in grapes and wines from different regions of France. American Journal of Enology and Viticulture, 51(2), 103–107.

[105] Ajtony, Z., Szoboszlai, N., Suskó, E.K., Mezei, P., György, K., & Bencs, L. (2008). Direct sample introduction of wines in graphite furnace atomic absorption spectrometry for the simultaneous determination of arsenic, cadmium, copper and lead content. Talanta, 76(3), 627–634.

[106] Komárek, M., Száková, J., Rohošková, M., Javorská, H., Chrastný, V., & Balík, J. (2008). Copper contamination of vineyard soils from small wine producers: a case study from the Czech Republic. Geoderma, 147(1), 16–22.

[107] Šelih, V.S., Šala, M., & Drgan, V. (2014). Multi-element analysis of wines by ICP-MS and ICP-OES and their classification according to geographical origin in Slovenia. Food Chemistry, 153, 414–423.

[108] European Union (1990). Commission Regulation No. 2676/90 of 17, September 1990. Official Journal, L272, 1–192.

[109] Pérez-Jordán, M.Y., Soldevila, J., Salvador, A., Pastor, A., & de la Guardia, M. (1999). Inductively coupled plasma mass spectrometry analysis of wines. Journal of Analytical Atomic Spectrometry, 14(1), 33–39.

[110] Burmakina, G.V., Mokh, N.S., Maksimov, N.G., Zimonin, D.V., Zhizhaev, A.M., & Rubaylo, A.I. (2016). Determination of manganese (II) in wines by stripping voltammetry on solid electrodes. Journal of Analytical Chemistry, 71(1), 71–76.

Chapter 21

D-O-C Stable Isotopes, ^{14}C Radiocarbon and Radiogenic Isotope Techniques Applied in Wine Products for Geographical Origin and Authentication

Paraskevi Chantzi, Anastasia-Elektra Poutouki and Elissavet Dotsika

Additional information is available at the end of the chapter

http://dx.doi.org/10.5772/64933

Abstract

Oxygen, deuterium and carbon isotopes were measured in wine products in two Greek vineyards, Amydaio (north) and Nemea (south). The enriched isotope values in Nemea reflects the higher evapo-transpiration rate and the more arid condition of Southern Greece. White wines were slightly more depleted than red wines of the same year and the same growing region probably due to different harvest periods. Further was pointed out the variety of isotope values with respect to vintage year indicating that the vintage year contributes to the development of isotopes in wine water. In both vineyards the trend lines intersect the oxygen and deuterium isotopes of irrigation water highlighting the source water and the initial isotopic composition of grape berries. $\delta^{13}C$ values of ethanol confirmed the origin of C3 plants and the authentication of wine products without detecting adulteration with industrial alcohol. The results of ^{14}C measurements in ethanol extracted from Greek wines follow the known pattern of ^{14}C variations in atmospheric CO_2. The homogeneity of $^{87}Sr/^{86}Sr$ and $^{144}Nd/^{143}Nd$ isotope values confirms that the territorial and geological signal is transferred through the vineyards in the final product, wine, certifying the exclusively provenance of the wine areas Amydaio and Nemea.

Keywords: oxygen isotope, carbon isotopes, radiocarbon, Greek wines, grape berries

1. Introduction

Wine is a product that has suffered serious damage from adulteration and mislabeling practices. The authenticity of a wine label, which confirms the alignment of the product with the relevant laws of alcohol production, is of great importance in many levels. First of all, the protection of the consumers against the globalization of food market, where the adulteration episodes are strongly existent, should be the dominant objective. On the other hand, the authentication in conjunction with the geographic origin maintains and defends traditional production techniques that influence the particular output features of wine and which are lost through the competition of mass production. In general, modern dietary habits governed by standardized products, as required by the food industry, resulting the lack of the link between products and their geographical origin influenced by the particular environment setting. There is also a great interest by wine producers and traders so that they can guarantee high quality wines offer a reliable label. However, fraud incentives persist and with upgraded methods. Therefore, there is a necessity for the upgrading of the authentication methods of the wine label. Toward this direction the use of stable isotopes ^{2}H, ^{13}C, and ^{18}O in diet samples started in the early 1970s and later enriched by the use of radiocarbon ^{14}C and radiogenic isotopes Sr and Nd. The recognition of isotopes as food authenticity indicators was a point of reference in the food authentication approach. In fact, stable isotope ratio analysis of wine isotope ratio mass spectrometry is official methods in the European Union for the detection of chaptalization, addition of water, sweetening with sugar, and authentication of geographic origin and year of harvest.

2. Factors that influence carbon and oxygen isotopic values

The nature ratio of stable isotopes of plant tissues is strongly influenced by raw materials of photosynthesis, water and carbon dioxide, as well as the fractionation process during the photosynthetic pathways. Regarding carbon dioxide, it should be noted that the atmospheric reservoir considered as constant where the $\delta^{13}C$ composition barley varies from −7% [1]. Since then the atmospheric pool supplies fixed carbon stable isotope values, water ("source water") and fractionation processes are responsible for the differentiation in $\delta^{13}C$ values in plants. According to the photosynthetic pathway that each plant follows three types are distinguished with characteristic range of $\delta^{13}C$ values: C4 (Hatch-Slack, C4-dicarboxylic acid path-way) plants generally range from −9 to −19% [2, 3] while for C3 (Calvin cycle) plants typically range from −20 (open areas exposed to water stress) to −35% (closed canopy) [2, 3]. The third photosynthetic pathway called crassulacean acid metabolism (CAM) presents $\delta^{13}C$ values between the end-members of C3 and C4 types. Generally, the CAM cycle corresponds to plants that store quantities of water in their tissues to use under demand conditions; therefore, the biggest distinction is limited to C3 and C4 photosynthetic pathways. The first represents the cycle that the first synthesized sugars are three carbon sugars while in C4 cycle first sugars fixed are four carbon sugars. Consequently, alcohol derived from grapes should reflect $\delta^{13}C$ values of the corresponding photosynthetic pathway that the grapes belong, which is the C3 type. The addition of different origin alcohol from C4 type (synthetic or beet) could be traced as the $\delta^{13}C$ values are more enriched than those of the C3 type. This intense discrimination is due to the fact that C3 cycle is "longer" enabling a greater isotopic segregation against the "short" C4 cycle which is closer related to the end of the products [4]. Moreover, since the plant interacts with atmospheric CO_2 any observed fluctuations should be justified by the particular environmental characteristics of the region where the vineyards grow. In more detail, the intake water source for vineyards owns a crucial role in plant growth by determining the sugar content in grapes and therefore the alcohol formation. The "water deficit" causes stress to plant and affects its function process which in an attempt to maintain the needed humidity preventing the loss though evaporation closes its leave stomata [2, 5, 6]. As a direct consequence to this is the limited interaction with environmental CO_2, which is the main factor in photosynthesis. The disruption of the system "leaf transpiration-carbon fixation per unit leaf area" results in a disruption in sugar formation in grape berries. Stress episodes lead to higher sugar level in fruits in contrast to those where water availability is stable and representative to the actual needs of the plant [7]. Consequently, $^{13}C/^{12}C$ isotope analysis was performed in wine samples. Ethanol samples which were extracted by number of wine samples were subjected in $^{18}O/^{16}O$, $^{2}H/^{1}H$, $^{13}C/^{12}C$, and radiocarbon (^{14}C) isotope analysis. As oxygen-18 and deuterium isotopes are affected by temperature, precipitation, humidity, altitude, latitude, and the distance from the sea [8, 9], the isotopic signature of vineyard's "source water" should be related to corresponding geographical region. The emphasis is given to oxygen isotopic values (^{18}O) for the characterization of the plant organic matter as they present a more normal behavior. However, the interpretation of their $\delta^{18}O$ values is more complicated than that of carbon as it does not present such a linear formation as CO_2 due to the several oxygen sources (precipitation, atmospheric CO_2, soil water) and the complexity of metabolic processes. The isotopic characteristics of feeding water, environmental conditions, photosynthetic fractionation, and the evapotranspiration effect on plant sap and leave stomata are considered the most dominant factors that gather the fixation of $\delta^{18}O$ values. However, we should further consider the cellular CO_2-H_2O equilibrium as well as possible contribution of CO_2 and/or H_2O to the organic oxygen in the cell [4, 10, 11].

3. Environmental and hydrogeological setting of study areas vineyards

Tradition of Greece in winemaking is pronounced with the unequivocal uniqueness of Greek wines resulted by the combination of the Greek terroir and the high quality of local wine varieties. Therefore, the authenticity control and origin assignment of Greek wine products have obtained remarkable importance. In the context of this project, two characteristic wine-producing areas of continental Greece: Amydaio in west Macedonia, north Greece, and Nemea in Corinthia, south Greece are selected. The first constitutes a significant Greek vineyard between Vermion and Borra mountains where it is dominated by the cultivation of Xinomauro variety while the second concerns vineyards with the exclusive cultivation of Agiorgitiko variety. So an attempt was made to trace the correspondence of "VQPRD (*Vins de Qualite Produits dans une Region Determinee*) Amydaio and Nemea" labels with their geographical

radiogenic analysis was also performed in selected wine samples. Rain samples were collected (11 samples for Nemea and 12 samples for Amydaio) by using 125 mL glass bottles. Ground water samples (10 samples for Nemea and 17 samples for Amydaio) were taken from irrigation water boreholes related to vineyards by using 125 mL glass bottles. In total 70 (54 samples for Amydaio and 16 samples for Nemea) different bottled vintage wines were chosen and prepared for stable isotopic analysis by extraction (approximately 40 mL from each bottle) through the cork using a 10 mL Hamilton Gastight Syringe with Gauge 26 Point Style 5 (side hole) needles. Finally, unfermented grape must samples (16 samples for Nemea and 33 samples for Amydaio) were collected immediately after harvest. Then they were centrifuged for approximately 1 hour to extract grape solids and larger yeasts, syringe-filtered with a 0.45 mm and 0.22 mm Cameo IV filter and stored into 20 mL glass bottles (without the addition of sulfur dioxide). Moreover, 10 samples of wine ethanol were collected by using a Cadiot column distillation system to quantitatively separate the ethanol, according to the procedure described in OIV MA-AS-311-05.

All isotopic analyses were performed in Laboratory of Stable Isotope and Radiocarbon of Nanoscience & Nanotechnology, Institute in N.C.S.R. Demokritos (Athens, Greece). Stable isotope analysis of $^{18}O/^{16}O$, D/H, and $^{13}C/^{12}C$ ratio in Greek wines, wine water, meteoric, and ground waters were carried out on a continuous flow Finnigan DELTA V plus (Thermo Electron Corporation, Bremen, Germany) stable isotope mass spectrometer [20, 21].

^{13}C, ^{18}O, and ^{2}H isotope analysis in ethanol with Thermo Scientific DELTA V Isotope Ratio Mass Spectrometers was performed according to Ref. [22]. Specifically, for $\delta^{13}C$ determination 1 μL of purified ethanol was injected into a small tin container for liquids and closed securely to prevent evaporation. The Flash Elemental Analyzer (FlashEA) with a single reactor system combining combustion and reduction in one reaction was used. The Thermo Scientific ConFlo interface diluted the CO_2 sample peak with 1 bar helium resulting in a split of about 1:12. For $\delta^{2}H$ and $\delta^{18}O$ determination 0.1 μL of pure ethanol was injected with a 0.5 μL syringe into a glassy carbon reactor in a high temperature carbon reduction system of a FlashEA. The autosampler was used and samples were stored in 2 mL vials with standard caps and septa.

The results are expressed in standard delta notation (δ) as per mil (‰) deviation from the standard V-SMOW as: $\delta = ((R_{sample} - R_{standard})/R_{standard}) \times 1000$, where R_{sample} and $R_{standard} = {}^{2}H/{}^{1}H$ or $^{18}O/^{16}O$ or $^{13}C/^{12}C$ ratios of sample and standard, respectively.

Measurement precision, based on the repeated analysis of internal standard waters, was 1.5, 0.5, and 0.2% for $\delta^{2}H$, $\delta^{18}O$, and $\delta^{13}C$, respectively.

All measurements were carried out according to laboratory standards that were periodically calibrated based on the international standards recommended by the IAEA.

^{14}C activity was determined by using liquid scintillation counting (LSC) on a Packard Tri-Carb TR/SL. Radiogenic isotope of Sr and Nd were measured with thermal ionization mass spectrometry (TIMS, Nd) and inductively coupled plasma mass spectrometry (ICP-MS, Sr).

5. Results and discussion

5.1. Isotopic composition of meteoric water

Twenty-nine water samples (ASt1 to ASt17 irrigation water samples and ASt18 to ASt29 precipitation water samples) were collected at the region of Amydaio in West Macedonia in North Greece, corresponding to the growing season of 2013. From Nemea, northeastern Peloponnese in southern Greece, a total of 21 water samples (NSt1 to NSt10 groundwater samples and NSt11 to NSt21 precipitation water samples) were collected, corresponding to the same growing season. The $\delta^{18}O$ and δD values of Amydaio and Nemea are shown in **Figure 2**. In the same figure are also pictured the Global Meteoric Water Line with correlation equation: [23]

$$\delta D = 8 \times \delta^{18}O + 10 \tag{1}$$

and the Local Meteoric Line that correspond to Greece (LMWL) with correlation equation: [24]

$$\delta D = 8.7\delta^{18}O + 19.5 \tag{2}$$

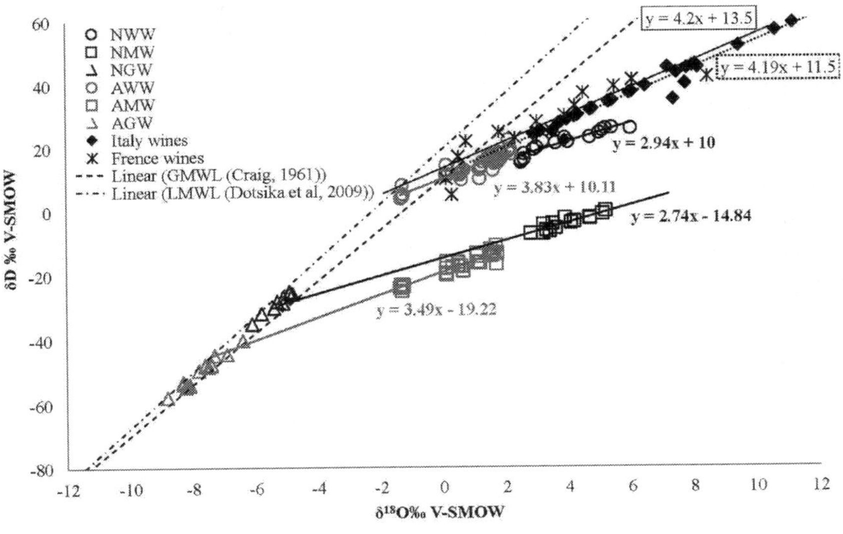

Figure 2. Equation correlation $\delta^{18}O$–$\delta^{2}H$ both in wine water and grape berries of Greece and Europe generally.

Generally, an isotopic correlation between $\delta^{18}O$ and $\delta^{2}H$ with a slope close to 8 indicates fresh precipitation of all types [23], as well as for surface water not subjected to excessive evaporation relative to input [24]. In contrast, slopes ranging from 6 to 4 reflect isotopic relations between $\delta^{18}O$ and $\delta\ ^{2}H$ for waters which are subjected to excessive evaporation relative to input [23].

Oxygen and deuterium isotope values for Amydaio precipitation water samples range from −9.9‰ to −9.6‰ and from −66.7‰ to −63.8‰, respectively, while for groundwater range from −8.8 to −6.4‰ and from −56.6‰ to −40.9‰, respectively. The correlation equation for Amydaio groundwater is $\delta D = 6.86*\delta^{18}O + 3.24$ where the slope 7.59 suggests that the samples have not undergone evapotranspiration process. For Nemea precipitation water samples of $\delta^{18}O$ and δD values range from −7.9‰ to −6.0‰ and from −48.1‰ to −35.2‰, respectively, while for groundwater samples range −6.1‰ to −4.8‰ and from −35.1‰ to −25.2‰, respectively. Evapotranspiration process seems not to have effected remarkably the irrigation water samples as their slope is 7.27 ($\delta D = 7.27*\delta^{18}O + 9.70$). Groundwater from both study areas do not exhibit remarkable variations in $\delta^{18}O$ and $\delta^{2}H$ values indicating the lack of seasonal and altitude impacts affecting the meteoric waters [25].

Similar equations with slightly greater slopes of 7 and 7.5were proposed for the northern part of Eastern Macedonia and for the Eastern Nemea, respectively [26]. A slope lower than 7 was reported for Central Macedonia [27], the southern part of Eastern Macedonia and Thrace [28], as well as the northern part of Epirus [29].

5.2. Isotopic composition of wine water

Amydaio vineyards are represented by 54 bottles of wine (AW1–AW54) consisting of 6 (AW1–AW6) Roditis (white wine), 24 (AW7–AW30) Xinomauro (red wine), and 24 ((AW31–AW54) Tannat (red wine). Nemea vineyards are represented by 16 bottles of wine (NW1–NW16) consisting of 11 (NW1–NW11) Agiorgitiko (red wine), 1 (NW12) Merlot-Agiorgitiko (red wine), 1 (NW13) Moschofilero-Rodidtis (white wine), and 3 (NW14–NW16) Roditis-Chardonnay (white wine).

The stable isotope values of Greek wine waters are presented in **Figure 2**. As Western and Southern Europe constitutes the most significant region, both in terms of vineyard area and quantity of production, French and Italian wines are also plotted in the same figure. Amydaio presents isotopic values ranging from −1.3‰ to 2.3‰ (mean value 1.1‰) for $\delta^{18}O$ and from 3.9 to 22.3‰ (mean value 14.2‰) for δD, while Nemea samples range from 2.5‰ to 6.1‰ (mean value 3.8‰) and from 15.2‰ to 26.1‰ (mean value 21.4‰) for $\delta^{18}O$ and δD, respectively. A first observation is the fact of isotope enrichment in the north-south direction. This trend has been also suggested by some authors for olive oils [30, 31]. $\delta^{13}C$ values also related to the north-south enrichment observation with wine water values to range from −29.4‰ to −26.1‰ (mean value −27.7‰) for Amydaio wines and from −26.2‰ to −24.7‰ (mean value −25.6‰) for Nemea wines. Moreover, Nemea wines present slightly more positive values with larger variations verifying the warmer and drier conditions than those of Amydaio where the conditions are cooler and the precipitation episodes more frequent and constant. Indeed, ambient conditions affect intermediate steps in plant functions [32, 33] and proceeds for sugar formation, therefore carbon isotope could trace deferent environmental conditions. The enzymes favor the ^{12}C isotope for the photosynthesis than ^{13}C as it is found in greater concentration in atmospheric CO_2 (98.9‰ for ^{12}C and 1.1‰ for ^{13}C). Moreover, this difference mass which cause a higher diffusion of $^{12}CO_2$ than that of $^{13}CO_2$ combined to the fact that ^{13}C forms stronger chemical compounds as it is heavier than ^{12}C induce in lower concentrations of ^{13}C in plants than atmosphere [3]. Especially in water deficit conditions leaves stomata close resulting in a reduced atmospheric CO_2 interchange, which leads to changes in $\delta^{13}C$ values among C3-photosynthesis plants such as grapes [2, 5].

Wine water and groundwater samples correlated well enough in both vineyards implying that the "source water" represented by groundwater could be linked to wine products. The isotopic composition of groundwater and precipitation are mainly related to latitude, distance from the sea, and altitude [27]. Groundwater samples from Amydaio and Nemea reflect their meteoric origin without remarkable variations (**Figure 2**) so it is assumed that environmental setting of the vineyards control the isotopic composition of wine products. Schmidt et al. have already reported the $\delta^{18}O$ correlation between carbohydrates and $\delta^{18}O$ of leaf water which are linked to the isotopic composition of groundwater with the evapotranspiration ratio to be influenced by humidity and temperature [9].

Amydaio wine samples are represented by two major vintage labels: 2009 and 2010. The correlation equation for all samples corresponding to both vintages is $\delta D = 3.86*\delta^{18}O + 10.01$ with characteristic slope of 3.86. Correlation equations for 2009 and 2010 are $\delta D = 3.75*\delta^{18}O + 10.38$ and $\delta D = 3.78*\delta^{18}O + 9.82$, respectively. Both vintage years retain their characteristic slope (3.75 for 2009 and 3.78 for 2010) with the vintage year of 2010 to present a better correlation against 2009 ($r^2 = 0.69$ for 2009 and $r^2 = 0.93$ for 2010). Nemea's wine samples present a correlation equation $\delta D = 2.79*\delta^{18}O + 10.71$ with characteristic slope of 2.79. Moreover, French and Italian wines exhibit slope of 4.19 and 4.20, respectively. The fact that Italian wines concern the north-west part of central Italy explains their similarity with French wines. Amydaio is also the most northerly of Greek vineyards, with vines growing at an altitude of 620–710 m and a significant network of lakes (Zazari, Cheimaditida, Petron, and Vegoritida lakes) to be presented; therefore, its slope is closer to those of French and Italy reflecting the strong moisture regime. The slope of Nemea wines is lower reflecting the more arid condition of Southern Greece. The slopes that both vineyards present reflect the different ambient conditions indicating the evapotranspiration process as the critical factor for the observed oxygen and deuterium isotope enrichment [34].

5.3. Isotopic effect of fermentation

Raco et al. demonstrated that for detection of geographical origin of wine $\delta^{18}O$ and δ^2H analysis should be carried out not only in wines but in grape berries and grape must as it clarifies the impact of fermentation process more precisely. The isotopic values of water from grape must from Amydaio and Nemea vineyards are presented in **Figure 2**. Oxygen and deuterium isotopic values of Amydaio grape must samples range from −1.4‰ to 1.7‰ and from −25.2‰ to −10.9‰, respectively, while Nemea grape must range from 2.5‰ to 5.2‰

and from −7.2‰ to −0.1‰, respectively. Moreover, in **Figure 2** are also reported the stable isotopic values of groundwater samples corresponding to vinification areas. Grape must samples are marked further away from groundwater-source water as during their maturation the transpiration process results in a water loss which leads to their isotopic enrichment. However, their transpiration lines with correlation equation

$$\delta D = 3.49 \times \delta^{18}O - 19.20 \ (r^2 = 0.89) \quad \text{Amydaio area} \tag{3}$$

$$\delta D = 2.75 \times \delta^{18}O - 14.84 \ (r^2 = 0.84) \quad \text{Nemea area} \tag{4}$$

intersect the Meteoric Water Lines in the area of groundwater samples highlighting the source water of vineyards and the initial isotopic composition of grape berries.

The wine line and transpiration line of grape must samples corresponding to Amydaio are differentiated about 0.37‰ for $\delta^{18}O$ and 29.21‰ for δD while the difference for Nemea's wines and grape must samples is about 0.03‰ for $\delta^{18}O$ and 25.55‰ for δD. The controlling factor of this isotopic setting is clearly the difference in deuterium isotopic values. The δD isotope values of grape must samples is referred to the member of the water molecule; however, the δD isotope values of wine water reflect hydrogen members of sugar and ethanol [35]. Refs. [44–48] demonstrated that during fermentation, deuterium (D) transfer from exchangeable hydroxyls or from nonexchangeable sites of sugars leading to significant differences in the "site-specific" hydrogen isotopic values of wine.

5.4. Influence of wine variety and vintage on isotopes

Isotopic values of $\delta^{18}O$ and δD for wine water in red wines produced in Amydaio range from 0.1‰ to 2.2‰ and from 9.8‰ to 22.3‰, respectively, while in white wines they range from −1.3 to −1.2‰ and from 3.9‰ to 8.1‰, respectively. For Nemea red wines the $\delta^{18}O$ and δD values of wine water range from 2.5‰ to 6.1‰ and from 15.2‰ to 26.1‰, respectively, while in white wines they range from 3.0‰ to 5.2‰ and from 19.9‰ to 25.9‰, respectively. In both vineyards the white wines are slightly more depleted than red wines of the same year and the same growing region. White and red wine grapes are harvested in different periods; red wine grapes usually harvested 1–3 weeks after white wine grapes. This interval leads to longer period of transpiration for red wine grapes resulting isotopically enriched red wine than white wine. However, it is revealed that the control factor that distinguishes the wine water values is their origin. Amydaio and Nemea wines clustered individually into two groups where the first concerns cooler-wetter climate and the second warmer-drier climate, respectively.

The distinct variation of wine water values based on their origin is also highlighted to the diagrams of $\delta^{18}O$ and δD versus vintage year (**Figures 3** and **4**). In the same figures, it is further pointed out the variety of isotope values with respect to vintage year indicating that the vintage year contributes to the development of isotopes in wine water. The special yearly weather variations play a critical role in establishing the basic isotope characteristics of wine on larger scale than geographical origin.

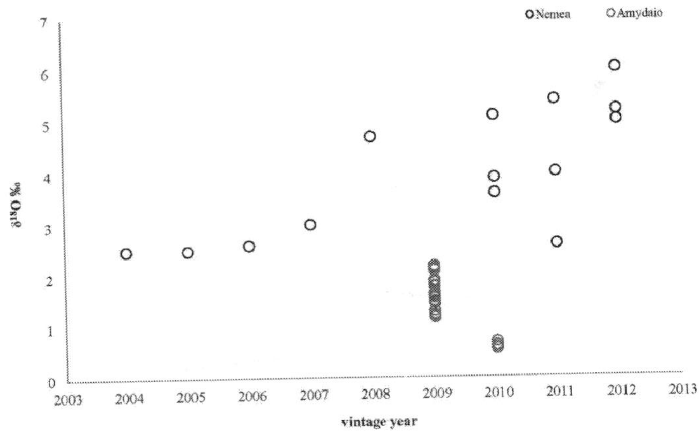

Figure 3. $\delta^{18}O$ % values of the analyzed wine samples versus the vintage year.

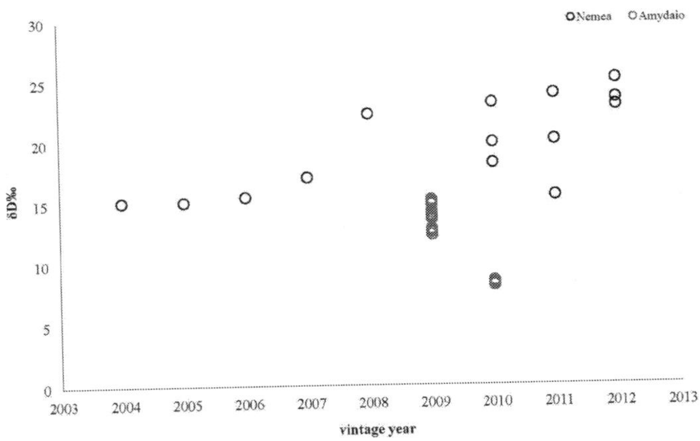

Figure 4. δD % values of the analyzed wine samples versus the vintage year.

5.5. Authentication: adulteration control

The formation of carbon isotope values in plants depends on both biotic and environmental factors. Photosynthetic pathways that referred to terrestrial plants are C3 (Calvin plants: first product of photosynthesis is a 3-carbon molecule), C4 (Hatch-Slack: first product of photosynthesis is a 4-carbon molecule, and crassulacean acid metabolism, which respond in different way to the atmospheric carbon dioxide (CO_2) concentration, to temperature oscillations, and water availability.

As has already been mentioned, the carbon isotopic signature of sugar in grape berries is totally correlated to photosynthetic pathway. As the isotopic discrimination that carbon offers depends on plant type (C3, C4, or CAM), the $\delta^{13}C$ values are a reliable tool for the detection of C3 or C4 sugar origin [3, 32]. The $\delta^{13}C$ values of C3 plants, where grape berries belong, range from −30‰ to −22‰ and for C4 plants from −14‰ to −10‰. So the carbon isotope values that referred to both Amydaio and Nemea vineyards confirm the C3 origin of their sugar and consequently the authentication of wine products. Although the addition of C4 plant sugars could be detected through carbon isotope values of wine water an adulteration through C3 plants would be harder to be traced. Such a challenge is addressed by the combined use of deuterium content of sugars and carbon isotopes of wines [34, 36, 37]. The equation D/Heth = {[(δ^2H/103) + 1]*155.7, where −54‰ is the fractionation of δ^2H of water with that of ethanol [38] was used in order to calculate the $(D/H)_I$ of wine ethanol. $(D/H)_I$ for Greek wines ranged between 102.4 and 105.8 ppm and they are consistent with the corresponding values given for wine (98–108 ppm). The extreme values correspond to only to the 3–5% of Greek wines. In fact, the upper limit of 98 ppm reflects wines originate from northern Greece for the years with strong precipitation episodes while the limit of 108 ppm reflects wines originate from the southern part of Greece (e.g., Crete) for the years that present increased temperature and almost absence of precipitation. In **Figure 5**, carbon isotope values versus deuterium contents of sugars that referred to Amydaio and Nemea wines are presented where reference alcohol from beet, cane [3], are given. Based on the adulteration triangle it is concluded that at the analyzed wine samples no sugar addition was detected. However, based on the $\delta^{13}C$ and $\delta^{18}O$ values with respect to the authentic wine of EU wine isotopic database (**Figure 6**) some suspicious samples are detected for Roditis 2010, Xinomavro 2010, and Syrah 2009 from Amydaio. These samples present low $\delta^{18}O$ values (−1.3‰ to 0.7‰) where this fact could reflect watering of wines and/or mixing with other varieties, respectively. However, it should be taken into account that Amydaio is characterized by purely continental climate with important annual variation in temperature due to the lack of significant bodies of water nearby. Negative oxygen values referred to white wines (Roditis) and as it has already been documented above their harvest were performed earlier than red wines with the evapotranspiration process to be shorter.

The concentration of Sr minerals in wines depends on many factors that are related to their geographic origin. Minerals found in the soil pass through the roots and transported to the plant with isotopic composition similar to that of soil. This way, the isotopic ratio $^{87}Sr/^{86}Sr$ can be used as a tracer of wine origin, where there is a strong correlation between the isotopic composition of the soil in the production area and the final wine that is produced [41–43]. The same approach underlies the use of neodymium isotopes $^{143}Nd/^{144}Nd$. Combined with $^{87}Sr/^{86}Sr$ provides useful information to geochemical analysis. Isotope ratios $^{87}Sr/^{86}Sr$ and $^{143}Nd/^{144}Nd$ reflect the isotopic composition of the soil that are affected by silicate formations in Amydaio area, while in Nemea carbonate minerals also have intense participation.

Finally, stable carbon isotope values of ethanol samples range between −29.2‰ and −25.9‰ reflecting the C3 origin of ethanol in Greek wines. The same conclusions, regarding the authenticity of Greek wines form Amydaio and Nemea, were reached as well by ^{14}C analysis in wine ethanol which compared with the atmospheric ^{14}C data. The results of ^{14}C measurements in ethanol extracted from Greek wines follow the known pattern of ^{14}C variations in atmospheric CO_2. A possible fraud episode with synthetic oil ethanol would result in lower values in the ^{14}C activity concentration and could therefore be detected.

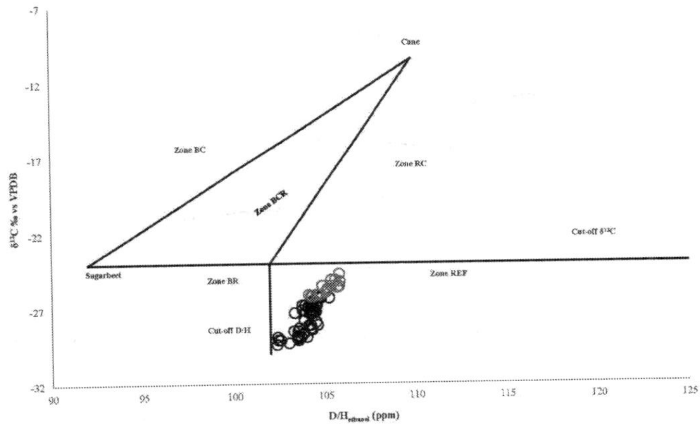

Figure 5. Correlation between $\delta^{13}C$ value of wine and alcoholic fraction of the isotope ratio of H ([39], as modified).

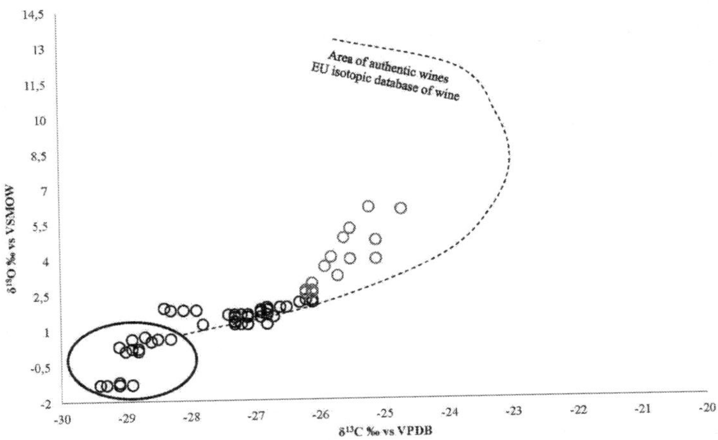

Figure 6. Correlation between $\delta^{18}O$ and $\delta^{13}C$ values of wine [40].

6. Conclusions

Stable isotopes (^{13}C, ^{18}O, ^{2}H), radiocarbon (^{14}C), and radiogenic isotopes ($^{87}Sr/^{86}Sr$, $^{143}Nd/^{144}Nd$) have been applied to determine the origin assignment and verify the geographical provenance of VQPRD Amydaio and Nemea wine products, which is considered as important characteristics both for consumers and the international regulations of wines. Stable isotope analyses of $^{18}O/^{16}O$, D/H, and $^{13}C/^{12}C$ ratio were performed in irrigation water, wine water, and grape must samples in order to detect the origin and the adulteration of wine. The $\delta^{13}C$ analysis of ethanol and wine water $\delta^{18}O$ underlines the importance of the photosynthetic pathway and the environmental conditions of wine (mean $\delta^{18}O_{wine}$ 1.5‰ with $\delta^{18}O_{rain}$ −8‰ for Amydaio and mean $\delta^{18}O_{wine}$ 4.5‰ with $\delta^{18}O_{rain}$ −6‰ for Nemea). The main factors that are responsible for the differentiation of the oxygen isotope ratios of wine water were discussed enhancing the relation of Amydaio and Nemea wine products with the corresponding vineyards. Data interpretation demonstrated the efficacy of $\delta^{18}O$ analysis in both wine and grape berry samples. The determination of $\delta^{18}O$ with the $\delta^{2}H$ isotope content of wine is proposed for achieving greater results on the detection of the geographical origin of wine. The range of $\delta^{13}C$ values and radiocarbon (^{14}C) analysis of ethanol corresponded to all wine labels of VQPRD Amydaio and Nemea vineyards confirming the C3 photosynthetic pathway origin which implies the authentication of wine product and not adulteration with industrial alcohol and/or sweetening agents (sugar beet or cane). The efficiency of direct and precise analysis of ethanol in liquid scintillation counter should be noted. $\delta^{18}O$ and $\delta^{13}C$ values of all wine labels of VQPRD Amydaio and Nemea vineyards are compared with an isotopic database of authentic European wines. The wine products of the zone VQPRD Amydaio and Nemea grouped with all the original European wines demonstrating the absence of adulteration episodes. A slight deviation of Amydaio wines from the EU database, due to more negative oxygen isotope values, attributed to the continental climate in area with important annual variation in temperature due to the lack of significant water bodies nearby. Moreover, $\delta^{13}C$ values of wine water and $D/H_{ethanol}$ values of ethanol extracted from wines of VQPRD Amydaio and Nemea define them as authentic products certifying nonadulteration process and an authenticity label. The homogeneity of $^{87}Sr/^{86}Sr$ and $^{144}Nd/^{143}Nd$ isotope values confirms that the territorial and geological signal is transferred through the vineyards in the final product, wine, certifying their exclusive provenance.

Author details

Paraskevi Chantzi[1*], Anastasia-Elektra Poutouki[1] and Elissavet Dotsika[1,2]

*Address all correspondence to: p.chantzi@inn.demokritos.gr

1 Laboratory of Stable Isotope and Radiocarbon, Institute of Nanoscience and Nanotechnology, NCSR Demokritos, Agia Paraskevi, Attiki, Greece

2 Institute of Geosciences and Earth Resources, Pisa, Italy

References

[1] Keeling, C.D., 1961: The concentration and isotopic abundances of carbon dioxide in rural and marine air. Cheochimica et Cosmochimica Acta, 24Q, 277–298.

[2] Ehleringer, J.R., Hall, A.E., & Farquhar, G.D., Eds., 1993: Stable Isotopes and Plant Carbon-Water Relations. San Diego, CA: Academic Press.

[3] O'Leary, M.H., 1988: Carbon isotopes in photosynthesis. BioScience, 38, 328–336.

[4] Hillaire-Marcel, G., 1986: Chapter 12—Isotopes and Food, In Handbook of Environmental Isotope Geochemistry, edited by P. Fritzj and C. Fontes. Amsterdam: Elsevier. Part B, pp. 507–548, The Terrestrial Environment, B, ISBN 9780444422255.

[5] Farquhar, G.D., Ehleringer, J.R., & Hubick, K.T., 1989: Carbon isotope discrimination and photosynthesis. Annual Review of Plant Physiology and Molecular Biology, 40, 503–537.

[6] Freeman, B.M., Kliewer, W.M., & Stern, P., 1982: Influence of windbreaks and climatic region on diurnal fluctuation of leaf water potential, stomatal conductance and leaf temperature of grapevines. American Journal of Enology and Viticulture, 31, 233–236.

[7] Roby, G., & Matthews, M.A., 2004: Relative proportions of seed, skin and flesh, in ripe berries from Cabernet-Sauvignon grapevines grown in a vineyard either well irrigated or under water deficit. Australian Journal of Grape and Wine Research, 10, 74–82.

[8] Clark, I., & Fritz, P., 1997: Environmental Isotopes in Hydrogeology. New York: Lewis Publishers. pp. 35–78.

[9] Rossmann, A., Reniero, F., Moussa, I., Schmidt, H.L., Versini, G., & Merle, M.H., 1999: Stable oxygen isotope content of water of EU data-bank wines from Italy, France and Germany. Zeitschrift für Lebensmittel-Untersuchung und Forschung A, 208 (5), 400–407.

[10] Epstein, S., Thompson, P., & Yapp, C.J., 1977: Oxygen and hydrogen isotopic ratios in plant cellulose. Science, 198 (4323), 1209–1215, doi: 10.1126/science

[11] Fehri, A., & Letolle, R., 1977: Transpiration and evaporation as the principal factors in oxygen isotope variations of organic matter in land plants. Physiologie Vegetale, 15, 363–370.

[12] Asche, S., Michaud, A.L., & Brenna, J.T., 2003: Sourcing organic compounds based on natural isotopic variations measured by high precision isotope ratio mass spectrometry. Current Organic Chemistry, 7 (15), 1527–1543.

[13] Ingraham, N.L., & Caldwell, E.A., 1999: Influence of weather on the stable isotopic ratios of wines: Tools for weather/climate reconstruction? Journal of Geophysical Research [Atmos.], 104 (D2), 2185–2194.

[14] Martin, G.J., & Martin, M.L., 2003: Climatic significance of isotope ratios. Phytochemistry Reviews, 2 (1–2), 179.

[15] Martin, M.L., & Martin, G.J., 1999: Site-specific isotope effects and origin inference. Analusis, 27 (3), 209–213.

[16] Zhang, B.L., Buddrus, S., Trierweiler, M., & Martin, G.J., 1998: Characterization of glycerol from different origins by H-2- and C-13-NMR studies of site-specific natural isotope fractionation. Journal of Agricultural and Food Chemistry, 46 (4), 1374–1380.

[17] Steenbrink, J., Hilgen, F.J., Krijgsman, W., Wijbrans, J.R., & Meulenkamp, J.E., 2006: Late Miocene to early Pliocene depositional history of the intramontane Florina–Ptolemais–Servia Basin, NW Greece: Interplay between orbital forcing and tectonics. Palaeogeography, Palaeoclimatology, Palaeoecology, 238 (1–4), 151–178, ISSN 0031-0182.

[18] Seger, M., & Alexander, J., 1993: Distribution of Plio-Pleistocene and modern coarse grained deltas south of the Gulf of Corinth, Greece. Special Publications International Association of Sedimentologists, 20, 37–48.

[19] Zelilidis, A., 2000: Drainage evolution in a rifted basin, Corinth graben, Greece. Geomorphology, 43, 253–262.

[20] Duhr, A., & Hilkert, A.W., 2004: Application note 30049: Automated H_2/H_2O Equilibration for δD Determination on Aqueous Samples Using Thermo Scientific GasBench II. Bremen, Germany: Thermo Fisher Scientific.

[21] Hilkert, A.W., & Avak, H., 2004: Application Note 30048: ^{18}O-Equilibration on Water, Fruit Juice and Wine Using Thermo Scientific GasBench II. Bremen, Germany: Thermo Fisher Scientific.

[22] Kracht, O., Hilkert, A.W., Racz-Fazakas, T., 2008: Application Note 30147: ^{13}C and Simultaneous ^{18}O and ^{2}H Isotope Analysis in Ethanol with Thermo Scientific DELTA V Isotope Ratio Mass Spectrometers. Bremen, Germany: Thermo Fisher Scientific

[23] Craig, H., 1961: Isotopic variation in meteoric waters. Science, 133, 1702–1703.

[24] Dotsika, E., Lykoudis, S., & Poutoukis, D., 2010: Spatial distribution of the isotopic composition of precipitation and spring water in Greece. Global and Planetary Change 71 (3e4), 141–149.

[25] Dotsika, E., Poutoukis, D., Michelot, J.L., & Raco, B., 2009: Natural tracers for identifying the origin of the thermal fluids emerging along the Aegean Volcanic arc (Greece): Evidence of Arc-Type Magmatic Water (ATMW) participation. Journal of Volcanology and Geothermal Research, 179 (1–2), 19–32, ISSN 0377-0273

[26] Leontiadis, I., Payne, B.R., Letsios, A., Papagianni, N., Kakarelis, D., & Chadjiagonakis, D., 1984: Isotope hydrology study of Kato Nevrokopi of drama. Proceedings of the Symposium on Isotope Hydrology 1983, International Atomic Energy Agency, Vienna. pp. 193–206.

[27] Christodoulou, T., Leontiadis, I.L., Morfis, A., Payne, B.R., & Tzimourtas, S., 1993: Isotope hydrology study of Axios River plain in northern Greece. Journal of Hydrology, 146, 391–404.

[28] Leontiadis, I.L., Vergis, S., & Christodoulou, T., 1996: Isotope hydrology study of areas in Eastern Macedonia and Trace, Northern Greece. Journal of Hydrology, 182, 1–17.

[29] Leontiadis, I.L., & Nikolaou, E., 1999: Environmental isotopes in determining groundwater flow systems, northern part of Epirus, Greece. Hydrogeology Journal, 7, 219–226.

[30] Angerosa, F., Bréas, O., Contento, S., Guillou, C., Reniero, F., & Sada, E., 1999: Application of stable isotope ratio analysis to the characterization of the geographical origin of olive oils. Journal of Agriculture and Food Chemistry, 47, 1013–1017. doi:10.1021/jf9809129.

[31] Bréas, O., Guillou, C., Reniero, F., Sada, E., & Angerosa, F., 1998: Oxygen-18 measurement by continuous flow pyrolysis/isotope ratio mass spectrometry of vegetable oils. Rapid Communications in Mass Spectrometry, 12(4), 188–192.

[32] O'Leary, M. H., 1995: Environmental effects on carbon isotope fractionation in terrestrial plants. In (E. Wada, T. Yoneyama, M. Minigawa, T. Ando & B. D. Fry, Eds) Stable Isotopes in the Biosphere. Kyoto, Japan: Kyoto University Press, pp. 78–91.

[33] Schmidt, H.-L., Werner, R.A., & Roßmann, A., 2001: ^{18}O pattern and biosynthesis of natural plant products. Phytochemistry, 58 (1), 9–32, ISSN 0031-9422.

[34] Dunbar, J., & Wilson, T., 1983: Oxygen and hydrogen isotopes in fruit and vegetable juices. Plant Physiology, 72, 725–727.

[35] Raco, B., Dotsika, E., Poutoukis, D., Battaglini, R., & Chantzi, P., 2015: O–H–C isotope ratio determination in wine in order to be used as a fingerprint of its regional origin. Food Chemistry, 168, 588–594, ISSN 0308-8146.

[36] Bréas, O., Guillou, C., Reniero, F., Sada, E., & Tanet, G., 1996: Deuterium/Hydrogen Measurements in Wines and Fruit Juices by Platinum Catalysed Equilibration Method. Rapid Commun. Mass Spectrometry, 10: 246–249.

[37] Dunbar, J., & Schmidt, H.-L., 1984: Fresenius Z. Analytical Chemistry, 317, 853.

[38] Bréas, O., Thomas, F., Zeleny, R., Calderone, G., Jamin, E., & Guillou, C., 2007: Performance evaluation of elemental analysis/isotope ratio mass spectrometry methods for the determination of the D/H ratio in tetramethylurea and other compounds-results of a laboratory inter-comparison. Rapid Communications in Mass Spectrometry, 21, 1555–1560.

[39] Ogrinc, N., Košir, I.J., Spangenberg, J.E., & Kidrič, J., 2003: The application of NMR and MS methods for detection of adulteration of wine, fruit juices, and olive oil. A review. Analytical and Bioanalytical Chemistry, 376, 424–430.

[40] Calderone, G., & Guillou, C., 2008: Analysis of isotopic ratios for the detection of illegal watering of beverages. Food Chemistry, 106, 1399–1405.

[41] Almeida, C.M., & Vasconcelos, M.T., 2004: Does the winemaking process influence the wine 87Sr/86Sr? A case study. Food Chemistry, 85, 7–12.

[42] Horn, P., Hoelzl, S., Todt, W., & Matthies, D., 1998: Isotope abundance ratios of Sr in wine provenance determinations in a tree-root activity study and of Pb in a pollution study on tree-rings. Isotopes in Environmental and Health Studies, 34, 31–42.

[43] Lancelot, J., Herrerias, J., Verdoux, P., Lurton, L., 1999: The use of strontium isotopes geochemistry for a high resolution identification of wines from the Rhone Valley. In: Fifth European Symposium of Food Authenticity, La Baule, 9–11 June, 1999.

[44] Martin, G.J., Zhang, B.L., & Martin, M.L., 1983: Application of quantitative deuterium NMR to the study of isotope fractionation in the conversion of saccharides to ethanols. Biochemical and Biophysical Research Communications, 111, 890–896.

[45] Martin, G.J., Zhang, B.L., Naulet, N., & Martin, M.L., 1986: Deuterium transfer in the bioconversion of glucose to ethanol studied by the specific isotope labeling at the natural abundance level. Journal of the American Chemical Society, 108, 5116–5122.

[46] Martin, G.J., Guillou, C., Martin, M.L., Cabanis, M.T., Tep, Y., & Aemy, J., 1988: Natural factors of isotope fractionation and the characterization of wines. Journal of Agricultural and Food Chemistry, 36, 316–322.

[47] Martin, G.J., Martin, M.L., & Zhang, B.L., 1992: Site-specific natural isotope fractionation of hydrogen in plant products studied by nuclear magnetic resonance. Plant, Cell & Environment, 15, 1037–1050.

[48] Zhang, B.L., Yunianta, V.C., & Martin, M.L., 1995: Site-specific isotope fractionation in the characterization of biochemical mechanisms. Journal of Biological Chemistry, 270, 16023–16029.